...ENTH EDITION

Business Data Networks and Telecommunications

LOAN

D1352272

...Hawaii at Manoa

PEARSON
Prentice
Hall

Pearson Education International

Executive Editor: Bob Horan
Editorial Director: Sally Yagan
Editor in Chief: Eric Svendsen
Product Development Manager: Ashley Santora
Editorial Project Manager: Kelly Loftus
Editorial Assistant: Mauricio Escoto
Marketing Manager: Anne Fahlgren
Marketing Assistant: Susan Osterlitz
PermissionsProject Manager: Charles Morris
Senior Managing Editor: Judy Leale
Production Project Manager: Debbie Ryan
Senior Operations Specialist: Arnold Vila
Operations Specialist: Carol O'Rourke
Art Director: Jayne Conte
Cover Designer: Studio Indigo
Composition: Aptara, Inc.
Full-Service Project Management: Puneet Lamba
Printer/Binder: R.R. Donnelley - Harrisonburg
Typeface: 8/10 New Baskerville

Pearson Education Ltd., London
Pearson Education Singapore, Pte. Ltd
Pearson Education, Canada, Inc.
Pearson Education–Japan
Pearson Education Australia PTY, Limited

Pearson Education North Asia, Ltd., Hong Kong
Pearson Educación de Mexico, S.A. de C.V.
Pearson Education Malaysia, Pte. Ltd
Pearson Education Upper Saddle River, New Jersey

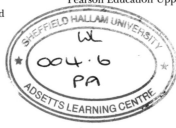

10 9 8 7 6 5 4 3 2 1
ISBN-13: 978-0-13-500939-0
ISBN-10: 0-13-500939-1

To Doug Engelbart, who foresaw all this in the 1960s.

Brief Contents

Contents

Preface for Teachers

THREE QUESTIONS

Teachers who are considering this book typically have three questions:.

1. Why select this book? In this preface, we will look at three reasons:

   Job-relevant information

   Great teacher support

   Pedagogy

2. How can I teach with it?

3. What's new since the last edition?

This preface walks through these questions, in order.

WHY SELECT THIS BOOK?

Job-Relevant Information

In designing this and previous editions, the author talked extensively with working networking professionals and managers. He also got extensive market data on the technologies that corporations actually use. Many textbooks seem to be surprisingly market blind, spending time on technologies that never were adopted by the market and on technologies that have not been sold since the previous century. As a result, they are not covering important new developments. For example, Ethernet, which is now the only wired LAN technology in almost every firm, is no longer a simple technology. Students need to know about priority, VLANs, the Rapid Spanning Tree Protocol for reliability, 802.1X, 802.1AE, 802.1af, 802.1at, and other advanced aspects of Ethernet.

Up-to-Date Content

Wireless   Of course, the content needs to be up-to-date. The material on wireless transmission—already strong in the last edition—has been strengthened throughout the book. Wireless material now starts earlier (in Chapter 3 on physical propagation), is the focus of Chapter 5 on wireless LANs chapters, and is taken into WANs in Chapter 6 (which discusses WiMAX and wireless access in general).

Security   Network managers now find themselves spending much of their time on security. The last edition was very strong on security. Now, security has been strengthened throughout the book, including in Chapters 4 (Ethernet LANs), Chapter 5 (Wireless LANs), Chapter 9 (the anchor chapter on security), and Chapter 11 (applications).

Voice Alternatives   Telephone bills in corporations are high, and technology is changing the way we use voice. Chapter 6 provides more in-depth discussions of VoIP and cellular telephony, as well as triple play bundles for voice, data, and television and the changes this will bring.

Market-Driven Content

Even more importantly, the book's content is strongly market-driven. Too many textbooks try to cover every technology that ever existed, even when its use today is almost nonexistent. As noted earlier, this leaves far too little time for today's critical technologies and emerging technologies. Students become historians, not market-ready graduates. In fact, some books seem to ignore market data. One recent textbook even called Frame Relay a new technology, despite the fact that Frame Relay revenues were almost equal to those of private line revenues in the WAN market in the late 1990s and that Frame Relay is now a legacy technology in decline. In contrast, here are some examples of the book's market-driven content:

➤ Wireless LANs. As just noted, the book responds to corporate calls for greater wireless knowledge.
➤ Security. As you would expect from the author, security is even more pervasive throughout the text.
➤ TCP/IP. Above the physical and data link layers used in LANs and WANs, TCP/IP is consolidating its hold on upper-layer technologies. In the seventh edition, TCP/IP is even more pervasive than in previous editions. Chapter 1 introduces several core concepts, Chapter 2 introduces the basic elements of TCP/IP, and Chapter 8 goes into detail on advanced aspects of TCP/IP standards, including how routers make routing decisions. Chapter 10 discusses key topics in TCP/IP management. Module A covers a number of even more advanced aspects of TCP/IP.

Job-Ready Detail

In the past, job interviewers often asked students to name the OSI layers and stopped at that. Today, however, even job interviewers for non-networking jobs grill job applicants on the details of networking. The days of feel-good "network appreciation" textbooks with far too little detail for job applicants should be over.

For example, in the Ethernet Chapter (Chapter 4), the book goes well beyond basic topology and switch operation to look at VLANs, link aggregation, the Rapid Spanning Tree Protocol, overprovisioning versus priority, and switch purchasing considerations. Detail has been beefed-up in other areas as well. The TCP/IP Chapter (Chapter 8) takes a detailed look at how routers operate, while many other books cover this critical topic superficially or even incorrectly.

Great Teacher Support

Teaching networking is very difficult, so textbooks must provide strong teacher support.

Ask the Author

I'm Ray@panko.com. If you have questions about the text, the material, or networking in general, please do not hesitate to contact me. Also, if you have suggestions for the next edition, please let me know. This book is on a two-year cycle, and it is always changing.

Detailed PowerPoint Lectures

The book has full PowerPoint lectures created and updated by the author—not just "a few selected slides." The PowerPoint lectures include builds for more complex figures and new information since the book went to press. Teachers can get annotated versions of the PowerPoint presentations to help them prepare and present lectures.

The PowerPoint slides are keyed directly to figures in the book. Almost all the important points in the book are covered in figures—including "study figures" that summarize the key points in the more complex sections.

Website

The book's website, http://www.prenhall.com/panko, created and updated by the author, is rich in teacher resources. This is where teachers can download answer keys, test item file questions, and the latest versions of the PowerPoint lectures (which are updated once or twice a year).

Flexibility: Have It Your Way

This book is not designed to be covered in its entirety, including all advanced modules. Rather, it is designed to give teachers options without their having to look for other material.

An 11-Chapter Core: A Complete Networking Course   The book has 11 core chapters that can each be covered in about one week. In a three-credit, one-semester course, this leaves one to two weeks for other material. However, many teachers prefer to limit their courses to the 11 core chapters, which form a complete networking course. This allows them to go through the material at a more deliberate pace. Also, staying with the 11 core chapters may be a good strategy for teachers working with the book for the first time or teaching a networking course for the first time.

Boxed Material within Chapters   Within chapters, some technical details are placed in boxes. Many teachers skip all boxed material. Others use them selectively.

"Letter" Chapters   The book has two types of material beyond the 11 core chapters: "letter" chapters, such as Chapter 1a, and advanced modules. These chapters and modules should be used judiciously. Even covering two or three letter chapters or one advanced module may be pushing it.

Several core chapters are followed by "letter" chapters. For example, Chapter 1a has a design exercise and hands-on exercises. Chapter 3a deals with cutting and connectorizing UTP wiring. Although neither 802.5 Token-Ring Networks nor FDDI networks are used in corporations today, Chapter 4a gives an overview of ring topologies and token-passing. Chapter 8a introduces WinDUMP and TCPDUMP for packet capture, and Chapter 10a has hands-on exercises for network management utilities.

Advanced Modules   There are three advanced modules. For teachers who really want to focus on TCP/IP, Module A has very detailed information about TCP/IP. Module C is designed for courses that focus on telecommunications. Chapter 6 looks at telecommunications from the viewpoint of corporate information systems (IS) staffs. Module C focuses on telecommunications from the carrier's point of view. Module B covers modulation in greater detail for teachers who feel that more information on telephone modems is needed.

Chapter Questions Are Tied to Answer Keys and Test Item File Questions

The chapters have test-your-understanding questions, roughly once per page, so that students can do a brain check on what they have just read. In addition, end-of-chapter thought questions, design questions, and troubleshooting questions help students attain higher-level mastery of the material. Answer keys for all questions are available to teachers.

Multiple-choice test item file questions are keyed to specific chapter questions. This allows teachers who wish to be selective to choose specific questions which students should master, then develop tests that reflect those selected questions.

Mailing List

A low-volume mailing list used a few times per year updates adopters on new developments—most commonly, new material at the website. The mailing list is also used to solicit adopter feedback on the text.

Pedagogy

Learning networking is difficult. Many students find that networking is the most conceptually difficult course in IS programs. Networking books need to have very strong pedagogy.

Clear Writing

All editions of this book have received accolades for clear writing—especially for its ability to teach difficult and complex topics. Every Chapter is classroom-tested.

Hands-On Opportunities

Students want opportunities to do things hands-on. With this book, they have those opportunities. The following sections are examples of hands-on training in this text:

- ➤ Chapters 1a and 10a, in their end-of-chapter reviews, present hands-on questions to reinforce concepts.
- ➤ Chapter 3a demonstrates how to cut and connectorize UTP. Teaching this material requires an investment of about $200, but undergraduate students love it.
- ➤ Chapter 8a shows students how to do packet capture and analysis.
- ➤ Other chapters have students go to the Internet to do specific tasks.

Chapter Questions

Frequent Test-Your-Understanding Questions   The book gives the student many opportunities to check his or her knowledge. Approximately once per page, there are Test-Your-Understanding questions to help the student see whether he or she has understood the material just read.

Meaty End–of-Chapter Questions   End-of-chapter questions help the student integrate the material in the chapter. Thought questions challenge the student to think more deeply about the material. Troubleshooting and design questions help students develop important skills that are critical in networking. Some chapters also have hands-on questions for students to do at home.

Coordinated Test Item File for Multiple Choice Questions   As noted earlier, test item file questions are keyed to specific Test-Your-Understanding and End-of-Chapter questions.

Up through the Layers/Familiar to the Unfamiliar

Like most books, the seventh edition takes an up-though-the-layers approach. However, this approach is significantly modified because most books that take this approach teach one layer at a time, in isolation. Only at the end of the book does the student see the whole picture. During the process, they have gained only a cursory framework within which to integrate chapter knowledge.

> ➤ This book is different. The book begins, in Chapters 1 through 3, with a strong framework to help students understand networking broadly so that when new knowledge appears, they understand its place. The difficult concept of layered network architectures is introduced early and is reinforced throughout the book.

> ➤ Chapters 4 through 7 deal with switched LANs, telecommunications, and switched WAN technologies. Every switched LAN and switched WAN technology is a mixture of Layer 1 (physical) and Layer 2 (data link) technologies. For this reason, this book covers Layer 1 and 2 technologies within the context of specific LAN and WAN technologies rather than individually (although Chapter 3 introduces specific physical layer information).

> ➤ Chapter 8 discusses internetworking, especially TCP/IP internetworking at Layer 3 (internet) and Layer 4 (transport). Once the student understands switched LAN and switched WAN technologies, they can appreciate the need to interconnect them.

> ➤ Chapters 9 and 10 cover material that cuts throughout the layers—security and network management. These topics are introduced early, but a full discussion has to wait until students have a solid understanding of layer technologies. These are anchor chapters that cement what was covered earlier and add detailed information.

> ➤ The last chapter, Chapter 11, deals with the application layer (in OSI, application layers). It might seem better to cover this information immediately after Chapter 8, but many schools cover applications in a separate course, and ending the course with applications is fun.

Synopsis Sections in Every Chapter

Each chapter ends in a synopsis section that summarizes key points. In classroom testing of the edition's chapters, these synopsis sections were very popular with students.

TEACHING WITH THIS BOOK

As noted earlier, this book has 11 core chapters. These form a complete course.

Junior and Senior Courses Information Systems

With courses for juniors and seniors, as noted earlier, covering the 11 core chapters (including "a" chapters that are case studies) will probably leave you with one or two semester weeks "free." As noted earlier, this leaves time for hands-on activities (discussed earlier), additional TCP/IP material (or other material in the advanced modules), a term project, or whatever you wish to cover. However, the book should not be covered in its entirety in a single term.

Community College Courses

For freshman and sophomore courses in community colleges, it is good practice to stay with the 11 core chapters, going over chapter questions in class. If you want to do hands-on material, it is advisable to cut some material from the core chapters.

Graduate Courses

Graduate courses tend to look a lot like junior and senior level courses, but with greater depth. More focus can be placed on end-of-chapter questions and novel hands-on exercises, such as OPNET simulations. It is also typical to have a term project.

CHANGES SINCE THE SIXTH EDITION

The seventh edition generally follows the same basic flow as the sixth. The following table lists some major and minor changes:

Seventh Edition	Remarks on the Seventh Edition, Relative to the Sixth Edition (6e)
Key Changes	Chapter 1 now jumps right into applications, quality of service, and network management, including operational management. This reflects the fact that networking is increasingly a managerial concern.
	Chapter 10 is much richer in network management and has been completely reorganized and heavily rewritten.
	The subject of wireless has been enhanced throughout the book. Propagation is moved up to Chapter 3, Chapter 5 on WLANs has been strengthened, and WiMAX is enhanced in Chapter 6.
	Security is strengthened in Chapter 4 on Ethernet, Chapter 5 on wireless LANs, and Chapter 9 (especially on management and stateful inspection firewalls).
	The short, but important, VoIP discussion in Chapter 6 has been strengthened.
Specific General Changes	The term "host" is used for all computers
	In the sixth edition, there was an important and pervasive distinction between single networks, which used switches, and internets, which used routers to connect single networks. In the seventh

	edition, these concepts have been changed to switched networks and routed networks (which are also called internets).
	In the sixth edition, the speed of WANs was set at 128 kbps to a few megabits per second. Due to speed increases, the range in the seventh edition has been raised to 256 kbps to about 50 Mbps. This is still far lower than LAN speeds.
	The terms "mobiles" and "mobile phones" usually replace the term "cellular phones."
	Perspective questions at the end of each chapter have students decide what was the most surprising thing to them in the chapter and what was the most difficult material for them in the chapter.
Chapter 1. An Introduction to Networking	This chapter has been almost completely rewritten to focus more heavily on applications and management.
	The chapter begins with a large section on applications, which are the things that users care about (including traditional Internet applications, IM, streaming audio and video, VoIP, file service, Web 2.0, P2P, and business-specific applications such as transaction processing).
	Rather than being delayed to Chapter 2, standards and the important distinction between proprietary and open standards (TCP/IP, OSI, etc.) are introduced in Chapter 1.
	The chapter has a strong quality of service section dealing with speed, cost, availability, error rates, latency and jitter.
	For cost, the system life cycle and the total cost of ownership (TCO) are introduced.
	The chapter ends with management, including operational management after the network is functioning. This section uses the OAM&P (operations, administration, management, and provisioning) framework.
	Security is shortened and focused on four key protections: authentication, cryptographic protections, firewalls, and host hardening. Authentication is developed throughout the book, including in Chapters 4, 5, and 9.
	The chapter discusses message switching versus packet switching and introduces the concepts of switched networks (single networks, in the sixth edition) and routed networks or internets (internets, in the sixth edition)
	The chapter emphasizes that LANs and WANs can be either switched or routed. Chapter 4 specifically ends with discussion of a routed LAN.
	The chapter introduces the Internet and key routed network concepts.
Chapter 1a. Introductory Design and Hands-On Exercises	This material was presented mostly in Chapter 1 in the sixth edition. There is a design exercise based on XTR, but simplified. Hands-On exercises that are largely from Chapter 1 in the previous edition.
Chapter 2. Layered Standards Architectures	This chapter is largely the same as in the sixth edition, although some material has been moved around or emphasized in the seventh edition.

(continued)

Material has been added on embedding and final frames, a topic that students typically find difficult. (See Figure 2-18.)

The text emphasizes that TCP and UDP are the only protocols at the transport layer.

Hybrid TCP/IP-OSI architecture (Figure 2-6) has been changed to begin with broad functions, then the five specific layers (Figure 2-20).

The Ethernet frame (Figure 2-11) has been changed in the new edition. See notes on Chapter 4, to follow, which discuss the changes.

Figure 2-13 "Why make only TCP/IP reliable?" was added to help students with this difficult, but critical, concept.

Chapter 3. Physical Layer Propagation	The section on digital signaling and baud rates has been rewritten, and the study figure has been extended. Most teachers ignore this box. The section on optical fiber has been almost completely rewritten. LAN fiber now comes before WAN fiber, which makes sense because only carriers use WAN fiber. Figure 3-24 visually contrasts the two types of fiber. In the seventh edition, the material on wireless propagation has been moved from Chapter 5 to Chapter 3, to bring all propagation information together (and to leave room for more wireless LAN discussion in Chapter 5). The section on decibels has been heavily rewritten. Small form factor connectors are introduced for fiber. The section on UTP quality standards has been rewritten to take recent developments into account. Figure 3-19 succinctly compares the fiber types.
Chapter 3a. Hands On: Cutting and Connectorizing UTP	No changes since the 6th edition.
Chapter 4. Ethernet which LANs	The seventh edition deletes the discussion of "fast Ethernet," is now slow, and "10/100 Ethernet," which is no longer relevant. The section on Ethernet standards has been changed to reflect the fact that the 802.3 Working Group has decided to produce both 40 Gbps and 100 Gbps versions as its next step. The section on Ethernet frame structure has been rewritten. The LLP subheader and PAD are not shown as part of the data field. This greatly simplifies the material the students need to understand about frame data fields holding packets. The section on introducing redundant links in Ethernet hierarchical networks has been rewritten to focus on RSTP (802.1w), not STP (802.1D), which is now obsolete. The section on VLANs has been streamlined, and security is given as an equal reason for using VLANs (a recent development in most firms). Overprovisioning is no longer an automatic win over priority, so statements saying that it is have been dropped. The Ethernet security section was strengthened, including the addition of artwork.

	The section on CSMA/CD has been rewritten somewhat, to help students learn the sequence of steps.
	The seventh edition drops switching matrix throughput, which is no longer a concern in networking.
	The power over Ethernet (POE) section has been completely rewritten, and the emerging POE Plus standard was added.
	There are now more design questions permitting easier student transitions in understanding from trivial LANs to complex LANs.
Chapter 4a. Token-Ring Networks and Early Ethernet Technologies	This chapter has been heavily rewritten to make early Ethernet equally important to token-ring networks. Again, this is material that most teachers will not want to cover.
Chapter 5. Wireless LANs (WLANs)	The chapter now starts with a discussion of 802.11 instead of the topic of new wireless technologies, which has been moved to the end of the chapter.
	Wireless propagation moved to Chapter 3.
	The material on 802.11 was heavily reorganized. Discussions of licensed and unlicensed radio bands appear earlier, and 802.11 standards are initially presented in the context of the two main unlicensed radio bands.
	The new edition lumps 802.11, 802.11a, and 802.11b in a section on early 802.1 standards and reduces their discussion to make room for today's dominant technologies. This edition is stronger than the sixth edition on 802.11g and even stronger on 802.11n and MIMO.
	Figure 5-18, the 802.11 WLAN standards table, now includes 802.11n.
	The new edition adds mesh networking and smart antennas to the 802.11 discussion because these are appearing in the marketplace.
	The crucial security section has been rewritten. WEP has been reduced in the discussion because it is no longer an issue, and a discussion on whether 802.11i can replace WPA has been added. VPNs and VLANs now are addressed in a separate section, and an Evil twin figure has been added.
	Software-defined radio is mentioned.
Chapter 6. Telecommunications	The seventh edition adds T1 TDM to the virtual circuits discussion. In cellular, more non-call features of mobile phones are needed.
	A stronger compare–and–contrast discussion has been added for WLAN with access points and cellular telephony in terms of hand-offs and roaming, and a figure has been added.
	The VoIP section has been heavily rewritten
	The last section, on residential Internet access, has been rewritten to be "rebuilding the last Kilometer" for better integration with the chapter. Telephone service and cable TV histories have been added, and the section on telephone modems has been reduced.
	There is now a distinct section on wireless access, including 3G and satellite service, and a strengthened WiMAX section.
	There is a section on triple-play services including voice, data, and TV.
	Fiber to the home is expanded because this offering is now appearing.

(continued)

The chapter notes the lag in U.S. speeds and high costs compared with international speeds and costs. The United States has a median download speed of 1.9 Mbps, while France has 17 Mbps, and Korea has 45 Mbps. The United States is now 16th in broadband speed.

Chapter 7. Wide Area Networks (WANs)	The chapter has been restructured to discuss carrier offerings in terms of Layer 1, 2, and 3 offerings.
	The typical speed range has been raised to 256 kbps to 50 Mbps for WANs—still much slower than LANs.
	This edition moves CSU/DSU devices forward to the discussion of leased lines, which better reflects the Layer 1 nature of CSU/DSU units.
	Frame Relay is reduced to reflect its declining use.
	The ATM cell concept has been eliminated from the text because it is confusing, is not fundamental, and is not relevant to corporate networking professionals.
	The discussion of IP WAN service has been heavily rewritten because Layer 3 carrier services are growing rapidly in importance, including IP carrier networks. In fact, carriers are beginning to set deadlines for clients switching from legacy Frame Relay service to IP carrier services.
	In the discussion of VPNs, the concept of host-to-host VPNs has been dropped. Now the book discusses only remote access and site-to-site VPNs. This sharpens the discussion.
	There is now a Test Your Understanding question on the implications for mesh versus hub-and-spoke designs, and there is a more complex design question on this at the end of the chapter.
Chapter 7a. Case Study: First Bank of Paradise's Wide Area Networks	No changes since the 6th edition.
Chapter 8. TCP/IP Internetworking	The section on router forwarding has been completely rewritten to streamline the presentation. ARP moved up to this section, and decision caching has been further emphasized.
	A box has been added with instructions for students on how to handle masks that do not break nicely at 8-bit boundaries and so must be handled at the bit level.
	The FIN section has been changed to note that after one side of a transmission initiates a close with a FIN, it will still send acknowledgments.
	The treatments of TCP and IP were moved up, and the treatments of other standards moved down to produce a smoother flow and to emphasize TCP and IP.
	The discussions of MPLS and DNS have been moved to Chapter 10 in order to have a complete section in Chapter 10 on TCP/IP management.
Chapter 8a. Hands On: Packet Capture and Analysis with WinDUMP and TCPDUMP	No changes since the 6th edition.

Chapter 9. Security	The chapter begins with a recap of Chapter 1.
	The chapter does not present CSI/FBI data.
	The section on attacks on individual consumers has been heavily modified for flow and content.
	The plan-protect-respond cycle has been introduced in the section on security management, and security planning principles have been emphasized.
	The stateful inspection firewall section has been completely changed to give a clearer concept of states and how stateful inspection relates to filtering.
	The DDoS figure has been simplified, and zombie changed to bot to reflect current practice.
	The topic of public key encryption for confidentiality has been left out to avoid confusion between the use of public keys in authentication and their use in confidentiality. Public key encryption for confidentiality is very rare and is better left to texts on security.
Chapter 10. Network Management	There is a completely new organization to the chapter.
	The chapter begins with a recap of management sections from earlier chapters.
	Within a planning section for technological infrastructure decisions, the treatment of traffic shaping has been improved, and a discussion of compression has been added.
	This chapter includes a dominant section on TCP/IP management, addressing IP subnet planning, NAT, MPLS, DNS, DHCP servers, and /SNMP. The NAT discussion is new and substantial. The section discusses authoritative DNS servers and how they are used.
	The final section deals with directory servers, including Active Directory.
Chapter 10a. Network Management Utilities and Router Configuration	Hands-on questions have been moved here from Chapter 10. Router configuration material has been moved here from Chapter 10.
Chapter 11. Networked Applications	Applications were introduced substantially in Chapter 1. E-commerce security has been left out of Chapter 11 because it is no longer a distinct issue.
	There has been a total rewrite of the Web Services (SOA) section, including WSDL.
Module A. More on TCP/IP	There are no changes in this module since the 6th edition.
Module B. More on Modulation	There are no changes in this module since the 6th edition.
Module C. More on Telecommunications	There are no changes in this module since the 6th edition.

Preface for Students

PERSPECTIVE

Initially, information systems (IS) graduates had a single career track: programmer–analyst–database administrator–manager. Today, however, many IS graduates find themselves on the networking career track—often to their surprise. This course is an introduction to the networking track.

Even programmers now need a strong understanding of networking. In the past, programmers wrote stand-alone programs that ran on a single computer. Today, however, most programmers write networked applications that work cooperatively with other programs on other computers.

LEARNING NETWORKING

Networking Is Difficult

Networking is an exciting topic. It is also a difficult topic. In programming, the focus is on creating and running programs. In networking, the critical skills are design, product selection, and troubleshooting. These rather abstract skills require a broad and deep knowledge of many concepts. Many IS students have a difficult time adjusting to these cerebral skill requirements.

Employers Are Growing More Demanding

In the past, many teachers tried to deal with the complexity of networking by selecting what was in essence a "network appreciation" book—a feel-good book that lacked the detailed knowledge needed for actual networking jobs.

Today, however, employers demand—and get—strong job readiness from new graduates. If you want to get a job in the IS field, you will need to have a competitive level of knowledge in every IS subject that you study. Even applicants for database jobs are grilled in networking knowledge (and networking applicants are grilled in database and other areas).

How to Study the Book

There are several keys to studying this book:

➤ Reading chapters once will not be enough. You will need to really study the chapters.

➤ Slow down for the tough parts. Some sections will be fairly easy, others difficult. Too many students study the harder stuff at the same speed they use to study the easier stuff.

➤ When you finish studying a section, do the Test Your Understanding questions immediately. If you don't get one of the questions, go back over the text. The understanding of networking is strongly cumulative, and if you skim over one section, you will have problems with other sections later. Multiple choice questions in the test item file are taken entirely from the Test-Your-Understanding questions and the End-of-Chapter questions.

➤ Later, in groups of a few students, go over the Test-Your-Understanding questions to see whether you got the correct answers.

➤ Study the figures. Nearly every key point in each chapter is covered in the figures. If there is something in a figure you don't understand, you need to study the corresponding section in the chapter.

➤ If several concepts are presented in a section or chapter, do not just study them individually. You need to know which one to use in a particular situation, and that requires compare/contrast knowledge. Study the figures that compare concepts, and make your own figures and lists or charts of features if the book does not provide them. Comparing different technologies in order to select the best one for corporate needs is a critical skill for all IT professionals.

➤ Study the synopsis section at the end of each chapter. The synopsis summarizes the core concepts in the chapter. Be very sure that you know them well. You might even study them before beginning the chapter, to get a broad understanding of the material.

Hands-On

One way to make networking less abstract to you is to do as many hands-on activities as possible.

➤ Be sure to do the hands-on exercises in Chapters 1a and 10a.

➤ To really understand TCP/IP, download WinDUMP and play with it. (See Chapter 8a.)

A NETWORKING CAREER

If you like the networking course and think you want a networking career, there are a number of steps you should take before graduation, even if your school does not have advanced networking courses.

➤ Most importantly, do a networking internship. Employers really want workers with job experience—often preferring it to an absurd degree over academic preparation.

➤ Learn systems administration (the management of servers). Learn the essentials of Unix and Windows Server. You can download a server version of Linux and install it on your home computer in order to play with Unix commands and network management functions.

➤ Learn about security. Security and networking are now inextricably intertwined.

➤ Consider getting one or more industry certifications. In networking, the low-level CompTIA Network+ certification should be obtainable with just a bit more study after you take your core networking course. Cisco's CCNA (Cisco Certified Network Associate) certification, which focuses on switching and routing, will require substantially more study. Microsoft server certification is also valuable. Employers like certifications, but they know that certifications are no substitutes for job experience.

About the Author

Ray Panko is a professor of IT management at the University of Hawai'i's College of Business Administration. Before coming to the university, he was a research physicist at Boeing, where he flew on an early flight test of the 747 prototype, and was a project manager at Stanford Research Institute (now SRI International), where he worked for Doug Englebart (the inventor of the mouse). He received his B.S. in Physics and his M.B.A. from Seattle University. He received his doctorate from Stanford University, his dissertation conducted under contract to the Office of the President of the United States. In his spare time, he collects die-cast models and competes in six-seat Hawai'ian outrigger canoe races.

An Introduction to Networking

Objectives

By the end of this chapter, you should be able to discuss the following:

- Networked applications.
- Network standards (protocols).
- Network quality-of-service (QoS) metrics.
- Network security.
- Switched networks.
- Routed networks (Internets).
- LANs and WANs.
- Network management.

NETWORKED APPLICATIONS

Meeting User Needs

The president of Stanley Works once told his board of directors, "Last year, we sold four million drills that nobody wanted." He went on to explain that what customers really want is *holes*. A drill is only a tool to produce holes, and it is not the only possible tool. To customers, tools are merely expensive requirements, not something desired. Only by focusing on what customers want could the company succeed in the marketplace.

What Is a Network?

In that spirit, we will begin with a preliminary definition of the term *network* in terms of user needs. Figure 1-1 shows that a **network** is a communication system which allows application programs on different hosts to work together. An application may be used directly by a person, as is the case for browsers. The application may also operate without direct human intervention, as is the case for webserver programs on World Wide Web server hosts that supply browsers with information.

> Our preliminary definition of a network is that a network is a communication system which allows application programs on different hosts to work together.

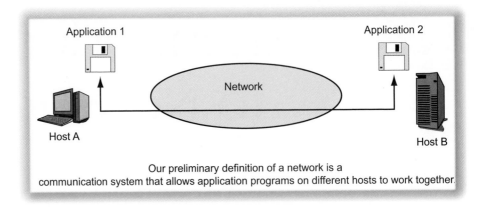

Figure 1-1   Black Box View of Networks

A Focus on Applications

Note that this definition focuses on *applications*. When you type the URL of a web-server into your browser, you don't care about how the Internet works—only that a webserver thousands of kilometers away can send you the information you need. To users, everything else is as exciting as plumbing.

We will spend most of our time in this course looking inside the *Network* bubble in Figure 1-1. As a networking professional, you will be on the team that keeps the bubble working invisibly to users. However, it is important to begin with your goal in mind. In networks, that means thinking about applications.

Hosts

Note that **host** is a general term for devices attached to a network. These devices can be large servers. They can also be small desktop PCs, notebooks, small personal digital assistants, and even cellular mobile telephones. In the future, even coffee makers[1] and many other small devices may be connected to the Internet.

TEST YOUR UNDERSTANDING

1. a) What is the book's preliminary definition of *network*? b) Why does the definition focus on applications? c) What is a host? d) Is your PC at home a host when you use it on a network? Explain. e) Is your mobile phone a host when you use it on a network? Explain.

---

[1] In fact, a protocol for controlling remote coffee pots has already been created. L. Masinter, *Hyper Text Coffee Pot Control Protocol (HTCPCP/1.0)*, RFC 2324 (informational), April 1, 1998. This is an informational request for comments, not a standards-track proposal. You should also note that the protocol was released on April 1—a date called April Fool's Day in many countries and celebrated as a time for playing jokes on people.

Networked Application Standards (Protocols)

Networked applications are applications that require networks to function. The most obvious example is the World Wide Web. Without the Internet, having a browser on your PC host would be useless.

Networked applications are applications that require networks to function.

Network standards govern the exchange of messages between hardware or software processes on different hosts. Network standards are also called **protocols.**[2]

Network standards, which are also called protocols, are sets of rules that govern the exchange of messages between hardware or software processes on different hosts.

If you want to call someone on the telephone, there have to be certain human behavioral protocols that govern the interactions. Most importantly, you both need to speak the same language. There are also subtle behavioral standards for turn-taking and other aspects of the conversation that you follow automatically through experience, but which would be rather complex to describe.

Human Protocols versus Network Protocols

Human communication protocols can be complex and ill-defined because people are intelligent and can cope with complexity, ambiguity, and speech errors. In contrast, computers are stupid (as you learned in your first programming course). Consequently, standards agencies must define network protocols very precisely and usually limit standards to a small number of commands and responses to commands.

HTTP Request–Response Cycle

The Hypertext Transfer Protocol (HTTP) that governs the World Wide Web is a very simple protocol at its core. As Figure 1-2 shows, the browser begins every HTTP interaction by sending an HTTP request message to the webserver application program. Typically, this message asks for a file. The webserver application sends back an HTTP response message, which delivers the file or gives an error code. This ends the HTTP **request–response cycle.**

In contrast to HTTP, many application standards, such as the Simple Mail Transfer Protocol (SMTP) standard for e-mail transmissions, typically require a dozen or more exchanges for a message transmission. Each exchange depends on previous exchanges in the session.

Client/Server Protocols

By the way, HTTP is called a **client/server** protocol because of the way its two application programs relate to each other. The browser is a client, in the sense that it receives services.

---

[2]Technically speaking, not all network standards are protocols. However, most network standards *are* protocols, and all of the standards in this book are protocols.

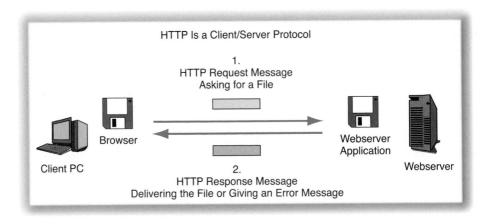

Figure 1-2 Hypertext Transfer Protocol (HTTP)

The webserver is a server because it provides service to clients. Most application program standards today are client/server protocols; but as we will see in Chapter 11, not all are.

In client/server applications, the server program provides services to the client application program. For the WWW, the browser is the client and the webserver is the server.

Many Application Standards

Each application has a different standard. There are many applications, so there are many application protocols. In fact, there are more standards for applications than there are for any other aspect of networking.

Proprietary Standards and Open Standards

There are two types of standards. **Proprietary standards** are created by a vendor to connect the vendor's own products together. Although a vendor may allow other manufacturers to use its proprietary standards for free or upon payment of royalty fees, the vendors that create proprietary standards control these standards, leaving other vendors vulnerable to changes in sharing policies.

In contrast, **open standards** are created by international standards agencies that are not under the control of a vendor. With open standards, vendors can develop products secure in the knowledge that a single vendor will not arbitrarily change those standards.

Open standards allow competition among vendors. Competition tends to reduce prices. It also protects the firm in case its usual vendor goes out of business. In the longer term, it fosters the development of features as vendors struggle to differentiate themselves. These features often are standardized in later versions of the protocol. HTTP is an open standard.

It's All about Standards

Standards are fundamental to networking. In the next chapter, we will look at network standards in some depth, including how they are developed and how standards agencies create *standards architectures* to bring order to the dozens or hundreds of standards they produce. In the chapters following that, we will see many protocols. In Chapter 11, after we have looked at protocols below the standards layer, we will return to application standards.

TEST YOUR UNDERSTANDING

2. a) What is a networked application? b) What is the book's definition of a *network standard?* c) What is a protocol? d) Why must network protocols be specified very precisely? Give a detailed answer. e) What are client/server applications? f) Are there few or many application standards? Explain. g) Distinguish between proprietary and open standards. h) What are the benefits of open standards?

The ARPANET and the Internet

Around 1970, the U.S. **Defense Advanced Research Projects Agency** (**DARPA**) created the **ARPANET**[3] to connect the researchers it funded. This network was restricted to people working on DARPA contracts, although this rule was not tightly enforced. In any case, DARPA funded many computer science researchers, so the ARPANET itself became widely used in computer science.

During the 1970s, several other research networks emerged, such as CSNET, which was open to all computer scientists, and BITNET in the world of social science and business research.[4] In 1980, DARPA changed the ARPANET into the backbone of a new international internet (network of networks), the **Internet.** Later, in the 1990s, commercial networks were allowed to connect to the Internet, and the Internet as we know it today emerged.[5] The Internet is an almost entirely commercial venture. There is almost no government money still flowing in to run the Internet.

Figure 1-3   The ARPANET and the Internet (Study Figure)

ARPANET
     Created by the Defense Advanced Research Projects Agency (DARPA) around 1970
Soon, Many Other Networks Appeared
     Tower of Babel—no interconnection
DARPA Created the Internet in 1980 to Connect Networks Together
     Initially used only by researchers
     Became commercial in the 1990s
     Today, the Internet is commercial
     Almost no government money flowing in to run the Internet

---

[3]Why not DARPANET? In the early years, the agency was simply the Advanced Research Projects Agency of the Department of Defense.

[4]*BIT* stood for "Because it's time!" Business researchers and social sciences wanted in on the action.

[5]Actually, there was an intermediate step. In 1988, the National Science Foundation reengineered and took over the backbone, calling it NSFNET. NSFNET specifically prohibited commercial traffic. This led to the creation of commercial ISPs, which are discussed later in the chapter. In April 1995, NSFNET was retired and the commercial ISPs ran the Internet in the United States. Similar government-to-commercial enterprise transitions quickly took place in other countries.

File Transfer Protocol (FTP)

E-Mail

The World Wide Web (WWW)

The Internet versus the Web
    The Internet is the transmission infrastructure
    The WWW is one of many applications that use the Internet for transmission

E-Commerce
    Buying and selling on the Internet

Figure 1-4   Traditional Internet Applications (Study Figure)

Traditional Internet Applications

Although the Internet is attractive because it allows users to reach millions of other hosts, the Internet became massive in large part because of the applications it offered in its infancy and that it has added over the years.

File Transfer Protocol (FTP)

One initial purpose of the ARPANET was to allow researchers in different locations to transfer large files to each other. Consequently, the ARPANET offered the **file transfer protocol (FTP).** FTP could transfer large files even if there were errors or breaks in the transmission. In addition, FTP could run on any host on the ARPANET, regardless of the computer's operating system. FTP is still popular today.

E-Mail

Before the ARPANET, individual servers had e-mail that allowed users of a single server to send messages to each other. Shortly after the ARPANET emerged, Ray Tomlinson of BBN decided on his own initiative to extend these e-mail so that users could send e-mail to users on different servers. On individual servers, your e-mail address was your user name. Tomlinson realized that Internet e-mail would also have to specify the host name. Looking at his keyboard, he saw a little-used character. He used that character to express a user's e-mail address as *username@mailserver*. Tomlinson's unauthorized e-mail system was soon used everywhere, leading to its rapid standardization. It is difficult to find a business card today that does not have the person's e-mail address.

The World Wide Web (WWW)

The most popular application on the Internet today is the **World Wide Web (WWW),** which uses HTTP as its communication protocol. Using only your browser, you can get access to over a hundred million webservers around the world. On the Web, you can get news, search for information, and even watch repeats of your favorite television

World Wide Web (Application)	E-Mail (Application)	Telnet (Application)	Other Applications
The Internet (Transmission System)			

Figure 1-5   The Internet versus the World Wide Web (and other applications)

shows. While FTP and e-mail were born in the 1970 s, Tim Berners-Lee did not create the WWW until the early 1990 s. However, its growth since then has been explosive.

The Internet versus the World Wide Web

The 'Web is so popular that some people mistake it for the entire Internet. However, many other applications use the Internet. The Internet is a transmission system. It delivers the messages of all applications without prejudice or even awareness of the application message content. The Internet, then, has a transmission level that is universal and an application level that is particular to each application—WWW, Telnet, e-mail, and so forth.

E-Commerce

Initially, the 'Web was simply a system for delivering files to browsers. Soon, however, it became the basis for **e-commerce**, which is buying and selling over the Internet. As we will see in Chapter 11, e-commerce builds on top of the World Wide Web, adding the functionality needed to handle catalogs, purchasing by credit cards, access to multiple other servers to get the information to satisfy the order, and many other things.

TEST YOUR UNDERSTANDING

3. a) Distinguish between the ARPANET and the Internet. b) What is the purpose of FTP? c) How did ARPANET e-mail extend traditional server e-mail? d) Distinguish between the World Wide Web and HTTP. e) When did the World Wide Web appear? f) Distinguish between the World Wide Web and the Internet. g) Distinguish between e-mail and the Internet. h) How are WWW service and e-commerce service related?

Newer Internet Applications

While the World Wide Web, e-mail, and other traditional core applications are still very widely used, the Internet today also offers many newer applications, some of which are already used heavily.

Instant Messaging (IM)

It is difficult to find a college student's PC that does not have an active **instant messaging (IM)** window. People in the user's circle of friends can send messages at any time, leading to an exchange of text messages or even a voice conversation. IM users can also

Instant Messaging (IM)

Streaming Audio and Video
    No need to wait until the entire file is downloaded before beginning to see or
    hear it

Voice over IP (VoIP)
    Telephony over the Internet or other IP networks

Web 2.0
    A hazy term that focuses on using the Internet to facilitate communication
    among users
    Including the creation of communities
    In addition, the users themselves typically generate the content
    Blogs, wikis, podcasts
    Community building sites such as MySpace and Facebook, video sharing sites
    such as YouTube, virtual words such as Second Life, and specific information
    sharing sites such as craigslist

Peer-to-Peer (P2P) Applications
    Growing processing power of PCs reduces the need for servers
    PCs can serve other PCs

Figure 1-6   Newer Internet Applications (Study Figure)

transfer files to one another.[6] This real-time message delivery stands in stark contrast to
e-mail, in which most people only check their mailboxes sporadically for new mail.

Streaming Audio and Video

Downloading audio and video to PCs is very popular today. If you just want to want to
listen to a song or watch a rerun of a television show, waiting for an entire file to download before you can hear or see it is inconvenient. With **streaming media**, you do not
have to wait for the entire file to be downloaded. After you download a few seconds of
the file, you begin to hear it or see it on your system. As you listen to or see the file,
more of it is downloaded in the background. The net result is that you hear or see the
entire file with only a brief lag at the beginning. Streaming media have long been used
in education and training on a limited basis. Now that many people are using streaming media applications, their use in education and training may increase.

Voice over IP (VoIP)

Another new popular application is **voice over IP (VoIP)**. Using a computer with multimedia hardware, users can place calls to people on the Internet and in some services

---

[6]In many cases, users multitask by communicating by IM while doing other things, such as talking on the
telephone. Many IM users feel that multitasking is NBD, but IMHO it is rude to IM while on the phone
with someone else, and there is 411 to back that up. Quite simply, humans have limited attention, and
TNSTAAFL. Yes, I know, WDALMIC?

to people on the traditional Public Switched Telephone Network (PSTN). Companies hope that VoIP can dramatically reduce the costs of long-distance and international telephone calling. As we will see in Chapter 6, many corporations are moving rapidly into VoIP for their corporate communication.

Web 2.0

The term **Web 2.0** was coined by O'Reilly Media in 2005. It is a fairly complex idea,[7] but one of its key elements is that users develop or enhance content, instead of the website owner controlling all of the content. Web 2.0 applications include blogs, wikis such as Wikipedia, podcasts, RSS feeds, social networking sites such as MySpace, video sharing sites such as YouTube, virtual worlds such as Second Life, and specific sharing sites such as craigslist.

One characteristic of Web 2.0 is that users develop or enhance content, instead of the website owner controlling all of the content.

Peer-to-Peer (P2P) Applications

In client/server applications, the user PC receives service from a centralized server. Traditionally, this was necessary because user PCs were underpowered and had to depend on servers. However, most PCs today have far more hard disk space than they can use, as well as spare processing power (especially when they are not being used by their owners). Instead of relying on servers, PC users can now provide services to one other. These PC-to-PC applications are called **peer-to-peer (P2P)** applications.

P2P applications have something of a bad name because the first widely used P2P applications were created for illegal file downloading—first for music files and later for movies. However, corporations are beginning to see the legitimate potential of P2P applications. This includes legitimate corporate file sharing applications. It also includes shared processing power. For instance, many financial firms (which do heavy mathematical modeling) spread the processing load across their idle PCs to contain costs. The unused processing power, storage, and network connectivity of user PCs collectively has been called the dark matter of IT. It may very well be that most processing power and storage in firms lies on user PCs. P2P applications can exploit this underused resource to save corporations money by reducing the need to purchase expensive servers.

TEST YOUR UNDERSTANDING

4. a) Distinguish between e-mail and IM. b) Distinguish between the traditional downloading of complete media files and streaming media. c) What is the benefit of streaming media? d) Describe VoIP. e) What benefits do companies hope to get from VoIP? f) Users create or enhance website content in _____ applications. g) Distinguish between client/server applications and P2P applications. h) What benefit do corporations see in peer-to-peer applications?

Corporate Networked Applications

So far, we have been looking at applications that you probably use as a student and in your personal life outside school. However, corporations also have specific applications

---

[7]For more information, see Tim O'Reilly, "What is Web 2.0," September 30, 2005. http://www.oreillynet.com/pub/a/oreilly/tim/news/2005/09/30/what-is-web-20.html

Applications Specific to Businesses
  Can consume far more corporate network resources than traditional and new
  Internet applications combined

Transaction-Processing Applications
  High-volume repetitive clerical transaction applications
  Accounting, payroll, billing, etc.

Enterprise Resource Planning (ERP) Applications
  Serve individual business functions while providing integration between functional modules

Organizational Communication Applications
  E-mail, etc.
  Groupware
    Integrate multiple types of communication, organize communication for retrieval, and provide multiple ways to disseminate and retrieve information

Converged Networks
  Voice and video have traditionally used different networks than data
  Convergence: Moving voice/video and data networks to a single network
  Can save the corporation a great deal of money by having only a single network
  Many technical issues remain

Figure 1-7　Corporate Network Applications (Study Figure)

that they need on both the Internet and on their internal networks. In most large organizations, these corporate network applications consume much more traffic than traditional and new Internet applications combined.

Transaction-Processing Applications

When corporate information systems began, the focus was on transaction-processing applications, which are clerical applications with high volumes of simple repetitive transactions, such as for payroll, billing, and accounts payable applications. Today, we call such applications **transaction-processing applications.** Although they are not exciting, they consume a good portion of a corporation's network resources and need to be addressed carefully in network planning. Poor network performance for transaction processing applications can be extremely costly to corporations in terms of lost productivity.

Enterprise Resource Planning (ERP) Applications

Originally, companies purchased "best of breed" transaction processing applications for individual business functions, such as payroll and inventory management. However, many corporate processes cut across functions, and if a corporation has selected applications on a function-by-functions basis, these applications probably will have a difficult time working together, if they can interoperate at all. Many firms now have or are beginning to install **enterprise resources planning (ERP)** applications that serve individual business functions while providing smooth integration between functional modules. These are also called **enterprise applications.**

Organizational Communication Applications

Organizations use e-mail and other traditional communication applications, but they also need more structured communication for project management and other disciplined processes. **Groupware** applications integrate multiple types of communication, organize communication for retrieval, and provide multiple ways to share information.

> Groupware applications integrate multiple types of communication, organize communication for retrieval, and provide multiple ways to share information.

Converged Voice/Data Networks

Most organizations today have two networks—one for voice and one for data. As we will see in Chapter 6, telephony traditionally has used very different technology than have data networks. However, the growth of VoIP has led to hope for **converged networks** that can serve both voice and data. Having only a single network should lower technology and management costs. However, voice and video have special needs which raise technical issues that have not been entirely solved.

TEST YOUR UNDERSTANDING

5. a) What are the characteristics of transaction processing applications? b) Why do companies purchase enterprise resource planning (ERP) applications? c) What are groupware applications? d) What is the potential benefit of voice/data convergence?

File Service

So far, we have looked at Internet applications and "big" corporate applications. However, there is another networked application that almost everyone uses in corporations: file service. As Figure 1-8 shows, **file service** is provided by a **file server;** in effect, it provides users with a shared hard drive that is accessible over a network.

> As Figure 1-8 shows, file service is provided by a file server; in effect, it provides users with a shared hard drive that is accessible over a network.

File Service for Data

On your PC's local hard drive, you store many data files—word-processing documents, spreadsheets, photographs, movies, and so forth. The figure shows that you can do the same thing on a file server. You are given a certain amount of space on the file server's hard drive, and file service manages your access to that space.

Backup   Your file server's systems administrator (the person who manages the file server) probably backs up the file server every night. (In contrast, you probably back up your own hard drive once a lifetime). It is a good idea to copy files you work on to the file server. In fact, you might never want to store files on your local hard drive—but always store them on the file server.

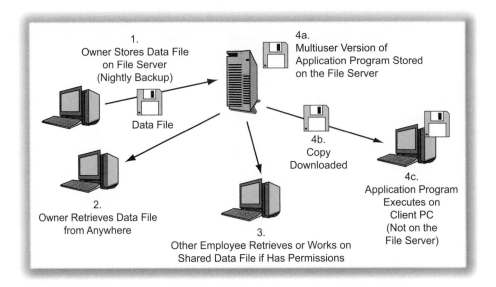

Figure 1-8   File Service

Access from Anywhere   When you are away from your office, you may need some of your office files. Most file servers allow remote access so that you can retrieve your files anywhere there is a network connection.

Sharing with Security   If you wish, you can allow other people to share some or all of your files. While you might be reluctant to do so for your main files, the systems administrator can set up directories for specific purposes, such as projects. You can then decide who may and who may not access that directory. You can also limit what authorized users can do in the directory. In some cases, you will want to give others read-only access so that they can only read files and cannot create new files, delete files, modify files, or take other actions. In other cases, you will allow them to take actions beyond mere reading.

File Service for Programs

Programs are merely files or collections of files, so you can store them on a file server.

Multiuser Software: Install Once   In fact, you can purchase a multiuser copy of a program and install it once on a file server, instead of having to install a copy on each PC individually. (Of course, multiuser copies are more expensive than the normal single-user copies sold in stores.) The labor savings from not doing multiple PC installations is substantial.

Execution on the Client PC   On your PC, when you execute a program saved on your local hard drive, that program is copied into RAM for execution. A file server, again, is like a remote hard drive. When you run a program stored on a file server, the file server sends your PC a copy of the program, and your PC stores it in RAM for execution.

This surprises many people. They assume that, because the file server is usually a more powerful computer than your desktop PC, the file server will do the "heavy work" of executing programs. Instead, your "little" PC has to do the processing. If this seems strange, remember that a file server is only a remote disk drive.

TEST YOUR UNDERSTANDING

6. a) To what PC component does a file server correspond? b) What are the three advantages of file service for data files? c) What is the advantage of storing multiuser versions of programs on a file server? d) When you run a program that is stored on a file server, where does the program execute—on the file server or on your PC?

Hints for Readers

KEYWORDS

Words shown in boldface are **keywords.** You will see a keyword printed in boldface on the page where the keyword is first defined or at least characterized. Pages where this occurs are numbered in boldface in the book's Index. When you look for a term in the Index, you should go to the boldface page number first. You can also go to the book's Glossary to get a definition of a keyword. There is a searchable version of the Glossary on the book's website.

KEY POINTS

In some cases, key points are set centered on the page, in smaller print, and with paragraph borders, as shown below. These are important points that you should stress when you study.

Points shown this way are important points that you should stress when you study.

TEST YOUR UNDERSTANDING QUESTIONS

At the end of many subsections, there are Test Your Understanding questions. Students who do the best in courses typically stop when they encounter Test Your Understanding questions and answer them before going on. In networking, concepts build on one another. If you understand each section as you go, you will be much more effective when you study later sections.

BOXES: ADVANCED MATERIAL

This material is being presented in a box. In this case, and in some other cases, the boxed material is designed to give you perspective or guidance. In most cases, however, the boxes contain advanced material. You can get a good understanding of each chapter without the advanced information in boxes, but the boxed material will give you a better understanding.

FOOTNOTES

This book uses footnotes to provide references or to provide explanations that are beyond the scope of the book, but that some readers may find interesting. None of the questions in the book's test bank asks about material in footnotes, but your instructor may include the material in footnotes in some of his or her own test questions.

> Quality of Service (QoS)
>   Indicators of network performance
>
> Metrics
>   Ways of measuring specific network quality-of-service variables

Figure 1-9   Network Quality of Service (QoS)

NETWORK QUALITY OF SERVICE (QoS)

In the early days of the ARPANET and the Internet, networked applications amazed new users. However, these users soon said, "Too bad this thing doesn't work better." Today, though, networking is a mission-critical service for corporations. If the network breaks down, much of the organization comes to a halt. Today, networks must work, and they must work *well*. Companies are concerned with network **quality-of-service (QoS)** measures, that is, indicators of network performance. Companies typically use a number of QoS **metrics** (measures) to quantify their quality of service so that they can set targets and determine whether they have met those targets.

TEST YOUR UNDERSTANDING

7. a) What are QoS measures? (Do not just spell out the acronym.) b) How are QoS metrics used?

Transmission Speed

There are many ways to measure how well a network is working. The most fundamental metric is speed. Just as the first question people ask about a newborn baby is whether it is a boy or a girl, the first thing that people ask when they encounter a new network is, "How fast is it?"

Bits per Second (bps)

Transmission speed[8] normally is measured in **bits per second (bps). A bit** is either a one or a zero. Obviously, a single bit cannot convey much information. Speeds today range from thousands of bits per second to trillions of bits per second. To simplify the writing of transmission speeds, professionals add metric prefixes to the base unit, bps. For example, Figrue 1-10 shows that in increasing factors of 1000 (not 1024 as with computer memory), we have **kilobits per second (kbps), megabits per second (Mbps), gigabits per second (Gbps),** and **terabits per second (Tbps).**

Speeds are measured in factors of 1000, not 1024.

---

[8]Purists correctly point out that *speed* is the wrong word to use to describe transmission rates. At faster transmission rates, bits do not physically travel faster. The sender merely transmits more bits in each second. Transmission rates are like talking faster, not running faster. However, transmission rates are called transmission speeds almost universally, so we will follow that practice in this book.

Speed
> Normally measured in bits per second (bps) *not* bytes per second
>> Metric suffixed in increasing units of 1,000 (not 1,024)
>> The metric abbreviation for kilo is lower-case k

1 kbps	1,000 bps	kilobits per second
1 Mbps	1,000 kbps	megabits per second
1 Gbps	1,000 Mbps	gigabits per second
1 Tbps	1,000 Gbps	terabits per second

> Sometimes, speed is measured in bytes per second, Bps, compared with bps
>> Bps usually is seen only in file transfers

Expressing Speed in Proper Notation

As Written	Places before Decimal Point	Properly Written
23.72 Mbps	2	23.72 Mbps
2,300 Mbps	4	2.3 Gbps
0.5 Mbps	0 (leading zeros do not count)	500 kbps

> There must be one to three spaces before the decimal point
>> Leading zeros do not count
> There must be a space between the number and the units
>> 12 Mbps is proper; 12Mbps is improper
> If the number is decreased by 1,000 (4523 becomes 4.523), then the suffix must be increased by 1,000 (Mbps to Gbps)
>> 4,523 Mbps becomes 4.523 Gbps
> If the number is increased by 1,000 (0.45 becomes 450), then the suffix must be decreased by 1,000 (Mbps to kbps)
>> 0.45 Mbps becomes 450 kbps

Rated Speed and Throughput
> Rated Speed
>> The speed a system should achieve according to vendor claims or to the standard that defines the technology
> Throughput
>> The data transmission speed a system actually provides to users
> Aggregate versus Rated Throughput on Shared Lines
>> The aggregate throughput is the total throughput available to all users
>> The individual throughput is an individual's share of the aggregate throughput

Figure 1-10   Transmission Speed (Study Figure)

Note that, consistent with metric notation, kilo is abbreviated as lower-case k instead of upper-case K. (In the metric system, K is the metric abbreviation for Kelvins, a measure of temperature.) Computer scientists often write K, but this is because they do not know the metric system. Networking people are smarter.

Bytes per Second

Although transmission speed is almost always measured in bits per second, it is occasionally measured in **bytes per second.** (A byte is eight bits.) While *bits per second* is written as bps, *bytes per second* is written as **Bps.** About the only time you will see Bps is in file transfers because file sizes normally are measured in bytes.

Writing Numbers in Proper Notation

The basic rule for writing speeds (and metric numbers in general) in proper notation is that there should be one to three places before the decimal point and that there should be a space between the number and the units. Figure 1-10 illustrates how to write speeds properly.

To write a speed in proper notation, there should be one to three places before the decimal point, and there should be a space between the number and the units.

➤ Given this rule, 23.72 Mbps is fine (two places before the decimal point).

➤ However, 2300 Mbps has four places before the decimal point (2300.00), so it should be rewritten as 2.3 Gbps (one place before the decimal point).

➤ Also, 0.5 Mbps has zero places to the left of the decimal point. (Leading zeros do not count.) It should be written as 500 kbps (three places).

Note also that there must be a space between the numerical prefix (23.72, etc.) and its metric suffix (Mbps, etc.). So 23.72 Mbps is proper, but 23.72Mbps is improper.

Suppose you have the speed 4,523 kbps. To get the number (4,523) right, you divide it by 1,000 to get 4.523. If you *divide the number* by 1,000, then you must *multiply the suffix* (kbps) by 1,000 to get Mbps. In proper notation, then, the number is 4.523 Mbps.

To give another example, suppose you have 0.45 Mbps. You need to *multiply the number* (0.45) by 1,000, getting 450. You then have to *divide the suffix* (Mbps) by 1,000 to give you kbps. The number in proper notation, then, is 450 kbps.

Rated Speed versus Throughput

Talking about transmission speed can be tricky. A network's **rated speed** is the speed it *should* achieve according to vendor claims or to the standard that defines the technology. For a number of reasons, networks often fail to deliver data at their rated speeds. In contrast to rated speed, a network's **throughput** is the data transmission speed that the network *actually* provides to users.

Note: Some students find the distinction between rated speed and throughput difficult to learn. However, we must use this distinction throughout the book, so be sure to take the time to understand it.

Network Demand, Budgets, and Decisions (see Figure 1-12)
    Network demand is growing explosively, while network budgets are growing slowly
    This creates a cost squeeze that affects every decision
    Overspending in one area will result in the inability to fund other projects

Systems Development Life Cycle Costs
    Hardware: Full price: base price and necessary options
    Software: Full price: base price and necessary options
    Labor costs: Networking staff and user costs
    Outsourcing development costs
    Total development investment

Systems Life Cycle (SLC) Costs
    System development life cycle versus system life cycle
    Total cost of ownership (TCO)
        Total cost over entire life cycle
        SDLC costs plus carrier costs
            Carrier pricing is complex and difficult to analyze
            Must deal with leases

Figure 1-11   Cost (Study Figure)

Throughput is the data transmission speed that a network **actually** provides to users.

Aggregate versus Individual Throughput

When a transmission line on a network is **multiplexed,** this means that several conversations between users will share the line's throughput. Consequently, it is important to distinguish between a line's **aggregate throughput,** which is the total it provides to all users who share it, and the **individual throughput** that single users receive as their shares of the aggregate throughput. As you learned as a child, sharing is theoretically good, but it means that you get less.

TEST YOUR UNDERSTANDING

8. a) In what units is transmission speed normally measured? b) Is speed normally measured in bits per second or bytes per second? c) Give the names and abbreviations for speeds in increasing factors of 1000. d) What is 55,000,000,000 bits per second with a metric suffix? e) Write out 100 kbps in bits per second (without a metric suffix). f) Write the following speeds properly: 0.067 Mbps, 23,000 kbps, 45.62 Gbps, and 13kbps.
9. a) Distinguish between rated speed and throughput. b) Distinguish between individual and aggregate throughput.

Cost
Network Demand, Budgets, and Decisions

Although speed is important, so is cost. Most of us would like to drive Ferraris. Few of us need a high-performance car, however, and fewer could afford one.

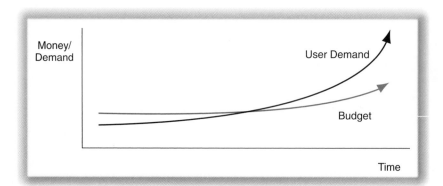

Figure 1-12   Network Demand and Budget

Figure 1-12 shows how demand and budget for network services are increasing over time. Demand, of course, is growing very rapidly. Everybody wants network applications and high-speed service. However, network budgets generally are growing slowly, if they are growing at all.

This creates a cost squeeze that governs every aspect of network management thinking. Network managers need to limit what they do and purchase only what they need. Overspending on one project will leave managers unable to do another project. In addition, corporations are requiring strong justification for new services and existing services alike.

In this book, you will encounter many questions asking you to compare alternative technologies that could be used to serve a need. In general, an answer to such a question will be wrong if it omits relative costs.

Systems Development Life Cycle (SDLC) Costs

You are undoubtedly familiar with the systems development life cycle (SDLC). When you begin a project, you need to consider the cost of the system during its development.

Hardware Costs   Consider what happens when you buy a personal computer. You first have to take hardware into account. When you look at the price of a computer, this may not include a display, and it usually does not include a printer. The **base price** is the price before adding components that will be needed in actual practice. In contrast, the **full price** of the hardware is the price of a complete working system. The distinction between base price and full price is also applicable to network hardware, including the switches and routers we will see later in this chapter.

Software Costs   After your first computer purchase, you realize that the software can be almost as expensive as the hardware. You have to consider the software you will need very carefully and understand the cost of that software. Individual software products, furthermore, often have misleading base prices that do not include all necessary components. Network product software decisions are similarly complex.

Labor Costs in Development   Although hardware and software costs are complex and difficult to measure, these problems pale before the problems involved in estimating labor costs in development. Planning, procurement, installation, configuration, testing, programming, and other labor costs can easily exceed hardware and software costs.

User costs should also be considered because the time that users spend on the system's development during requirements definition and later development states is substantial and is not free to the company, any more than network staff time is free.

Outsourcing Development Costs   If the company outsources some or all of the development costs, then outsourcing costs need to be considered in the overall picture.

Total Development Investment   To evaluate potential projects, the networking staff must forecast the total development investment—the total of hardware, software, and labor costs during development. These expenditures truly are investments that should pay off over the life of the project.

Systems Life Cycle (SLC) Costs

The preoccupation of information systems professionals with the systems development life cycle has always puzzled networking professionals, who note that most costs come *after* the SDLC has finished. Training IS (information system) students in SDLC costs, but not in operational work and costs afterward, is like training doctors only in obstetrics and ignoring care after birth.

It is important to consider **system lifecycle costs,** which are costs over a system's entire life—not just during the systems development period. The cost of a system over the entire life cycle is called the **total cost of ownership (TCO).**

Operating and management costs usually are very important over the system life cycle. When making equipment and software purchases, it is important to consider how much labor is involved in operating and managing the equipment and software. These costs must be considered very carefully in product selection.

One new factor in systems life cycle analysis is carrier costs. If you must deal with a communications carrier to carry your signals from one corporate site to another, then you also have to consider carrier pricing. This is rarely easy to do, and it is even harder to compare the prices of alternative carriers offering roughly the same service, because of the wording in their contracts. In addition, you usually have to sign equipment leases or service agreements that lock you in for various periods of time, sometimes up to several years.

TEST YOUR UNDERSTANDING

10. a) Compare network demand trends and network budget trends. b) What are the implications of these trends? c) What period of a network's life does the SDLC cover? d) Why are hardware and software base prices often misleading? e) List the four categories of SDLC costs. f) Why must user costs be considered?

11. a) Distinguish between the systems development life cycle and the system life cycle. b) What is the total cost of ownership (TCO)? c) Why should operating and management costs be considered, in addition to hardware, software, and transmission costs, in purchasing decisions? d) What additional cost factor comes into SLC costs, compared with SDLC costs? e) Why must carrier contracts be entered into carefully?

Other Quality-of-Service Metrics

Availability

Although speed and cost are fundamental, there are several other important measures for how a network is working. One of these is metrics is **availability,** which is the percentage of time that the network is available for use. In contrast, **downtime** is the percentage of time that the network is not available.

Ideally, systems would be available 100% of the time, but that is impossible in reality. On the Public Switched Telephone Network, the availability target usually is 99.999%. This is known as the "five nines." Data networks generally have lower

Figure 1-13   Other Quality-of-Service Metrics (Study Figure)

Speed

Cost

Availability
    The percentage of time a network is available for use
    Downtime is the amount of time a network is unavailable (minutes, hours, days, etc.)

Error Rates
    Packet error rate: the percentage of packets lost or damaged
    Bit error rate: the percentage of bits lost or damaged

Latency and Jitter
    Latency
        Delay, measured in milliseconds
    Jitter
        Variation in latency between successive packets
        Makes voice sound jittery

Service Level Agreements
    Guarantees for performance
    Penalties if the network does not meet its service metrics guarantees
    Guarantees specify worst cases (no worse than)
        Lowest speed (e.g., no worse than 1 Mbps)
        Maximum latency (e.g., no more than 125 ms)
    Often written on a percentage basis
        No worse than 100 Mbps 99.5% of the time

availability, but are under pressure to improve their availability, given the cost of network downtime to firms today.[9]

Error Rates

We will see later in this chapter that hosts send data in small messages called packets. Ideally, all packets would arrive intact, but this does not always happen. The **packet error rate** is the percentage of packets that are lost or damaged during delivery. The **bit error rate,** in turn, is the percentage of bits that are lost or damaged.

Most networks today have very low average error rates. However, when network traffic is very high, error rates can soar because the network has to drop the packets it cannot handle. Companies must measure error rates when traffic levels are high in order to have a good understanding of error rate risks.

Latency and Jitter

When packets move through the network, they will encounter some delays. The amount of delay is called **latency.** Latency is measured in **milliseconds (ms).** A millisecond is a thousandth of a second. When latency reaches about 125 milliseconds, turn-taking in telephone conversations becomes difficult.

A related concept is **jitter,** which Figure 1-14 illustrates. Jitter occurs when the latency between successive packets varies. Some packets will come too far apart in time, others too close in time. While jitter does not bother most applications, VoIP and streaming media are highly sensitive to jitter. If the sound is played back without adjustment, it will speed up or slow down hundreds of times per second. These speed changes make voice sound jittery.

Service Level Agreements

When you buy some products, you receive a guarantee promising that they will work and specifying penalties if they do not work. In networks, service providers often

Figure 1-14   Jitter

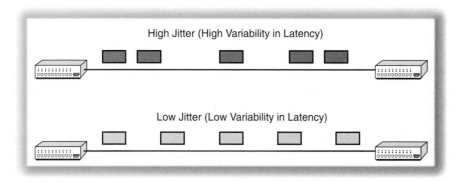

High Jitter (High Variability in Latency)

Low Jitter (Low Variability in Latency)

---

[9]On a more detailed basis, availability can be discussed in terms of the mean time to failure (MTTF) and the mean time to repair (MTTR). The former asks how frequently downtime occurs. The latter asks how long service is down after a failure begins. Frequent short failures may be preferable to infrequent, but very long, outages.

provide **service level agreements (SLAs),** which are contracts that guarantee levels of performance for various metrics such as speed and availability. If a service does not meet its SLA guarantees, the service provider must pay a **penalty** to its customers.

SLAs guarantees are expressed as *worst cases*. An SLA might guarantee that speed will be *no lower* than a certain amount, that the latency will be *no higher* than a certain value, and so forth.

SLAs guarantees are expressed as **worst cases.**

In addition, SLAs have percentage-of-time elements. For instance, an SLA on speed might guarantee a speed of at least 480 Mbps 99.9% of the time. This means that the speed will nearly always be at least 480 Mbps, but may be slower 0.1% of the time without incurring penalties.

TEST YOUR UNDERSTANDING

12. a) What is availability? b) What is downtime? c) What are the "five nines"? d) Does corporate network availability usually meet the five nines expectation of the telephone network? e) What are packets? f) Distinguish between the packet error rate and the bit error rate. g) When should error rates be measured? Why? h) What is latency? i) In what units is latency measured? j) What is jitter? k) For what applications is jitter a problem?

13. a) What are service level agreements? b) Does an SLA measure the best case or the worst case? c) Would an SLA measure the highest latency or the lowest latency? d) Would an SLA guarantee a lowest availability or a highest availability? e) What happens if a carrier does not meet its SLA guarantee? f) If carrier speed falls below its guaranteed speed in an SLA, under what circumstances will the carrier *not* have to pay a penalty to the customers?

SECURITY

In addition to network quality of service, corporations today are extremely concerned with network security. Attacks by external and internal adversaries can be extremely expensive for corporations. We will look at network security in depth in Chapter 9. For now, however, we will discuss a few security concepts that we will see in this book before we reach Chapter 9.

Authentication

In a famous Peter Steiner cartoon in *The New Yorker*, a dog is sitting at a computer keyboard. The dog brags to another dog in the room: "On the Internet, nobody knows you're a dog."[10] In a network, a host cannot see the user trying to log into it. To **authenticate** the user—that is, have the user prove his or her identity—the server needs proof of identity, most commonly, a password.

In general, a person trying to establish his or her identity is called the **supplicant,** and the device attempting to authenticate the user is called the **verifier.** The supplicant sends **credentials** (proofs of identity) to the verifier, and the verifier checks these credentials. If authentication is done well, impostors will not be able to pose as legitimate users.

---

[10]*The New Yorker*, vol. LXIX, no. 20, July 5, 1993, p. 61.

Security
        Attacks can be extremely expensive
        Companies need to install defenses against attacks
        Chapter 9 discusses network security in depth
Authentication
        Supplicant attempts to prove its identity to a verifier
        Example: user logging into a server is a supplicant; the server is a verifier
        Proofs of identity are called credentials
        If authentication is done well, impostors will not be able to pose as legitimate users.
Cryptographic Protections
        Eavesdroppers may intercept your messages
                Read and even change messages
                Send new messages impersonating the other side
        Cryptography is the use of mathematics to protect information in storage or in transit
                Also when in storage
        Encryption for confidentiality
                An eavesdropper cannot read encrypted messages
                Legitimate receiver, however, can decrypt the message
Firewalls
        Examines each passing packet
        Drops and logs provable attack packets
        It lets other packets get through
Host Hardening
        Some attacks will inevitably get past safeguards and reach hosts
        Hosts must be hardened to withstand attacks
        Hardening is a set of protections we will see in Chapter 9
                Example: installing antivirus software on the host
                Example: downloading security updates

Figure 1-15   Network Security (Study Figure)

Cryptographic Protections

When you transmit packets, there is a danger that an eavesdropper will intercept them. The eavesdropper will then be able to read your sensitive information and even change packets to send false information to the receiver. The interceptor will also be able to send new messages in your name. To address these threats, network planners turn to **cryptography,** which is the use of mathematics to protect information in storage or in transit.

Cryptography is the use of mathematics to protect information in storage or in transit.

The most obvious cryptographic protection is **encryption for confidentiality.** Encryption for confidentiality effectively scrambles messages so that interceptors cannot read them. The legitimate receiver, of course, can **decrypt** messages and read them.

Firewalls

Another security tool is the firewall. A **firewall** examines each packet passing through it. If a packet is a **provable attack packet,** the firewall drops and logs the packet. Otherwise, it lets the packet through.

Host Hardening

Although firewalls will stop most attacks, some attack packets will inevitably get through to clients and servers. In Chapter 9, we will see a number of ways to **harden** hosts so that they can withstand attacks. An obvious example of host hardening is installing antivirus software on desktop and notebook PCs. Another example is downloading security updates for software running on the host.

TEST YOUR UNDERSTANDING

14. a) What is authentication? b) Why is authentication needed? c) Describe these authentication terms: supplicant, verifier, and credentials. d) What is cryptography? e) What does confidentiality mean in transmission? f) What do firewalls do? g) What types of packets do firewalls drop? (Be very specific.) h) Why is host hardening necessary?

SWITCHED NETWORKS

Now that we have looked at networked applications and quality-of-service metrics, we can finally begin to look at how networks do their jobs. We will look first at simple *switched* networks and then at more complex *routed* networks, which are also called internets.

Ethernet Switching with a Single Switch

There are several switched network technologies. We will focus on Ethernet switching, which most corporations use for networking within their buildings. In the smallest Ethernet networks, all host computers connect to a single Ethernet switch, which handles the transmission of messages between hosts. Figure 1-16 shows one of these small Ethernet networks.

A switch has multiple **ports** (connection points for transmission lines). The company must make a switching decision when a frame comes into one port. The decision is what port to use to send the frame back out. How does it make this **switching decision**? The answer is that it uses the destination address that must appear in each message.

In Ethernet, each host has a unique **Ethernet address,** which looks something like C3–2D–55–3B–A9–4F.[11] (We will look at Ethernet addresses more closely in Chapter 4.) When a source host sends an Ethernet message, the message has a destination Ethernet address, which specifies the Ethernet address of the destination host. This is like the address on a postal envelope.

When the switch receives a message, the switch looks at the destination address. In this example, the destination Ethernet address is C3–2D–55–3B–A9–4F. As the figure shows, the switch has a switching table that tells the switch what port to use to send the frame back out. For C3–2D–55–3B–A9–4F, the associated port is port 15. The switch sends the frame out through that port, to the destination host.

---

[11]As we will see in Chapter 4, Ethernet addresses are also called MAC addresses and, sometimes, physical addresses.

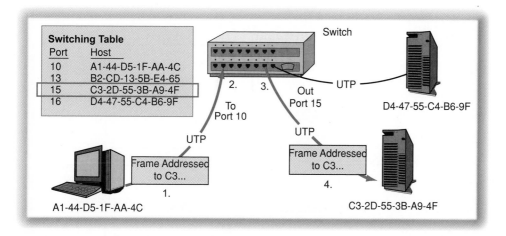

Figure 1-16   Ethernet Switching

TEST YOUR UNDERSTANDING

15   a) On a switch, what is a port? b) What is a switching decision? c) On what do Ethernet switches base switching decisions? d) What does an Ethernet switch do after it reads the destination Ethernet address in an arriving frame?

Workgroup Switches and Core Switches

In all but the smallest Ethernet networks, there are multiple Ethernet switches. Each switch sends a message closer to its final destination. Think of a railroad boxcar that passes through several switching yards along the way to its destination. Boxcars are analogous to messages, and rail yards are analogous to switches.

Figure 1-17 shows a switched network in a multistory building. The figure shows a workgroup switch on each floor. A **workgroup switch** is one that connects PCs and servers to the network. There is a core switch down in the basement equipment room. **Core switches** connect switches to other switches and to other devices (called routers) that we will see later in this chapter.

➤   When the wired client on Floor 1 transmits a message to the server on Floor 2, the client sends the message to the workgroup switch on its floor.

➤   That workgroup switch realizes that it cannot deliver the message directly to a computer attached to it. The workgroup switch then forwards the message to the core switch in the basement equipment room.

➤   The core switch sends the message to the workgroup switch on Floor 2.

➤   The workgroup switch sends the message on to the server.

TEST YOUR UNDERSTANDING

16.   a) Distinguish between core and workgroup switches. b) In Figure 1-17, how many workgroup switches are there? c) How many core switches? d) Suppose that there is a server connected to Workgroup Switch 1. Through what switches will messages travel if they are sent by the wireless client on that floor?

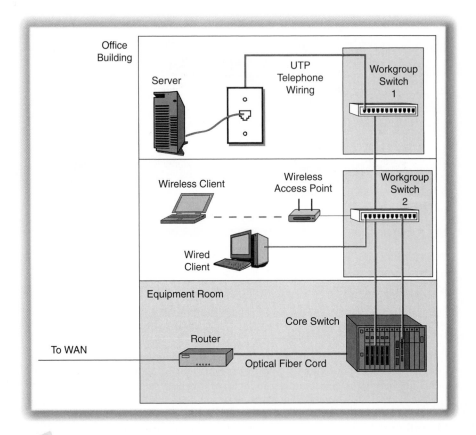

Figure 1-17   Switched Network in a Multistory Building

Access Lines, Trunk Lines, and Multiplexing

Access Lines

The devices in Figure 1-17 are connected by transmission lines. The transmission lines *between hosts and workgroup switches* are called **access lines** because they give a host access to the network. In the figure, the horizontal lines on the first and second floors of the building are access lines.

Most access lines use **4-pair unshielded twisted pair (UTP)** copper wiring, which is illustrated in Figure 1-18. This type of wiring looks like your home telephone wiring but is somewhat thicker and terminates in thicker connectors. The building in Figure 1-17 also augments its wired network with a wireless access point that allows wireless PCs to connect to the building's server on the wired network.

Trunk Lines

In turn, transmission *between switches* takes place over **trunk lines.** These are the vertical lines in Figure 1-17. As we will see in Chapters 3 and 4, trunk lines in local area

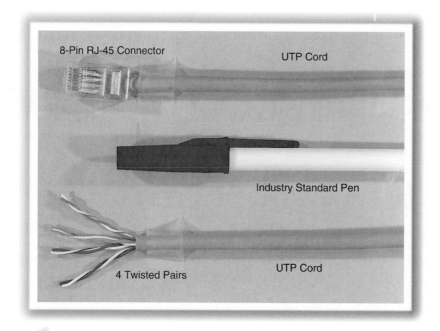

Figure 1-18   Four-Pair Unshielded Twisted Pair (UTP) Copper Wiring

networks can either use 4-pair UTP or optical fiber (which transmits signals as light pulses through a thin flexible glass rod).

Access lines typically serve a single host. However, trunk lines must carry the transmissions of many hosts. Consequently, trunk lines usually need to have much higher transmission speeds than do access lines.

TEST YOUR UNDERSTANDING

17. a) Distinguish between access lines and trunk lines. b) Which type of line needs higher speeds—access lines or trunk lines? Why? c) What type of line is the connection between the Server host and Workgroup Switch 2 in Figure 1-17? d) What type of line is the connection between Workgroup Switch 1 and the Core Switch in Figure 1-17? e) What transmission media do most access lines use? f) Is the 4-pair UTP cord coming out of your PC and plugging into the network wall jack a trunk line or an access line? Explain.

Packet Switching

All switches today, except many of those used for telephony, use a process called packet switching. Packet switching grew from an earlier technology, **message switching.** In message switching, entire messages were sent through switched networks.

Unfortunately, messages varied from short to very long. This created two problems. First, it created problems for trunk lines. Trunk lines **multiplex** (mix) the traffic of many conversations between pairs of hosts. It is much cheaper to have messages

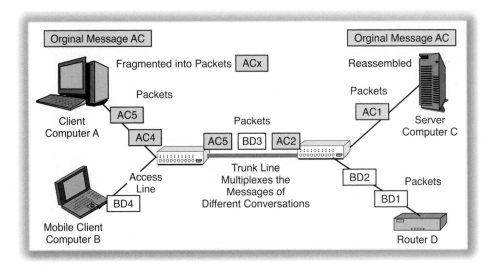

Figure 1-19   Packet Switching and Multiplexing

share a single large line than to have a different line for each message. (Think how expensive it would be if every car on a freeway had its own lane.) Unfortunately, for obscure statistical reasons,[12] it is difficult to fill trunk lines efficiently when messages have very different lengths and when some of them are very long.

A second problem with message switching is that, if there was even a single-bit error, the entire message had to be retransmitted. In long messages, the probability of an error was significant, and frequent retransmissions added to transmission inefficiency.

Figure 1-19 shows that most data networks today use a more sophisticated transmission method called **packet switching.** In packet switching, the source host breaks long application messages into a number of short messages. We have seen earlier that these short messages are generically called **packets.** The host sends each packet out separately. (Think of cars leaving individually to go to a restaurant after a football game.)

Statistically, packet switching can fill multiplexed trunk lines more efficiently than message switching can, driving down trunk line costs. In addition, if there is a transmission error, only the single short packet containing the error needs to be transmitted—not the entire original message. This also reduces trunk line costs by avoiding long retransmissions.

Now for something really confusing: In switched networks, packets are called **frames.** In other words, switched networks do packet switching, but they call their packets frames. This is done to confuse networking students, and it usually succeeds.

In switched networks, packets are called frames. So switched networks do use packet switching, but they call their packets frames.

---

[12]To give a very loose comparison, you can fill a jar more fully with sand than with rocks.

TEST YOUR UNDERSTANDING

18. a) What is multiplexing? b) What is the benefit of multiplexing? c) What is packet switching? d) What does packet switching do to application messages? e) Distinguish between message switching and packet switching. f) Why is packet switching better than message switching? g) What are packets called in switched networks?

ROUTED NETWORKS (INTERNETS)
Routers and Routed Networks

By the 1980s, there were many switched network technologies. They were deeply incompatible, and it was impossible for a host on one type of switched network to communicate with a host on another type of switched network. Even networks that used the same switching technology were rarely connected together. The situation was a networking Tower of Babel.

The solution to this inability to communicate across networks was developed by Vint Cerf and Bob Kahn.[13] As Figure 1-21 shows, their solution was to add another level of transmission functionality on top of switched networks by connecting networks with devices called **routers.** Cerf and Kahn originally called routers **gateways**—a term that still sees some use. Networks like the Internet, then, are **routed networks.** They are also called internets because they are networks of networks.

Figure 1-20   Routed Networks (Study Figure)

The 1980s: A Switched Tower of Babel

Routers and Routed Networks
  Routers connect different switched networks together
  Routed networks are also called internets
  Routers are more complex (and expensive) than switches
    Designed to work no matter how complex the network
    Require more hands-on administration than switches

IP Addresses
  IP addresses
  Switched network addresses
  Two addresses for each host

Packets and Frames
  Packets are called frames in switched networks
  Packets are called packets in routed networks
  A packet is carried in a frame within each switched network

---

[13]V. Cerf and R. Kahn, "A Protocol For Packet Network Intercommunication," *IEEE Transactions on Communication,* vol. C-20, no. 5, May 1974, pp. 637–648.

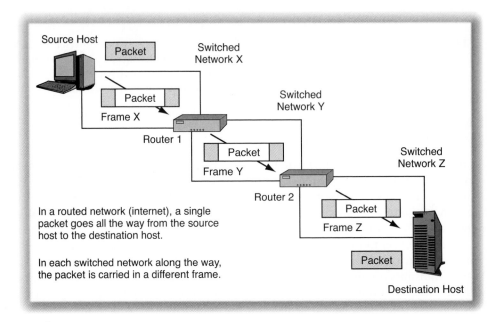

Figure 1-21   Routed Network (Internet)

This leads to the following definition of a routed network: A routed network (internet) is a group of networks connected by routers so that any application on any host on any switched network in the internet can communicate with any application on any other host on any other switched network in the internet.

A routed network (internet) is a group of networks connected by routers so that any application on any host on any switched network in the internet can communicate with any application on any other host on any other switched network in the internet.

Routed networks perform packet switching. However, while packets are called frames in switched networks, packets are called packets in routed networks.

How are switching and routing different? Most importantly, switching usually is simple and therefore inexpensive. In contrast, we will see in Chapter 8 that router forwarding decisions are complex, which translates into more processing power per packet handled and therefore higher cost. In addition, while switches rarely need management attention once installed, routers need frequent management work. Networking professionals say, "Switch where you can; route where you must." However, routed networks can be extremely large and still be manageable. This is not always the case for switched networks.

TEST YOUR UNDERSTANDING

19. a) Give a definition for *routed network*. b) Is there a distinction between the terms *routed network* and *internet*? If so, what is it? c) In an internet, what device connects networks

together? d) Routed networks do packet switching. On routed networks, what are packets called? e) When is a packet called a packet, and when is a packet called a frame? f) Why do networking specialists say, "Switch where you can; route where you must"? g) For what kinds of networks are routed internets especially good?

IP Addresses

Switched Network Addresses

Before routed networks appeared, hosts already had addresses on their switched networks. For instance, on Ethernet switched networks, hosts had 48-bit addresses that were expressed for people in hexadecimal notation. An example is A1–BB–1F–F1–33–CE. Different switched networks had different addressing schemes. For instance, most Frame Relay switched networks had 10-bit addresses called data link control indicators (DLCIs). There was no way to translate between addresses on different types of existing switched networks, and new switched networks (with new addressing schemes) were still emerging.

IP Addresses

Cerf and Kahn addressed this problem by giving each host a second address—a globally unique **IP address.** (IP is the abbreviation for the Internet Protocol, which is the standard created to deliver messages across the networks in an internet.) An IP address is a string of 32 bits, but people have difficulty remembering or writing a stream

Figure 1-22   The Internet (Study Figure)

Hosts
>All computers on any internet are called hosts, including client PCs, personal digital assistants, mobile phones, etc.

Internet Service Providers (ISPs)
>Provide access to the Internet
>Carry your traffic
>Network access points (NAPs) connect the ISPs together
>>Smaller ISPs must pay settlement charges
>You and organizations pay for the ISP operations

Internet Access Lines
>Connect you to your ISP

How the Internet Is Financed
>Through ISP subscriber payments
>Like the telephone network
>Almost no government money involved

The TCP/IP Standards
>The set of protocols that governs the Internet
>Standards for both applications and packet delivery
>Created by the Internet Engineering Task Force (IETF)

of bits. Routers work directly with 32-bit strings, but inferior biological entities express IP addresses in **dotted decimal notation,** which consists of four numerical segments separated by dots. An example is 128.171.17.13.

Two Addresses for Each Host

Consequently, every host on a routed network has two addresses—an address on its individual switched network and a universal IP address that is unique across all switched networks within the routed network.

TEST YOUR UNDERSTANDING

20. a) What is a host's address on the Internet or an internet? b) What is a host's address on an Ethernet network? c) How many bits long is an IP address? d) How are IP addresses presented for human reading? e) Why did Cerf and Kahn create a second address for each host instead of just using the host's existing address on its switched network as its routed network address?

Packets and Frames

Packets

We saw earlier that most switched networks today use packet switching, but call their packets *frames*. Routed networks also use packet switching but call their packets *packets*. If you are confused and somewhat outraged by this apparently stupid inconsistency, you are to be commended for your understanding.

Frames and Packets

Figure 1-20 shows the relationship between packets in a routed network and frames in a switched network. The packet travels end-to-end between the source and destination hosts. Routers pass the packet from one to another along the way. As the packet goes from one router to another, it must travel through a switched network.

The figure shows that the packet travels within a different frame in each switched network. Consequently, when a host transmits a packet to another host, there will be only one packet, but there may be many frames, one in each switched network along the way. If five switched networks separate the source host from the destination host, there will be one packet along the way, but there will be five frames.

TEST YOUR UNDERSTANDING

21. a) What are messages called in internets? b) Distinguish between frames and packets. c) In an internet, the source and destination hosts are separated by five networks (including their own networks). When the source host transmits, how many packets will travel through the internet? d) How many frames?

The Internet Today

The largest routed network (internet) today, of course, is the global Internet you use every day. In this book, we will spell *internet* in lower case to refer to any routed network, and we will spell *Internet* in upper case to refer to the global Internet.

In this book, we will spell **internet** in lower case to refer to any routed network, and we will spell **Internet** in upper case to refer to the global Internet.

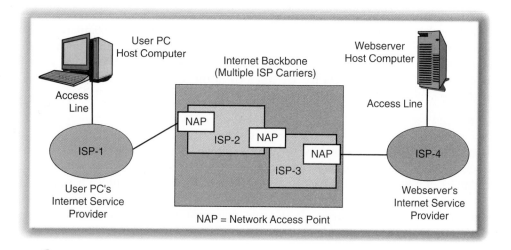

Figure 1-23   The Internet

Internet Service Providers (ISPs)

Figure 1-22 shows the Internet today. It shows that the Internet consists of many commercial carriers called **Internet service providers (ISPs)**. Each ISP is a large routed network. In fact, several ISPs span a dozen or more countries.

To connect to the Internet, you *must* connect to an ISP. ISPs are correctly called the on-ramps to the Internet. ISPs also carry your messages to the destination host and the destination host's messages back to your PC.

If the source and destination hosts use different ISPs, their packets (messages) are exchanged between ISPs at **network access points (NAPs)**.[14] It is common for packets to pass through several NAPs along their journeys. This may seem like an awkward way of operating the Internet, but it allows for competition and flexibility. It is also how the worldwide telephone network operates, as we will see in Chapter 6.

Internet Access Lines

You also need an access line to your ISP. This can be your regular telephone line, a cable television system connection, or even a wireless connection. Sometimes, ISP fees include access line fees. In other cases, they do not.

Financing the Internet

Who pays for all of this? The answer is that you do. The fees that subscribers pay fund their own ISPs. For home networks, the monthly fee is only about $10 to $70. Larger organizations, however, such as universities and major corporations, pay several million dollars per year to their ISPs.

---

[14]ISPs of comparable traffic exchange volume will transfer packets without any payments to each other. ISP pairs that send different amounts of traffic have settlement fees.

The TCP/IP Standards

The Internet is governed by a set of protocols called the **TCP/IP standards.** These standards govern both applications and the way that the Internet delivers packets from source hosts to destination hosts. The TCP/IP standards are created by the **Internet Engineering Task Force (IETF).**[15] We will look at the TCP/IP standards in Chapter 2 and more extensively in Chapters 8 and 10.

Host Names and the Domain Name System (DNS)

As we saw earlier, every host on a routed network needs an IP address. Unfortunately, an IP address, even in dotted decimal notation, is difficult for people to remember. Consequently, many hosts are given **host names,** which are easier for people to remember. Some examples of host names are google.com, www.msn.com, and panko.shidler.hawaii.edu.

Although host names are easy to remember, the IP address is still the host's official address. If you have a host's host name, you must learn its IP address before you can send it packets. To give an analogy, if you want to call someone whose name you know, you must learn their telephone number.

To continue the telephone example, you can call the directory information number and tell the operator the person's name. The operator will look up the name in the directory database and read you the person's telephone number.

Similarly, if your computer needs to look up a host name's IP address, it contacts a **domain name system (DNS)** server, as Figure 1-24 shows. It sends the server the host name. The DNS server looks up this host name and sends back the IP address. The host that sends the DNS request message can now communicate with the named host. In Chapter 10, we will see how DNS works in more detail.

Figure 1-24   Domain Name System (DNS)

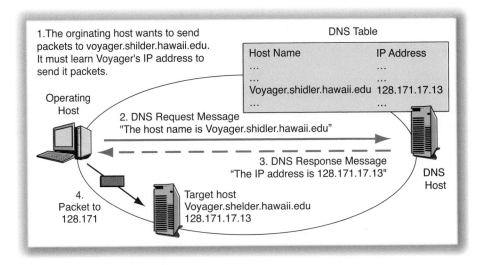

---

[15]Originally, DARPA paid Bolt Beranek and Newman (now BBN Technologies) to create standards for the operation of the Internet. Stewardship of these standards was then passed to the IETF.

TEST YOUR UNDERSTANDING

22. a) What are the two basic services offered by ISPs on the Internet? b) What are NAPs? c) Why are NAPs crucial to universal connectivity on the Internet? d) What kind of line is needed to connect to an ISP? e) What set of protocols governs Internet transmission? f) What standards agency creates these standards? g) How is the Internet funded?

22. a) What is a host's official address on the Internet? b) Why are host names used? c) When you type a host name for a computer in a URL, what does your computer have to do? d) What type of server is needed?

LANS AND WANS

We have distinguished between switched networks and routed networks (internets). Both types of networks should be further divided into local area networks (LANs) and wide area networks (WANs). Consequently, there can be switched LANs, routed LANs, switched WANs, and routed WANs. The Internet is a routed WAN.

Local Area Networks (LANs)

Local area networks (LANs) are networks that operate on the **customer's premises—** the land and buildings owned by the LAN user. The premises may be a single house or

Figure 1-25   LANs and WANs (Study Figure)

Category	Local Area Networks	Wide Area Networks
Abbreviation	LAN	WAN
Can use switched network technology?	Yes	Yes
Can use routed network technology?	Yes, especially in large LANs	Yes, in fact, that is what the Internet is
Distance span	Customer premises (apartment, office, building, campus, etc.)	Between sites within a corporation or between different corporations
Implementation	Self	Carrier with rights of way
Ability to choose technologies	High	Low
Need to manage technologies	High	Low
Cost per bit transmitted	Low	High with arbitrary changes
Therefore, typical transmission speed	Usually 100 Mbps to 10 Gbps	About 256 kbps to 50 Mbps

apartment. They may also be a small business, an office building, or a university campus. The customer's premises are also called the customer's site.

LANs are networks that operate on the customer's premises (site).

Due to the fact that the LAN operates on the customer's premises, the user company can select whatever technology it wishes to use. Of course, everything you own ends up owning you, and companies also need to install, operate, and maintain their LANs themselves.

Small LANs are likely to be switched LANs. Larger LANs are likely to use routers to divide the LAN into smaller switched networks that are linked together into a local internet.

Wide Area Networks (WANs)

While LANs carry traffic *within* sites, **wide area networks (WANs)** carry traffic *between* sites. These sites might be sites of the same company, sites of different businesses, or company sites and the access sites of Internet service providers.

WANs connect different sites.

Companies do not have legal **rights of way** to lay wires outside of their sites. (Imagine how your neighbors would feel if you started laying wires across their yards.) Consequently, for wide area networking, companies must use **carriers,** which are organizations to which the government gives transmission rights of way. In return for receiving these rights of way, carriers agree to be regulated by the government. This regulation affects both service offerings and prices.

For WANs, companies must use carriers with rights of way.

Carriers are likely to offer only a few services. Consequently, customers must adapt their network plans to the services offered by carriers to which the customers have access. In addition, carrier prices tend to be high and to change rapidly in ways that are only slightly (if at all) related to changes in technology and costs. Finally, carriers often require contracts lasting months or even years. This can lock the firm into a high-priced contract (think of most mobile phone contracts), but at least this gives predictability in pricing.

Cost per Bit and Transmission Speed

When you place a long-distance call, you pay more per minute than you do when you place a local call. Consequently, when you place long-distance calls, you tend to make fewer and shorter calls. In general, when the unit price rises, unit demand decreases.

The same economic logic works for networking. In local area networks, the cost per bit transmitted is far lower than it is in wide area networks. Consequently, companies tend to build LANs that give speeds ranging between 100 Mbps and 10 Gbps, while they tend to limit WAN speeds to between 256 kbps and about 50 Mbps.

Strategic Network Management
  As far as possible, build a coherent roadmap
  Pay special attention to decisions that lock you in for long periods of time
  Legacy technologies are technologies selected previously that limit services today
    For upgrading, service benefits must exceed update costs

Product Selection with Multicriteria Decision Making
  The entire systems development life cycle (SDLC) must be followed
  For network products, almost ways buy instead of make network elements
    Must use multicriteria decision making

Ongoing Management (OAM&P)
  The most important (and expensive) part of the systems life cycle
  Operations
    Moment-by-moment traffic management
    Network operations center (NOC) using SNMP
  Maintenance
    Fixing things that go wrong
    Preventive maintenance
    Should be separate from the operations staff
  Provisioning (Providing Service)
    Includes physical installation
    Includes setting up user accounts and services
    Reprovisioning when things change
    Deprovisioning when accounts and services are no longer permitted
    Collectively extremely expensive
  Administration
    High end: planning
    Middle: analysis of operations to indicate needed changes
    Low: paying bills, managing contracts, etc.

Figure 1-26   Network Management (Study Figure)

Companies tend to build LANs that give speeds ranging between 100 Mbps and 10 Gbps, while they tend to limit WAN speeds to between 256 kbps and about 50 Mbps.

TEST YOUR UNDERSTANDING

24. a) What are LANs? b) What is a WAN? c) When are carriers needed? d) What are rights of way? e) Does a company have more control over LANs or WANs? f) Compare typical LAN and WAN speeds. g) Why are typical WAN speeds slower than typical LAN speeds? Give a complete and logical answer.

NETWORK MANAGEMENT

Although technology is the most visible aspect of networking, management is the most important aspect of networking. The best-managed networks provide far better service to users at much lower costs than do the worst-managed networks.

Strategic Network Planning

Network technology is changing rapidly. Business requirements are changing even more rapidly. In some cases, the network administrator can only react to unexpected events. However, as far as possible, the network administrator must engage in **strategic network planning** that anticipates changes and builds a roadmap to guide the network in a coherent way for several years.

In strategic planning, an important goal is to identify *current decisions that will lock the firm in for a long period of time.* For instance, if a company pursues a particular technology, this might lock the firm into a particular vendor or at least a particular standard for several years. Although all strategic planning decisions are important, lock-in decisions are the most crucial and should receive the greatest amount of attention. A wrong choice can have dire consequences.

In strategic planning, an important goal is to identify current decisions that will lock the firm in for a long period of time.

Another important focus for strategic networking planning is on the firm's existing legacy technologies. Legacy technologies are technologies that were selected by your predecessor. Although they probably were good choices at the time, they are now obsolete and impose limits on the services that the networking function can provide. Updating legacy technologies is attractive from a service viewpoint, but the cost of the upgrade must justify the added services. No firm can afford to upgrade all of its legacy network systems immediately.

TEST YOUR UNDERSTANDING

25. a) Why is it important to carefully consider decisions that will lock the firm in for a long period of time? b) What are legacy technologies? c) Should legacy technologies be upgraded if they interfere with services?

Product Selection with Multicriteria Decision Making

Once a project is selected and initiated, the network staff must go through the traditional systems development life cycle to implement the project. Given that almost all readers know about the systems development life cycle, we will not discuss it in detail.

In software development projects, there usually is a **make-versus-buy decision.** Should the programming staff create the software itself, or should the company purchase the software? In networking projects, this decision rarely makes sense. User companies like banks and retail stores do not have the technical expertise to make their own switches and routers. Instead, they must *select* and *buy* these technologies. Consequently, in this book, we will look at the factors you need to understand when you make purchasing decisions involving several alternative technologies.

Criterion	Criterion Weight (Max: 5)	Product A		Product B	
		Product Rating (Max: 10)	Criterion Score	Product Rating (Max: 10)	Criterion Score
Functionality	5	9	45	7	35
Availability	2	7	14	7	14
Cost	5	4	20	9	45
Ease of Management	4	8	32	6	24
Electrical Efficiency	1	9	9	8	8
Total Score			120		126

Figure 1-27   Multicriteria Decision Making in Purchase Decisions

When making purchasing decisions, companies tend to use **multicriteria decision making,** which Figure 1-27 illustrates. In this approach, the company decides what product characteristics will be important in making the purchase. Things that are important in the purchasing decision are called **criteria.**

Of course, costs are important—both purchase costs and ongoing costs. However, other decision criteria are also important. In Figure 1-27, the criteria for the product are functionality, availability, cost, ease of management, and electrical efficiency.

Next to each criterion is the **criterion weight.** This weight gives the relative importance of each criterion compared with those of other criteria. Here, weights range from 1 to 5. Note that cost and functionality have the largest weights (5), emphasizing their importance.

For each product (there are only two in the figure), the evaluation team gives the product a **rating** for each decision criterion. In this example, the ratings range from 1 to 10, with higher values indicating higher value. More functionality is better, so higher numbers in ratings reflect greater functionality. In contrast, for cost, lower cost is better, so higher rated values must indicate lower cost.

After filling in the ratings on all criteria for all products, the network staff computes the **criterion score** for each product. To do this, the staff multiplies the criterion weight times the rating for that product in that criterion. It then totals the criterion scores into a **total score.**

In Figure 1-27, Product A has a total score of 120, while Product B has a total score of 126. Speaking simplistically, Product B appears to be a better choice. However, the two total scores are very close. Numbers must never drive our thinking. A closer look shows that Product A has very good functionality and ease of management, although its cost is high. Product B has poorer scores on functionality and ease of management. It may be possible to negotiate a lower price on Product A and redo the analysis.

TEST YOUR UNDERSTANDING

26. a) What is the make-versus-buy decision? b) For routers and switches, do firms usually make or buy? c) We are considering products A, B, and C. Our criteria are price, performance, and reliability, with weights of 20%, 40%, and 40%, respectively. Product A's evaluation scores on these three criteria are 8, 6, and 6, respectively. For B, the values are 6, 8, and 8. For C, they are 7, 7, and 7. Present a multicriteria analysis of the decision problem, in tabular form and showing all work. Interpret the table.

Operational Management

After a network component is in place, it probably will be used for many years. During its **operational life,** there will be substantial labor costs. We will classify these costs in a way that telecommunications carriers have traditionally done—in terms of **operations, administration, maintenance, and provisioning (OAM&P).**[16]

Operations and the Simple Network Management Protocol (SNMP)

You probably have seen pictures of **network operations centers (NOCs)** for major telecommunications carriers. These are large rooms with dozens of monitors showing the conditions of various parts of the network. Most corporations also have network operations centers. These corporate NOCs are smaller, usually having only about a half dozen monitors.

For remote device management, most NOCs use the **simple network management protocol (SNMP),** which is illustrated in Figure 1-28. In the NOC, there is a computer that runs a program called the **manager,** which manages a large number of **managed devices,** such as switches, routers, servers, and PCs.

Actually, the manager does not talk directly with the managed devices. Rather, each managed device has an **agent,** which is hardware, software, or both. The manager talks to the agent, which in response talks to the managed device.

The network operations center constantly collects data from the managed devices, using SNMP **Get** commands. It places this data in a **managed information base (MIB).** Data in the MIB gives NOC managers a detailed picture of the traffic flowing through the network. This can include failure points, links that are approaching their capacity, or unusual traffic patterns that may indicate attacks on the network.

In addition, the manager can send **Set** commands to the switches and other devices within the network. Set commands can reroute traffic around failed equipment or transmission links, reroute traffic around points of congestion, or turn off expensive transmission links during periods when less expensive links can carry the traffic adequately.

Normally, the manager sends a command and the agent responds. However, if the agent senses a problem, it can send a **trap** command on its own initiative. The trap command gives details of the problem.

Maintenance

You have undoubtedly seen telephone company maintenance trucks on their way to downed transmission lines, broken transformers, or other trouble spots. In addition to

---

[16]Many firms describe their ongoing work in terms of the ISO Telecommunications Management Network Model. In this model, these activities are called FCAPS, which stands for *fault management, configuration management, accounting management, performance management,* and *security management.*

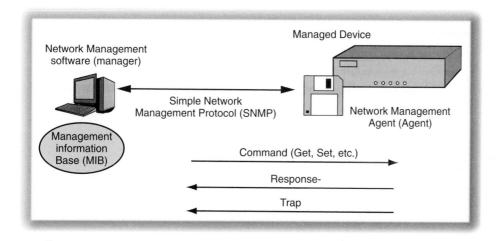

Figure 1-28   Simple Network Management Protocol (SNMP)

fixing equipment failures, telephone companies do preventative maintenance to prevent future failures.

In the same way, companies often have to fix their internal corporate network switches and other physical components. They also have to handle software problems. Although the network operations center can fix some problems remotely, most firms have separate NOC and maintenance staffs. The NOC staff usually is heavily occupied with the moment-by-moment operation of the network, and it makes sense to have other networking professionals focus on maintenance.

Provisioning

If you get cable television service, the cable company has to **provision** your residence—that is, set up service. This includes physical setup (running the coaxial cable into your home). It also involves setting up your account on the company's computers. The cable company also has to reprovision customers when they change their service by adding channels, dropping optional services, or switching pricing plans.

Within a corporate network, provisioning may involve the installation of additional switches, networks, and transmission lines to serve new users. In networks, every time a new user joins the firm, the company has to provision service for that user. In fact, provisioning has to be done for every user account on every server and access point on the network.

Furthermore, once a user is provisioned for a particular resource, he or she may have to be **reprovisioned** if his or her authorizations change—say, if he or she is upgraded from read-only data access to full read/write access. The user also has to be reprovisioned if he or she changes jobs within a firm, joins project teams, or does many other things. Users also have to be **deprovisioned** when they leave project teams or leave the company entirely. Contractors and other outside organizations also have to be provisioned, reprovisioned, and deprovisioned when they start to work, change the way they work, or stop working with a company.

Administration

Operations, maintenance, and provisioning involve real-time work to keep the network running. Administration tasks include "everything else."

➤ At the high end, administration includes network planning and project management.

➤ In the middle, administration includes the collection of performance data, maintenance data, and other data on the network for use in planning.

➤ At the low end, it includes paying bills to vendors and telephone companies, monitoring proposals and contracts, doing network budgeting, comparing network budgets with actual costs, and doing other "business" work.

TEST YOUR UNDERSTANDING

27. a) List the main elements in an SNMP network management system. b) Does the manager communicate directly with the managed device? Explain. c) Distinguish between Get and Set commands. d) Where does the manager store the information it receives from Get commands? e) What kinds of messages can agents initiate?

28. a) For what is OAM&P an abbreviation in ongoing management? b) Distinguish between operations and maintenance. c) What is provisioning? d) When may reprovisioning be necessary? e) When may deprovisioning be necessary? f) List at least one administrative task at the high end, in the middle, and at the low end.

The Conclusion Section

The Conclusion section wraps up the chapter and helps you reflect on what you have learned.

SYNOPSIS

The Synopsis gives the highlights of the chapter. Reading it after you study the body of the chapter will help you see how things fit together. Studying only the Synopsis, CliffsNotes® style, may seem attractive, but if you don't understand the chapter well, you won't understand many of the key points. Also, while the points in the synopsis are especially important, they represent only a fraction of what you will have to know on an exam.

END OF CHAPTER QUESTIONS

Test Your Understanding questions are limited to helping you understand basic concepts. At the end of the chapter, there are questions that help you reflect on and integrate what you have learned. Do

these only after you have mastered the individual concepts in the chapter.

THOUGHT QUESTIONS

Thought Questions are general questions that require you to integrate what you have learned. Learning individual facts in isolation will not be enough to answer thought questions (or to prepare you for your career). Thought questions that include the words, "what do you think" generally have no right or wrong answers. They require you to come to a reasoned opinion, hopefully after considering all sides of the issue.

TROUBLESHOOTING QUESTIONS

Troubleshooting Questions help you apply what you have learned to real-world situations. Networks are complex aggregations of devices and transmission lines. When problems occur, you will need to use your detailed knowledge and your understanding of the situation to come up with several possible causes and then systematically eliminate causes until you have found the correct one.

DESIGN QUESTIONS

Design Questions give you a description of a situation and ask you to design a solution that meets a person's or organization's needs. This is the real litmus test for whether you have understood the material in the chapter. Design requires you to put together all of the bits and pieces in the chapter. Keep in mind that designers are governed by their worst moments. If you leave out something important, the client will be faced with a system that will not work without an additional unplanned investment. Clients recoup their additional expenses through lawsuits against designers.

PERSPECTIVE QUESTIONS

Perspective Questions help you to reflect on your experience in working through the chapter. What surprised you? What was hardest for you? These questions help you as you restudy the material later.

PROJECTS

Some chapters have projects that involve research into a topic.

HANDS-ON EXERCISES

Hands-On Exercises ask you to use your computer or another device to accomplish something specific, such as to see whether your computer's connection to the network is working or to figure out why you cannot reach a webserver that you normally can reach.

CONCLUSION

Synopsis

The chapter began with a focus on applications. Networks exist to allow application programs on different hosts to work together. (The term *host* is used for any computer attached to a network, regardless of the host's size.) In this chapter, we looked at a number of traditional Internet applications, newer Internet applications, and other corporate applications. The discussion was intended to emphasize the wide variety of applications that networks must serve.

One theme in the discussion of applications was network standards, which are also known as protocols. Open standards allow products from different vendors to work together. This spurs competition and product advancement. We will look more closely at standards in Chapter 2.

We spent almost half of the chapter looking for applications of various types. We looked at the traditional and new Internet applications that you are familiar with as an individual. We also looked briefly at corporate applications, such as transaction processing and converged network applications. We also saw the advantages of file service for both data files and program files.

Today, networks are mission-critical corporate infrastructures. They must work, and they must work well. We looked at several quality-of-service (QoS) metrics, including speed, cost, availability, error rate, latency, jitter, and security. We also looked at service level agreements (SLAs), which are vendor warranties for QoS. SLAs specify the maximum amount of time that a worst-case condition (low speed, high latency, etc.) may exist.

Packet switching involves fragmenting a long message into smaller messages called packets and then sending these packets individually. Packet switching creates very efficient multiplexing, greatly reducing transmission costs.

We looked at the technology of switched networks, which use simple devices called switches to forward packets. One oddity of terminology is that in switched networks, packets are called *frames*. We saw that each frame carries the switched address of the destination device for the frame—a destination host or a router within the switched network. Switches along the way read the destination address to decide which port to use to send the frame back out.

During the 1980s, numerous switched networks appeared, using many different technologies. The solution to the chaos this created was to integrate multiple switched networks into routed networks, also called internets. This was made possible by the invention of sophisticated devices called routers. Routers are much more expensive than switches because of the need to move messages across multiple switched networks. In routed networks, messages actually are called packets. One packet travels from the source host to the destination host. In each switched network along the way, the switch is carried in a different packet. In a routed network, every host has two addresses. One is its address on its own switched network. The other is its globally unique IP address on the Internet.

We looked at the Internet very briefly. We saw that to use the Internet, you need an access line and an Internet service provider (ISP). Your ISP gives you access to the Internet and carries your messages. If the other server you are trying to reach is served by a different ISP, that is fine because ISPs interconnect with one another at network access points (NAPs).

We noted that local area networks (LANs) and wide area networks (WANs) could be either switched or routed. LANs operate on the customer premises. This allows the user organization to choose any technology it wishes. WANs carry traffic between customer sites. Companies do not have rights of way to lay transmission lines beyond their premises, so WAN transmission is done by carriers. The cost per bit transmitted is higher in WANs than in LANs, so WAN speeds usually are much lower than LAN speeds.

We ended the chapter with a discussion of network management, including strategic network planning, multicriteria decision making for product selection, and OAM&P (operations, administration, maintenance, and provisioning).

End-of-Chapter Questions

THOUGHT QUESTIONS

1. a) The telephone system has an availability of 99.999 percent. How much downtime is that per year? b) With an availability of 99.9 percent, how much downtime is that per year? c) With an availability of 99 percent, how much downtime is that per year?

2. Is minimizing the cost per bit transmitted more important in LANs or WANs? Justify your answer.

3. Create a table comparing switched and routed networks across the various dimensions discussed in the chapter (name of message, etc.).

TROUBLESHOOTING QUESTIONS

Troubleshooting is identifying and fixing problems. Troubleshooting is an important skill, and we will see troubleshooting questions throughout this textbook. Research has shown that people often make fundamental mistakes when they do troubleshooting. Most fundamentally, they usually consider only one or two possible causes for their problem. Often, the one or two possible causes they consider are incorrect. Consequently, they often waste time trying to solve the wrong problem. Only later do they realize that they need to consider additional possibilities. Premature focusing on one or two possible causes tends to extend downtime needlessly and sometimes leads to "solutions" that fail to fix the real problem.

In troubleshooting questions, you will be expected to create multiple hypotheses, not just one or two. It is almost always best to draw a diagram of all of the components of a system to broaden your perspective. After you develop multiple possible causes of the problem, you can then use logic or experimentation to prioritize them and eliminate false causes.

1. Here is a sample troubleshooting problem for you to solve. You have been using a telephone modem to access the Internet. The modem's rated download speed is 56 kbps. You switch to a cable modem, which should allow you to receive at 3 Mbps. In general, your download speed for webpages is faster than it was with your telephone modem; however, your actual download rates usually vary from only 500 kbps to 1.5 Mbps.

    a) List likely reasons for your not being able to get a full 3 Mbps. **Do NOT just come up with one or two possible explanations**. *Hint:* Consider Figure 1-24, which shows the Internet.

    b) Assess the likelihood of each alternative, given the facts in the problem description and any other analysis you can consider.

2. Your DSL line has a listed speed of 500 kbps. However, when you make downloads, a speed counter tells you that you are receiving only 50 kBps. Can you explain this apparent inconsistency?

PERSPECTIVE QUESTIONS

1. What was the most surprising thing you learned in this chapter?

2. What was the most difficult material for you in this chapter?

PROJECTS

1. Do a report on streaming video formats.

2. Do a report on some aspect of Web 2.0 (its definition, wikis, blogs, etc.).

GETTING CURRENT

Go to the book website's New Information and Errors pages for this chapter to get new information since this book went to press and to correct any errors in the text.

Introductory Design and Hands-on Exercises

Objectives

By the end of this chapter, you should be able to do the following:

- Design a small LAN to be connected to the Internet.
- Use some simple network management commands and tools.

DESIGN EXERCISE

XTR Consulting

Consider a network for XTR Consulting, a fictional small environmental consulting company with 16 professionals (including managers) and a secretary. The company has an office suite in a large downtown office building. Its leaders wish to have an internal Ethernet LAN.

The Initial Situation

We will begin with the initial situation in the company:

➤ When planning for the network began, XTR's 16 professionals and the secretary (17 staff members in all) each had a good personal computer.

➤ Each staff member also had a printer. Victor Chao, XTR's president, had a laser printer. So did Ann Jacobs (the secretary) and one of the consultants. The other consultants had color inkjet printers, because the cost of giving each a laser printer was seen as prohibitive.

➤ When a consultant with an inkjet printer needed a laser printout, he or she saved the file onto a floppy disk and gave it to Ann to print. Similarly, when consultants wanted to share a file, one had to save it to disk and walk the disk over to the other consultant. Walking files around was jokingly called "**sneakernet**." It wasted time and created confusion over who had the most current version of each file.

➤ The consultants, who traveled frequently, had to copy files that they would need on the road to the firm's one loaner notebook computer. If they had an unanticipated need on the road for another file, there was no way to retrieve it.

➤ Although all PCs had modems, only one person could dial into the Internet at a time, and telephone access was very slow—especially for downloading the large maps they often needed.

➤ Although one of the employees, Kumiko Touchi, is very good with computers, there were many problems that Kumiko could not fix. Also, pulling Kumiko away from highly paid consulting work to fix a computer problem was absurd financially.

Broad Network Design

Given this situation, the firm naturally wanted to network its computers together and to connect the internal network to the outside world. Victor Chao hired a network consultant, Robert Blanco, to create a broad design for the network. Your job will be to flesh out Blanco's broad design, including its initial cost. His design is shown in Figure 1a-1.

Labor Costs

XTR will hire a company to do the actual installation. The company will charge $75 per person hour.

Figure 1a-1   Broad XTR Design

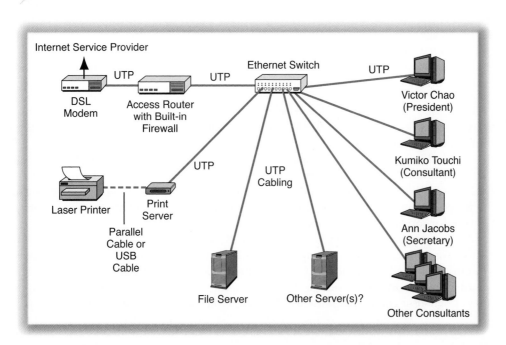

Switch

As shown in the figure, the firm will have an Ethernet switch to connect all of the elements. The switch will operate at 100 Mbps and will need quite a few ports. This will be a very simple switch that can simply be plugged into a power outlet, with no installation work beyond physical installation, which will take about 30 minutes.

Wires

A UTP wire cord will run between the switch and each device. The company will run patch cords (precut UTP cords of approximately the correct length for each run). Patch cords will average $40 each.

The cords will run under carpets rather than neatly through false ceilings or walls. Neater, but more expensive, installation will not be done, because the firm may be moving soon and a neat installation would take more than four hours per UTP connection to the switch in XTR's building. Running the cords under the carpets, but leaving the cords otherwise exposed, will take only about 30 minutes per UTP connection to the switch.

Network Interface Cards (NICs)

The firm will have to install an NIC in three of the firm's current desktop PCs, now renamed client PCs. The other desktops came with adequate NICs when they were purchased. NICs, like the switch, will operate at 100 Mbps. Installation of an NIC will take about 15 minutes. Setting up each PC to run on the network will take an average of 30 minutes.

Dedicated Server(s)

In Mr. Blanco's design, the firm will purchase one or more new PCs to be used as dedicated servers. These will run the Windows Server 2003 network operating system instead of the desktop versions of Windows that the client computers use (Vista and XP).

Pricing for the Microsoft Windows server depends upon the number of users who will share the server. Look up the price for Windows Server 2003 online, and be sure you have enough CALs for the number of employees.

In addition, each server will require six hours of installation time for the hardware setup and operating system setup.

File Service   One server task will be to provide **file service,** meaning that the consultants can store their files on the server. Consequently, the main server is called a file server. **File service** is attractive because the server is backed up nightly.

In addition, one user can save a file to a directory on a file server. Other users can later retrieve the file and can even change it and resave it if they are authorized to do so. However, unauthorized users cannot even see the file. Assigning and changing access permissions in various directories to the various users of a file server are constant chores for server administrators.

The company can also install its client PC software on the file server—including Microsoft Office, which the company uses extensively. Of course, the company cannot just install a standard version of Office on the file server and let all users share it. They must install a multiuser version and purchase a volume license. This will cost much more than single-user copies. Assume that Microsoft Office will require two hours to set up.

E-Mail   The firm will need e-mail service. This mail service will communicate over the Internet with mail hosts in other companies. The company can either run the mail service software on the file server or purchase a separate mail server. The e-mail software will take about five hours to install and configure, including setting up mail accounts for XTR's employees. The e-mail vendor charges an annual license fee of $500. There will also be a monthly charge per user on an ongoing basis. Setting up the e-mail program and user accounts will take eight hours.

Dedicated Print Servers

XTR will replace the firm's diverse printers with the three existing high-capacity laser printers that will be spread around the office area for easy access by users. Larger firms use dedicated **print servers,** as shown in Figure 1a-1. When client users print, their printouts go to a print server that then feeds the print job to the attached printer.

Although print servers are called servers, they are not full computers. They are simple electronic devices that cost only $50 to $300 and have the processing power and software needed just to receive print jobs and feed them to the printer attached to them. A print server also has a built-in NIC to enable it to talk to the switch.

Each print server connects to a switch with a UTP cord and to its printer through an ordinary parallel cable or USB cord. Parallel cables and USB cords are very short—only one or two meters. UTP cords, in contrast, can be up to 100 meters long. Consequently, print servers sit right next to the printer, and there usually is a long UTP wiring run to the switch that connects the print server to the network.

Each print server takes about 30 minutes to set up, including configuration. To configure the print server, the installer runs an installation program on a client PC. This program on the client PC communicates across the network with the print server.

Internet Access

For Internet access, the firm will replace its slow telephone access with high-speed DSL service (see Chapter 6) from the local telephone company. Several client PCs can use this access line simultaneously. This will require the purchase of an access router. One end of the access router will plug into the DSL modem via a UTP cord, and the other end will plug into the switch via another UTP cord.

In order to be configured, the access router has a built-in webserver. To configure the access router, the installer opens a browser on a client PC and then types the IP address of the access router to connect the browser to the built-in webserver. The actual configuration employs a graphical user interface. Physical installation and configuration will take about 30 minutes for the DSL modem and access router.

Firewall

The access router has a built-in firewall. Setting up the firewall will require about five hours. The major part of this will be working with XTR employees to determine what traffic should be blocked and what traffic should be permitted. Setup time to implement the plan after firewall installation will take approximately 12 hours of labor.

Your Detailed Design

With this information, you will flesh out the network design for the company. To guide you, answer the following questions and hand the answers in to your teacher:

1. List the requirements for the network.
2. List the items that need to be purchased and installed.

3. Should the company use separate file and e-mail servers, or should it use one server for both? Justify your answer.

4. Do you think that the company should purchase the server or servers you listed in the previous question, or should it save money by taking away existing PCs from employees and using them as dedicated servers (assuming that the existing PCs are powerful enough to be used as servers)?

5. How many print servers should the company use—one per printer or just one for all three printers? Justify your number.

6. How many ports will be needed on the switch? Justify your number.

7. How many UTP cords will be needed? Justify your number.

8. Switches can be purchased with 12, 24, 36, and 48 ports. Which switch should the company purchase? Justify your decision.

9. Cost out the hardware for your PC server or servers, using an online source such as Dell.com. Provide a detailed listing of your server's or servers' features. Specify the version of Windows Server. The server should have enough CALs for XTR's employees. You can purchase these separately. Show details!

10. For the volume license for Microsoft Office, just assume that a volume license will cost 10 times as much as a single-user version. (Volume licensing is rather complex and situation-specific.)

11. Cost out the full installation expense that XTR must face. (Do not include network access.) Use a spreadsheet model. Cite specific products (including product numbers and technical details) and sources of data. This is a really big task, but also a very realistic one. It is the kind of thing you will have to do often as a network professional. Do it right.

12. What are the least expensive components of this network's cost?

13. What are the most expensive components of this network's cost?

14. Repeat parts 10–12 for the neat wiring installation.

HANDS-ON EXERCISES

1. Binary and Decimal Conversions Using the Microsoft Windows Calculator

It is relatively easy to convert 32-bit IP addresses into dotted decimal notation if you use the Microsoft Windows Calculator. Go to the *Start* button, then to *Programs* or *All Programs*, then to *Accessories*, and then click on *Calculator*. The Windows Calculator will pop up. Initially, it is a very simple calculator. Choose *View* and click on *Scientific* to make Calculator an advanced scientific calculator.

Binary to Decimal

To convert eight binary bits to decimal, first divide the 32 bits into four 8-bit segments. Click on the *Bin* (binary) radio button and type in the 8-bit binary sequence you wish to convert. Then click on the *Dec* (decimal) radio button. The decimal value for that segment will appear.

Note that you cannot convert the whole 32-bit IP address at one time. You have to do it in four 8-bit segments.

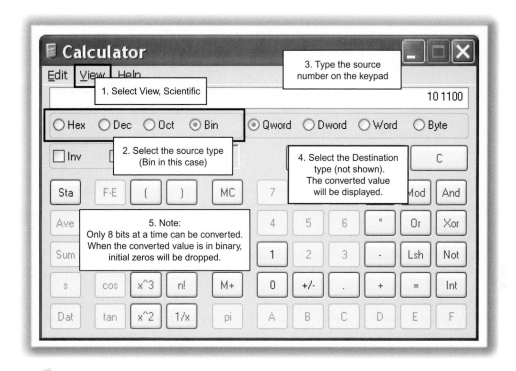

Figure 1a-2   Windows Calculator

Once you have the four decimal segment values, write them in order with dots between them. It will look something like 128.171.17.13. You have now converted the 32-bit IP address to dotted decimal notation.

Decimal to Binary

To convert decimal to binary, go to *View* and choose *Scientific* (if you have not already done so). Click on *Dec* to indicate that you are entering a decimal number. Type the number. Now click on *Bin* to convert this number to binary.

One subtlety is that Calculator drops initial zeros. So if you convert 17, you get 10001. You must add three initial zeros to make this an 8-bit segment: 00010001.

Another subtlety is that you can convert only one 8-bit segment at a time.

1. a) What is 11001010 in decimal?
   b) Express the following IP address in binary: 128.171.17.13. (Hint: 128 is 10000000. Put spaces between each group of 8 bits.)
   c) Convert the following address in binary to dotted decimal notation: 11110000 10101010 00001111 11100011. (Spaces are added between bytes to make reading easier.) (Hint: 11110000 is 240 in decimal.)

2. Test Your Download Speed

How fast is your Internet connection? Test your download speed at *both* http://www.pcpitstop.com/internet/bandwidth.asp   *and*   http://webservices.zdnet.com/zdnet/bandwidth. If you can, test your bandwidth during periods of light and heavy use.

2. a) What kind of connection do you have (telephone modem, cable modem, LAN, etc.)?
    b) What was your download speed during the test for pcpitstop?
    c) What was your download speed during the test for zdnet?

3. Working with the Windows Command Line

Windows offers a number of tools from its command line prompt. Network professionals need to learn to work with these commands.

Getting to the Command Line

To get to the command line, click on the *Start* button and choose *Run*. Type either *cmd* and or *command*, depending on your version of Windows, and then hit *OK*.

Command Line Rules

At the command line, you need to type carefully because even a single-letter error will ruin the command. You also need to hit *Enter* at the end of each line. You can clear the command line screen by typing **cls** and then pressing *Enter*.

3.   Go to the command line. Clear the screen.

4. See Your Configuration

In Windows, you can find information about your own computer with ipconfig or winipconfig. In newer versions of windows, type the command **ipconfig/all** and then press *Enter*. Older versions of windows require you to type the command **winipconfig** and *Enter*. This will give you your IP address, your physical address (your Ethernet address), the IP addresses of your organization's or ISP's DNS hosts, and other information—some of which we will see in Chapter 8.

4.   Use ipconfig/all or winipconfig. a) What is your computer's IP address? b) What is its Ethernet address? c) What are the IP addresses of your DNS hosts?

5. Ping and Tracert

Ping

To find out whether you can reach a host and to see how much latency there is when you contact a host, use the **ping** command. You ping an IP address or host name much as a submarine pings a target to see whether it exists and how far away it is. To use the command, type **ping *hostname*** and press *Enter*, or type **ping *IPaddress*** and press *Enter*. Ping may not work if the host is behind a firewall, because firewalls typically block pings.

5.   a) Ping a host whose name you know and that you use frequently. What is the latency? If this process does not work because the host is behind a firewall, try pinging other hosts until you succeed.

Ping 127.0.0.1 (PC, Call Home)

Ping the address 127.0.0.1. This is your computer's **loopback address.** In effect, the computer's network program sends a ping to itself. If your PC seems to have trouble communicating over the Internet, type **ping 127.0.0.1** and *Enter*. If the ping fails, you know that the problem is internal and you need to focus on your network software's configuration. If the ping succeeds, then your computer is talking to the outside world at least.

> b) Ping 127.0.0.1. Did it succeed?

Tracert

The Windows **tracert** program is like a super ping. It lists not only latency to a target host, but also each router along the way and the latency to that router. Actually, tracert shows three latencies for each router because it tests each router three times. To use tracert, type **tracert** *hostname* and press *Enter*, or type **tracert** *IPaddress* and press *Enter*. Again, hosts (and routers) behind firewalls will not respond.

> c) Do a tracert on a host whose name you know and that you use frequently. You can stop the tracert process by hitting Control-C.
> d) What is the latency to the destination host?
> e) How many routers are there between you and the destination host? If this does not work because the host is behind a firewall, try reaching other hosts until you succeed.
> f) Distinguish between the information that ping provides and the data that tracert provides.

6. To Get Your IP and Ethernet Addresses with Windows XP

Follow these steps if you have an XP computer (this exercise will give you configuration information without your having to go to the command line): Choose *Start*, then *Control Panel*, then *Network and Internet Connections*, then *Network Connections*. In the window that appears, click on a network connection icon to select it. Choose *File, Status*. You will see the *Connection Status* dialog box for that connection. Select the *Support* tab to see your IP address. Click on *Details* while in the *Support* tab to see more information, including the physical (Ethernet) address of your NIC.

> 6. If you have a Windows XP computer, find your IP address and the physical (Ethernet) address of your computer.

7. To Get Your IP and Ethernet Addresses with Windows Vista

Do the following if you have a Microsoft Windows Vista computer:

> ➤ Click on *Start*.
> ➤ Click on *Network and Internet*.
> ➤ Click on *Network and Sharing Center*.
> ➤ Under *Connection*, choose *View Status*.
> ➤ Click *Details*.

> 7. If you have a Windows Vista computer, find your IP address and the physical (Ethernet) address of your computer.

Chapter 2

Network Standards

Learning Objectives

By the end of this chapter, you should be able to discuss the following:

- How network standards define message ordering, semantics, and syntax.
- Reliable versus unreliable service and connection-oriented versus connectionless service.
- Layered standards architectures.
- Standards at Layers 1 and 2 (the physical and data link layers) and how Ethernet works at these layers.
- Layer 3 (internet layer) standards and how IP works at this layer.
- Layer 4 (transport layer) standards and how TCP and UDP work at this layer.
- Layer 5 (application layer) standards and how HTTP works at this layer.
- Vertical communication among layer processes on the same device.
- Common layered standards architectures and the dominance of the hybrid TCP/IP–OSI standards architecture.

STANDARDS GOVERNING MESSAGE EXCHANGES

Chapter 1 introduced the concept of network standards, which are also called protocols. In later chapters, we will see many specific network standards. This chapter looks at standards somewhat more broadly, focusing on types of standards and their relationships with one another.

Network Standards (Protocols)

We saw in Chapter 1 that network standards, which are also called protocols, are sets of rules that govern the exchange of messages between hardware or software processes on different hosts. Chapter 1 also noted that open standards bring competition because anyone can build products that interoperate with products from larger vendors. Competition generally reduces prices and increases innovation. In this chapter, we will focus on open network standards.

In order to be more precise, we will extend the definition of network standards that we saw in Chapter 1 to the following: **Network standards** are rules that govern the

Network Standards
>   Also known as protocols
>   Network standards govern the exchange of messages between hardware or software processes on different host computers, including message order, semantics, syntax, reliability, and connection orientation

Network Standards Govern
>   Message order
>>   Turn taking, order of messages in a complex transaction, who must initiate communication, etc.
>   Message semantics (meaning)
>>   HTTP request message: "Please give me this file"
>>   HTTP response message: "Here is the file" (Or, "I could not comply for the following reason:")
>   Message Syntax (organization)
>>   Like human grammar, but more rigid
>>   Header, data field, and syntax (Figure 2-2)

Figure 2-1   Network Standards (Study Figure)

exchange of messages between hardware or software processes on different host computers, including messages (ordering, semantics, and syntax), reliability, and connection orientation.

Network standards are rules that govern the exchange of messages between hardware or software processes on different host computers, including messages (ordering, semantics, and syntax), reliability, and connection orientation.

Message Order, Semantics, and Syntax

This new definition adds several concepts to the preliminary definition in Chapter 1. We will look first at the three major topics of message ordering, semantics, and syntax. We will look at reliability and connection orientation in the next section.

Message Ordering

In medicine and many other fields, a *protocol* is a prescribed series of actions to be performed in a particular order. In cooking, for example, recipes work this way. If you do not put together the ingredients of a cake in the right order, the cake is not likely to turn out very well. In this same way, network standards govern **message ordering.** For the Hypertext Transfer Standard that we saw in the last chapter, the protocol is very simple. The client sends an HTTP request message, and the server sends back an HTTP response message. A server will not send an HTTP response message unless that client has first sent an HTTP request message. Many protocols, including the Transmission Control Protocol (TCP) standard that we will see in this chapter and in Chapter 8, involve many messages being passed around in a precise message order.

Semantics

Computers are not intelligent, as you learned in your first programming course. Neither are switches, routers, hosts, or any other network devices. Consequently, protocols usually define only a few message types, and these types usually have only a few options. Consequently, network protocols greatly limit the **semantics** (meaning) of their messages.

Syntax

In addition, while human grammar is very flexible (even when properly used), network messages have very precise **syntax,** that is, message arrangement options. In HTTP, to give another example using that protocol, we will see in Chapter 11 that most lines in HTTP request and response messages have the syntax "keyword colon value." For instance, the HTTP request message usually has the line "Host: *hostname.*" This line gives the name of the host to which the HTTP request or response should be sent.

TEST YOUR UNDERSTANDING

1. a) Give the definition of network standards that this chapter introduced. b) What three things about message exchanges do network standards govern? c) Give an example not involving networking in which the order in which you do things can make a big difference. d) Distinguish between syntax and semantics.

Syntax: General Message Organization

Figure 2-2 shows that message syntax in general may have three parts—a header, a data field, and a trailer. The header and trailer may be further divided into smaller units called fields.

Data Field

The **data field** contains the content delivered by a message. In an HTTP response message, the data field contains the file that the message is delivering.

The data field contains the content delivered by a message.

Header

The message **header,** quite simply, is everything that comes before the data field. For the HTTP request message, the entire message is the header. There is no data field. In the HTTP response message, the header is everything before the file that the client requested.

The message header, quite simply, is everything that comes before the data field.

Trailer

Some messages also have **trailers,** which consist of everything coming after the data field. HTTP messages do not have trailers. They have nothing after the data field.

The message trailer is everything that comes after the data field.

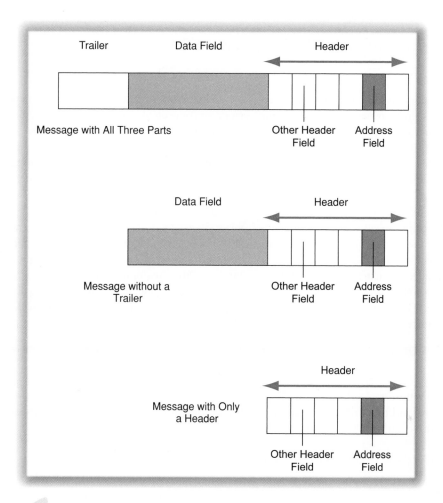

Figure 2-2   General Message Organization

Not All Messages Have All Three Parts

HTTP messages demonstrate that only a header is present in all messages. Data fields are not always present but are very common. Trailers are not common.

Fields in Headers and Trailers

The header and trailer usually contain smaller sections called **fields.** For example, a frame or packet has a destination address header field, which allows switches or routers along the way to pass on the frame or packet they receive.

The header and trailer usually contain smaller sections called fields.

Learning Standards Concepts Is Difficult

Networking, in general, is difficult to learn. There are many concepts, and most concepts are accompanied by TLAs and FLAs (three-letter algorithms and four-letter algorithms). Network standards certainly exemplify this.

Network standards also raise a much worse barrier to learning. There are not only many network standards concepts, but these concepts are highly related. Cognitive scientists call this high degree of interrelatedness *integrative complexity*, meaning that it is difficult to understand anything until you know everything. It is extremely important to learn each section well before going on. It is also important after you have studied the chapters to go back over the material again and again.

The newness and abstractness of network standards concepts also make learning difficult. You will find that you are constantly relearning what you already learned and understood because abstractness prevents your brain from storing information efficiently. Most students have to go through the chapter several times to really get it down.

The bottom line is that this chapter requires especially heavy study. It is tempting to bypass this heavy study and have only a "general idea" about network standards. However, network standards are the foundation of networking. If you don't get a solid base understanding in this chapter, this will be a very long term for you. The effort is worth the pain.

When we look at network standard messages in this chapter and in later chapters, we will be concerned primarily with header fields and trailer fields because these are the defining characteristics of standards.

TEST YOUR UNDERSTANDING

2. a) What are the three general parts of messages? b) What does the data field contain? c) What is the definition of a header? d) Is there always a data field in a message? e) What is the definition of a trailer? f) Are trailers common? g) Distinguish between headers and header fields.

RELIABILITY AND CONNECTIONS

Two general characteristics apply to all standards. The first is whether the standard is reliable or unreliable. The second is whether the standard is connection-oriented or connectionless.

Reliability

Protocols are either reliable or unreliable. **Reliable protocols** resend lost or damaged messages. **Unreliable protocols** do not correct errors.

Reliable protocols correct errors by resending lost or damaged messages. Unreliable protocols do not correct errors.

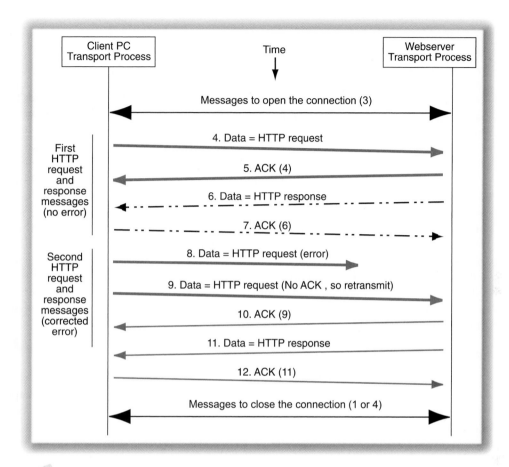

Figure 2-3 Reliable Transmission Control Protocol (TCP) Session

Unreliable Protocols

HTTP is an **unreliable** protocol. It does not do error correction.

Reliable Protocols

Figure 2-3 shows how an important standard called the *Transmission Control Protocol (TCP)* implements reliability.

Acknowledging Correctly Received Segments   Every time a TCP process receives a correct TCP segment (TCP messages are called **TCP segments**), it sends back an **acknowledgment** (ACK) segment. If the original sender receives the acknowledgment for a segment, it knows that its segment arrived correctly. See Segments 4 and 5 in Figure 2-3 for an example of a transmission and an acknowledgment.

Not Acknowledging Incorrect or Not-Received Segments   However, the receiving TCP process does not send an acknowledgment if the segment was damaged or (obviously) if the segment never arrived (see Segment 8). If the sending TCP process does not receive an acknowledgment for a segment promptly, it knows that the segment did not arrive at the other side. It resends the segment (see Segment 9). This process corrects the transmission error, giving reliability. Note that the TCP process that originally sends a TCP segment—not the receiving TCP process—decides whether to resend it.

The Expense of Reliability

Reliability gives clean information. This certainly seems like a fundamentally good thing. (After all, who wants to be called, "unreliable?") However, reliability is extremely expensive. Consider the computer processing cycles that are needed for reliability. The heaviest aspect of reliability is doing the calculation needed to determine whether an error has occurred. The sender treats the entire message as a very large binary number and does a lengthy computation on it. This computation yields a brief number that is included with the message. The receiver redoes the computation and compares its calculation with the transmitted number. Reliability requires far more expensive processing cycles than any other aspect of protocol functioning.

TEST YOUR UNDERSTANDING

3. a) What is reliability? b) How does TCP implement reliability? c) In TCP, what is the receiver's role in reliability? d) In TCP, what is the sender's role in reliability? e) What is the disadvantage of reliability?

Connection-Oriented and Connectionless Protocols

Another general characteristic of protocols is whether they are connection-oriented or connectionless.

Connection-Oriented Protocols

When you call someone on the telephone, you do not just begin speaking when the other party picks up the phone. There is at least a tacit initial negotiation to check that the called person is willing to speak. In addition, at the end of a conversation, it is rude simply to hang up without both sides indicating that they wish to end the call. In networking, as Figure 2-4 illustrates, **connection-oriented** standards work this way too. They have explicit openings and closings. (In Figure 2-3, note that TCP is a connection-oriented protocol.)

Connection-oriented standards have explicit openings and closings.

Connectionless Protocols

As Figure 2-4 also shows, **connectionless** protocols do not establish a connection before transmitting. When you send someone an e-mail, for instance, you do not call him or her ahead of time to ask whether you may send the e-mail message or call him or her afterward to ask whether you can stop sending messages.

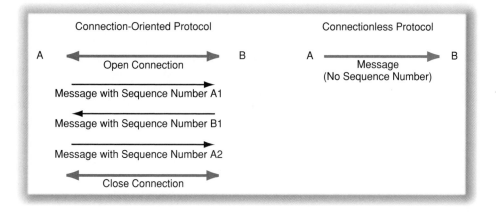

Figure 2-4   Connection-Oriented and Connectionless Protocols

Advantages of Connection-Oriented Protocols

Figure 2-4 also shows that each message that is transmitted is given a **sequence number** to indicate its transmission order during the connection. This has several benefits:

➤ Thanks to sequence numbers, the parties can tell when a message is lost. (There will be a gap in the sequence numbers.)

➤ Acknowledgments can refer to specific messages according to the sequence numbers of those messages. This allows a sender to transmit a number of messages without having to wait for a reply to each one.

➤ Long messages can be fragmented into many smaller messages that can fit inside packets. The fragments will be given sequence numbers so that they can be assembled at the other end. **Fragmentation** followed by **reassembly** is an important concept in networking.

Disadvantages of Connection-Oriented Protocols

Although connection-oriented protocols (thanks largely to their sequence numbers) are useful, they have one major disadvantage. They place a heavy burden on the network and on the computers attached to the network. Opens, closes, acknowledgements, and other supervisory messages can consume a good deal of bandwidth, and doing the work needed to create and use sequence numbers consumes a good deal of processing power. Connection-oriented protocols are referred to as **heavyweight protocols.**

In contrast, connectionless protocols do not create comparable burdens on networks and on computers attached to the network. Connectionless protocols are **lightweight protocols.**

Connectionless and Unreliable Protocols Dominate

Given the need to keep costs low, almost all of the protocols we will see in this book are connectionless and unreliable. The main exception to this is the connection-oriented

Advantages

Connection-oriented protocols give each message a sequence number

Thanks to sequence numbers, the parties can tell when a message is lost (There will be a gap in the sequence numbers)

Error messages, such as ACKs, can refer to specific messages according to the sequence numbers of these messages

Long messages can be fragmented into many smaller messages that can fit inside of packets

The fragments will be given sequence numbers so that they can be assembled at the other end

Fragmentation followed by reassembly is an important concept in networking

Messages can refer to earlier messages by sequence number

Important in database-based transaction processes where several messages must be exchanged to make a purchase, record a transaction, or do some other common business task

Disadvantages

Connection-oriented protocols place a heavy load on networks and on computers connected to the Internet

Figure 2-5   Advantages and Disadvantages of Connection-Oriented Protocols (Study Figure)

Transmission Control Protocol, which we will see again later in this chapter. (We will also see why it is the exception to the pattern.)

Given the need to keep costs low, almost all of the protocols we will see in this book are connectionless.

TEST YOUR UNDERSTANDING

4. a) Distinguish between connectionless and connection-oriented protocols. b) Which can have sequence numbers? c) What are the advantages that sequence numbers bring to connection-oriented protocols? d) Explain fragmentation and reassembly. e) What is the disadvantage of connection-oriented protocols? f) Are most protocols connectionless or connection-oriented? g) Are most protocols reliable or unreliable?

LAYERED STANDARDS ARCHITECTURES

Up to now, we have been talking about the characteristics of individual standards. In networking, there are dozens of standards that do different things, just as different rooms in a house have different functions.

Architectures

To continue the house example, you do not design a house by building one room, then another, and then another, without any plan. Rather, you first create an **architecture**— a broad plan that specifies what rooms the house will have and how these rooms will relate to each other in terms of uses and traffic flow. The architecture ensures that the rooms will collectively provide everything the owner needs. Only after the architecture is finished will the architect begin to design individual rooms in detail.

TCP/IP–OSI

Similarly, data network standards are not created in isolation. A **network architecture** is a broad plan that specifies everything necessary for two application programs on a switched or routed network to be able to work together effectively. Figure 2-6 illustrates

Figure 2-6   Hybrid TCP/IP-OSI Architecture

Broad Function	Layer	Name	Specific Function
Interoperability of application programs	5	Application	Application layer standards govern how two applications work with each other, even if they are from different vendors.
Transmission across an internet (routed network)	4	Transport	Transport layer standards govern aspects of end-to-end communication between two end hosts that are not handled by the internet layer. These standards also allow hosts to work together even if the two computers are from different vendors or have different internal designs.
	3	Internet	Internet link layer standards govern the transmission of packets across an internet—typically by sending them through several routers along the route. Internet layer standards also govern packet organization, timing constraints, and reliability.
Transmission across a single switched network	2	Data Link	Data link layer standards govern the transmission of frames across a single switched network—typically by sending them through several switches along the data link. Data link layer standards also govern frame organization, timing constraints, and reliability.
	I	Physical	Physical layer standards govern transmission between adjacent devices connected by a transmission medium.

the most popular standards architecture for networking today, the **Hybrid TCP/IP–OSI Architecture.** Later in this chapter, we will discuss its odd name. For now, we will focus on its organization.

A network architecture is a broad plan that specifies everything needed for two application programs on a switched or routed network to be able to work together effectively.

Layers

The architecture has a series of layers. Each **standards layer** provides services to the layer immediately above it. For example, when you travel by car, the road provides service for your tires (support), and tires provide support for your car. Your car, of course, provides support for you.[1]

In a layered standards architecture, each layer provides services to the layer above it.

Why Layers?

Why divide networking standards into five pieces (layers)? The first answer is that whenever you have a major task, it is a good strategy to break it up into smaller pieces and attack the pieces individually. This strategy also works in standards development, which is a very large task.

In addition, if a team is doing the work, team members should receive individual tasks that suit their skills. In networking standards development, for instance, electrical engineers usually create physical layer standards, while specialists in a particular application usually create application layer standards for that application.

Another benefit of layering is that development at one layer can ignore concerns at other layers. Application standards designers can design new application standards without having to worry about how to get messages between the two hosts. Physical layer standards designers, in turn, do not have to worry about which applications will use a particular physical link.

A fourth reason for layering is that it allows standards to be updated or changed at various layers independently. If you change from one switched LAN standard to another—say, from Ethernet to an 802.11 wireless LAN or from an old version of Ethernet to a newer version—you do not have to change standards above the data link layer. In addition, if you switch from handling e-mail to browsing webservers (these are application layer functions), you do not need to use different lower-layer (Layer 1–4) standards.

TEST YOUR UNDERSTANDING

5. a) What is a network architecture? b) What is the most popular network architecture today? c) In layered standards architectures, to what layer or layers does a layer provide service?

6. Why do standards architectures break down the standards development process into layers?

---

[1]To give another example, if you have a cat, food provides sustenance for the cat, and the cat provides services for you. Okay, bad example.

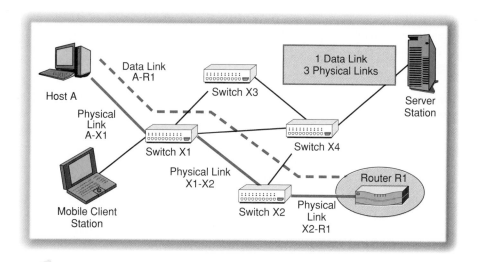

Figure 2-7   Physical and Data Link Layer Standards in a Switched Network

Layer 1 and Layer 2 Standards for Switched Networks (Switched LANs and WANs)

The bottom two layers (physical and data link) govern transmission across a switched network—either a switched LAN or a switched WAN. Figure 2-7 compares these two layers. It shows a switched network. In this network, Host A and Router R1 are separated by two switches—X1 and X2.

Physical Links

Physical layer (Layer 1) standards govern transmission between adjacent devices connected by a transmission medium. In Chapter 1, we saw that the main transmission media were UTP, optical fiber, and wireless (radio) transmission. Such connections are **physical links.** In the figure, there are three physical links between Host A and Router R1: from Host A to the first switch, from the first switch to the second switch, and from the second switch to Router R1. The physical layer provides transmission between devices so that higher layers do not have to worry about this.

Physical layer (Layer 1) standards govern transmission between adjacent devices connected by a transmission medium.

Data Links

The **data link layer (Layer 2)** standards, in turn, govern the transmission of frames all the way from Host A to Router R1 across two switches (X1 and X2). The path that a frame takes across the network is the frame's **data link.** There is only one data link between the source and destination computers on a switched network.

Data link layer (Layer 2) standards govern the transmission of frames across a single switched network—typically by sending them through several switches along the data link. Data link standards also govern frame organization, reliability, and other matters.

> Data link layer (Layer 2) standards govern the transmission of frames across a switched network—typically by sending them through several switches along the data link. Data link layer standards also govern frame organization, reliability, and other matters.

Many Switched Network Standards

There are many types of switched network standards, including Ethernet for switched LANs and Frame Relay for switched WANs, to name just two. Each switched network standard specifies both physical layer standards and data link standards. There are many types of switched networks, so most of the network standards we will see in this book will be physical and data link standards.

> Switched networks (switched LANs and switched WANs) are governed by physical and data link layer standards.

TEST YOUR UNDERSTANDING

7. a) What devices does a physical link connect? b) What is a data link? c) Five switches separate two computers on a switched network. How many physical links are there between the two computers? d) How many data links are there between them? e) What do data link layer standards govern? f) Which layers govern switched LAN transmission? g) Which layers govern switched WAN transmission?

Standards for Routed Network Transmission (Layers 3 and 4)

Initially, there were standards only for switched networks. However, as switched networks began to proliferate, companies needed new standards to link two or more switched networks together into what are called routed networks, or internets. In Chapter 1, we saw that these groups of switched networks are connected by routers so that any application on any host on any switched network can communicate with any application on any other host on any other switched network in a routed network (internet). Standards at the internet and transport layers collectively govern transmission across a routed network.

> Routed networks are also called internets.

The Internet and Data Link Layers

Many people have a hard time differentiating between the data link layer and the internet layer. An example may help clarify this. In Figure 2-8, there are three switched networks in an internet. Host A in Switched Network X is transmitting to Host B in Switched Network Z. The connection linking the two hosts involves three networks, so there are three data links (A to R1, R1 to R2, and R2 to B).

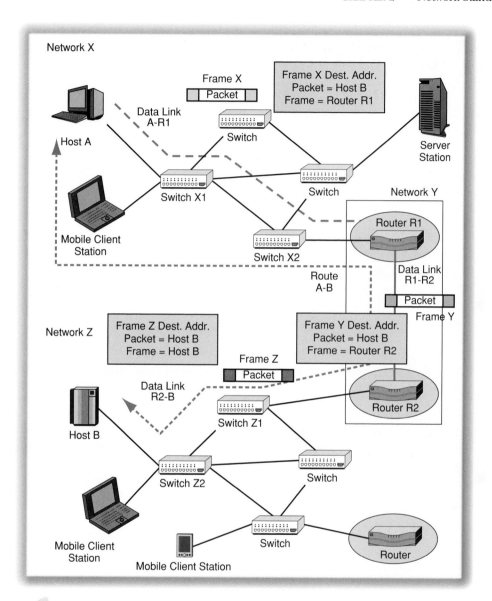

Figure 2-8   Internet and Data Link Layers in a Routed Network (Internet)

In turn, the path that the packet takes across its routed network is its *route*. There is only one route between the source and destination hosts (Host A and Host B). The main standard for internet transmission is the *Internet Protocol* (IP).

Standards at the **internet layer (Layer 3)** govern the transmission of packets from one router to another, *across an entire routed network.* Standards at the internet layer specify what each router along the way does with packets. (Routers act individually in sequence to forward packets to the destination host.) Standards at the internet layer govern how each router along the way passes packets on to the next router on the route to the destination host. There is only one route across the routed network in the figure.

> Standards at the internet layer (Layer 3) govern the transmission of packets from one router to another, across an entire routed network.

Note that we do not capitalize *internet* when describing the internet layer or when we refer to specific routed networks. In general, when *internet* is spelled with a lowercase *i*, this indicates either an internet in general or the internet layer. When *Internet* is spelled with an uppercase *I*, this designates the global Internet. (At the beginning of sentences or in titles, however, *internet* is always capitalized.)

> When internet is spelled with a lowercase i, this indicates either an internet in general or the internet layer. As noted earlier, when Internet is spelled with an uppercase I, this designates the global Internet. (At the beginning of sentences or in titles, however, internet is always capitalized.)

The Internet Layer and Transport Layer

The Internet Layer   Figure 2-9 reiterates that the internet layer governs the transmission of packets from one router to another, across an entire routed network. Many routers are involved in the transmission of a packet over a large internet.

The Transport Layer   In turn, **transport layer (Layer 4)** standards govern the aspects of end-to-end communication between the two end hosts that are *not* handled by the internet layer. The many routers along the way are not involved with the transport layer. Only the two hosts are involved at this layer.

Figure 2-9   Internet and Transport Layers Standards

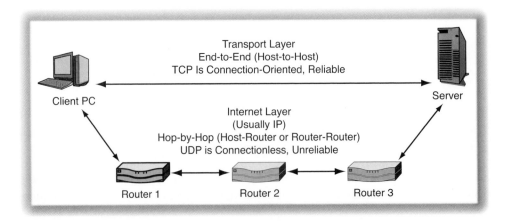

Transport layer (Layer 4) standards govern the aspects of end-to-end communication between the two end hosts that are not handled by the internet layer.

Typically, the transport layer standard (usually, TCP) is reliable, fixing any errors created at the transport layer or lower layers and delivering clean data to the application program. However, later in this chapter, we will see in the discussion of UDP that transport layer protocols are not always reliable.

TEST YOUR UNDERSTANDING

8. a) What do the internet and transport layers do collectively? b) Distinguish between what the internet and transport layer standards govern. c) What is the main internet layer standard? d) What errors does the transport layer usually fix? e) What does it mean in this book if *internet* is spelled with a lower-case *i*?

Standards for Applications

At the highest layer, **application layer (Layer 5)** standards govern how two applications work with each other, even if they are from different vendors. For example, a Microsoft Internet Explorer browser can work with an open-source Apache webserver application program via the HTTP protocol.

Application layer (Layer 5) standards govern how two applications work with each other, even if they are from different vendors.

In Chapter 1, we saw the Hypertext Transfer Protocol (HTTP) application layer standard for the World Wide Web. HTTP is only one of many application protocols used on the Internet. Other applications have their own protocols, for example, E-mail (SMTP POP, etc.), database (ODBC), and FTP (FTP). There are many applications, and there are more application protocols than there are standards at any other layer.

Figure 2-10   Application Layer Standards (Study Figure)

Application Layer Standards
    Govern how two applications work with each other, even if they are from different vendors

Many application layer standards because there are many applications
    World Wide Web (HTTP)

    E-Mail (SMTP, POP, etc.)

    FTP (FTP)

    Database (ODBC)

    Etc.

There are more application layer standards than any other type of standards

TEST YOUR UNDERSTANDING

9. a) What do application layer standards govern? b) Which layer has the most standards? Why is this the case?

LAYERS 1 (PHYSICAL) AND 2 (DATA LINK) IN ETHERNET

Having looked at layering in general, we will now begin to look at individual layers in more detail, beginning with Layers 1 and 2, which together govern switched LANs and WANs.

Ethernet Physical Layer Standards

At the Physical Layer (Layer 1), the bits of the messages are translated into signals to be sent down the communications medium. The physical layer standard usually translates the bits into sequences of voltage changes, flashes of light, or changes in radio waves. The physical layer translates the bits of a message individually or in small groups that are unrelated to the message's logical organization. At the receiving end, the signals are translated back into the bits of the frame. Ethernet has many physical layer standards. We will see the most important of these standards in Chapter 4.

TEST YOUR UNDERSTANDING

10. What does the sending physical layer process do with the bits of the frame?

Ethernet Frames

Messages at the data link layer are called **frames.** Data link standards dictate the semantics and syntax of frame transmission within a single switched network. As we saw in Chapter 1, switches work with Ethernet frames. Figure 2-11 illustrates the standard **Ethernet frame,** which, formally speaking, is the **802.3 MAC layer frame.** In Chapter 4, we will look at Ethernet frames in detail. For now, we will focus on only a few fields.

Messages at the data link layer are called frames.

Octets

The Ethernet header and trailer are divided into fields. Field lengths can be measured in bits. Another common measure for field lengths in networking is the octet. An **octet** is a group of eight bits. Isn't that a byte? Yes, exactly. *Octet* is just another name for *byte.* The term is widely used in networking, however, so you need to become familiar with the it. *Octet* actually makes more sense than *byte,* because *oct* means "eight." We have octopuses, octagons, and octogenarians.[2]

An octet is a group of eight bits.

Ethernet Addresses

48-Bit Ethernet Addresses   Ethernet frames have Ethernet data link layer source addresses and destination addresses that identify the sending and receiving computers,

---

[2] What is the eighth month? (Careful!)

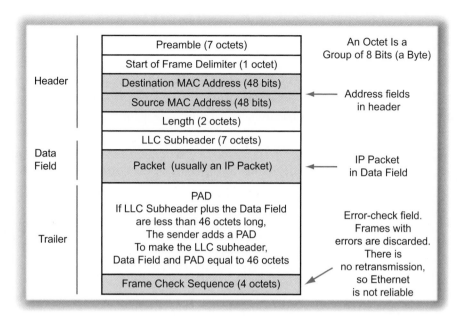

Figure 2-11   Ethernet Frame

respectively. Each Ethernet network interface card (NIC) has a unique 48-bit **Ethernet address** set at the factory before the card ships.

Hexadecimal Notation   As Chapter 4 discusses, Ethernet addresses often are written in hexadecimal (Base 16) notation. In "hex" notation, a typical 48-bit **Ethernet address** is A7-BF-23-D4-33-99.

Ethernet Switches and Ethernet Addresses   As we saw in Chapter 1, switches between the source and destination hosts read the destination Ethernet address in each arriving frame. They look up the Ethernet address in the switching table. The line containing this address contains a port number. The switches send the frame back out this indicated port number.

The Ethernet Data Field

An Ethernet frame usually (but not always) contains an IP packet. In the first chapter, we saw that frames encapsulate packets. Figure 2-11 shows how Ethernet implements this encapsulation by showing the position of the packet in the Ethernet frame.

Frame Check Sequence Field: Unreliable Operation

The last field in an Ethernet **frame is the frame check sequence field.** This four-octet field holds a binary number that the sending NIC calculates on the basis of the bits in other fields in the frame.

The receiving NIC recomputes the number and compares its result with the value contained in the frame check sequence field. If the two do not match, an error must have occurred during transmission.

If there is an error, the receiving NIC's data link process simply discards the frame. There is no request for retransmission. Having no means of error correction, the Ethernet protocol at the data link layer is unreliable, despite detecting errors. Error detection is not error correction, which is necessary for reliability.

Error detection is not error correction, which is necessary for reliability.

TEST YOUR UNDERSTANDING

11. What is an octet?

12. a) How many bits long are Ethernet addresses? b) When are Ethernet addresses set on NICs? c) In what notation are Ethernet addresses typically written for human reading? d) What device in an Ethernet network besides the destination host reads the Ethernet address? e) What is its purpose in reading the Ethernet destination address?

13. Where is the IP packet carried in an Ethernet frame?

14. a) How many *bits* long is the Ethernet frame check sequence field? b) What is the purpose of the Ethernet frame check sequence field? c) How does the receiving NIC use the value in the frame check sequence field? d) What happens if a receiving NIC detects an error? e) Does this error detection and discarding process make Ethernet a reliable standard? Explain.

Ethernet Is Unreliable and Connectionless

In addition to being unreliable (as we have just seen), Ethernet is connectionless. NICs send Ethernet frames without opens, closes, acknowledgments, or sequence numbers.

TEST YOUR UNDERSTANDING

15. a) Is Ethernet connectionless or connection-oriented? Explain. b) Is Ethernet reliable or unreliable? Explain.

LAYER 3: THE INTERNET PROTOCOL (IP)

Internet layer standards govern the transmission of messages called packets across a routed network. As noted earlier, the Internet Protocol (IP) is the most common standard at Layer 3, the internet layer.

Layer 2 versus Layer 3

Figure 2-9 illustrated how the data link layer and the internet layer are related. Put simply, the data link is the path a frame takes through a single switched network. In turn, the internet layer path, called a route, is the path a packet takes from the source host to the destination host across an internet.

Figure 2-8 also showed how frames and packets are related. As a packet travels through a routed network, it travels in the data field of a frame within each switched network. If three switched networks are involved in the transmission from the source to the destination host, there will be one packet that will travel in three different frames. To give an analogy, if you mail a letter, the letter goes all the way from you to the receiver, like a packet. Along the way, however, it will travel in a series of trucks and airplanes, which are analogous to frames.

The figure showed how addressing is done in frames and packets. The packet always has the same Layer 3 internet destination address: the IP address of Host B.

However, the three frames have Layer 2 data link layer destination addresses that are the endpoints in their individual switched networks.

➤ The destination address in Frame X in Switched Network X is the data link layer address of Router R1 because Router R1 is the final destination of Frame X.

➤ The destination address in Frame Y in Switched Network Y is the data link address of Router R2 because Router R2 is the final destination of Frame Y.

➤ The destination address in Frame Z in Switched Network Z is the data link address of the destination host, Host B, because Host B is the final destination of Frame Z.

TEST YOUR UNDERSTANDING

16. a) Four switched networks are involved in transmissions from the source to the destination host. How many packets will there be along the way when the source host transmits a packet? b) How many frames will there be along the way? c) How many routes will there be along the way? d) How many data links will there be along the way? e) How many destination IP addresses will there be? f) How many data link layer destination addresses will there be?

The IP Packet

In Figure 2-11, we saw an Ethernet frame at Layer 2. We saw that an Ethernet frame typically carries an IP packet in its data field. Figure 2-12 shows the organization of an **Internet Protocol (IP)** packet.

Illustrated with 32 Bits per Line

An IP packet, like an Ethernet frame, is a long string of bits (1s and 0s). Unfortunately, drawing the packet this way would require a page several meters wide. Instead, Figure 2-12 shows an IP packet as a series of rows with 32 bits per row. In binary counting, the first bit is zero. Consequently, the first row is bits 0 through 31. The next row is bits 32 through 63. This is a different way of showing syntax than we saw with the Ethernet frame and with HTTP messages, but it is a common way of showing syntax, so you need to be familiar with it.

32-Bit Source and Destination Addresses

Like an Ethernet frame, an IP packet has source and destination addresses. These **IP addresses** are 32 bits long. (In contrast, we saw earlier in this chapter that Ethernet addresses are 48 bits long.) While an Ethernet address gives a host's address on its switched network, an IP address gives a host's internet layer address on a routed network consisting of multiple switched networks.

For human comprehension, it is normal to express IP addresses in **dotted decimal notation.** As we saw in Chapter 1, a typical IP address in dotted decimal notation is 128.171.17.13—four numbers separated by dots. The numbers have to be between 0 and 255.

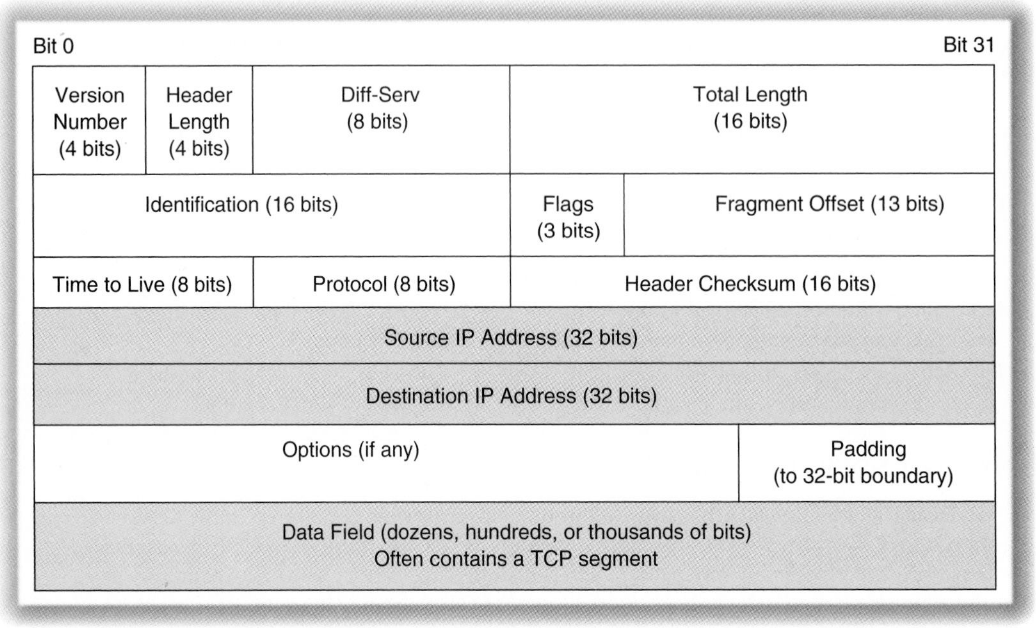

Bit 0			Bit 31

Version Number (4 bits)	Header Length (4 bits)	Diff-Serv (8 bits)	Total Length (16 bits)	
Identification (16 bits)		Flags (3 bits)	Fragment Offset (13 bits)	
Time to Live (8 bits)	Protocol (8 bits)	Header Checksum (16 bits)	
Source IP Address (32 bits)			
Destination IP Address (32 bits)			
Options (if any)		Padding (to 32-bit boundary)	
Data Field (dozens, hundreds, or thousands of bits) Often contains a TCP segment			

Notes: Bits 0–3 hold the version number   Bits 16–31 hold the total length value
Bits 4–7 hold the header length   Bits 32–47 hold the identification value
Bits 8–15 hold the Diff-Serv information

Figure 2-12   Internet Protocol (IP) Packet

Ethernet switches within a switched Ethernet network read the Ethernet destination address in the Ethernet header in a frame to learn where to send the frame. Similarly, routers along the way read the IP address in the destination IP address field of each packet. On the basis of this address, the router forwards the IP packet to the next router or, if the destination host is on a switched network connected to the router, to the destination host itself.

TEST YOUR UNDERSTANDING

17. a) How many octets long is an IP header if there are no options? (Look at Figure 2-12.) b) What is the bit number of the first bit in the destination address field? (Remember that the first bit in binary counting is bit 0.) c) How long are IP addresses? d) You have two addresses: B7-23-DD-6F-C8-AB and 217.42.18.248. Specify what kind of address each address is. e) What device in an internet besides the destination host reads the destination IP address? f) What is this device's purpose in doing so?

IP Characteristics

Like Ethernet, IP is a connectionless protocol. What about reliability? As Figure 2-12 shows, the IP packet has a header checksum field,[3] which the receiver uses to check

---

[3] The figure actually shows the packet header for IP Version 4, which is the dominant version of the Internet Protocol in use today. The newer IP Version 6, which is beginning to spread, does not have a header checksum. It does neither error detection nor error correction.

for errors in the IP header (but not its body). As in Ethernet, if the receiver detects an error, it simply discards the packet. There is error detection but no error correction, so IP is an unreliable protocol.

TEST YOUR UNDERSTANDING

18. a) Is IP connectionless or connection-oriented? b) Is IP reliable or unreliable?

LAYER 4: THE TRANSPORT LAYER

In this section, we will take a closer look at the transport layer. As noted earlier, transport layer (Layer 4) standards govern the aspects of end-to-end communication between the two end hosts that are *not* handled by the internet layer. The internet and transport layers work together to implement internetworking, just as the physical and data link layers work together to implement transmission through a single switched network.

The internet and transport layers work together to implement internetworking, just as the physical and data link layers work together to implement transmission through a switched network.

Layers 3 and 4

Figure 2-9 showed the relationship between the internet and transport layers.

Hop-by-Hop Layers

Note that the internet layer is a **hop-by-hop layer.** Internet layer standards govern the communication between the source host and the first router, between every pair of routers along the way, and between the last router and the destination host. Although packets get all the way from the source host to the destination host, protocol operation at the internet layer is still hop by hop between routers. Internet processes, which implement internet protocols, operate the source host on each intermittent router and on the destination host. Similarly, data link layer standards govern hop-by-hop transmission across multiple switches.

End-to-End Layers

In contrast, the transport layer is an **end-to-end layer.** Its protocols govern communication directly between the transport process on the source host and the transport process on the destination host. The application layer that we will see after this section also is an end-to-end layer. There are only two transport and application processes in operation—one on each host.

TEST YOUR UNDERSTANDING

19. Ten routers separate two hosts. a) How many internet layer processes will be active on the two hosts and the routers between them? b) How many transport layer processes will be active? (The answer is not directly in the book. You will have to think about this one a little.) c) Which layers are hop-by-hop layers? (The physical layer is not considered to be either a hop-by-hop or an end-to-end layer.) d) Which layers are end-to-end layers? (The physical layer is not considered to be either a hop-by-hop or an end-to-end layer.)

TCP: A Reliable Protocol

Most protocols are unreliable. However, TCP, which operates at the transport layer, is reliable, as we saw earlier. As sFigure 2-6 showed, the transport layer is the highest layer apart from the application layer. Making TCP reliable means that any errors made at the transport layer or at lower layers will be caught and corrected by TCP. This gives the application layer clean data.

Caution: The following discussion is difficult but extremely important.

Why not simply make all layers reliable? The answer is that reliability is expensive. As we saw earlier in this chapter, reliability requires considerable processing and storage requirements on each device to hold copies of outgoing messages and to decide whether to resend a message. In fact, error correction consumes far more processor cycles per message than any other process in switching or routing.

Reliability is expensive. In fact, error correction consumes far more processor cycles per message than any other process in switching or routing.

TCP is a very complex protocol. We will look at it in detail in Chapter 8. However, we have already seen that extensive traffic is generated by openings, closings, acknowledgments, and retransmissions. TCP places heavy burdens on the sending and receiving hosts and on the internet's traffic.

Figure 2-13   Why Not Make All Layers Reliable? (Study Figure)

Reliability Is Expensive

> Where errors are rare (in hops between routers and switches), the cost is not justified
>
> Switches and routers would be much more expensive if they did hop-by-hop error correction
>
> There are many switch and router hops, so doing error correction between hops would be very expensive
>
> Error correction at the transport layer corrects errors made at lower layers, making correction at lower layer unnecessary as well as expensive

Error Checking Makes Sense at the Transport Layer

> There are only two transport processes: one on the source host, one on the destination host
>
> So error correction has to be done only once
>
> The transport process is just below the application layer
>
> So doing error correction at the transport layer frees the application layer from doing error correction

There are two reasons to make TCP reliable. The first is that TCP works at the transport layer, which is immediately below the application layer. Error correction at this layer automatically corrects errors at all lower layers. Doing error correction at only one layer—the highest layer before the application layer—provides error-free data to the application program at minimal cost, compared with doing error correction at all layers.[4]

The second reason for doing error correction only at the transport layer is that the transport layer is an end-to-end layer. This means that error correction is done only on the two hosts. In contrast, if error correction occurred on each hop between switches at the data link layer and on each hop between routers at the internet layer, the costs of switches and routers would be far higher than they are. Furthermore, incurring this cost at each switch and router would be extremely wasteful because transmission errors are rare today.

TEST YOUR UNDERSTANDING

20. a) Why are most standards unreliable? b) For what two reasons is making TCP reliable a good choice?

The User Datagram Protocol (UDP)

TCP is widely used at the transport layer, but it is a heavyweight protocol. It consumes many computer processing cycles when doing the calculations needed for error correction. It also consumes extensive traffic capacity because it sends many supervisory messages (ACK segments, opening messages, closing messages, and so forth).

For applications that do not need error correction, there is another popular protocol at the transport layer. This is the **User Datagram Protocol (UDP).** Figure 2-14 compares these two protocols.

The figure shows that both protocols work at the transport layer. They are complementary, and application developers can specify either TCP or UDP for use at the transport layer with their applications. While TCP is connection-oriented, reliable, and burdensome on the two hosts and on the routed network, UDP is connectionless and unreliable, so it places a lighter load on the two hosts and the routed network.

What applications use UDP? Voice over IP (VoIP) uses UDP because voice packets must arrive with minimal delay. There is no time to wait for retransmissions if there is an error.

The Simple Network Management Protocol (SNMP), which we saw briefly in Chapter 1, also uses UDP to transport SNMP application messages. SNMP must send queries constantly to many devices in a routed network. Using UDP reduces the traffic load on the internet. In addition, the occasional loss of a few messages simply means that the management program's information about a few of the network's devices will be slightly out of date. Error correction is simply not worth the cost in SNMP.

---

[4]It is also possible to make the application layer the only layer where error correction is done. However, by making TCP reliable at the transport layer, there is no need for each application program writer to have to develop error correction code. As we will see later, another transport layer protocol, UDP, does *not* do error correction. This makes UDP ideal for applications that do not need error correction or that prefer to do their own error handling.

	TCP	UDP
Layer	Transport	Transport
Connection-orientation?	Connection-Oriented	Connectionless
Reliable?	Reliable	Unreliable
Burden on the two hosts	High	Low
Traffic burden on the network	High	Low

Figure 2-14   TCP and UDP at the Transport Layer

TEST YOUR UNDERSTANDING

21. Compare TCP and UDP in terms of layer of operation, connection-orientation, reliability, and burden (traffic and processing on devices).

The Only Protocols at the Transport Layer

In the TCP/IP transport layer, there are only two protocols—TCP and UDP. Most other layers have far more protocols.

TEST YOUR UNDERSTANDING

22. At the transport layer, what are the only TCP/IP protocols?

LAYER 5: HTTP AND OTHER APPLICATION STANDARDS

The highest layer is the application layer (Layer 5). Standards at this layer govern how application programs talk to one another. In our examples so far, we have used HTTP most of the time. However, as noted earlier, there are many application layer standards—more than there are standards at any other layer. There are application layer standards for e-mail, database queries, and every other application. After network professionals master the network and internetwork standards that this course presents, they spend much of the rest of their careers mastering application standards.

HTTP is a very simple application layer protocol. The browser sends HTTP request messages, and the webserver sends back HTTP response messages. However, many application protocols are much more complex and involve the exchange of many messages. Although we used HTTP in our early examples in this chapter, HTTP is not the only application layer standard. Unfortunately, some students become fixated on examples and lose sight of general principles.

TEST YOUR UNDERSTANDING

23. a) Is the application layer standard always HTTP? b) Which layer has the most standards? c) At which layer would you find standards for instant messaging? (The answer is not explicitly in this section.)

VERTICAL COMMUNICATION ON HOSTS, SWITCHES, AND ROUTERS

In this chapter so far, we have looked at horizontal communication at a single layer, between processes on different hosts, switches, and routers. Indeed, this kind of horizontal communication will be the focus of most of this book. However, communication also needs to take place within a single host, router, or switch. More specifically, a layer process on a device often needs to communicate vertically with the process one layer above it (Layer $N + 1$) and the process one layer below it (Layer $N - 1$). So a Layer 3 protocol has to communicate with Layer 4 ($N + 1$) and Layer 2 ($N - 1$) processes on the same computer.

Layered Communication on the Source Host

Figure 2-15 looks at vertical communication within a single host computer. Here the computer is a client PC with a browser.

At the Application Layer

The browser creates an HTTP request message intended for the webserver process on a webserver destination computer. The two application layer processes are on different computers, so they cannot communicate directly. Rather, they must use processes at other layers on their own computers and on intermediate switches and routers to carry messages between the two application programs.

Figure 2-15 shows that immediately after the browser at the application layer creates the HTTP message, it passes this application message down to the next-lower layer, which is the transport layer.

At the Transport Layer

The transport layer process on the client PC then creates a TCP segment. It places the HTTP message in the body of the TCP segment and adds a header designed to be read by the transport process on the other host. A TCP segment carrying an application message, then, consists of a TCP header and the application message. As noted earlier in the chapter, placing a message in the data field of another message is *encapsulation*. The transport layer, in other words, encapsulates the HTTP message in the data field of a TCP segment. The transport layer then passes the TCP segment down to the internet layer.

At the Internet Layer

The internet layer process creates an IP packet by making the TCP segment the data field and adding an IP header. This packet now contains an HTTP message, a TCP header, and an IP header. Continuing the pattern, the internet layer process passes the IP packet down to the data link layer.

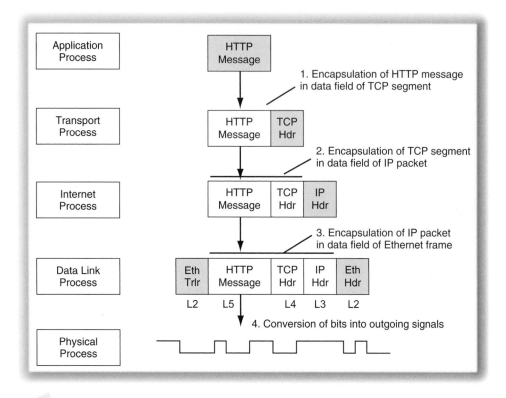

Figure 2-15   Layered Communication on the Source Host

At the Data Link Layer

The data link layer takes the IP packet as its data field and adds a data link header and perhaps a data link trailer. The final frame, then, consists of a data link header, an IP header, a TCP header, the HTTP message, and a data link trailer. If the switched network is an Ethernet network, the data link header is an Ethernet header and the data link trailer is an Ethernet trailer. This data link layer frame is the final message created on the source host.

A Study Hint

Figure 2-15 lists the final frame with the layer numbers of its headers, messages, and trailers. There is a data link layer (L2) header, an IP (L3) header, a TCP (L4) header, an application layer (L5) message, and a data link layer (L2) trailer. Note the regular pattern—L2, then L3, then L4, then L5. The final element (L2) breaks the sequence. If you are confused about which protocol is used at each layer, remember the L2, L3, L4, L5, and L2 pattern.

To create a final frame, begin with the highest layer message or header. Then add headers and perhaps trailers until the frame is complete.

➤ In this example, the highest layer is Layer 5, which creates the HTTP message.

➤ Next, the L4 header is added, which is either TCP or UDP. In this case, it is a TCP header. We now have the HTTP message (L5) and the TCP header (L4).

➤ Next comes an L3 header. At Layer 3, IP is almost always the standard, so the header is an IP header. We now have the HTTP message (L5), a TCP header (L4), and an IP header (L3).

➤ Finally, the Layer 2 header and trailer are added. We now have the final frame—an L2 trailer, the HTTP message (L5), a TCP header (L4), an IP header (L3), and an L2 header.

At the Physical Layer

The data link layer passes the data link layer frame to the physical layer. The physical layer converts the ones and zeros of this frame into signals and transmits these signals to the next device, usually a switch.

A Simple Process, with Repetition

Some students have a difficult time with layered communication on source (and destination) hosts. If you are one of them, just keep in mind that there is a simple process repeated multiple times.

➤ When a layer N creates its message, the layer process passes the message down to the $N - 1$ layer.

➤ The $N - 1$ layer encapsulates the message in the data field of the Layer $N - 1$ message and adds a Layer $N - 1$ header and (at the data link layer only) perhaps a layer $N - 1$ trailer. The Layer $N - 1$ process then repeats the cycle.

The only exception is the physical layer, which receives data link layer frames, converts them into signals, and transmits them. And, of course, the layer that begins the transmission (in the example above, the application layer) does not encapsulate any messages from higher layers.

A Postal Analogy

To give an analogy, when you write a letter, you encapsulate it in an envelope, which the postal service delivers. Along the way, the postal service will encapsulate your message in mailbags for delivery between post offices.

At the other end, the post office decapsulates the envelope from its mail bag and delivers the envelope to your intended receiver. The recipient decapsulates the letter from the envelope and reads it.

TEST YOUR UNDERSTANDING

24. a) When a layer creates a message, what does it usually do immediately afterward? b) What does the layer below it usually do after receiving the next-higher-layer message? c) What is encapsulation? d) With Web communication using HTTP, what message does IP encapsulate in packet data fields?

25. a) What are the two steps after a layer process creates its layer message? b) What is the final frame if SMTP (an e-mail protocol that requires TCP) is used at the application layer and if Frame Relay (which has a header and a trailer) is used instead of Ethernet at the data link layer? c) What is the final frame if SNMP (which requires UDP) is used at

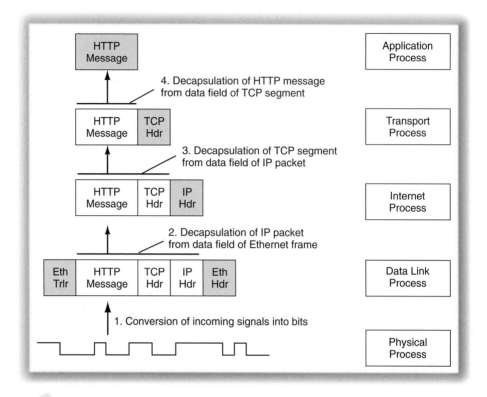

Figure 2-16   Decapsulation on the Destination Host

the application layer and if the ATM data link layer protocol (which has a header but no trailer) is used instead of Ethernet at the data link layer?

On the Destination Host

When the signal of the last frame finally reaches the destination host, the process of **decapsulation** on the destination host is the reverse of the encapsulation process on the source host, as Figure 2-16 shows. For this example, we will use the frame carrying an HTTP message, which is illustrated in Figure 2-15.

➤   The physical layer turns the signals into the bits of the frame and passes the frame to the data link layer.

➤   The data link layer checks the frame for errors. If there are no errors, it decapsulates the IP packet from the frame and passes the packet up to the internet layer process.

➤   The internet layer checks the IP packet header for errors. If there are no errors, it decapsulates the TCP segment from the IP packet and passes the segment up to the transport layer process. (It also sends an acknowledgment.)

➤   The transport layer checks the TCP segment for errors. If there are no errors, it decapsulates the HTTP message and passes this message up to the application layer process.

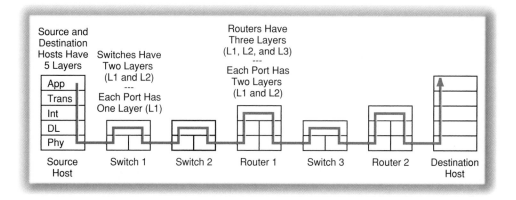

Figure 2-17 Layered End-to-End Communication

What if there is an error at any layer? If there is an error (apart from the application layer), the layer process simply discards the message and does not send an error message. This is true for all layers above the physical layer. This is even true for the transport layer, although at the transport layer, the message is re-sent by the original sender.

TEST YOUR UNDERSTANDING

26. a) Which host decapsulates—the sending host or the receiving host? b) Describe what each layer's process does on the receiving host when the host receives an Ethernet frame containing an HTTP message.

On Switches and Routers along the Way

So far, we have looked at layered communication only on the two hosts. However, layer processing also takes place on all intermediate switches and routers, as Figure 2-17 shows. Note that the highest layer on switches is the data link layer and that the highest layer on routers is the internet layer. This is why networking specialists call switches Layer 2 devices and call routers Layer 3 devices.

TEST YOUR UNDERSTANDING

27. a) Why are switches called Layer 2 devices? b) Why are routers called Layer 3 devices? c) Do routers first encapsulate or decapsulate? (The answer is not explicit in the text. Look at Figure 2-17.)

Layering on a Source Router

In the example we just worked through, the application layer process began the message transmission. However, this is not always the case. For instance, in the ICMP standard that we will see in Chapter 10, if a router cannot deliver a packet, it may send an

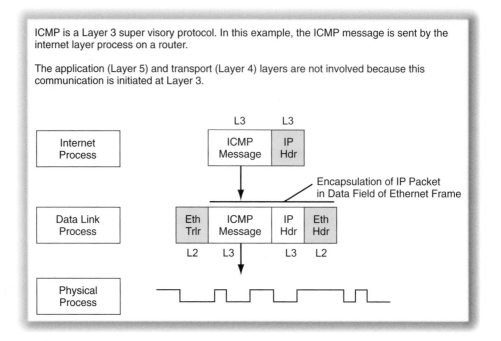

ICMP is a Layer 3 super visory protocol. In this example, the ICMP message is sent by the internet layer process on a router.

The application (Layer 5) and transport (Layer 4) layers are not involved because this communication is initiated at Layer 3.

Figure 2-18   Layered Message Exchange Initiated at the Internet Layer

ICMP message back to the internet layer process on the source host, giving an indication of what the problem was.

Figure 2-18 shows what a router will do to send an ICMP message back to the source host that sent the problem packet. The router's internet layer process is the router's highest layer process involved in ICMP. The router's internet layer process creates the ICMP message to be sent to the source host. Note that because the highest layer involved in the process is the internet layer (L3), there is no L4 or L5 message or header.

The internet layer process puts the ICMP message into an IP packet by adding an IP header. The IP header is a Layer 3 header, so we have the odd situation of one L3 message (the ICMP message) being encapsulated in the data field of an L3 packet.

The internet layer process then passes the packet to the data link layer on the port through which the message will go out. If this is an Ethernet port, then the final frame will include an Ethernet header (L2), an IP header (L3), an ICMP message (L3, because it is a Layer 3 supervisory protocol carried in the IP header's data field), and an Ethernet trailer (L2). The layering sequence will be L2-L3-L3-L2, where ICMP and the packet it contains are both at Layer 3.

To see how to build this frame, begin with the highest layer (L3). At this layer we have the ICMP message. Next add an IP header (L3). Finally, work down to the next-lower-layer, Layer 2. Ethernet adds an L2 header before the packet and an L2 trailer after the packet.

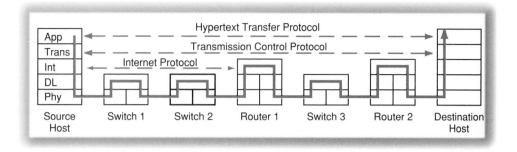

Figure 2-19   Combining Horizontal and Vertical Communication

TEST YOUR UNDERSTANDING

28. a) Why is there no transport or application content to the final frame in Figure 2-18?
    b) To create the frame in the figure, at what layer do you begin? c) At what layer is an ICMP message?

Combining Vertical and Horizontal Communication

In earlier parts of this chapter, we looked at *horizontal* communication between hardware or software processes at the *same layer* but on two *different* devices. In this section, we looked at *vertical* communication between *adjacent layer* processes on the *same* device. Figure 2-19 shows how the two views are related.

Vertical communication is necessary because hosts, switches, and routers connect only at the physical layer. For a process at any other layer, the only way to communicate with its peer process on another device is to pass its message down to the physical layer for transmission. The vertical communication process we have been examining does this in an orderly manner. In the examples shown in Figure 2-15 and Figure 2-16, the application process transmitted an entire message to its peer. Other layers simply added communication messages for their peers in the headers and perhaps trailers that they added during encapsulation. To continue the postal analogy, the address on an envelope is intended for the postal service to read, while the letter contained in the envelope is intended only for the recipient of the letter to read.

TEST YOUR UNDERSTANDING

29. To what software process is the transport layer message addressed?

MAJOR STANDARDS ARCHITECTURES

It might be nice if there were only one set of standards that governed all network equipment. The reality, however, is that there are several standards architectures, which are families of related standards that collectively allow an application program on one machine on a routed network to communicate with another application program on another machine on that same internet.

> Standards architectures are families of related standards that collectively allow an application program on one machine on a routed network to communicate with another application program on another machine on that same internet.

Unfortunately, hardware and software processes cannot talk to each other if they use standards from different architectures. Consider one non-network example of multiple standards architectures and incompatibility: Different countries have different standards for their electrical systems (voltages, cycles per second, plug design, etc.). If you take your computer from one country to another, it may operate at the wrong voltage and with the wrong number of cycles per second. Plugging your PC into a wall socket might damage the computer beyond repair. Fortunately, you probably would not even be able to plug your computer into wall sockets because your power cord would not fit the wall jack.

TCP/IP and OSI Architectures

Although there are several major network standards architectures, two of them dominate actual corporate use: OSI and TCP/IP. Figure 2-20 illustrates both. Although they often are described as competitors, we will see that they actually work together in most corporate networks. What corporations really use today is primarily a hybrid (combined) TCP/IP–OSI architecture.

Figure 2-20   The Hybrid TCP/IP-OSI Architecture

Broad Purpose	TCP/IP	OSI	Hybrid TCP/IP-OSI
Applications	Application	Application (Layer 7)	Application (Layer 5)
		Presentation (Layer 6)	
		Session (Layer 5)	
Internetworking	Transport	Transport (Layer 4)	TCP/IP Transport Layer (Layer 4)
	Internet	Network (Layer 3)	TCP/IP Internet Layer (Layer 3)
Communication within a single switched LAN or WAN	Use OSI Standards Here	Data Link (Layer 2)	Data Link (OSI) Layer (Layer 2)
		Physical (Layer 1)	Physical OSI Layer (Layer 1)

Notes:
The Hybrid TCP/IP-OSI Architecture governs the Internet and dominates internal corporate networks.
OSI standards dominate the physical and data link layers (which govern communication within individual networks) almost exclusively.
TCP/IP dominates the internet and transport layer in internetworking and governs 80% to 90% of all corporate traffic above the data link layer.

	OSI	**TCP/IP**
Standards Agency or Agencies	ISO (International Organization for Standardization) ITU-T (International Telecommunications Union–Telecommunications Standards Sector)	IETF (Internet Engineering Task Force)
Dominance	Nearly 100% at physical and data link layers	80% to 90% at the internet and transport layers
Documents Are Called	Various	Mostly RFCs (requests for comments)

Notes:
Do not confuse OSI (the architecture) with ISO (the organization).
The acronyms for ISO and ITU-T do not match their names, but these are the official names and acronyms.

Figure 2-21   OSI and TCP/IP

TEST YOUR UNDERSTANDING

30. a) What is a standards architecture? b) What are the two dominant network standards architectures? c) Are they competitors?

OSI
OSI is the "Reference Model of Open Systems Interconnection." *Reference model* is another name for *architecture.* An open system is one that is open to communicating with the rest of the world. In any case, OSI is rarely spelled out, which is merciful.

Standards Agencies: ISO and ITU-T
Standards architectures are managed by organizations called **standards agencies.** Figure 2-21 shows that OSI has two standards agencies.

➤ One is the **International Organization for Standardization (ISO),** which generally is a strong standards organization for manufacturing, including computer manufacturing. By the way, do not confuse OSI and ISO. OSI is an architecture. ISO is a standards agency.

➤ The other is the **International Telecommunications Union–Telecommunications Standards Sector (ITU-T).**[5] Part of the United Nations, the ITU-T oversees international telecommunications.

Although ISO or the ITU-T must *ratify* all OSI standards, other organizations frequently *create* standards for inclusion in OSI. For instance, we will see in Chapter 4 that the IEEE creates Ethernet standards. These standards are not official until the ITU-T or ISO ratifies them, although this has always been a mere formality.

---

[5]No, the names and acronyms do not match for ISO and ITU-T, but these are the official names and acronyms for these two organizations.

Layer Number	OSI Name	Purpose	Use
1	Physical	Physical connections between adjacent devices.	Nearly 100% dominant
2	Data Link	End-to-end transmission in a single switched network. Frame organization. Switch operation.	Nearly 100% dominant
3	Network	Generally equivalent to the TCP/IP internet layer. However, OSI network layer standards are not compatible with TCP/IP internet layer standards	Rarely used
4	Transport	Generally equivalent to the TCP/IP transport layer. However, OSI transport layer standards are not compatible with TCP/IP transport layer standards	Rarely used
5	Session	Initiates and maintains a connection between application programs on different computers.	Rarely used
6	Presentation	Designed to handle data formatting differences and data compression and encryption. In practice, a category for general file format standards used in multiple applications.	Rarely used as a layer. However, many file format standards are assigned to this layer.
7	Application	Governs remaining application-specific matters.	Some OSI applications are used

Figure 2-22   OSI Layers

OSI's Dominance at Lower Layers (Physical and Data Link)

Although OSI is a seven-layer standards architecture (see Figure 2-20), standards from its five upper layers are rarely used. Figure 2-22 describes the seven OSI layers.

However, at the two lowest layers—the physical and data link layers—corporations use OSI standards almost universally in their networks. These two layers govern transmission within a switched network. Almost all switched networks—both switched LANs and switched WANs—follow OSI standards at the physical and data link layers, regardless of what upper-layer standards they use. OSI standards are almost 100 percent dominant at the bottom two layers.

Almost all switched networks—both switched LANs and switched WANs—follow OSI standards at the physical and data link layers, regardless of what upper-layer standards they use.

Other standards agencies, recognizing the dominance of OSI at the physical and data link layers, simply specify the use of OSI standards at these layers. They then create standards only for internetworking and applications.

OSI Network and Transport Layers

The network layer functionality of OSI corresponds closely to the internet layer functionality of TCP/IP that we saw earlier in this chapter. The transport layer functionality of OSI, in turn, is very similar to the transport layer functionality of TCP/IP. However, while *functionality* may be similar between OSI and TCP at these layers, actual OSI and TCP/IP *standards* at these layers are completely incompatible. More importantly, OSI standards are rarely used at the network or transport layers by real organizations.[6]

OSI Session Layer

The **OSI session layer (OSI Layer 5)** initiates and maintains a connection between application programs on different computers. For instance, suppose that a single transaction requires a number of messages. If there is a connection break, the transmission can begin at the last session layer checkpoint. For example, if communication fails during a database transaction, the entire transaction does not have to be done over—only the work since the last rollback point.

OSI Presentation Layer

The **OSI presentation layer (OSI Layer 6)** is designed to handle data formatting differences between the two computers. For example, most computers format character data (letters, digits, and punctuation signs) in the ASCII code. In contrast, IBM mainframes format them in the EBCDIC code.

The OSI presentation layer is also designed to be used for compression and data encryption for application data. In practice, however, the presentation layer in practice is rarely used for either data format conversion or compression and encryption.

Rather, the presentation layer has become a category for general file format standards used in multiple applications, including MP3, JPEG, and many other general OSI file format standards.

OSI Application Layer

The **OSI application layer (OSI Layer 7)** governs remaining application-specific matters that are now covered by the session and presentation layers. The OSI application layer, freed from session and presentation matters, focuses on concerns specific to the application in use.

TEST YOUR UNDERSTANDING

31. a) What standards agencies are responsible for the OSI standards architecture? Just give the acronyms. b) At which layers do OSI standards dominate usage? c) Name

---

[6]Although OSI physical and data link layer standards are dominant, and while many OSI application layer standards are used, almost no systems implement OSI standards at the network, transport, session, or presentation layers. Then why, you may ask, do you need to be able to describe these layers? The answer is that you need to get a job. In a very large percentage of all job interviews, an interviewer, noting that you have taken a networking course, asks you to describe the OSI layers. I kid you not.

and describe the functions of OSI Layer 5. d) Name and describe the intended use of OSI Layer 6. e) How is the OSI presentation layer actually used? f) Beginning with the physical layer (Layer 1), give the name and number of the OSI layers.

TCP/IP

The **TCP/IP** architecture is mandatory on the Internet at the internet and transport layers. TCP/IP is also widely used at these layers by companies for their internal corporate internets.

The TCP/IP architecture is named after two of its standards, TCP and IP, which we have looked at briefly in this chapter. However, TCP/IP also has many other standards, including the UDP standard we have already seen. This makes the name TCP/IP rather misleading. Another confusing point about names is that TCP/IP is the *standards architecture*, while TCP and IP are individual *standards* within the architecture.

Note that TCP/IP is the standards architecture, while TCP and IP are individual standards within the architecture.

The Internet Engineering Task Force (IETF)

TCP/IP's standards agency is the **Internet Engineering Task Force (IETF).** Traditionally, the IETF has been viewed as being in competition with ISO and ITU-T for standards development.[7] However, in recent years, the IETF and these other organizations have begun to cooperate in standards development. For instance, the IETF is working closely with ITU-T for VoIP transmission standards.

The IETF historically has been rather informal. Its committees traditionally focus on consensus rather than on voting, and technical expertise is the source of most power within the organization. Although corporate participation has somewhat "tamed" the IETF, it remains a fascinating organization, although its members might argue that "organization" is too strong a word.

A great deal of the IETF's success is due to the fact that the IETF typically produces simple standards, then adds to their complexity over time. In fact, IETF standards often have the word *simple* in their name—for instance, the Simple Mail Transfer Protocol. Consequently, IETF TCP/IP standards are developed quickly. In addition, TCP/IP products can be developed quickly and inexpensively because they are simple. "Inexpensive and fast to market" is almost always a good recipe for success. In contrast, OSI standards often take a very long time to be developed and often are so bloated with functionality that they are uneconomical.

Note that the success of TCP/IP is not primarily due to its use on the Internet. Corporations had already shifted many of their networks to TCP/IP before the Internet became a dominant force in the 1990s.

---

[7]In 1992, IETF member Dave Clark summarized the situation this way: "We reject kings, presidents, and voting. We believe in rough consensus and running code." His last point is that the IETF normally will not create a standard unless there are working products that implement it. This avoids a frequent problem with OSI standards, which often are developed before any products exist—a tendency for standards implementation to be difficult because things that the standards developers believed to be very clear turn out to be ambiguous or even wrong during implementation.

Requests for Comments (RFCs)

Most documents produced by the IETF have the rather misleading name **requests for comments (RFCs).** Every few years, the IETF publishes a list of which RFCs are **Official Internet Protocol Standards.** Each list of standards adds some RFCs to the list and drops previously listed standards.

Dominance at the Internetwork Layers (Internet and Transport)

As noted earlier, physical and data link layer standards govern the transmission of data within a *switched network.* We saw earlier that OSI standards are completely dominant at these layers.

TCP/IP internet and transport layer standards, in turn, govern transmission across an entire internet, ensuring that any two host computers can communicate. TCP/IP application standards ensure that the two application programs on the two hosts can communicate as well.

TCP/IP is dominant in the internetworking layers (internet and transport). However, it is less dominant at these layers than OSI standards are at the bottom two layers. In most organizations, TCP/IP standards govern 80 percent to 90 percent of internet and transport layer traffic, and this dominance is growing. In a few years, the use of other architectures at the internet and transport layers in new products will be increasingly rare.

TCP/IP is dominant in corporate networking at the internet and transport layers, although less dominant than OSI is at the physical and data link layers. In most organizations, TCP/IP standards govern 80 percent to 90 percent of internet and transport layer traffic.

TEST YOUR UNDERSTANDING

32. a) Which of the following is an architecture: TCP/IP, TCP, or IP? b) Which of the following are standards: TCP/IP, TCP, or IP? c) What is the standards agency for TCP/IP? d) Why have this agency's standards been so successful? e) What are most of this agency's documents called? f) At which layers is TCP/IP dominant? g) How dominant is TCP/IP today at these layers, compared with OSI's dominance at the physical and data link layers?

The Application Layer

OSI is completely dominant at the physical and data link layers, and TCP/IP is very dominant at the internet and transport layers. What about the application layer? The answer here is highly complex.[8] Overall, it seems best to say that no standards agency

---

[8]Many application protocols come from the IETF. These include such popular standards as e-mail protocols (SMTP, POP, IMAP, etc.), the FTP standards, and the Simple Network Management Protocol (SNMP) in network management.

Some application standards come directly from OSI. This is particularly true of graphics file format standards. However, many OSI standards were too complex for widespread use. The IETF then produced simpler versions of these OSI standards. A good example is the Lightweight Directory Access Protocol (LDAP) for access to directory servers. LDAP evolved from the OSI directory access protocol standard.

Other standards agencies are also producing application layer standards. HTTP and HTML standards come from the World Wide Web Consortium (W3C), although the IETF is producing some WWW standards. Most confusingly, incompatible Service Oriented Architecture web services standards are being produced by several competing standards agencies.

or architecture dominates at the application layer, although the IETF is particularly strong, especially for popular standards such as e-mail and FTP.

Although many applications do not come from the IETF, they almost all run over TCP/IP standards at the internet and transport layers. This is true even for standards created by ISO and ITU-T.

At the application layer, in fact, there is growing cooperation between ISO, the ITU-T, and the IETF. In voice over IP, the ITU-T and the IETF have harmonized several key standards. ISO and the IETF, in turn, have been cooperating in file format standards.

TEST YOUR UNDERSTANDING

33. a) Is any standards architecture dominant at the application layer? b) Do almost all applications run over TCP/IP standards at the internet and transport layers?

TCP/IP and OSI: The Hybrid TCP/IP–OSI Standards Architecture

Although people sometimes view TCP/IP and OSI as competitors, most organizations use them together. The most common standards pattern in organizations is to use OSI standards at the physical and data link layers and TCP/IP standards at the internet and transport layers. This is very important for you to keep in mind because this **hybrid TCP/IP–OSI standards architecture,** shown in Figure 2-20, will form the basis for most of this book.

TEST YOUR UNDERSTANDING

34. a) What layers of the hybrid TCP/IP–OSI standards architecture use OSI standards? b) What layers use TCP/IP standards? c) Do switched LAN standards come from OSI or TCP/IP? Explain. (The answer is not explicitly in this section.) d) Do switched WAN standards come from OSI or TCP/IP? Explain. (Again, the answer is not explicitly in this section.)

A Multiprotocol World at Higher Layers

At the same time, quite a few networking products (especially legacy products) in organizations follow other architectures, as shown in Figure 2-23. Real corporations live and will continue to live for some time in a multiprotocol world in which network administrators have to deal with a complex mix of products following different architectures above the data link layer. In this book, we focus on OSI and TCP/IP because they are by far the most important and are becoming ever more so. However, in a typical organization, 10 to 20 percent of all upper-layer traffic still uses protocols from other standards architectures.

IPX/SPX

The most widely used non-TCP/IP standards architecture found at upper layers in LANs is the **IPX/SPX architecture.** Older, Novell NetWare file servers required this architecture. Many NetWare users are switching to TCP/IP.

Systems Network Architecture (SNA)

IBM mainframe computers traditionally used the Systems Network Architecture (SNA) standards architecture, which actually predates OSI and TCP/IP. Most firms,

IPX/SPX
> Used by older Novell NetWare file servers for file and print service
>
> Sometimes used in newer Novell NetWare file servers for consistency with older NetWare servers

SNA (Systems Network Architecture)
> Used by IBM mainframe computers

AppleTalk
> Used by Apple Macintosh desktops and notebooks to talk to Macintosh servers

Figure 2-23   Other Major Standards Architectures

however, are transitioning or have already transitioned their mainframe communications from SNA to TCP/IP.

AppleTalk

Until OSX, Macintosh® desktop and notebook computers were designed to use Apple®'s proprietary **AppleTalk® architecture** when they talk to Macintosh servers. Macintoshes are rare in corporations today, so AppleTalk is not widely seen, and most recent Apple products speak TCP/IP.

TEST YOUR UNDERSTANDING

35. a) Under what circumstances might you encounter IPX/SPX standards? b) SNA standards? c) AppleTalk standards?

CONCLUSION

Synopsis

In this chapter, we looked broadly at standards. Most of this book (and the networking profession in general) focuses on standards, which are also called protocols.

Standards govern message exchanges. More specifically, they place constraints on message semantics (meaning) and message syntax (format).

Standards are connection-oriented or connectionless. In connection-oriented protocols, there is a distinct opening before content messages are sent and a distinct closing afterward. There also are sequence numbers, which allow fragmentation and are used in supervisory messages (such as acknowledgments) to refer to specific messages by sequence numbers. In connectionless protocols, there are no such openings and closings. Connectionless protocols are simpler than connection-oriented protocols, but they lose the advantages of sequence numbers.

Layer	Protocol	Connection-Oriented or Connectionless?	Reliable or Unreliable?
5 (Application)	HTTP	Connectionless	Unreliable
4 (Transport)	TCP	Connection-oriented	Reliable
4 (Transport)	UDP	Connectionless	Unreliable
3 (Internet)	IP	Connectionless	Unreliable
2 (Data Link)	Ethernet	Connectionless	Unreliable

Figure 2-24   Characteristics of Protocols Discussed in This Chapter

In turn, reliable protocols do error correction, while unreliable protocols do not (although unreliable protocols may do error detection without error correction). In general, standards below the transport layer are unreliable in order to reduce costs. The transport standard usually is reliable; this allows error correction processes on just the two hosts to correct errors at the transport layer and at lower layers, giving the application clean data. Figure 2-24 compares the main protocols we have seen in this chapter in terms of connection orientation and reliability.

Standards are created within broad plans called standards architectures. Most real-world corporate networks use a five-layer architecture.

➤ The physical layer governs transmission between adjacent devices. There are no messages at the physical layer. Physical layer standards almost always come from OSI.

➤ The data link layer governs transmission between two devices on a switched network. Messages at the data link layer are called frames. The path a frame takes through a switched network is called a data link. Data link layer standards govern the operation of switches within a single switched network. Data link layer standards almost always come from OSI.

➤ The internet layer governs transmission between the source and destination hosts on a routed network. Messages at the internet layer are called packets. The path a packet takes through an internet is called a route. Internet layer standards govern the operation of packets within a routed network. Note that messages at the data link layer are frames, while messages at the internet layer are packets. If there are ten switched networks involved in a transmission between two hosts, there will be ten frames and ten data links along the way but only one packet and one route. Internet layer standards usually come from TCP/IP.

➤ The transport layer governs communication between the source and destination hosts in an internet. It usually corrects errors at lower layers. Transport layer standards usually come from TCP/IP.

➤ The application layer governs interactions between application programs. Application standards come from TCP/IP, OSI, and other standards architectures.

Transmission in single switched networks—both switched LANs and switched WANs—is governed by physical and data link layer standards, which almost always come from OSI. Ethernet is a fairly simple standard for physical and data link layer transmission in switched LANs. Ethernet is connectionless and unreliable. This simplicity leads to low costs, and low costs have brought Ethernet to dominance in switched LANs. Representations of the Ethernet frame's syntax usually show each field's length in octets. Ethernet addresses are 48 bits long. They are often expressed in hexadecimal notation for human reading.

Internetworking involves the internet and transport layer standards. At the internet layer, the Internet Protocol (IP) is a fairly simple connectionless and unreliable protocol. IP depends on TCP for error correction and even for the correct ordering of received packets. The IP packet's syntax usually is shown as a series of lines with 32 bits on each line. IP addresses are 32 bits long. Humans usually express them in dotted decimal notation.

The Transmission Control Protocol (TCP) is a complex transport layer protocol that is connection-oriented and reliable. The receiving transport process acknowledges every correct TCP segment. If a segment is not acknowledged promptly, the sender retransmits it. TCP corrects errors at all lower layers as well as at the transport layer, so it gives the application program clean data.

The User Datagram Protocol (UDP) is an alternative to TCP at the transport layer. It is a lightweight protocol that places a lower burden than TCP on devices and networks. However, it is connectionless and unreliable. It is up to the application layer program to deal with errors that occur.

HTTP is a simple connectionless and unreliable protocol that depends on TCP at the transport layer for reliable data transmission. HTTP has a simple text-based syntax for its headers.

Standards processes at each layer also communicate vertically with the process in the layer directly above them and the process in the layer directly below them on the same computer. On the sending device, a process usually creates a message and then passes the message down to the next-lower-layer process, which encapsulates the message in the data field of its own message. On the receiving process, the reverse process (decapsulation) is used. Hosts are Layer 5 devices, while switches are Layer 2 devices and routers are Layer 3 devices.

Standards are created by standards agencies, which first create standards architectures that guide the creation of individual standards. Standards from different architectures are incompatible. The TCP/IP architecture is managed by the IETF, while the OSI architecture is managed by two organizations, ISO and ITU-T. The dominant standards architecture used in organizations today is the hybrid TCP/IP–OSI architecture, with OSI standards being used at the bottom two (physical and data link) layers and TCP/IP standards being used at the upper (internet and transport) layers.

Again, application layer standards come from TCP/IP, OSI, and other standards architectures. Although many applications do not come from the IETF, they almost all run over TCP/IP standards at the internet and transport layers. This is true even for standards created by ISO and ITU-T.

Firms use OSI standards almost universally at the physical and data link layers. At higher layers, most firms are multiprotocol environments in which 20–30 percent of all traffic at upper layers comes from standards architectures other than TCP/IP— usually, IPX/SPX (for some Novell NetWare file servers), SNA (for IBM mainframe communications), and AppleTalk (for older Macintosh computers).

End-of-Chapter Questions

BASIC THOUGHT QUESTIONS

1. To open a TCP connection to the transport layer process on Host B, the TCP process on host A sends a TCP SYN segment. What will the final frame be that Host A sends if Host A is on a Frame Relay (FR) switched network? FR frames have both headers and trailers. The application layer process is not at all involved in the connection-opening attempt.

2. Figure 2-11 shows the fields in an Ethernet frame. Ethernet is the dominant standard for switched LANs. However, there are many other data link standards. One example is the Point-to-Point Protocol (PPP). This protocol is used at the data link layer to connect two routers with a point-to-point leased line from the telephone company. PPP frames begin and end with a one-octet flag field containing the content 01111110. These unambiguously signal the start of a new frame and the end of that frame, respectively. The second two octets always have the values 11111111 and 00000011. These are the address and control fields, respectively. They exist for historical reasons that are no longer important. Obviously, there is no need for an address field in a point-to-point connection, and the function of the control field has been replaced by the advanced use of the data field for supervisory communication. The next two octets form the protocol field, which describes the contents of the data field. If the PPP frame is delivering a packet, the protocol field contains the value 8021h. In PPP, this data field is called the information field. It can be up to 1500 octets long, although a shorter value can be negotiated. Next comes the frame check sequence field. As in Ethernet, this field is used to detect errors. If the receiver detects an error, it simply discards the frame. There are no acknowledgments. a) Which fields form the PPP header? b) Which fields form the PPP trailer? c) Figure 2-15 shows the final frame when an HTTP application transmits an application message over an Ethernet switched network. Give the final frame if PPP is used instead of Ethernet. Just say *PPP header* and *PPP trailer*. Do not give details regarding the fields that make up the PPP header and trailer. d) Give the final frame if the application is SNMP, which requires UDP at the transport layer. The protocol is still PPP at the data link layer. Again, just say *PPP header* and *PPP trailer*. Do not give details regarding the fields that make up the PPP header and trailer. e) When TCP sends a pure acknowledgment, it transmits a TCP message that has only a header. The application layer is not involved at all in the acknowledgment process. Give the final frame when the packet travels over an Ethernet LAN.

3. a) In Figure 2-19, how many switched networks, physical links, data links, and routes are shown?

   b) In Figure 2-19, how many data link, internet, transport, and application **processes** in total are involved in the transmission?

4. Ethernet stations need Ethernet addresses. Do Ethernet switches need to have Ethernet addresses too when they forward frames? Explain your reasoning.

EVEN HARDER THOUGHT QUESTIONS

1. Normally, only the transport layer standard is reliable. However, in Chapter 5, we will see that 802.11 wireless LAN standards at the data link layer are reliable. Why do you think this is so? (Hint: Review the logic in Figure 2-13 for clues.)

2. How do you think TCP would handle the problem if an acknowledgment were lost, so that the sender retransmitted the unacknowledged TCP segment, therefore causing the receiving transport process to receive the same segment twice?

3. You can place both TCP/IP clients and servers and IPX clients and servers on the same Ethernet network, and each client will talk to its server. How do you think this is possible? (Hint: Consider the Ethernet frame in Figure 2-11.)

4. How can you make a connectionless protocol reliable? (Try to answer this one, but you may not be able to do so.)

5. Spacecraft exploring the outer planets need reliable data transmission. However, the acknowledgments would take hours to arrive. This makes an ACK-based reliability approach unattractive. Can you think of another way to provide reliable data transmission to spacecraft? (Try to answer this one, but you may not be able to do so.)

PERSPECTIVE QUESTIONS

1. What was the most surprising thing you learned in this chapter?

2. What was the most difficult material for you in this chapter?

GETTING CURRENT

Go to the book website's New Information and Errors pages for this chapter to get new information since this book went to press and to see any errors in the text

Physical Layer Propagation

Learning Objectives

By the end of this chapter, you should be able to discuss the following:

- Signals and propagation effects.
- Binary data representations for important types of data.
- Binary signaling (on/off signaling or using two possible voltage states).
- (In the box on digital signaling) Binary versus digital signaling, including the difference between bit rates and baud rates.
- Unshielded twisted pair (UTP) wiring, including relevant propagation effects that must be controlled by limiting cord length and by limiting the untwisting of pairs during connectorization.
- The differences between serial and parallel transmission, including the speed advantage of parallel transmission.
- Optical fiber cabling, including relevant propagation effects and different types of optical fiber cabling and signaling.
- Wireless radio transmission and its many difficult propagation effects.
- Network topologies, including point-to-point connections, stars, extended stars (hierarchies), rings, meshes, and buses.

INTRODUCTION: THE PHYSICAL LAYER

Characteristics of the Physical Layer

Chapter 2 presented an overview of layered standards. Most of that chapter focused on the data link, internet, transport, and application layers. In this chapter, we will go down to Layer 1, the physical layer, which differs from upper layers in two ways:

> ➤ It is the only layer that does not deal with messages. It takes the bits of frames and turns them into signals.
> ➤ It alone deals with propagation effects, which change signals when they travel over transmission lines.

This chapter covers physical layer signaling and three major transmission media: UTP, optical fiber, and wireless (radio) transmission. It also introduces the physical layer concept of topology.

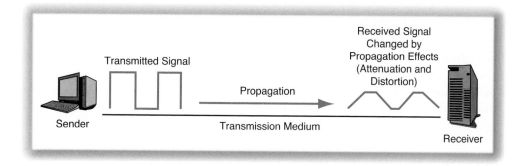

Figure 3-1 Signal and Propagation

Signals and Propagation
Propagation
Suppose that you and a friend are standing a few meters apart. You stretch a rope tautly between you. Now you close your eyes. Your friend jiggles the rope. The disturbance **propagates** (travels) down the rope. When it arrives, you feel it.

Signals
Something like this happens in network transmission. As Figure 3-1 shows, the transmitter creates a disturbance in a transmission medium—wire, optical fiber, or radio. This disturbance is the signal. A **signal,** then, is a disturbance that propagates down a transmission medium to the other side, which reads the signal.

A signal is a disturbance that propagates down a transmission medium to the other side, which reads the signal.

Propagation Effects
Note that the received signal differs from the transmitted signal. The changes in this figure are due to **propagation effects**—that is, changes in the signal during propagation. In the figure, the signal has **attenuated** (weakened). It also has been **distorted** (its shape has been changed). If propagation effects are too large, the receiver will not be able to interpret the signal correctly. In this chapter, we will see several propagation effects, most of which are important only in certain transmission media.

TEST YOUR UNDERSTANDING

1. a) What is a signal? b) What is propagation? c) What are propagation effects? d) Why are propagation effects bad?

BINARY DATA REPRESENTATION

The messages that we saw in the previous chapter were long strings of bits. This type of data has only two possible values (ones and zeros), so it is called **binary data** (from the Greek word for "two").

In binary data, there are only two values.

Physical layer signaling converts ones and zeros into signals. Consequently, before information can be sent out, the sender must convert word, text, graphics images, video files, and other information to be transmitted into ones and zeros. This is done primarily at the application layer, rather than at the physical layer. However, to understand physical layer operation, you need some understanding of binary data representation.

Inherently Binary Data

As just noted, certain types of data are inherently binary. For instance, IP addresses are 32-bit binary strings, while Ethernet addresses are 48-bit binary strings.

Binary Numbers

Some message fields contain numbers. Senders represent whole numbers (integers) as simple **base 2** or **binary numbers.** Figure 3-2 illustrates binary counting and arithmetic.

With binary numbers, counting begins with 0. This is a source of frequent confusion.

In counting, add 1 to each binary number to give the next binary number. There are four simple rules for addition in binary, as Figure 3-2 illustrates.

➤ If you add 0 and 0, you get 0.

➤ If you add 0 and 1, you get 1.

➤ If you add 1 and 1, you get 10 (carry the one).

➤ If you add 1, 1, and 1, you get 11 (carry the one).

Figure 3-2   Arithmetic with Binary Numbers

Counting begins with 0, not 1 So the first three items are 0, 1, and 10 (10 is 2 in binary)						
Basic Rules					Examples	
				1	1000	8
0	0	1	1	+1	+1	+1
+0	+1	+0	+1	+1	=1001	=9
=0	=1	=1	=10	=11	+1	+1
					=1010	=10
					+1	+1
					=1011	=11
					+1	+1
					=1100	=12

These rules are simple to use, as the figure illustrates.

➤ For example, 8 is 1000.
➤ The number 9 adds a 1 to the final 0, giving a 1, so 9 is 1001.
➤ Adding 1 to the final 1 gives us 10, so 10 is 1010.
➤ The number 11 adds a 1 to the final 0, giving 1011.
➤ The number 12 adds a 1 to the final 1. With carries, this gives 1100.

Decimal numbers can also be converted into binary representations. There are several ways to do this. Generally, the representation contains two parts: the number itself and an indication of where the decimal point is.[1]

Encoding Alternatives

Sometimes, a field represents one of several different **alternatives,** such as site names in a corporation or product numbers. How many possible alternatives can a field represent? The answer depends on the field's length. If a field is N bits long, it can represent 2^N possible alternatives. This is illustrated in Figure 3-3.

If a field is N bits long, it can represent 2^N possible alternatives.

➤ If the field is only one bit long, it can represent only 2^1 (2) alternatives. For instance, a 1-bit gender field might represent *female* by 1 and *male* by 0.
➤ If the field is two bits long, it can represent 2^2 (4) possibilities, representing each by 00, 01, 10, or 11. A 2-bit season field might represent *Spring* by 00, *Summer* by 01, *Autumn* by 10, and *Winter* by 11.

Figure 3-3   Binary Encoding for a Number of Alternatives

Number of Bits in Field	Number of Alternatives That Can Be Encoded[1]	Specific Bit Sequences	Example
1	$2^1 = 2$	0, 1	Yes or No, Male or Female, etc.
2	$2^2 = 4$	00, 01, 10, 11	North, South, East, West
4	$2^4 = 16$	0000, 0001, 0010, ...	Top 10 security threats (6 values go unused)
8	$2^8 = 256$	00000000, 00000001, ...	ASCII text representation (128 values go unused)

[1]When encoding alternatives, the number of alternatives = $2^{\text{number of bits in the field}}$

---

[1]In Chapter 1, we saw how to use the Microsoft Windows Calculator in scientific mode to convert between decimal, binary, and hexadecimal. You can also use it to check your binary calculations (obviously, not on tests). First click on the Bin button to put the calculator in binary mode. Then use the calculator normally. For example, to add two binary numbers, click on Bin, enter the first binary number, hit the plus button, type the second binary number, and click on the equal sign to see the total.

➤ A 1-octet (one-byte) field can represent 2^8 (256) alternatives.

➤ A 2-octet field can represent 2^{16} (65,536) alternatives.

➤ A 3-octet field can represent 2^{24} (over 16 million) alternatives.

➤ A 4-octet field can represent 2^{32} (about 4 billion) alternatives.

Remember that binary counting begins with 0. Therefore, if you have eight bits, there are 256 possibilities. These begin with 0 and end with 255.

Text (ASCII and Extended ASCII)

Web pages use the **ASCII code,** whose individual symbols are each seven bits long, but are usually stored as whole bytes. Seven bits gives 128 possibilities. This is enough for

➤ capital letters (The letter *A* is 1000001)

➤ lowercase letters (The letter *a* is 1100001)

➤ digits (The number 3 is 0110011)

➤ punctuation and other characters (A *period* is 0101110)

➤ printing control (A carriage return is 0001101, and a *line feed* is 0001010)

Figure 3-4   ASCII and Extended ASCII (Study Figure)

Purpose
    To represent text (A, a, 3, $, etc.) as binary data for transmission

ASCII
    Traditional code to represent text data in binary
    Seven bits per character
    2^7 (128) characters possible
    Sufficient for all keyboard characters (including shifted values)
        Capital letters (*A* is 1000001)
        Lower-case letters (*a* is 1100001)
        Digits (0 through 9) (*3* is 0110011)
        Punctuation and other special characters (a *period* is 0101110)
        A space is 00100000
        Printing control (a *carriage return* is 0001101, and a *line feed* is 0001010)
        Eighth bit in data bytes normally is not used

Extended ASCII
    Used on PCs
    8 bits per character
    2^8 (256) characters possible
    Extra characters can represent formatting in word processing, etc.

Text-to-ASCII and Text-to-Extended ASCII Calculators
    Readily available on the Internet

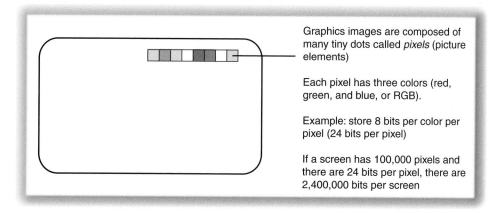

Graphics images are composed of many tiny dots called *pixels* (picture elements)

Each pixel has three colors (red, green, and blue, or RGB).

Example: store 8 bits per color per pixel (24 bits per pixel)

If a screen has 100,000 pixels and there are 24 bits per pixel, there are 2,400,000 bits per screen

Figure 3-5   Graphics Image and Conversion to Binary

Seven-bit ASCII is sufficient for typing all keys on a PC keyboard. During transmission, the eighth bit of each byte usually is not used.[2]

However, PCs internally use **extended ASCII,** which uses all eight bits to represent text more richly. Going from seven to eight bits gives 128 extra character codes. Different application programs use these extra character codes differently. For instance, word processing programs typically use these extra codes to represent formatting.

If you go to a search engine, you can easily find converters to represent characters in both ASCII and extended ASCII.

Raster Graphics

In **raster graphics,** the screen is divided into a grid of dots called *pixels,* as Figure 3-5 illustrates. Graphics programs represent each pixel by one to three bytes. Using a single byte per pixel allows 256 colors. This option, used by **GIF** files, is not attractive to the human eye. In contrast, **JPEG** uses three octets per pixel—one octet each for red, green, and blue. This permits over 16 million colors. This is more than the human eye can distinguish and so gives very satisfying color.

In addition, JPEG and other graphics file mechanisms do file compression to reduce the size of files. Therefore, compressed graphics files allow the sender to transmit fewer than eight bits per pixel (GIF) or twenty-four bits per pixel (JPEG).

Computer screens have at least a half million pixels. Good consumer digital cameras create pictures with about 5 million pixels.

Overall, a graphics file can be thought of as a long string of bytes.

---

[2]Early systems used the eighth bit as a "parity bit" to detect errors in transmission. This could detect a change in a single bit in the byte. At today's high transmission speeds, however, transmission errors normally generate multibit errors rather than single-bit errors. Consequently, parity is useless and is ignored.

TEST YOUR UNDERSTANDING

2. a) Give the binary representations for 13, 14, 15, 16, and 17 by adding one to successive numbers (12 is 1100). b) Rounding off, about how many possible addresses can you represent with 32-bit IP addresses? (You probably will need a spreadsheet program to answer this question.) c) If you have four bits, how many possibilities can you represent? d) With four bits, what is the smallest nonnegative binary number you can represent? e) What is its decimal equivalent? f) What is the largest binary number you can represent if you have four bits? g) What is its decimal equivalent? h) If you need to represent three alternatives—fraud, error, and hacking—how many bits will you need per entry? i) If you need to represent 40 courses offered in a department, how many bits will you need per entry?

3. a) How many bits does ASCII use to represent keyboard characters? b) If you transmit "Hello World!" (without the quotation marks) in ASCII, how many bytes must you transmit? c) Your computer screen has a resolution of 800 pixels horizontally and 600 pixels vertically. How many bytes will a JPEG screen image be if there is no compression? Express the file size in proper notation. Use *B* for bytes. d) How long will it take to transmit the file over a 30 kbps line? e) How large would the file be with 20:1 compression (which is common)? f) How long will it take to transmit the file over a 30 kbps line?

SIGNALING

Converting Data to Signals

So far, we have looked at how to encode *data* as a string of bits. For transmission, the sender also must convert these bits into *signals* that will represent the bits, as Figure 3-6 illustrates. Note that most data must be converted twice—once into binary data and once into signals.

Most data must be converted twice—once into binary data and once into signals.

On/Off Signaling

The simplest way to signal is to divide time into brief **clock cycles** (periods of time) and to have each clock cycle represent one bit. As Figure 3-7 shows, the sender turns a

Figure 3-6   Data Encoding and Signals

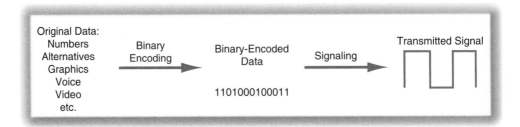

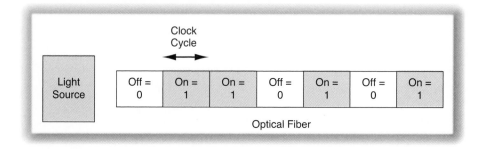

Figure 3-7   On/Off Signaling

signal on for a 1 during the clock cycle or off for a 0. (Think signal.) Optical fiber generally uses this type of **on/off signaling.**

Binary Signaling

In binary *data,* there are only two possible states. In **binary signaling,** there also are two possible states. It is easy to map binary data into binary signals. On/off signaling is an example of binary signaling, which uses **states** (on or off) to represent binary data (1 or 0).

Binary signaling uses two possible states to represent information or data (1s and 0s).

Binary Voltage Signaling

When signals travel over wires, it is common to use two different voltages to represent 1 and 0, as Figure 3-8 illustrates. This is also binary signaling.

232 Serial Ports

In Figure 3-8, a high voltage is anything between 3 and 15 volts. A low voltage, in turn, is anything between negative 3 and negative 15 volts.

Now comes the strange part: A high voltage represents a 0, while a low voltage represents a 1. No, this is not a misprint. This is the way a **232 serial port** works on your PC.[3]

Constancy during Each Clock Cycle

As in on/off signaling, time is divided into brief clock cycles. The sender holds the signal constant within each clock cycle.[4] At the end of each clock cycle, the signal either stays the same or changes to the other voltage level (state).

---

[3]Sometimes called RS-232-C serial ports, they actually follow the newer ANSI/TIA/EIA-232-F standard. In Europe, equivalent ports are specified by three standards: V.24, V.28, and ISO-2110.
[4]This constancy allows the receiver to read the voltage at any time during the clock cycle. Even if the receiver is slightly off on the timing of its read for a bit, the receiver will still read the value correctly.

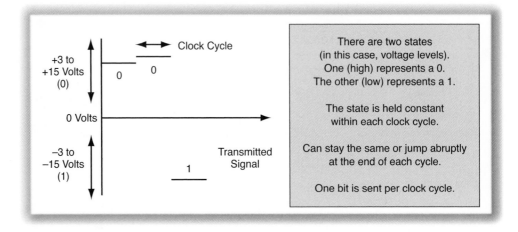

Figure 3-8   Binary Voltage Signaling in 232 Serial Ports

Relative Immunity to Attenuation Errors

Binary signal transmission is attractive because it is relatively immune to attenuation errors (losses in signal intensity). Suppose that, as in Figure 3-9, a signal starts at 12 volts and attenuates to 6 volts (a 50 percent loss). It will still be read correctly as a high voltage and, therefore, a 0. Binary signaling also is relatively immune to other types of propagation error. Of course, large propagation effects still cause errors, but with careful

Figure 3-9   Relative Immunity to Errors in Binary Signaling

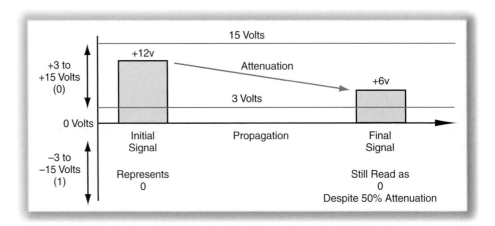

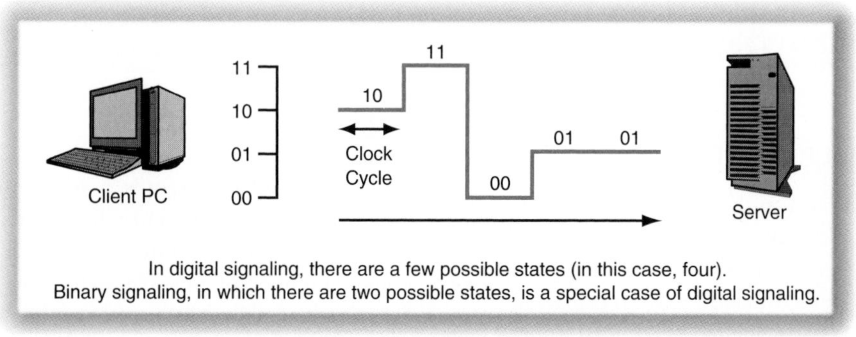

In digital signaling, there are a few possible states (in this case, four).
Binary signaling, in which there are two possible states, is a special case of digital signaling.

Figure 3-10   Four-State Digital Signaling

design, the **bit error rate (BER),** which is the percentage of errors per thousand or million bits transmitted, will be very low.

Binary Signaling versus Digital Signaling

Binary signaling is often called digital signaling. Although many professionals now consider the terms equivalent, the two concepts are not completely the same, as discussed in the box "Multistate Digital Signaling."

Concisely, while binary signaling is limited to two states (on/off, high/low voltage, etc.), **digital signaling** can have two states, four states, eight states, or more.[5] Figure 3-10 illustrates four-state digital signaling. The box notes that multistate digital signaling can transmit more bits per clock cycle at the cost of a higher error rate.

Almost all signaling systems today are binary, and because digital signaling can involve just two states (two qualifies as few), it is correct to call all binary transmission systems *digital transmission systems.*

TEST YOUR UNDERSTANDING

4. a) In binary 232 serial port transmission for transmission over a binary transmission line, how are 1 and 0 represented? b) How does this give resistance to transmission errors? c) A signal is sent at 9 volts in a 232 serial port. What fraction of its strength can it lose before it becomes unreadable? d) What is the advantage of multistate digital signaling? e) What is the disadvantage of multistate digital signaling? f) Today, almost all signaling is binary. Is it correct to call all binary signaling digital signaling? g) Is it correct to call all digital transmission systems binary transmission systems?

---

[5]The number of states is always given by 2^N, where $\underline{N}$ is a small integer.

Multistate Digital Signaling

A *FEW* POSSIBLE STATES

Figure 3-10 illustrates a slightly more complex form of signaling: digital signaling. In binary signaling, there are only two possible states (in the case of 232 serial ports, voltage levels) to represent information. In **digital signaling,** in contrast, there are a *few* possible states; the sender holds the state (voltage level in this example) constant during the clock cycle. The figure specifically shows digital signaling with four possible states. However, digital signaling can use eight, sixteen, thirty-two, or occasionally (but rarely) more states.

Note also that binary signaling is a special case of digital signaling. If *few* is *two*, then digital signaling is binary.

In digital signaling, there are a **few** possible states. Binary signaling is a special case of digital signaling in which there are exactly **two** states.

WHY MORE THAN TWO STATES?

Why use more than two possible states? The answer is that you can send multiple bits per clock cycle if you have more than two possible states. Binary transmission sends only one bit per clock cycle (a 1 or a 0). With four possible states, however, there can be four possibilities. If we let the four voltage levels represent 00, 01, 10, and 11, then we can send two bits per clock cycle—doubling the transmission rate.

Adding more possible states allows the sender to transmit even more bits per clock cycle. However, there are diminishing returns. Each doubling in the number of possible states allows the sender to transmit only one more bit per clock cycle. For instance, having eight possible states allows only three bits per clock cycle, and having sixteen possible states allows only four bits per clock cycle.

If there are too many possible states, furthermore, the possible states will be very close together. If the signal changes even slightly during transmis-

sion, the receiver will record the wrong state. This is why digital signaling uses only a *few* possible states.[6] In essence, sending more bits increases transmission rates, but results in more errors. Beyond a small number of possible states, the net transmission rate actually falls because of the need to retransmit incorrect transmissions.

BIT RATES AND BAUD RATES

Bit Rates

In digital data transmission, the rate at which we transmit data is called the **bit rate.** It is measured in bits per second. This is what users care about.

Baud Rate

In turn, the **baud rate** is the number of clock cycles the transmission system uses per second. If there are 1000 clock cycles per second, the baud rate is 1000 baud (not 1000 bauds per second). Looked at another way, if each clock cycle is one millionth of a second, then the baud rate is 1 megabaud.

The bit rate is the rate at which we transmit data. The baud rate is the number of clock cycles the transmission system uses per second.

Combining the Two Concepts

The bit rate and the baud rate are connected. The bit rate is just the baud rate times the number of bits transmitted in each clock cycle. If the baud rate is 10 kbaud and the sender transmits one bit per clock cycle (binary transmission), then the bit rate is 10 kbps. However, if the baud rate is 10 kbaud and the sender transmits 4 bits per baud, then the bit rate is 40 kbps.

Looked at another way, if you wish to transmit 80,000 bits per second and the baud rate is 10,000, then you must transmit 8 bits per clock cycle. Or, if you wish to transmit 80,000 bits per second and the baud rate is 20 kbaud, then you must transmit 4 bits per baud.

---

[6]Originally, digital signaling used 10 states; it was called *digital* because our 10 fingers are called digits.

Digital Signaling

    Clock cycles

    Signal is held fixed during each clock cycle

    Binary signaling: two states

    Digital signaling: a few states (two or more)

    Up to about 256 states

Why More than Two States?

    With more than two states, can send more than one bit per clock cycle

    Two states = 1 bit per clock cycle (1 or 0)

    Four states = 2 bits per clock cycle (00, 01, 10, 11)

    Eight states = 3 bits per clock cycle, etc.

    Each doubling of states gives one more bit per clock cycle

Problem of Multiple States

    As the number of states increases, the difference between states decreases

    There is less tolerance for changes in the signal

    This is why there is a limit of a few states (256 maximum and usually much less)

Concepts

    Bit rate: Number of bits sent per second

    Baud rate: Number of clock cycles per second

        If 1,000 clock cycles per second, 1 kbaud

        If each clock cycle is 1/1,000 second = 1,000 clock cycles/second = 1 kbaud

    Bits per baud: Number of bits that can be sent per clock cycle

        1 if two states

        2 if four states

        …

Computing the Bit Rate

    Know the baud rate and the number of bits per baud

    Multiply them

    If baud rate is 10,000 baud (not bauds)

    If two bits per clock cycle

    Then bit rate is 2 × 10,000, or 20,000 bps = 20 kbps

Computing the Bit Rate

    Know the baud rate and the number of states

    Compute the number of bits from the number of states

    Multiply the bits per clock cycle (per baud)

    If baud rate is 10,000 baud (not bauds)

    If four states, can send 2 bits per clock cycle

    Then bit rate is 2 × 10,000, or 20,000 bps = 20 kbps

Computing the Required Number of States

    Know the required bit rate and baud rate

    Divide the bit rate by the baud rate to get the bits per baud

    Compute the required number of states

    Required bit rate is 4 Mbps

    Baud rate is 1 Mbaud

    Bit rate/baud rate = 4 bits per clock cycle

    4 bits per clock cycle are required

Figure 3-11   Multistate Digital Signaling (Study Figure)

Bits per Baud

The number of bits per clock cycle is determined by the number of possible states per clock cycle. In binary transmission, there are two possible states. Figure 3-10 shows that if you have four possible states, then you can transmit two bits per clock cycle. Each doubling of the number of states allows you to transmit one bit per clock cycle (one bit per baud). So if you wish to transmit three bits per clock cycle, you will need eight possible levels. Looked at the other way, if you have eight possible states, then you can transmit four bits per clock cycle.

Adding more states per clock cycle to send more bits per clock cycle has rapidly diminishing returns. Five bits requires 32 states, six bits requires 64 states, seven bits requires 128 states, and eight bits requires 256 states. The number of required states grows very rapidly compared with the number of bits per clock cycle. The difference between two states becomes so small beyond a few bits per clock cycle that the slightest propagation effects will cause the signal to be misread as a different state.

TEST YOUR UNDERSTANDING

5. a) Distinguish between binary and digital transmission. b) Is all binary transmission digital? c) Is all digital transmission binary? d) What is desirable about having multiple possible states instead of just two? e) What is undesirable about having multiple possible states? f) How many more bits can you send per clock cycle every time you double the number of possible states?

6. a) Distinguish between the bit rate and the baud rate. b) When are the two equal? (You will have to think about this one.) c) If you have 10000 clock cycles per second, what is the baud rate? d) If you have 10000 clock cycles per second and you transmit in binary, what is the bit rate? e) If instead you have 10000 clock cycles per second and use 16 voltage levels for digital signaling, what is the bit rate? f) To transmit 30 kbps over a 10 kbaud line, how many possible states will you need?

UNSHIELDED TWISTED PAIR (UTP) COPPER WIRING

Having looked at propagation effects and signaling in general, we will now see how these concepts apply to UTP transmission. Later, we will see how they apply to optical fiber transmission.

4-Pair UTP and RJ-45

The 4-Pair UTP Cable

Ethernet networks typically use **4-pair unshielded twisted pair (UTP)** wiring.[7] The TIA/EIA-568 standard governs UTP wiring in the United States. In Europe, the comparable standard is ISO/IEC 11801.

This name "4-pair unshielded twisted pair (UTP)" sounds complicated, but the medium is very simple. Figure 3-13 illustrates a 4-pair UTP cord with its four wire pairs showing.

➤ A length of UTP wiring is a **cord**.

➤ Each cord has eight copper wires.

---

[7]We will see later that some types of wiring use shielding, in which a metal mesh is placed around each pair of wires and around the jacket. This type of wiring is called shielded twisted pair wiring (STP). STP reduces interference, which is discussed later in this chapter. However, shielding is expensive, so almost all UTP wiring used in organizations today is unshielded, although some USB cables are shielded.

4-Pair UTP Cable
> The TIA/EIA-568 standard governs UTP wiring in the United States
> In Europe, the comparable standard is ISO/IEC 11801

Cord Organization
> A length of UTP wiring is a cord
> Each cord has eight copper wires
>> Each wire is covered with dielectric (nonconducting) insulation.
> The wires are organized as four pairs
>> Each pair's two wires are twisted around each other several times per inch
> There is an outer plastic jacket that encloses the four pairs

Connector
> RJ-45 connector is the standard connector
> Plugs into an RJ-45 jack in a NIC, switch, or wall jack

Characteristics
> Inexpensive and easy to purchase and install
> Rugged: Can be run over with chairs, etc.
> Dominates media for access links

Figure 3-12   Unshielded Twisted Pair (UTP) Wiring (Study Figure)

> ➤ Each wire is covered with dielectric (nonconducting) **insulation**.[8] This prevents short circuits between the electrical signals traveling on different wires.
> ➤ The wires are organized as four pairs.
> ➤ Each pair's two wires are twisted around each other several times per inch.
> ➤ There is an outer plastic **jacket** that encloses the four pairs.

RJ-45 Connectors

At the two ends of a UTP cord, the wires must be separated and placed within an 8-pin **RJ-45 connector,** which also is shown in Figure 3-13. The RJ-45 connector at each end of a 4-pair UTP cord snaps into an **RJ-45 jack** (port) in the NIC, the switch, or the wall jack.[9]

Easy, Inexpensive, and Rugged

UTP is inexpensive to purchase, easy to **connectorize**[10](add connectors to), and relatively easy to install. It is also rugged, so if a chair runs over it accidentally, it probably

---

[8]Benjamin Franklin coined the terms *conductor, insulation,* and many other terms in electricity (including *positive* and *negative*). He also created the theory of electricity flow, although he thought that electricity flowed from the positive end of a battery to the negative end. Today, we know that electrons flow from negative to positive.

[9]Home telephone connections use a thinner RJ-11 connector and jack. They were designed to terminate six wires but usually terminate only a single pair.

[10]Yes, *connectorize* is a really ugly term. Hey, I don't make up these names!

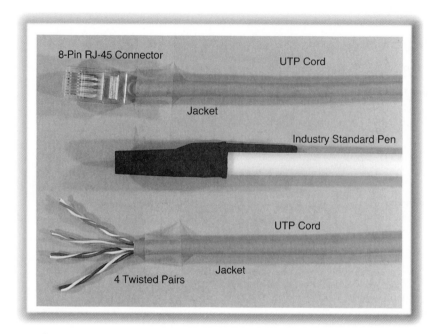

Figure 3-13   A 4-Pair Unshielded Twisted Pair (UTP) Cord with RJ-45 Connector

will survive undamaged. Four-pair UTP dominates corporate usage in access links from the NIC to the first switch because of UTP's low cost and durability. (Access links are frequently exposed to harsh treatment in the office.)

TEST YOUR UNDERSTANDING

7. a) What standards govern UTP? b) What is a length of UTP wiring called? c) In 4-pair UTP, how many wires are there in a cord? d) How many wire pairs are there in a cord? e) What surrounds each wire? f) How are the two wires of each pair arranged? g) What is the outer covering called?

8. Why is 4-pair UTP dominant in LANs for the access line between a NIC and the switch that serves the NIC?

Attenuation and Noise Problems

UTP signals change as they travel down the wires. If they change too much, they will not be readable. As noted earlier in this chapter, there are several types of propagation effects. We will now look at the most important propagation effects for UTP transmission, beginning with attenuation and noise.

Attenuation

As Figure 3-14 illustrates, when signals travel, they attenuate (grow weaker). To give an analogy, as you walk away from someone who is speaking, his or her voice will grow

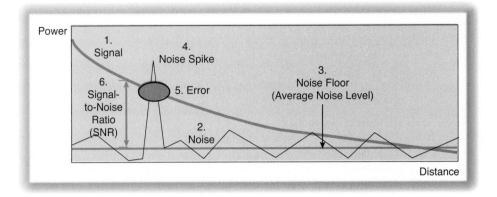

Figure 3-14   Attenuation and Noise

fainter and fainter. As noted earlier in this chapter, if a signal attenuates too much, the receiver will not be able to recognize it.

Attenuation is measured relative to the original strength of the signal. If a signal starts at the power level P_1 and falls to P_2, then we say that the signal has attenuated to P_2/P_1 of its initial strength. To give an example, if the power begins at 20 milliwatts (mW) and falls to 5 mW, then P_1 is 20 mW and P_2 is 5 mW. This signal had decreased to 25 percent (5/20) of its original value. Expressed another way, the signal has attenuated by 75 percent.

Expressing Power Ratios in Decibels

If there is attenuation, the signal power that reaches the receiver will decrease. If P_1 and P_2 are the initial power and final power of the signal, respectively, then P_2 will be smaller than P_1. The final power's ratio to the initial power will be P_2/P_1. This is the **power ratio.** The final power will be only P_2/P_1 of the original power. For example, if P_2/P_1 is 0.25, then the final power will be only 25 percent of the original power.

As a consequence of attenuation, a signal can decline to 1/4 of its original power or even to 1/10000 of its original power. When physical processes span a great range, engineers often express them in logarithms. In the case of attenuation, ratios are often expressed in **decibels (dB).** The equation for decibels in power ratios is dB = $10 \log_{10} (P_2/P_1)$.

$$dB = 10 \log_{10} (P_2/P_1) \qquad \text{(Equation 1)}$$

To give an example, suppose that the initial power is 100 milliwatts (mW) and the final power after propagation is 37 mW.

➤   Then the power is reduced to P_2/P_1 of its original value or is 37/100 (0.37).

➤   From Excel, LOG10(0.37) is −0.4318.

➤   Multiplying this by 10 and rounding off gives −4.3 dB. The signal power ratio has decreased to 4.3 dB of its original value.

Power Ratios

The signal power ratio is P_2/P_1.

   P_1 is the initial power and P_2 is the received power.

The final received power is P_2/P_1 of the original power

Example: Power starts (P_1) at 200 milliwatts (mW) and falls to (P_2) 100 mW

   $P_2/P_1 = 100 / 200 = 0.5 = 50\%$

Power Ratios Expressed in Decibels

Power reduction ratios vary widely in attenuation

   When there is a wide range of values, engineers often express them in logarithms

The equation for power ratios in decibels is dB = 10 log10 P_2/P_1)

   Where P_1 is the initial power and P_2 is the final power after transmission

   If P_2 is smaller than P_1, then the answer will be negative

In calculations, the Excel LOG10 function can be used

Example

Over a transmission link, power drops to 37 percent of its original value

$P_2/P_1 = 37\%/100\% = .37$

From Excel: LOG10(0.37) = −0.4318

10*LOG10(0.37) = −4.3 dB (the negative indicates power reduction through attenuation)

Two Useful Approximations

−3 dB loss is a power ratio of 1/2

   −6 dB is a power ratio of 1/4

   −9 dB is a power ratio of 1/8

−10 dB loss is a power ratio of 1/10

   −20 dB is a power ratio of 1/100

   −30 dB is a power ratio of 1/1000

When to Use Ratios as Fractions or as Decibels

Expressed in decibels only for human reading

In equations, the power ratio fraction is used

Figure 3-15   Expressing Power Ratios in Decibels (Study Figure)

Approximately Three Decibels Is a Halving in Power   While Equation 3 will give you an exact answer, you often can estimate power ratios in decibels. You can do this with two useful facts. First, cutting a power ratio in half is −3 dB. Each additional halving is an additional −3 dB. For example, a decrease to approximately 1/4 of the original power would give a ratio of −6 dB. A −9 dB power ratio would be a decline to approximately 1/8 of the original power.

In the example given previously, the signal power declined to 37 percent of its initial value. This is a little more than a third, so the decibel value of the power ratio should be a little worse than −3 dB. In fact, we saw that it was −4.3 dB.

Approximately 10 Decibels Is a Reduction to 1/10 of the Original Power   Second, a decline to 1/10 of the initial power is a −10 dB power ratio, which is a 10 dB loss. So falling to 1/100 of the original power would be a ratio of approximately 20 dB. Incredibly, a 20 dB decline in power can occur in UTP transmission without the signal becoming unintelligible.

Which Measure to Use?   You may be confused about when to use simple power ratios (P_2/P_1) and when to use decibels. The answer is that equations that have power ratios as an input always use P_2/P_1. Decibel measures are used to make it easier for users to understand power ratios. Just as computers use 32-bit IP addresses while people use dotted decimal notation, equations use P_2/P_1, while people use decibels.

Noise

Electrons within a wire are constantly moving, and moving electrons generate random electromagnetic energy. This random electromagnetic energy is **noise.** Noise energy adds to the signal energy, so the receiver actually sees the total of the signal plus the noise.

> Random electromagnetic energy is noise.

The mean of the noise energy is the **noise floor**—despite the fact that it is an average and not a minimum, as the name *floor* would suggest. Figure 3-14 shows a noise floor.

> The mean of the noise energy is the noise floor.

As a consequence of noise being a random process, there are occasional **noise spikes** that are much higher or lower than the noise floor. As Figure 3-14 shows, if a noise spike is about as large as the signal, the combined signal and noise may be unrecognizable by the receiver.

Noise, Attenuation, and Propagation Distance

If a signal is far larger than the noise floor, then we have a high **signal-to-noise ratio (SNR).** With a high SNR, few random noise spikes will be large enough to cause errors. However, as a signal attenuates during propagation, it falls ever closer to the noise floor. Noise spikes will equal the signal's strength more frequently, so errors will become more frequent. In other words, even if the noise level is constant, longer propagation distances create attenuation that results in a lower SNR and therefore more noise errors.

> Even if the noise level is constant, longer propagation distances result in a lower SNR and therefore more noise errors.

Limiting UTP Cord Distance to Limit Attenuation and Noise Problems

Fortunately, installers can control attenuation and noise by limiting the length of UTP cords. The Ethernet standard currently limits UTP propagation distances to 100 meters at all speeds up to 1 Gbps. If UTP cords are restricted to 100 meters, the signal still will be comfortably larger than the noise floor when it arrives at the receiver, so there will be few noise errors.

> The Ethernet standard limits UTP cords to 100 meters at all speeds up to 1 Gbps.

TEST YOUR UNDERSTANDING

9. a) Describe the attenuation problem and why it is important. b) A signal is 1/16 of its initial power when it arrives at the receiver. What is the power ratio in decibels? c) A signal is 1/100 of its initial value when it reaches the receiver. What is this power ratio in decibels? d) Describe the noise problem. e) As a signal propagates down a UTP cord, the noise level is constant. Will greater propagation distance result in fewer noise errors, the same number of noise errors, or more noise errors? Explain.

10. a) What problem or problems is(are) reduced to acceptable limits by limiting the length of a UTP cord in Ethernet? b) What is the limit on UTP cord length in Ethernet standards?

Electromagnetic Interference (EMI) in UTP Wiring

General EMI

Noise is unwanted electrical energy within the propagation medium. In turn, **electromagnetic interference (EMI)**—or more simply, **interference**—is unwanted electrical energy coming from external devices, such as electrical motors, fluorescent lights, and even nearby UTP cords (which always radiate some of their signal). Like noise energy, interference energy adds to the signal energy and can make the received signal unreadable.

Using Twisted Pair Wiring to Reduce Interference

Fortunately, there is a simple way to reduce EMI to an acceptable level. This is to twist each pair's wires around each other several times per inch, as Figure 3-16 illustrates.

Consider what happens over a full twist. Over the first half of the twist, the interference might add to the signal. Over the other half, however, this same interference

Figure 3-16   Electromagnetic Interference (EMI) and Twisting

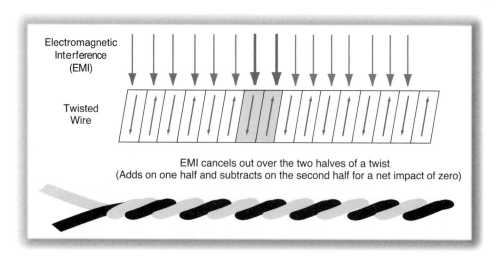

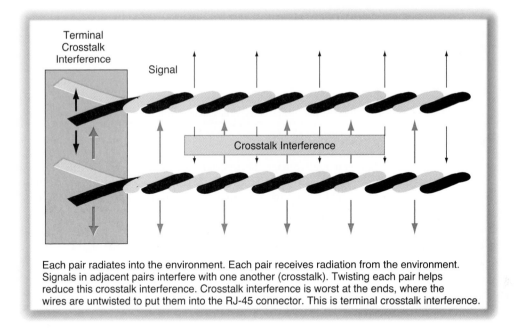

Each pair radiates into the environment. Each pair receives radiation from the environment. Signals in adjacent pairs interfere with one another (crosstalk). Twisting each pair helps reduce this crosstalk interference. Crosstalk interference is worst at the ends, where the wires are untwisted to put them into the RJ-45 connector. This is terminal crosstalk interference.

Figure 3-17   Crosstalk Interference and Terminal Crosstalk Interference

would subtract from the signal. The interference on the two halves would cancel out, and the net interference would be zero.

Does twisting really work this perfectly? No, of course not. However, twisting the wiring dramatically reduces interference, limiting it to an acceptable level. As a historical note, Alexander Graham Bell himself patented twisted-pair wiring as a way to reduce interference in telephone transmission.

Crosstalk Interference

As Figure 3-17 shows, individual pairs in a cord will radiate some of their energy, producing electromagnetic interference in other pairs within the cord. This mutual EMI among wire pairs in a UTP cord is **crosstalk interference.** It is always present in wire bundles and must be controlled. Fortunately, the twisting of each pair normally keeps crosstalk interference to a reasonable level.

Terminal Crosstalk Interference

Unfortunately, when a UTP cord is connectorized, its wires must be untwisted to fit into the RJ-45 connector, as shown in Figure 3-17. The eight wires are now parallel, so there is no protection from crosstalk interference. Crosstalk interference at the ends of the UTP cord, which is **terminal crosstalk interference,** usually is much larger than the rest of the crosstalk interference over the entire rest of the cord.

Installers must be careful not to untwist UTP wires more than 1.25 cm when adding connectors. This precaution will not completely eliminate terminal crosstalk interference, but it will limit crosstalk interference to an acceptable level.

Installers must be careful not to untwist UTP wires more than 1.25 cm when adding connectors. This precaution will not completely eliminate terminal crosstalk interference, but it will limit crosstalk interference to an acceptable level.

TEST YOUR UNDERSTANDING

11. a) Distinguish between electromagnetic interference (EMI), crosstalk interference, and terminal crosstalk interference. b) How is EMI controlled? c) How is terminal crosstalk interference controlled in general? Explain. d) Does this precaution eliminate terminal crosstalk interference?

Serial and Parallel Transmission

Figure 3-18 shows an important distinction in wire communication: serial versus parallel transmission.

Serial Transmission

In the next chapter, we will look at Ethernet standards. In slower versions of Ethernet that run at 10 Mbps and 100 Mbps, a single pair in a UTP cord transmits the signal in each direction. Two-way transmission, then, uses two of the four pairs—one in each direction. The other two pairs are not used, although the standard calls for them to be

Figure 3-18　Serial versus Parallel Transmission

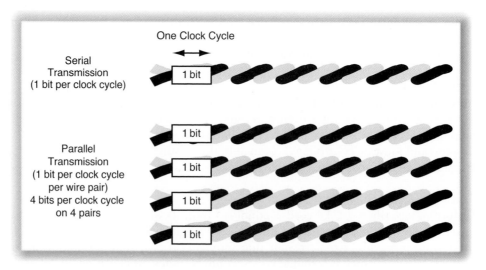

Note: Parallel means that transmission occurs on multiple pairs, not just four pairs.

present. If one pair of wires is used to send a transmission, this is **serial transmission** because the transmitted bits must follow one another in series.

Parallel Transmission

For gigabit Ethernet, however, when the NIC or switch transmits, it transmits on all four pairs in each direction.[11] This means that it can transmit four bits at a time instead of just one. Sending data simultaneously on two or more transmission lines in the same direction is **parallel transmission.** The benefit of parallel transmission is that it is faster than serial transmission for the same clock cycle duration. Speed is the key benefit of parallel transmission.

Sending data simultaneously on two or more transmission lines in the same direction is parallel transmission.

Note that parallel transmission does not mean the use of four transmitting pairs (or grounded wires). It means using more than a single pair of wires or one wire and a ground wire. For example, your computer's internal bus, which carries signals among components in your system unit, has about 100 wires for high-density parallel transmission. To give another example, parallel cables used to connect PCs to printers use eight wires to carry data in each direction.

Trend

In general, serial transmission has largely replaced parallel transmission in practice. Reducing the clock cycle in serial transmission has the same benefit as using more transmission paths, and it is generally cheaper. Parallel transmission survives only in some short-distance transmission applications such as gigabit Ethernet and PC parallel buses.

TEST YOUR UNDERSTANDING

12. a) Distinguish between serial and parallel transmission. b) What is the main benefit of parallel transmission? c) In parallel transmission, how many pairs (or single wires and ground wires) are used in each direction? d) Is serial or parallel transmission more widely used today?

Wire Quality Standard Categories

UTP cords vary in transmission quality. If wire quality is too low, signals will not be able to travel far at a given speed or may not be able to travel at all. The **TIA/EIA/ANSI-568** standard defines wiring quality levels as **category** numbers. The term *category* is often abbreviated as cat. Figure 3-19 shows the relationship between wiring quality, speed, and maximum transmission propagation distances in Ethernet. Quite simply, as transmission speeds have increased, higher-quality wiring has become necessary.

Cat 3 and Cat 4

The first categories of wiring that were defined were **Cat 3** and **Cat 4** wiring. Both were sufficient for 10 Mbps Ethernet transmission over 100 meters. Some old buildings still

---

[11]What if both sides transmit at the same time? Each knows what signal it sent. It subtracts this signal from the signal it hears on the wire. The remaining signal is the signal sent by the other side.

Category	Technology	Maximum Speed	Maximum Ethernet Distance at This Speed
1	UTP	Never defined	Not applicable
2	UTP	Never defined	Not applicable
3	UTP	10 Mbps	100 meters
4	UTP	10 Mbps	100 meters
5	UTP	1 Gbps	100 meters
5e	UTP	1 Gbps	100 meters
6	UTP	10 Gbps	55 meters
6A	UTP	10 Gbps	100 meters
7	STP[1]	10 Gbps+	100 meters

[1]STP is shielded twisted pair. There is foil around each pair and a metal mesh around the four pairs.

Figure 3-19   Wire Quality Standards

have Cat 3 and Cat 4 wiring. If they do, they cannot bring more than 10 Mbps to the desktop.[12]

Cat 5 and Cat 5e

Most buildings today, however, have **Cat 5** or **Cat 5e** (enhanced) wiring. Both categories can bring not only 100 Mbps to the desktop, but even 1 Gbps. In addition, both can support these speeds over distances up to 100 meters. This is more than sufficient for most corporate needs for the foreseeable future for desktop access lines.

The 10 Gbps Problem

Within data centers and switching centers, however, speeds of 10 Gbps are becoming important over short distances. Although optical fiber can be used at these distances, UTP wiring is less expensive, so new categories have been defined for twisted-pair wiring. The first category in this class was **Cat 6.** However, Cat 6 was not as good as originally planned. It can only carry 10 Gbps Ethernet traffic up to 55 meters. Consequently, augmented category 6 (**Cat 6A** or AC6) wiring was defined to allow UTP to carry 10 Gbps up to the usual 100 meters.

Another way to carry 10 Gbps traffic up to 100 meters is to use **Category 7** wiring. In contrast to other wiring categories, Cat 7 wiring uses **shielded twisted pair (STP)** wiring instead of UTP. STP wiring places a metal foil shield around each twisted pair and also a wire mesh around the four pairs. This almost completely eliminates electromagnetic interference. However, STP is more difficult and expensive to install than UTP. In addition, Cat 7 STP wiring requires a new type of connector. Companies are likely to be very reluctant to use Cat 7 STP unless they have high-EMI environments.

---

[12]The wiring in the author's college is "Cat 3 on a good day."

Perspective

Although you need to understand the several wiring quality standards and their limitations, the situation in corporations today is fairly simple. Overall, companies have generally been content with the Cat 5e wiring that they have had in place for many years. Up to a gigabit per second, which is a very high speed to bring to desktops, Cat 5e has no problem.

The main problem to date has been the speed limits created by very old Cat 3 and Cat 4 wiring plants. If a 10 Mbps limit to the desktop is sufficient, then there is no problem. However, firms that have legacy Cat 3 and Cat 4 wiring generally are upgrading to high-quality wiring.

To reach 10 Gbps, higher-quality UTP will be needed. However, few firms envision bringing 10 Gbps to the desktop in the foreseeable future. Copper connections for 10 Gbps will take place mostly within server rooms or switching rooms, as a cheaper alternative to optical fiber.

TEST YOUR UNDERSTANDING

13. a) What is the maximum length of UTP cords in Ethernet standards up to and including 1 Gbps? b) What wiring characteristic do UTP categories standardize? c) In an older building with Cat 3 wiring in place, how fast can you send Ethernet traffic? d) What two UTP quality categories dominate sales in the marketplace today? e) Cat 5e and Cat 6 are sufficient for Ethernet transmission speeds up to _____. f) What new categories of wiring are being developed for 10 Gbps Ethernet transmission? g) How far can Cat 6, Cat 6A, and Cat 7 carry 10 Gbps Ethernet signals?

OPTICAL FIBER

Light through Glass

In the 1840s, scientists discovered that light would follow water flowing out of a hose. Where the water stream bent, the light would follow. This raised the possibility that light signals could be sent through glass rods—a technology that allows more controlled transmission than through water. Figure 3-20 illustrates transmission through glass. This technology is called **optical fiber** or, more simply, **fiber.**

Optical fiber can carry signals much farther than UTP at any given speed. While UTP transmission is limited to about 100 meters, optical fiber can easily span distances of 200 to 300 meters. This is optical fiber's main benefit today.

Some texts say that optical fiber can also carry signals faster than UTP. However, both UTP and fiber can carry signals up to 10 Gbps, which is the highest speed of Ethernet today. In addition, this range satisfies the needs of almost all corporations. Optical fiber is for going farther, not going faster.

That being said, wiring normally has an expected life of 5 to 7 years—sometimes much longer. For wiring that is installed within walls or conduits, and at speeds of 10 Gbps, optical fiber probably provides growth potential for the next major speed plateau at 40 Gbps or 100 Gbps.

Although optical fiber can carry signals very far, optical fiber is more expensive to purchase and install than UTP cabling. Fiber is the champagne of transmission media, and like champagne, it is expensive.

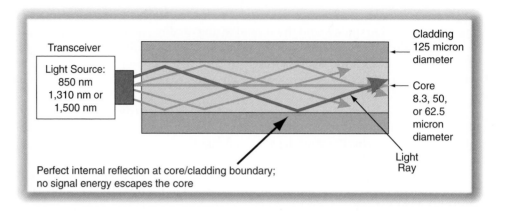

Figure 3-20 Optical Fiber Transceiver and Strand

TEST YOUR UNDERSTANDING

14. a) How does optical fiber signaling represent 1s and 0s? b) What is the main benefit of optical fiber compared with UTP today? c) Which costs less to buy and install—UTP or optical fiber?

The Roles of Fiber and Copper

Although optical fiber use is growing rapidly, it is not completely replacing copper UTP wiring. Rather, as Figure 3-21 shows, the two play different roles in most LANs, so they are not direct competitors.

➤ UTP is dominant for access lines because UTP is less expensive than fiber and because access line distances normally are well within UTP's usual 100-meter distance limit and speed limits. Also, UTP is somewhat more rugged than fiber, and this can be important in office areas, where optical fiber deals with its traditional mortal enemy—moving chairs.

➤ In contrast, while UTP may be used for some trunk lines, most trunk lines are optical fiber. Trunk lines tend to span longer distances and so can justify the higher cost of optical fiber.

TEST YOUR UNDERSTANDING

15. In LAN transmission, what are the typical roles of UTP and optical fiber?

Optical Fiber Construction and Operation

Figure 3-20 illustrates how an optical fiber strand is constructed and how it carries signals.

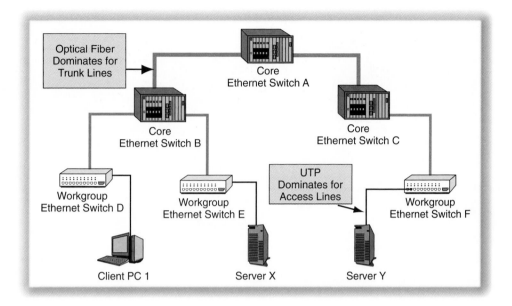

Figure 3-21 Roles of UTP and Optical Fiber in LANs

Core

To send signals, the **transceiver** (sender/receiver) injects light into a very thin glass rod called the **core.** Typical core diameters are 8.3, 50, or 62.5 microns (millionths of a meter). In comparison, an average human hair is about 75 microns thick.[13]

Cladding

As noted earlier, surrounding the core is a thicker glass cylinder called the **cladding.** The cladding normally is 125 microns in diameter, regardless of the core diameter. Consequently, an optical fiber strand with a 50 micron core diameter is often called 50/125 fiber.

As Figure 3-20 shows, if a light ray enters at an angle, it hits the cladding and reflects back into the core with **perfect internal reflection** so that no light escapes into the cladding or beyond.[14] In contrast, UTP tends to radiate energy out of the wire bundle, causing rapid attenuation.

---

[13]Actually, human hair varies in thickness from about 40 microns to 120 microns. Hair less than 60 microns thick is considered fine hair. Hair more than 80 microns thick is called thick or coarse.

[14]This is based on Snell's Law. The cladding has a slightly lower index of refraction than the core. This difference in index of refraction creates perfect internal reflection. By the way, if the fiber is bent too much, the angle between the core and cladding will become wrong for the light waves, and a great deal of light will be lost into the core. Of course, if optical fiber is bent even farther, it will crack. Glass is glass.

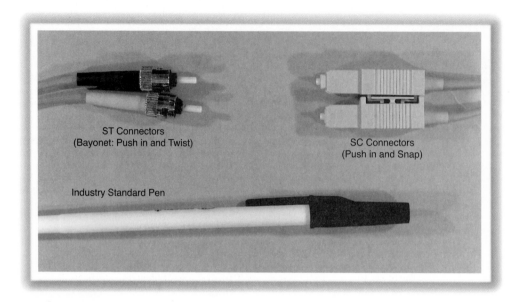

Figure 3-22   Full-Duplex Optical Fiber Cord with SC and ST Connectors

Strand Thickness

An opaque coating surrounds the cladding to keep out light and to strengthen the fiber. Part of this coating includes strands of yellow Aramid (Kevlar[15]) yarn to strengthen the fiber. The coating and outer jacket bring the outer diameter of a LAN fiber strand to about 900 microns (0.9 mm). This usually is the case for fiber, regardless of the core diameter.

Two Strands for Full-Duplex Operation

Note in Figure 3-22 that an **optical fiber** cord normally has two **strands** of fiber—one for transmission in each direction. This gives **full-duplex communication** (simultaneous two-way communication).

Connectors

In UTP, there is only one type of connector—the RJ-45 connector. However, there are several types of optical fiber connectors. The most popular are the **SC** and **ST** connectors, which are shown in Figure 3-22. There also are several smaller connectors, which are called **small form factor (SFF)** connectors. These smaller connectors allow more ports to be placed on a switch of any given size. This diversity of connector types somewhat complicates fiber selection, but not too much, because it is possible to put

---

[15]Yes, Kevlar is the stuff used to make bulletproof vests. If you cut optical fiber with a wire cutter built for UTP, you will dull the wire cutter very quickly.

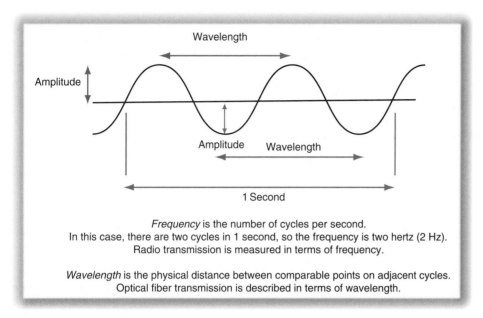

Frequency is the number of cycles per second.
In this case, there are two cycles in 1 second, so the frequency is two hertz (2 Hz).
Radio transmission is measured in terms of frequency.

Wavelength is the physical distance between comparable points on adjacent cycles.
Optical fiber transmission is described in terms of wavelength.

Figure 3-23   Frequency and Wavelength

different types of connectors at the two ends of a fiber cord so that a single cord can connect to ports on two different switches or routers with different types of plugs.

Transceiver Frequency and Wavelength

In optical fiber, a transceiver (transmitter/receiver) transmits light into the core and accepts light arriving from the core.

Figure 3-23 shows that light waves can be described in terms of **frequency,** which is the number of times the wave goes through a complete cycle per second. Frequency is measured in **hertz (Hz).** High frequencies are expressed in metric notation.

The **wavelength,** in turn, is the distance between comparable parts on two successive cycles—peak to peak, trough to trough, and so on. In light, wavelengths are measured in **nanometers (nm).** A nanometer is one billionth of a meter.

In optical fiber transmission, laser light usually is reported in wavelengths. In fiber, wavelengths come in three main "windows" centered at 850 nm, 1310 nm, and 1550 nm (see Figure 3-20). These windows are approximately 50 nm wide. Within these windows, attenuation is very low compared with attenuation at other wavelengths.

Longer wavelengths bring longer propagation distances at higher speeds. However, longer wavelengths also increase the transceivers' costs. Consequently, the goal is to select the shortest wavelength that will provide the speed and distance needed. For LANs, this is almost always 850 nm. For WANs, which must span long distances, 1310 nm is the most widely used wavelength, although 1550 nm is also used.

Longer-wavelength light can travel farther and faster, but longer-wavelength transceivers are more expensive. For LAN fiber, 850-nm transceivers usually are sufficient. WAN fiber normally uses 1310-nm or even 1550-nm transceivers.

The third characteristic of a wave is *amplitude,* which is the strength of the wave. Amplitude usually is measured as power.

TEST YOUR UNDERSTANDING

16   a) In optical fiber, what are the roles of the core and the cladding? b) What is the ability to transmit in both directions simultaneously called? c) Why does a fiber cord normally need two fiber strands? d) Does optical fiber have a single connector type? e) What are the two most common fiber connector types? f) What are small optical fiber connectors called? g) Is optical fiber transmission usually expressed in terms of wavelength or frequency? h) The amplitude of a radio or light wave usually is expressed in terms of___.

LAN Fiber versus Carrier WAN Fiber

Corporations use optical fiber to create LANs. Carriers use optical fiber to carry signals over long WAN distances. This is a book about corporate networking, so we will focus on the types of fiber that most corporations encounter: fiber for local area networks. Figure 3-24 illustrates the main differences between LAN and carrier WAN fiber.

Wavelengths

First, LAN fiber almost always uses 850 nm lasers. There is no need to use more expensive 1310 nm or 1550 nm lasers to carry signals over LAN distances. Carrier fiber, in contrast, normally uses 1310 or 1550 nm lasers because it needs longer-wavelength light for the distances carriers need to span (usually tens of kilometers).

Figure 3-24   LAN Fiber versus Carrier WAN Fiber

	LAN Fiber	Carrier WAN Fiber
Required Distance Span	200 to 300 m	1 to 40 kilometers
Transceiver Wavelength	850 nm	1,310 nm (and sometimes 1,550 nm)
Type of Fiber	Multimode (thick core)	Single mode (thin core)
Core Diameter	50 microns or 62 microns	83 microns
Primary Distance Limitation	Modal dispersion	Absorptive attenuation
Quality Metric	Modal bandwidth (MHz.km)	NA

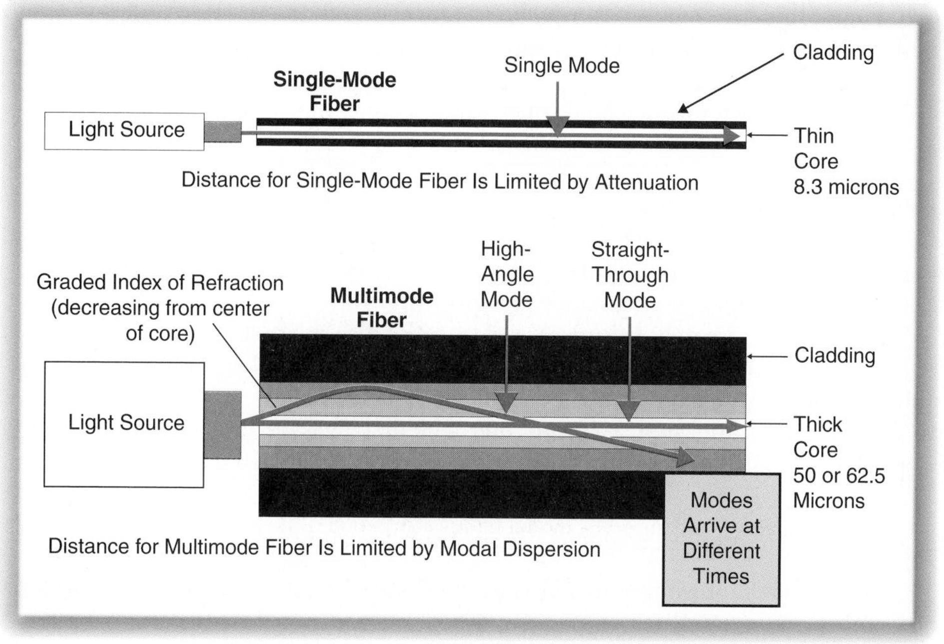

Figure 3-25   Multimode Fiber and Single-Mode Fiber

Multimode Fiber

LANs use **multimode fiber,** which has a relatively thick core, as Figure 3-25 shows. Initially, North American multimode fiber had a core diameter of 62.5 microns. In Europe, which developed fiber slightly later, the core diameter was a little thinner—only 50 microns in diameter. A thinner core can carry signals farther. Although 50-micron fiber is no more expensive than 62.5-micron fiber today, most U.S. organizations continue to standardize on 62.5-micron fiber, while European companies primarily use 50-micron fiber.

As Figure 3-25 shows, because multimode fiber has a thick core, light rays can enter over a wide range of angles. Light traveling straight through the core arrives fastest. In contrast, light entering at high angles will zigzag through the core many times. It will travel much farther and so will arrive later.

This difference in arrival times causes a problem called **modal dispersion,** in which light from successive light pulses overlaps. If the cord length is too great, light from successive light pulses will overlap so much that the signal will be unreadable. Modal dispersion is a limiting factor for distance in multimode fiber.

Carrier Single-Mode Fiber

In contrast, carriers use single-mode fiber. This fiber has a core diameter of only 8.3 nm. With such a thin diameter, only a single mode can travel through the fiber. There is no modal dispersion to limit distance. For single-mode fiber, the only distance limitation is **absorptive attenuation,** with the light being absorbed by glass molecules.

Fortunately, this attenuation is very low, especially at longer wavelengths. Consequently, single-mode fiber can carry signals much farther than multimode fiber.

The limiting factor in multimode fiber is modal dispersion. The limiting factor in single-mode fiber is absorptive attenuation.

Multimode Fiber Quality

Just as UTP wiring has quality categories, multimode LAN fiber varies in quality. Simply put, higher-quality multimode fiber is better at limiting modal dispersion.[16] This allows a longer propagation distance at a given speed.

We saw that wire quality is given as a series of discrete quality levels (Cat 3, Cat 4, etc.). Multimode optical fiber quality, in contrast, is measured as a *continuous* variable called **modal bandwidth.** Modal bandwidth is measured as **MHz.km.** Usually, two modal bandwidths are listed, as in 200/500. The first gives the modal bandwidth of the fiber at 850 nm. The second gives the modal bandwidth at 1310 nm.

In fiber, quality is measured by modal bandwidth, which is measured as MHz.km.

The modal bandwidth concept is too difficult to explain in an introductory network course (fortunately for you). Quite simply, more modal bandwidth is better for signal distance.

Wavelength, Core Diameter, Modal Bandwidth, and Distance in Multimode Fiber

To increase LAN propagation distance, one can increase the light wavelength, decrease the core diameter, and buy better-quality fiber having greater modal bandwidth. In practice, however, LAN fiber uses only 850 nm light because longer wavelengths would be too expensive.

1000BASE-SX with 62.5/125 Fiber

For example, gigabit Ethernet, which carries signals at 1 Gbps, is dominated by the 1000BASE-SX standard, which transmits light at 850 nm. (The S stands for short wavelength to denote 850 nm transceivers.) With a 62.5 micron core diameter and 160 MHz.km modal bandwidth, 1000BASE-SX can transmit gigabit Ethernet signals over a distance of 220 m (the length of two football fields). With 200 MHz.km fiber, the distance rises to 275 m.

1000BASE-SX with 50/125 Fiber

Few companies need longer distances than 62.5/125 multimode fiber can provide for their trunk lines. However, if they do, they can use 50/125 fiber rather than 62.5/125 fiber.

---

[16]As one way to reduce modal dispersion, fiber manufacturers today only build graded-index multimode fiber, in which the index of refraction decreases from the center of the core to the core's outer edge. This slows light traveling down the center, compared with light farther out in the radius. Consequently, light going straight through along the center is slowed down, while light zigzagging through the core is speeded up during its time away from the center. This reduces the time lapse between the direct mode and other modes, thus reducing modal dispersion.

Wavelength	Core Diameter	Modal Bandwidth	Maximum Propagation Distance
850 nm	62.5 microns	160 MHz.km	220 m
850 nm	62.5 microns	200 MHz.km	275 m
850 nm	50 microns	500 MHz.km	550 m

Note: Modal Bandwidth Is a Measure of Multimode Optical Fiber Quality. Not applicable to single-mode fiber.
[1]The ISO/IEC 11801 OM1 multimode fiber standard has a modal bandwidth of 200 MHz.km at 850 nm.
[2]The ISO/IEC 11801 OM2 multimode fiber standard has a modal bandwidth of 500 MHz.km at 850 nm.

Figure 3-26   Wavelength, Core Diameters, Modal Bandwidth, and Maximum Propagation Distance for Ethernet 1000BASE-SX

Thinner fiber offers higher modal bandwidth. With a modal bandwidth of 400/1000 MHz.km, 50/125 fiber can carry 1000BASE-SX signals up to 500 meters.

TEST YOUR UNDERSTANDING

17. a) Do LANs normally use multimode fiber or single-mode fiber? b) What is the main physical characteristic distinguishing multimode fiber from single-mode fiber? c) Why is multimode fiber used in LANs? (Give a complete explanation.) d) What are typical core diameters for multimode fiber? e) What is modal dispersion? f) What is the limiting propagation problem for multimode fiber transmission distance? g) Of what is modal bandwidth a measure? h) In what units is modal bandwidth expressed? i) What is the core diameter of single-mode fiber? j) What is the limiting factor for transmission distance in single-mode fiber? k) In what two ways can businesses select multimode fiber to transmit Ethernet 1000BASE-SX LAN signals farther?

Noise and Electromagnetic Interference

In fiber, attenuation and modal dispersion are important limitations for different types of fiber. However, noise and electromagnetic interference, which are serious problems in UTP, are *not* problems for fiber.

There is no noise energy in fiber. Electrons randomly moving around within the transmission medium generate electromagnetic energy called noise. However, electrons do not generate light energy that adds to the light signal.

In UTP, EMI is an important concern. In optical fiber, this concern vanishes. The only type of electromagnetic interference applicable to fiber is light coming into the fiber from the outside. Fiber prevents this completely by having an opaque coating around the cladding.

TEST YOUR UNDERSTANDING

18. Are noise and interference major propagation problems for optical fiber propagation? Explain.

RADIO SIGNAL PROPAGATION

Frequencies

Wireless radio signals propagate as waves, as we saw in Figure 3-23. While optical fiber waves are measured in terms of wavelength, radio waves are described in terms of *frequency*. Useful radio frequencies for WLAN transmission (and data networking in general) come in the high **megahertz (MHz)** to low **gigahertz (GHz)** range. As with transmission speeds, metric designations increase by a factor of 1000, rather than 1024.

Frequency is used to describe the radio waves used in WLANs.

TEST YOUR UNDERSTANDING

19. Is wireless radio transmission usually expressed in terms of wavelength or frequency?

Antennas

Radio transmission requires an antenna. Figure 3-27 shows that there are two types of radio antennas: omnidirectional antennas and dish antennas.

➤ **Omnidirectional antennas** transmit signals equally strongly in all directions and receive incoming signals equally well from all directions. Consequently, the antenna does not need to point in the direction of the receiver. However, because the signal spreads in all directions, only a small fraction of the energy transmitted by an omnidirectional antenna reaches the receiver.

➤ **Dish antennas**, in contrast, point in a particular direction, which allows them to focus stronger outgoing signals in that direction for the same power and to receive weaker incoming signals from that direction.

Figure 3-27   Omnidirectional and Dish Antennas

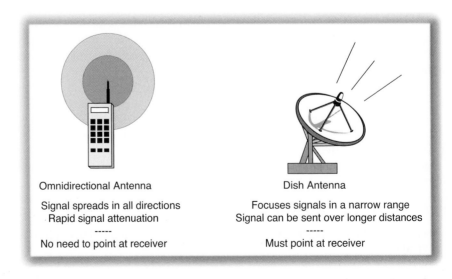

Omnidirectional Antenna

Signal spreads in all directions
Rapid signal attenuation

No need to point at receiver

Dish Antenna

Focuses signals in a narrow range
Signal can be sent over longer distances

Must point at receiver

Dish antennas are good for long distances because of their focusing ability, but they need to know the direction of the other party. Also, omnidirectional antennas are easier to use. (Imagine if you had to carry a dish with you whenever you carried your cellular phone. You would not even know where to point the dish!)

WLANs use omnidirectional antennas almost exclusively. Distances are short and users do not always know the location of the nearest access point.

TEST YOUR UNDERSTANDING

20. a) Distinguish between omnidirectional and dish antennas in terms of operation. b) Under what circumstances would you use an omnidirectional antenna? c) Under what circumstances would you use a dish antenna? d) What type of antenna normally is used in WLANs? Why?

Wireless Propagation Problems

Although wireless communication gives mobility, wireless transmission is not very predictable, and there often are serious propagation problems. Figure 3-28 illustrates four common wireless propagation problems.

Inverse Square Law Attenuation

Compared with signals sent through wires and optical fiber, radio signals attenuate very rapidly. When a signal spreads out from any kind of antenna, its strength is spread over the area of a sphere. (In omnidirectional antennas, power is spread equally over the sphere, while in dish antennas, it is concentrated primarily in one direction on the sphere.)

The area of a sphere is proportional to the square of its radius, so signal strength in any direction weakens by an **inverse square law** $(1/r^2)$, as Equation 5–1 illustrates.

Figure 3-28   Wireless Propagation Problems

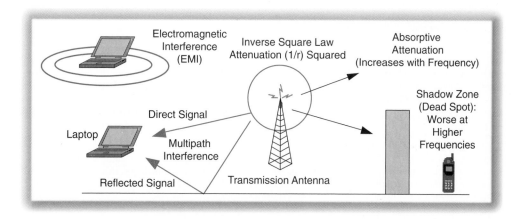

Here, S_1 is the signal strength at distance r_1, and S_2 is the signal strength at a farther distance, r_2.

$$S_2 = S_1 \times (r_1/r_2)^2 \qquad \text{(Equation 5-1)}$$

To give an example, if you triple the distance ($r_1/r_2 = 1/3$), the signal strength (S_2) falls to only one-ninth ($1/3^2$) of its original strength (S_1). With radio propagation, you have to be relatively close to your communication partner unless the signal strength is very high, an omnidirectional antenna is used, or both.

Absorptive Interference

As a radio signal travels, it is partially absorbed by the air molecules, plants, and other things it passes through. This **absorptive attenuation** is especially bad in moist air, and office plants are the natural enemies of wireless transmission. Absorptive attenuation is especially bad for longer-distance outdoor propagation.

Shadow Zones (Dead Spots)

To some extent, radio signals can go through and bend around objects. However, if there is a large or dense object (such as a brick wall), blocking the direct path between the sender and receiver, the receiver may be in a **shadow zone (dead spot),** where it cannot receive the signal. If you have a cellular telephone and often try to use it within buildings, you probably are familiar with this problem.

Multipath Interference

In addition, radio waves tend to bounce off walls, floors, ceilings, and other objects. As Figure 3-28 shows, this may mean that a receiver will receive two or more signals—a direct signal and one or more reflected signals. The direct and reflected signals will travel different distances and so may be out of phase when they reach the receiver. (For example, one may be at its highest amplitude while the other is at its lowest, giving an average of zero.) This **multipath interference** may cause the signal to range from strong to nonexistent within a few centimeters (inches).[17] If the difference in time between the direct and reflected signal is large, some reflected signals may even interfere with the next direct signal. Multipath interference is the most serious propagation problem at WLAN frequencies.

Multipath interference is the most serious propagation problem at WLAN frequencies.

Electromagnetic Interference (EMI)

A final common propagation problem in wireless communication is *electromagnetic interference (EMI)*. As we saw earlier in this chapter, electrical motors and many other devices produce EMI at frequencies used in data communications. Among these devices are cordless telephones, microwaves, and, especially, devices in other nearby wireless networks.

---

[17]Wireless access points often have two antennas to combat multipath interference. Many wireless NICs, in turn, have an antenna that looks like a fan. This really is a "rake" antenna, which gets this name because it has many small antennas side by side, like the tines in a leaf rake. Later in this chapter, we will see that the 802.11n standard actually uses multipath interference to *improve* transmission.

Frequency-Dependent Propagation Problems

Two propagation problems are affected by frequency.

➤ First, higher-frequency waves attenuate more rapidly with distance than lower-frequency waves because they are absorbed more rapidly by moisture in the air, leafy vegetation, and other water-bearing obstacles. Consequently, as we will see later in this chapter, WLAN signals around 5 GHz attenuate more rapidly than signals around 2.4 GHz.

➤ Second, shadow zone problems grow worse with frequency. As frequency increases, radio waves become less able to go through and bend around objects.

TEST YOUR UNDERSTANDING

21. a) Which offers more reliable transmission characteristics—UTP or radio transmission? b) Which attenuate more rapidly with distance—signals sent though wired media or radio signals? c) If the signal strength from an omnidirectional radio source is 8 milliwatts (mW) at 30 meters, how strong will it be at 120 meters, ignoring absorptive attenuation? Show your work. d) How are shadow zones (dead spots) created? e) Why is multipath interference very sensitive to location? f) What is the most serious propagation problem in WLANs? g) List some sources of EMI. h) What propagation problems become worse as frequency increases?

NETWORK TOPOLOGIES

The term **network topology** refers to the physical arrangement of a network's computers, switches, routers, and transmission lines. Topology, then, is a physical layer concept. Different network (and internet) standards specify different topologies. Figure 3-29 shows the major topologies found in networking. Some are seen only in older legacy LANs using obsolete technology.

Network topology is the physical arrangement of a network's computers, switches, routers, and transmission lines.

Point-to-Point Topology

The simplest network topology is the **point-to-point topology,** in which two nodes are connected directly. Although some might say that a point-to-point connection is not a network, WANs often are built of point-to-point private leased lines provided by the telephone service, and dial-up telephone connections effectively provide point-to-point connections between users and the Internet.

Star Topology and Extended Star (Hierarchy) Topology

Modern versions of Ethernet, which is the dominant LAN standard, use the star and extended star topologies. In a **simple star topology,** all wires connect to a single switch. In an **extended star** (or **hierarchy**) **topology,** there are multiple layers of switches organized in a hierarchy. We will see Ethernet hierarchies in the next chapter.

Mesh Topology

In a mesh topology, there are many connections among switches, so there are many alternative routes to get from one end of the network to the other. We will see **mesh topologies** with ATM and Frame Relay in Chapter 7 and with routers in Chapter 8.

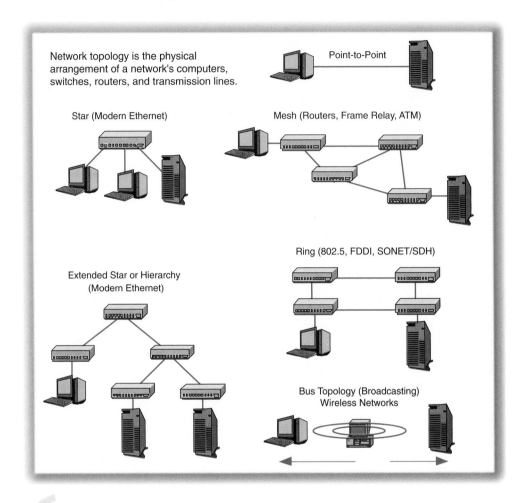

Figure 3-29   Major Topologies

Ring Topology

In a **ring topology,** computers are connected in a loop. Messages pass in only one direction around the loop. Eventually, all messages pass through all computers. In LANs, the obsolete 802.5 Token-Ring Network and FDDI network technologies discussed in Chapter 4a used a ring topology. In addition, the worldwide telephone network increasingly uses rings to connect its switches via the SONET/SDH technology, a topic that is discussed in Module B.

Bus Topologies

In a **bus topology,** when a computer transmits, it broadcasts to all other computers. Wireless LANs, which we will see in Chapter 5, use a bus topology by broadcasting signals. So do Ethernet hubs, which we will see in Chapter 4. The obsolete Ethernet 10BASE5 and 10BASE2 technologies discussed in Module D also used a bus topology.

TEST YOUR UNDERSTANDING

22. a) What is a network topology? b) At what layer do we find topologies? c) List the major network topologies, and give the defining characteristics for each one. Present each topology in a separate paragraph. d) What technology is associated with each topology?

CONCLUSION

Synopsis

Chapter 2 discussed standards in general. It focused on the data link layer through the application layer because these layers all operate in the same general way—by sending messages. The physical layer, which was the focus of this chapter, is very different. It sends signals instead of messages, and it alone governs transmission media and propagation effects.

When signals propagate, they endure propagation effects such as attenuation. If propagation effects are too large, the receiver will not be able to read the signal. In this chapter, we looked at three transmission media and their major propagation effects: unshielded twisted pair copper wiring (4-pair UTP), optical fiber, and wireless (radio) transmission.

Before data can be transmitted as signals, the data must be represented as strings of bits (1s and 0s).

➤ Whole numbers (integers) typically are represented as binary numbers. Binary counting begins with 0. You should know how to count in binary by adding 1 to each previous value.

➤ You should also know that a field with N bits allows you to represent 2^N alternatives.

➤ For text, 7-bit ASCII and 8-bit extended ASCII normally are used to represent characters in binary.

➤ Other forms of data, including graphics and the human voice, can also be represented in binary.

After conversion, these binary data streams must then be converted into signals to propagate down the transmission medium. In simple on/off signaling in optical fiber, the sender turns on the light source during a clock cycle for a 1 or turns off the light source for a 0. On/off signaling is binary signaling, in which there are two possible states or line conditions (on and off). Binary signaling can also use two voltage ranges on a wire. One voltage range represents a 0; the other, a 1.

In the box on digital signaling, we saw that in digital signaling there are a few possible states (2, 4, 8, 16, etc.). Adding more possible states allows the sender to transmit more bits per clock cycle, but doing so decreases immunity to attenuation propagation errors. Binary transmission is a special case of digital transmission.

In unshielded twisted pair wiring, a cord consists of four twisted wire pairs. There is an RJ-45 connector at each end. UTP is rugged and inexpensive and so dominates the access links that connect computers to workgroup switches.

The two wires of each pair in a UTP cord are twisted around each other to reduce electromagnetic interference (EMI) problems. Two simple installation expedients keep propagation problems to acceptable levels. First, restricting UTP cord lengths to 100 meters usually prevents serious attenuation and noise errors. Second, limiting the untwisting of wires to no more than 0.5 inches (1.25 cm) usually keeps terminal crosstalk interference to an acceptable level.

Earlier versions of Ethernet transmitted serially—sending on only one wire pair in each direction. Gigabit Ethernet transmits in parallel, sending on all four wire pairs when it transmits. Other forms of parallel transmission use more than four transmission paths. Parallel transmission is faster than serial transmission for a given clock cycle, but is more expensive, so serial transmission now dominates in the marketplace.

There are different grades (called *categories*) of UTP quality. Almost all UTP sold today is Category 5e or Category 6 wiring. This is sufficient for up to 1 Gbps in Ethernet. To achieve 100-meter distances at higher speeds over UTP, companies need to install newer Cat 6A or Cat 7 wiring, although if they only need a distance of up to 55 meters, they can send 10 Gbps Ethernet over Cat 6 wiring.

For trunk lines, optical fiber dominates in LAN transmission and will grow even more dominant as speed requirements increase. In fiber, a transceiver injects signals into a thin glass core that is 62.5 microns, 50 microns, or 8.3 microns in diameter. There are three major propagation windows centered around 850 nm, 1310 nm, and 1550 nm. Attenuation is lower at longer wavelengths, so carrier WAN fiber normally uses 1310 nm or 1550 nm light. However, 850 nm transceivers are much less expensive, and they can easily span typical LAN distances, so LAN fiber uses 850 nm light.

The transceiver uses simple binary on/off light signaling. An optical fiber cord has two strands for full-duplex communication. Fiber can use several connectors, including SC, ST, and newer small form factor (SFF) attenuation.

LAN fiber almost always uses multimode fiber, which has a "thick" diameter of 62.5 or 50 microns. This is less expensive than single-mode fiber, which has a core diameter of only 8.3 microns. Although multimode cannot carry signals as far as single-mode fiber, it can easily carry light 200 meters or more. This is more than adequate for nearly all LAN trunk lines. Multimode fiber propagation is limited by modal dispersion. Higher-quality multimode fiber, which is characterized by a higher modal bandwidth (measured as MHz.km), can carry signals farther than lower-quality multimode fiber. In addition, multimode fiber with a thinner core diameter (50 microns instead of 62.5 microns) can carry signals farther.

Carriers use single-mode fiber, which is very expensive but can carry high-speed signals over the very long distances needed by carriers (tens of kilometers). Single-mode fiber is limited by attenuation, which is very low but is the limiting propagation effect over long distances.

Wireless radio transmission is becoming popular in wireless LANs, and carriers are even beginning to offer it in WAN transmission. While optical fiber waves are measured in wavelengths, radio waves are measured in frequency. Mobile wireless transmission uses omnidirectional antennas, while wireless transmission to and from fixed sites typically uses dish antennas. Dish antennas focus signals and so give better reception than omnidirectional antennas, but you must know where to point the antenna. For mobile users, only omnidirectional antennas make sense.

Both UTP and fiber propagation effects can be largely neutralized by choosing the right-quality wire and fiber and respecting distance limits. However, we saw that radio propagation suffers from many propagation problems, including inverse square law attenuation, absorptive attenuation, electromagnetic interference, multipath interference, and shadow zones. In addition, there is no general way to control these propagation effects. While UTP and optical fiber transmissions are highly predictable, wireless transmission is always problematic.

A network's topology describes the way that transmission lines connect computers, switches, and routers. Examples of topologies are the point-to-point, star, hierarchy (extended star), mesh, ring, and bus topologies. As you work through this book, you will encounter these topologies and their implications for performance and reliability.

End-of-Chapter Questions

THOUGHT QUESTIONS

1. In binary, 35 is 100011. Compute 36 through 40 in binary. (The number 40 is 101000.)

2. a) (If you read the box on bit rates and baud rates) A field is 8 bits long. How many values can it represent?

   b) Repeat for 1, 4, and 16. Do not use a spreadsheet program. (Hint: Each bit doubles the number of possible alternatives.)

3. (If you read the box on bit rates and baud rates) The clock cycle is one millionth of a second. There is eight-level digital signaling.

   a) What is the baud rate?

   b) How many bits are sent per clock cycle?

   c) What is the bit rate?

   d) If your clock cycle is 1/10,000 of a second and you want a bit rate of 50 kbps, how many possible states will you need per clock cycle?

4. If power falls to 40 percent of its initial value through attenuation, how many decibels is this loss?

5. What type of interference is most likely to create problems in UTP transmission?

6. When a teacher lectures in class, is the classroom a full-duplex communication system or a half-duplex communication system?

7. In the author's college, the wiring has been described as "Cat 3 on a good day." What Ethernet speeds can it carry?

8. a) How many possible paths are there between any two computers in a point-to-point topology?

   b) In a hierarchical topology? (Hint: Draw a picture.)

   c) In a mesh topology with four switches (without back-tracking)? (Hint: Draw a picture.)

TROUBLESHOOTING QUESTIONS

1. A tester shows that a UTP cord has too much interference. What might be causing the problem? Give at least two alternative hypotheses, and then describe how to test them.

2. What kinds of errors are you likely to encounter if you run a length of UTP cord 200 meters? (Recall that the standard calls for a 100-meter maximum distance.)

HANDS-ON EXERCISE

Chapter 3a discusses how to connectorize bulk UTP cabling. To try it out, you will need a box of bulk UTP cabling, a wire cutter, a wire stripper, a crimper, a bag of RJ-45 connectors, and a tester (because only about half of connections done by novices work). All of this will set you back about $300. When you know this, the price of UTP patch cables, which are cut, connectorized, and tested at the factory, seems more reasonable, doesn't it?

PERSPECTIVE QUESTIONS

1. What was the most surprising material for you in this chapter?

2. What was the most difficult thing for you in this chapter?

PROJECT

Write a one-page research report on Category 7 STP or small form factor optical fiber connectors.

GETTING CURRENT

Go to the book website's New Information and Errors pages for this chapter to get new information since this book went to press and for corrections to any errors in the text.

Hands-On: Cutting and Connectorizing UTP[1]

INTRODUCTION

Chapter 3 discussed UTP wiring in general. This chapter discusses how to cut and connectorize (add connectors to) solid UTP wiring.

SOLID AND STRANDED WIRING

Solid-Wire UTP versus Stranded-Wire UTP

The TIA/EIA-568 standard requires that long runs to wall jacks use **solid-wire UTP,** in which each of the eight wires really is a single solid wire.

However, patch cords running from the wall outlet to a NIC usually are **stranded-wire UTP,** in which each of the eight "wires" really is a bundle of thinner wire strands. So stranded-wire UTP has eight bundles of wires, each bundle in its own insulation and acting like a single wire.

Relative Advantages

Solid wire is needed in long cords because it has lower attenuation than stranded wire. In contrast, stranded-wire UTP cords are more flexible than solid-wire cords, making them ideal for patch cords—especially the one running to the desktop—because they can be bent more and still function. They are more durable than solid-wire UTP cords.

---

[1]This material is based on the author's lab projects and on the lab project of Prof. Harry Reif of James Madison University.

Solid-Wire UTP
> Each of the eight wires is a solid wire
> Low attenuation over long distances
> Easy to connectorize
> Inflexible and stiff—not good for runs to the desktop

Stranded-Wire UTP
> Each of the eight "wires" is itself several thin strands of wire within an insulation tube
> Flexible and durable—good for runs to the desktop
> Impossible to connectorize in the field (bought as patch cords)
> Higher attenuation than solid-wire UTP—Used only in short runs
>> From wall jack to desktop
>> Within a telecommunications closet (see Chapter 3)

Figure 3a-1   Solid-Wire and Stranded-Wire UTP (Study Figure)

Adding Connectors

It is relatively easy to add RJ-45 connectors to solid-wire UTP cords. However, it is very difficult to add RJ-45 connectors to stranded-wire cords. Stranded-wire patch cords should be purchased from the factory precut to desired lengths and preconnectorized.

In addition, when purchasing equipment to connectorize solid-wire UTP, it is important to purchase crimpers designed for solid wire.

CUTTING THE CORD

Solid-wire UTP normally comes in a box or spool containing 50 meters or more of wire. The first step is to cut a length of UTP cord that matches your need. It is good to be a little generous with the length. This way, bad connectorization can be fixed by cutting off the connector and adding a new connector to the shortened cord. Also, UTP cords should never be subjected to pulls (strain), and adding a little extra length creates some slack.

STRIPPING THE CORD

Now the cord must be stripped at each end using a **stripping tool** such as the one shown in Figure 3a-2. The installer rotates the stripper once around the cord, scoring (cutting into) the cord jacket (but not cutting through it). The installer then pulls off the scored end of the cord, exposing about 5 cm (about two inches) of the wire pairs.

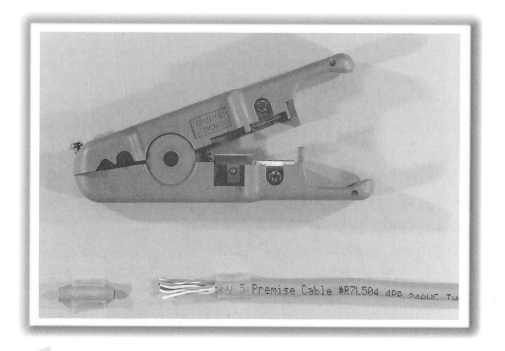

Figure 3a-2   Stripping Tool

It is critical not to score the cord too deeply, or the insulation around the individual wires may be cut. This creates short circuits. A really deep cut also will nick the wire, perhaps causing it to snap immediately or later.

WORKING WITH THE EXPOSED PAIRS

Pair Colors

The four pairs each have a color: orange, green, blue, or brown. One wire of the pair usually is a completely solid color. The other usually is white with stripes of the pair's color. For instance, the orange pair has an orange wire and a white wire with orange stripes.

Untwisting the Pairs

The wires of each pair are twisted around each other several times per inch. These must be untwisted after the end of the cord is stripped.

Ordering the Pairs

The wires now must be placed in their correct order, left to right. Figure 3a-3 shows the location of Pin 1 on the RJ-45 connector and on a wall jack or NIC.

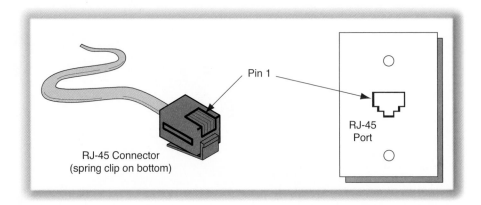

Figure 3a-3   Location of Pin 1 on an RJ-45 Connector and Wall Jack or NIC

Which color wire goes into which connector slot? The two standardized patterns are shown in Figure 3a-4. The T568B pattern is much more common in the United States.

The connectors at both ends of the cord use the same pattern. If the white-orange wire goes into Pin 1 of the connector on one end of the cord, it also goes into Pin 1 of the connector at the other end.

Figure 3a.4   T568A and T568B Pin Colors

Pin*	T568A	T568B
1	White-Green	White-Orange
2	Green	Orange
3	White-Orange	White-Green
4	Blue	Blue
5	White-Blue	White-Blue
6	Orange	Green
7	White-Brown	White-Brown
8	Brown	Brown

Note: Do not confuse T568A and T568B
pin colors with the TIA/EIA-568 Standard.

Cutting the Wires

The length of the exposed wires must be limited to 1.25 cm (0.5 inch) or slightly less. After the wires have been arranged in the correct order, a cutter should cut across the wires to make them this length. The cut should be made straight across, so that all wires are of equal length. Otherwise, they will not all reach the end of the connector when they are inserted into it. Wires that do not reach the end will not make electrical contact.

ADDING THE CONNECTOR

Holding the Connector

The next step is to place the wires in the RJ-45 connector. In one hand, hold the connector, clip side down, with the opening in the back of the connector facing you.

Sliding in the Wires

Now, slide the wires into the connector, making sure that they are in the correct order (white-orange on your left). There are grooves in the connector that will help. Be sure to push the wires all the way to the end or proper electrical contact will not be made with the pins at the end.

Before you crimp the connector, look down at the top of the connector, holding the tip away from you. The first wire on your left should be mostly white. So should every second wire. If they are not, you have inserted your wires incorrectly.[2]

Some Jacket Inside the Connector

If you have shortened your wires properly, there will be a little bit of jacket inside the RJ-45 connector.

CRIMPING

Pressing Down

Get a really good **crimping tool** (see Figure 3a-5). Place the connector with the wires in it into the crimp and push down firmly. Good crimping tools have ratchets to reduce the chance of your pushing down too tightly.

Making Electrical Contact

The front of the connector has eight pins running from the top almost to the bottom (spring clip side). When you **crimp** the connector, you force these eight pins through the insulation around each wire and into the wire itself. This seems like a crude electrical connection, and it is. However, it normally works very well. Your wires are now connected to the connector's pins. By the way, this is called an **insulation displacement connection (IDC)** because it cuts through the insulation.

---

[2]Thanks to Jason Okumura, who suggested this way of checking the wires.

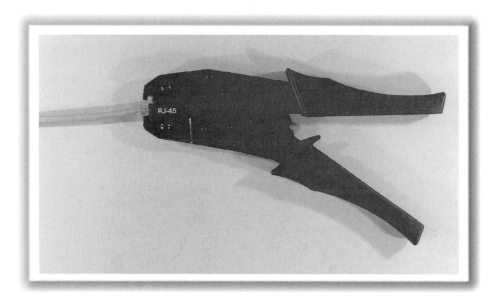

Figure 3a-5   Crimping Tool

Strain Relief

When you crimp, the crimper also forces a ridge in the back of the RJ-45 connector into the jacket of the cord. This provides **strain relief,** meaning that if someone pulls on the cord (a bad idea), they will be pulling only to the point where the jacket has the ridge forced into it. There will be no strain where the wires connect to the pins.

TESTING

Purchasing the best UTP cabling means nothing unless you install it properly. Wiring errors are common in the field, so you need to test every cord after you install it. Testing is inexpensive compared to troubleshooting subtle wiring problems later.

Testing with Continuity Testers

The simplest testers are **continuity testers**, which merely test whether the wires are arranged in correct order within the two RJ-45 connectors and are making good electrical contact with the connector. They cost only about $100.

Testing for Signal Quality

Better testers cost $500 to $2,000 but are worth the extra money. In addition to testing for continuity problems, they send **test signals** through the cord to determine whether the cord meets TIA/EIA-568 signal quality requirements. Many include **time domain**

reflectometry (TDR), which sends a signal and listens for echoes in order to measure the length of the UTP cord or to find if and where breaks exist in the cord.

TEST YOUR UNDERSTANDING

1. a) Explain the technical difference between solid-wire UTP and stranded-wire UTP. b) In what way is solid-wire UTP better? c) In what way is stranded-wire UTP better? d) Where would you use each? e) Which should only be connectorized at the factory?

2. If you have a wire run of 50 meters, should you cut the cord to 50 meters? Explain.

3. Why do you score the jacket of the cord with the stripping tool instead of cutting all the way through the jacket?

4. a) What are the colors of the four pairs? b) If you are following T568B, which wire goes into Pin 3? c) At the other end of the cord, would the same wire go into Pin 3?

5. After you arrange the wires in their correct order and cut them across, how much of the wires should be exposed from the jacket?

6. a) Describe RJ-45's insulation displacement approach. b) Describe its strain relief approach.

7. a) Should you test every cord in the field after installation? b)For what do inexpensive testers test? c) For what do expensive testers test?

Ethernet LANs

Learning Objectives

By the end of this chapter, you should be able to discuss the following:

■ Ethernet physical layer standards and how they affect network design.

■ The Ethernet data link layer and the Ethernet MAC layer frame.

■ Basic Ethernet data link layer switch operation.

■ Advanced aspects of Ethernet switch operation.

■ Ethernet switch purchasing criteria.

■ Ethernet security.

INTRODUCTION

And the Winner Is . . .

According to Nortel Networks, 95 percent of wired LAN ports worldwide were Ethernet ports in 2002. That percentage is even higher today. Indeed, when people talk about LANs without referring to technology, they usually mean Ethernet. In this chapter, we will look at Ethernet in detail. In the next chapter, we will look at the other main LAN technology in corporations today—wireless LAN technology.

Ethernet became the dominant LAN technology because of its simple, and therefore inexpensive, data link layer operation. This cost advantage, coupled with adequate performance, has driven competing technologies out of the corporate LAN market.

Ethernet became the dominant LAN technology because of its simple, and therefore inexpensive, data link layer operation combined with adequate performance.

TEST YOUR UNDERSTANDING

1. a) What is the dominant LAN technology today? b) Why did it become dominant?

A Short History of Ethernet Standards

Prehistory: Xerox, Intel, and Digital Equipment

Metcalfe and Boggs created Ethernet technology at the Xerox Palo Alto Research Center in Palo Alto, California, in the mid-1970s.[1] In the early 1980s, Xerox, Intel, and Digital

---

[1]Bob Metcalfe has noted that he got the idea for Ethernet by visiting the University of Hawai'i's packet radio Alohanet project. The Alohanet project sent packets over radio and handled the problem of controlling when stations could transmit. Metcalfe realized that the same could be done over coaxial cable. See Bob Metcalfe, "Internet Fogies Reminisce and Argue at Interop Confab," *Infoworld*, September 21, 1992, p. 45.

Equipment Corporation teamed up to produce the first two commercial standards for Ethernet: Ethernet I and Ethernet II.

The 802 Committee

After creating Ethernet II, the three companies passed responsibility for Ethernet standards to the newly created **802 LAN/MAN Standards Committee** of the **Institute for Electrical and Electronics Engineers (IEEE)**. The "**802 Committee**," as everybody calls it, is broadly responsible for creating local area network standards and metropolitan area network standards.

Figure 4-1   A Short History of Ethernet Standards (Study Figure)

Early History of Ethernet Standards

> Developed at the Xerox Palo Alto Research Center by Metcalfe and Boggs
>
> Standardized by Xerox, Intel, and Digital Equipment Corporation
>
> Developed the Ethernet I and Ethernet II standards in the early 1980s

The 802 Committee

> Development passed to the Institute for Electrical and Electronics Engineers (IEEE)
>
> IEEE created the 802 LAN/MAN Standards Committee for LAN standards
>
> This committee is usually called the 802 Committee
>
> The 802 Committee creates working groups for specific types of standards.
>
>> 802.1 for general standards
>>
>> 802.3 for Ethernet standards
>>
>> 802.11 for wireless LAN standards
>>
>> 802.16 for WiMax wireless metropolitan area network standards

The 802.3 Working Group

> This group is in charge of creating Ethernet standards
>
> The terms *802.3* and *Ethernet* are interchangeable today
>
> Figure 4-2 shows Ethernet physical layer standards
>
> Ethernet also has data link layer standards (frame organization, switch operation, etc.)

Ethernet Standards are OSI Standards

> Layer 1 and Layer 2 standards are almost universally OSI standards
>
> Ethernet is no exception
>
> ISO must ratify them
>
> In practice, when the 802.3 Working Group finishes standards, vendors begin building compliant products

The 802.3 Ethernet Working Group

The 802 Committee delegates the actual work of developing standards to specific working groups. The 802 Committee's **802.3 Working Group**, for example, creates Ethernet-specific standards. We will use the terms *Ethernet* and *802.3* interchangeably in this book.

We will use the terms *Ethernet* and *802.3* interchangeably in this book.

Other Working Groups

The 802 Committee has several other working groups. For example, the 802.11 Working Group creates the wireless LAN standards that we will see in the next chapter. In turn, the 802.16 Working Group is creating the WiMax standards for wireless subscriber access within a city. The 802.1 Working Group creates general standards.

Ethernet Standards Are OSI Standards

Ethernet standards are LAN standards, so they are Layer 1 (physical) and Layer 2 (data link) standards. Recall from Chapter 2 that standards at the lowest two layers are always OSI standards. Although the 802.3 Working Group creates Ethernet standards, these standards are not official OSI standards until ISO ratifies them later. In practice, however, as soon as the 802.3 Working Group releases an 802.3 standard, vendors begin building products based on the specification.

TEST YOUR UNDERSTANDING

2. a) What working group creates Ethernet standards? b) To what committee does this working group report? c) In what organization is this committee? d) Are there other working groups? e) If so, what do they do? f) Does this book use *Ethernet* and *802.3* interchangeably? g) Why would you expect Ethernet standards to be OSI standards? h) When do vendors begin developing products based on Ethernet standards?

ETHERNET PHYSICAL LAYER STANDARDS

Ethernet defines standards at the physical and data link layers. We will look first at Ethernet physical layer standards.

Major Ethernet Physical Layer Standards

As just noted, Ethernet is a LAN technology, so standards must be set at both the physical and data link layers. We will look at 802.3 physical layer standards first.

When the 802.3 Working Group first created physical layer Ethernet standards, it used technologies that are no longer in use. Figure 4-2 shows the Ethernet physical layer standards that have been ratified.

Venders have stopped manufacturing products based on the oldest standards, and for higher speeds, vendors are manufacturing products for only some standards. In addition, some vendors provide optical fiber of a higher quality than standards specify. These can carry Ethernet signals over longer distances. If you are designing an Ethernet network, you will need to conduct timely research on what is available.[2]

---

[2]In Chapter 3, we saw standards for optical fiber. Ethernet standards build on these optical fiber standards, adding light source standards and signaling methods.

Physical Layer Standard	Speed	Maximum Run Length	Medium
UTP			
10BASE-T	10 Mbps	100 meters	4-pair Category 3 or higher
100BASE-TX	100 Mbps	100 meters	4-pair Category 5 or higher
1000BASE-T	1,000 Mbps	100 meters	4-pair Category 5 or higher
10GBASE-T	10 Gbps	55 meters or 100 meters	4-Pair Category 6: Maximum length 55 meters 4-Pair Category 6A or Category 7: length 100 meters
Optical Fiber			
1000BASE-SX	1 Gbps	220 m	62.5/125 micron multimode, 850 nm. 160 MHz-km modal bandwidth.
1000BASE-SX	1 Gbps	275 m	62.5/125 micron multimode, 850 nm. 200 MHz-km modal bandwidth.
1000BASE-SX	1 Gbps	500 m	50/125 micron multimode, 850 nm. 400 MHz-km modal bandwidth.
1000BASE-SX	1 Gbps	550 m	50/125 micron multimode, 850 nm. 500 MHz-km modal bandwidth.
1000BASE-LX	1 Gbps	550 m	62.5/125 micron multimode, 1,310 nm.
1000BASE-LX	1 Gbps	5 km	9/125 micron single mode, 1,310 nm.
10GBASE-SR/SW	10 Gbps	65 m	62.5/125 micron multimode, 850 nm.
10GBASE-LX4	10 Gbps	300 m	62.5/125 micron multimode, 1,310 nm, wave division multiplexing.
10GBASE-LR/LW	10 Gbps	10 km	9/125 micron single mode, 1,310 nm.
10GBASE-ER/EW	10 Gbps	40 km	9/125 micron single mode, 1,550 nm.
40 Gbps Ethernet 100 Gbps Ethernet	40 Gbps or 100 Gbps	Under Development	To be determined

Notes:

For 10GBASE-x, LAN versions (X and R) transmit at 10 Gbps. WAN versions (W) transmit at 9.95328 Gbps for carriage over SONET/SDH links (see Chapter 6 and Module C).

The 40 Gbps and 100 Gbps Ethernet standards are still under preliminary development.

Figure 4-2   Ethernet Physical Layer Standards

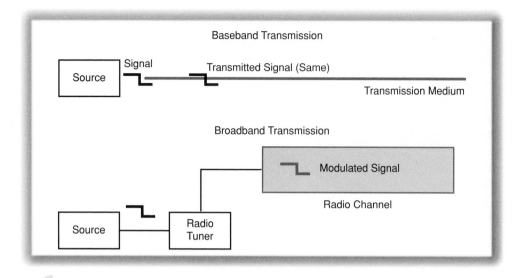

Figure 4-3   **Baseband versus Broadband Transmission**

Baseband Transmission

As Figure 4-2 shows, most Ethernet physical layer standards have *BASE* in their names. This is short for *baseband*. In **baseband** transmission, the transmitter simply injects signals into the transmission medium, as Figure 4-3 illustrates. For UTP, signals consist of voltage changes that propagate down the wires. For fiber, signals are light pulses.

Broadband Transmission

In contrast, in **broadband** transmission, the sender transmits signals in radio channels. However, radio-based broadband transmission is more expensive than baseband transmission. Although the 802.3 Working Group came up with early broadband LAN standards, these standards did not thrive because of their high cost.

Transmission Speed

The names of Ethernet physical layer standards also indicate transmission speeds— 10 Mbps, 100 Mbps, or more. These are the speeds of NICs and switch ports. For example, a 12-port gigabit Ethernet switch will be able to send and receive at 12 Gbps across all of its ports.

10 Mbps Physical Layer 802.3 Standards

The slowest Ethernet physical layer standards in use today operate at 10 Mbps. The 802.3 10BASE-T standard normally uses 4-pair UTP wiring. The 10BASE-T standard is no longer commonly used today.

100 Mbps Physical Layer 802.3 Standards

Next, the 802.3 Working Group produced 100 Mbps standards for UTP (T) and optical fiber (F). The **100BASE-TX** standard is the dominant Ethernet standard for connecting stations to switches today. **100BASE-FX**, in contrast, has almost entirely disappeared in favor of faster fiber-based Ethernet standards.[3]

Standard 100BASE-TX is the dominant Ethernet standard for connecting stations to switches today.

Gigabit Ethernet (1000BASE-x)

Advancing by another factor of 10, the 802.3 Working Group then produced **gigabit Ethernet (1000BASE-x)**. One gigabit per second is the dominant speed for connecting switches to switches, switches to routers, and routers to routers. However, companies are increasingly using gigabit Ethernet to connect servers and some desktops to the switches that serve them.

A UTP version, **1000BASE-T**, can be used to bring gigabit Ethernet to the desktop. It can also be used to connect switches to other switches and switches to routers. Optical fiber connections tend to be three times as expensive as UTP connections, so 1000BASE-T is attractive.

However, most firms use fiber versions of gigabit Ethernet. Although fiber is more expensive to install, today's fiber should be able to carry much faster future versions of Ethernet. There are two fiber versions of gigabit Ethernet. The **1000BASE-SX** (short wavelength) version transmits at 850 nm, using inexpensive laser signaling and multimode fiber. This standard has a maximum transmission distance of 220 meters for older 62.5 micron fiber with a modal bandwidth of 160 MHz-km, and 275 m for newer 62.5 micron fiber with a modal bandwidth of 200 MHz-km. In addition, some fiber vendors offer ultra-high modal bandwidth fiber that can extend 1000BASE-SX to nonstandard but reliable lengths of 300–500 meters. Overall, 1000BASE-SX is the dominant gigabit Ethernet standard and is the most widely used standard for connecting switches to other switches in organizations.[4]

Standard 1000BASE-SX is the most widely used standard for connecting switches to other switches in organizations.

10GBASE-x

The next step, 10 Gbps technology, uses optical fiber almost exclusively. Although the 802.3 Working Group is developing copper wiring standards,[5] the 10 Gbps Ethernet standards in widespread use today are fiber standards. Figure 4-2 shows that the 802.3 Working Group created quite a few 10 Gbps fiber standards.

Ethernet at 40 Gbps, 100 Gbps, or Both

The 802.3 Working Group is now working on the next generation of Ethernet speed standards. Traditional Ethernet experts want to continue the pattern of

---

[3]Why TX and FX instead of T and F? The 802.3 Working Group created other 100 Mbps Ethernet standards, but only the TX and FX standards saw market acceptance.
[4]The 1000BASE-LX (long wavelength) version uses more expensive lasers to transmit at 1,310 nm. It can send data twice as far as SX using multimode fiber. With single-mode fiber, it can send data 5 km. However, 1000BASE-SX is used far more widely than 1000BASE-LX.
[5]The 802.3 Working Group is developing the 10GBASE-CX4 standard for runs of less than 15 meters. The 10GBASE-CX4 standard uses eight wires, but these are not arranged as twisted pairs and there is metal shielding around the wiring. This will be useful only for switch-to-switch connections in wiring cabinets and server-to-switch connections in data centers. 10GBASE-CX4 uses adaptations of the Infiniband connectors and cabling created for storage area networks.

increasing speed by a factor of 10. This would give the next generation of Ethernet a speed of 100 Gbps.

However, wide area network people know that transmission will probably take place over the SONET/SDH physical layer standard for carrier fiber. For 10 Gbps Ethernet, this was not a problem because SONET/SDH offers a speed of 9.95328 Gbps, so it was comparatively easy to create versions that ran at a slightly different speed. However, the SONET/SDH increment beyond 10 Gbps is 40 Gbps instead of 100 Gbps. Consequently, Ethernet specialists who are focused on metropolitan area networking want the next version to run at Gbps.

Initially, the speed of 100 Gbps was selected as the next Ethernet speed. However, the 40 Gbps group won the concession that standards will be developed for both speeds. However, the development process is still in an early stage.

TEST YOUR UNDERSTANDING

3. a) At what layer is the 100BASE-TX standard? b) What can you infer from the name *100BASE-TX*? c) Distinguish between baseband and broadband transmission. d) Why does baseband transmission dominate for LANs? e) What is the most widely used 802.3 physical layer standard for connecting stations to switches today? f) What is the most widely used 802.3 physical layer fiber standard for connecting switches to other switches today? g) What are the two likely speeds for the next iteration of Ethernet speed standards?

Link Aggregation (Trunking)

Ethernet transmission capacity usually increases by a factor of 10. What should you do if you only need somewhat more speed than a certain standard specifies? For instance, suppose that you have gigabit Ethernet switches and need a switch-to-switch link of 1.5 Gbps instead of 10 Gbps.

Figure 4-4 illustrates that sometimes two or more trunk lines connect a single pair of switches. This requires the switches to implement the **802.3ad** standard. The IEEE calls this **link aggregation**. Networking professionals also call this **trunking** or **bonding**.

Link aggregation allows you to increase trunk speed incrementally by a factor of two or three, instead of by a factor of ten. This incremental growth uses existing ports and usually is inexpensive compared with upgrading a switch to the next higher Ethernet speed. In contrast, moving to the next higher Ethernet speed usually involves purchasing new switches or at least new port modules for existing switches.

However, after two or three aggregated links, the company should compare the cost of link aggregation with the cost of a 10-fold increase in capacity by moving up to the next Ethernet speed. Going to a single faster trunk line will also give more room for growth.

The incremental growth that link aggregation brings usually is inexpensive compared with upgrading a switch to the next higher Ethernet speed. However, after two or three aggregated links, the company should compare the cost of link aggregation with the cost of a 10-fold increase in capacity by moving up to the next Ethernet speed.

TEST YOUR UNDERSTANDING

4. a) What is link aggregation (trunking or bonding)? b) Why may link aggregation be more desirable than installing a single faster link? c) Why may link aggregation not be desirable if you will need several aggregated links to meet capacity requirements?

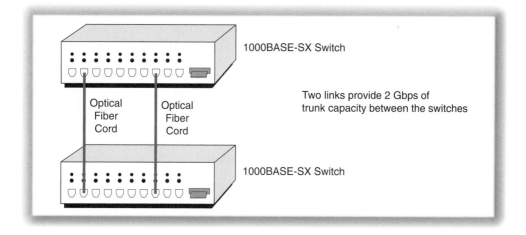

1000BASE-SX Switch

Two links provide 2 Gbps of
trunk capacity between the switches

Optical   Optical
Fiber   Fiber
Cord   Cord

1000BASE-SX Switch

Figure 4-4   Link Aggregation (Trunking or Bonding)

Ethernet Physical Layer Standards and Network Design

Using Figure 4-2

Note that if you know the speed you need (100 Mbps, 1 Gbps, and so forth), and if you know what distance you need to span, the information in Figure 4-2 will show you what type of transmission link you can use. If link aggregation is also available with your switches, you have even more choices.

For instance, if you need a speed of 1 Gbps and if your two switches are 130 meters apart, you would select 1000BASE-SX multimode fiber to minimize cost. Although 1000BASE-LX would also do the job, 1000BASE-SX is more than sufficient. Generally speaking, picking the shortest-distance standard that will do the job will minimize cost.

Alternatively, if you are designing a network from scratch, say, for a new facility, the options presented in Figure 4-2 will allow you to consider alternative placements for your switches. If you can place your switches farther apart on average, for instance, you can reduce the total number of switches, and this can save money—although each of the switches in a more dispersed network will cost more to purchase because each will have to handle more traffic.

As noted earlier, however, Figure 4-2 lists standards that have been ratified. Some of these standards have been ignored by vendors, and some vendors have products that use superior fiber and so can cover long distances on the basis of the same standard.

Switches Regenerate Signals to Extend Distance

The 100-meter limit for UTP and the longer distance limits for fiber shown in Figure 4-2 apply only to connections *between a single pair of devices*—for example, between a station and a switch, between two switches, or a between switch and a router.

The 100-meter limit for UTP and the longer distance limits for fiber shown in Figure 4-2 apply only to connections *between a pair of devices*, not to end-to-end connections between stations across multiple switches.

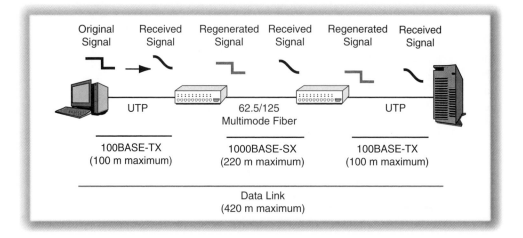

Figure 4-5   Data Link Using Multiple Switches

What should you do if a longer distance separates the source host and the destination host? Figure 4-5 shows a data link with two intermediate switches. In addition to the two 100-meter maximum length UTP access links, there is a 220-meter maximum length 1000BASE-SX optical fiber link (using 62.5/125-micron 160-MHz-km modal bandwidth fiber) between the two switches. This setup can span a maximum of 420 meters.

Each switch along the way **regenerates** the signal. If the signal sent by the source host begins as a 1, it is likely to be distorted before it reaches the first switch. The first switch recognizes it as a 1 and generates a clean new 1 signal to send to the second switch. The second switch regenerates the 1 as well.

The key point is that Figure 4-2 shows maximum distances between pairs of devices, not end-to-end transmission distances. To deliver frames over long distances, intermediate switches regenerate the signal. There is no maximum end-to-end distance between pairs of stations in an Ethernet network. Although cumulative delay might be a problem with a dozen or more intermediate switches, this rarely is a problem in real networks.

There is no maximum end-to-end distance between pairs of stations in an Ethernet network.

TEST YOUR UNDERSTANDING

5. a) How could you use the information in Figure 4-2 in network design? b) If more than one type of Ethernet standard shown in Figure 4-2 can span the distance you need, what would determine which one you choose? c) In Figure 4-2, is the distance the maximum for a single physical link or for the end-to-end distance between two stations across multiple switches? d) At what layer or layers is the 802.3 100BASE-TX standard defined—physical, data link, or internet? e) How does regeneration allow a firm to create LANs that span very long distances? f) If you need to span 300 meters by using 1000BASE-SX, what options do you have? (Include the possibility of using an intermediate switch.) g) How would you decide which option to choose?

THE ETHERNET FRAME

Layering

The Logical Link Control Layer

When the 802 Committee assumed control over Ethernet standardization, it realized that it would have to standardize non-Ethernet LAN technology as well. Consequently, the 802 Committee divided the data link layer into two layers, as Figure 4-6 illustrates.

➤ The lower part of the standard—the media access control layer—is specific to the particular LAN technology. For example, there are separate media access control layer standards for Ethernet and 802.11 wireless networks.

➤ The upper layer—the **logical link control (LLC)** layer—adds some functionality on top of technology-specific functionality. Unfortunately, time has proven the added functionality of the LLC layer to be of little value, so it is now largely ignored. As we will see a little later, the sender adds an LLC subheader to each 802.3 frame. There is only a single LLC layer standard, **802.2.**

The 802.3 MAC Layer Standard

As just noted, the lower part of the data link layer is the MAC layer. *MAC* stands for **media access control**. The MAC layer defines functionality specific to a particular LAN technology.

Note in Figure 4-6 that while Ethernet (802.3) has many physical layer standards, it only has a single media access control layer standard, the **802.3 MAC Layer Standard**. This standard defines Ethernet frame organization and NIC and switch operation.

Figure 4-6   Layering in 802 Networks

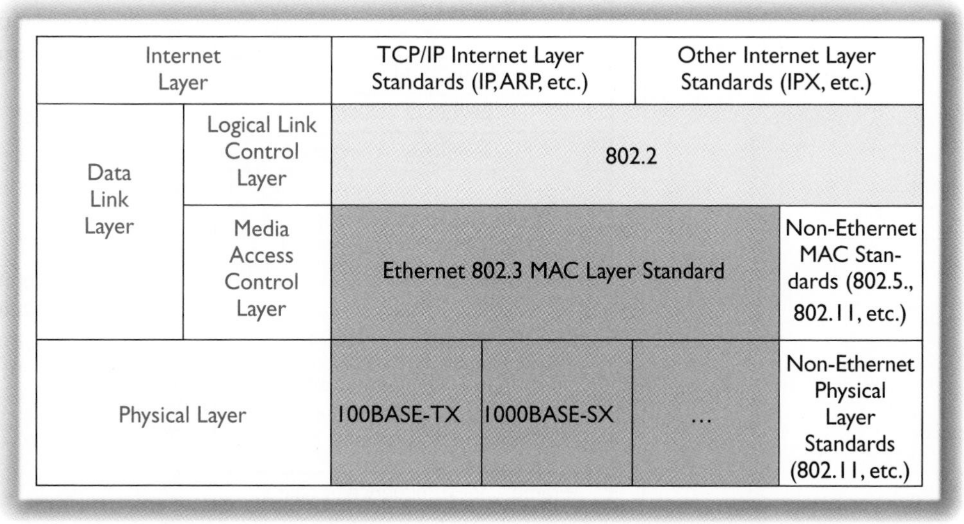

Internet Layer		TCP/IP Internet Layer Standards (IP, ARP, etc.)		Other Internet Layer Standards (IPX, etc.)	
Data Link Layer	Logical Link Control Layer	802.2			
	Media Access Control Layer	Ethernet 802.3 MAC Layer Standard			Non-Ethernet MAC Standards (802.5., 802.11, etc.)
Physical Layer		100BASE-TX	1000BASE-SX	...	Non-Ethernet Physical Layer Standards (802.11, etc.)

TEST YOUR UNDERSTANDING

6. a) Distinguish between the MAC and LLC layers. b) Does Ethernet have multiple physical layer standards? c) Does Ethernet have multiple MAC layer standards? d) What is the name of Ethernet's single MAC standard?

The Ethernet Frame's Organization

Figure 4-7 shows the Ethernet MAC layer frame, which we saw briefly in Chapter 2. We will now look at the Ethernet frame in more depth.

Preamble and Start of Frame Delimiter Fields

Before a play in American football, the quarterback calls out something like "Hut one, hut two, hut three, hike!" This cadence synchronizes all of the offensive players.

In the Ethernet MAC frame, the **preamble** field (7 octets) and the **start of frame delimiter** field (1 octet) synchronize the receiver's clock to the sender's clock. These fields have a strong rhythm of alternating 1s and 0s. The last bit in this sequence is a 1 instead of the expected 0, to signal that the synchronization is finished.

Source and Destination Address Fields

Hex Notation   We saw in Chapter 2 that the source and destination Ethernet address fields are 48 bits long and that while computers work with this raw 48-bit form, humans normally express these addresses in Base 16 **hexadecimal (hex) notation**.

Figure 4-7   The Ethernet MAC Layer Frame

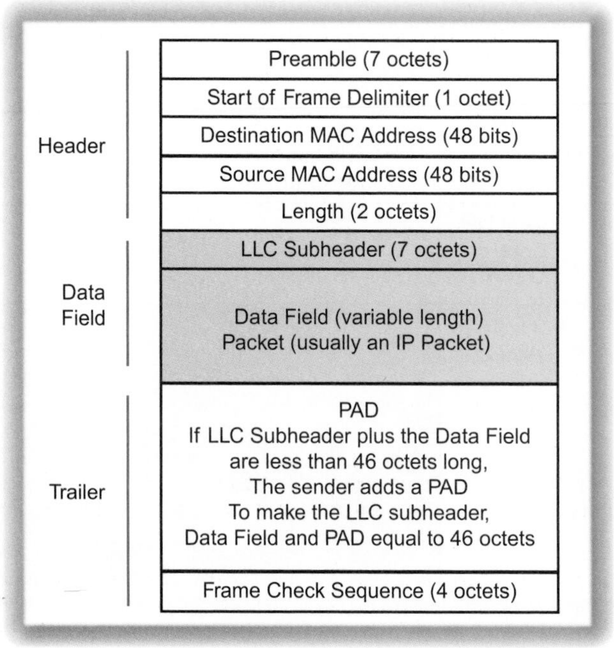

4 Bits*	Decimal (Base 10)	Hexadecimal (Base 16)	4 Bits*	Decimal (Base 10)	Hexadecimal (Base 16)
0000	0	0 hex	1000	8	8 hex
0001	1	1 hex	1001	9	9 hex
0010	2	2 hex	1010	10	A hex
0011	3	3 hex	1011	11	B hex
0100	4	4 hex	1100	12	C hex
0101	5	5 hex	1101	13	D hex
0110	6	6 hex	1110	14	E hex
0111	7	7 hex	1111	15	F hex

*Note: $2^4 = 16$ combinations
For example, A1-34-CD-7B-DF-47 hex begins with 10100001 for A1.

Figure 4-8   Hexadecimal Notation

➤ First, divide the 48 bits into twelve 4-bit units, which computer scientists call nibbles.

➤ Second, convert each nibble into a hexadecimal symbol, using the method shown in Figure 4-8.

➤ Third, write the symbols as six pairs with a dash between each pair—for instance, B2-CC-67-0D-5E-BA. (Each pair represents one octet.)

MAC Layer Addresses   Ethernet addresses exist at the MAC layer, so Ethernet addresses are **MAC addresses**. They are also called **physical addresses** because physical devices (NICs) implement Ethernet at the physical, MAC, and LLC layers.

Length Field

The **length** field contains a binary number that gives the length of the data field (not of the entire frame) in octets. The maximum length of the data field is 1500 octets. There is no minimum length for the data field.

The Data Field   The data field contains two subfields: the LLC subheader and the packet that the frame is delivering.[6]

LLC Subheader   The **logical link control layer (LLC) subheader** is eight octets long.[7] The purpose of the LLC subheader is to describe the type of packet contained in the data field. For instance, if the LLC subheader ends with the code 08-00 hex (Base 16), then the data field contains an IP packet.[8]

---

[6]Why does the data field have two parts? The answer is that the data field of the MAC layer frame actually is the LLC layer frame, which has a header (the LLC subheader) and a data field consisting of the packet being carried in the LLC frame. However, to avoid damaging neurons, it is best simply to think of the MAC layer data field as having two parts.

[7]There are some exceptions, but they are rare.

[8]The LLC subheader has several fields. In the SNAP version of LLC, which is almost always used, the first three octets are always AA-AA-03 hex. The next three octets are almost always 00-00-00 hex. The final two octets constitute the Ethertype field, which specifies the kind of packet in the data field. Common hexadecimal Ethertype values are 0800 (IP), 8137 (IPX), 809B (AppleTalk), 80D5 (SNA services), and 86DD (IP version 6).

The Packet The **data** field also contains the packet that the MAC layer frame is delivering. The packet usually is far longer than all other fields combined. The packet encapsulated in the data field usually is an IP packet. However, it could also be a packet from another standards architecture, for instance, an IPX packet. As long as the source and destination stations understand the packet format, there is no problem.

PAD field

The **PAD** field is unusual because it does not always exist. Although there is no minimum length for 802.3 MAC layer frame data fields, if the data field is less than 46 octets long, then the sender must add a PAD field so that the total length of the data field and the PAD field is exactly 46 octets long. For instance, if the data field is 26 octets long, the sender will add a 20-octet PAD field. If the data field is 46 octets long or longer, the sender will not add a PAD field.

There is no minimum length for the data field, but if the data field is less than 46 octets long, then a PAD field must be added to bring the total length of the data and pad fields to 46 octets.

Frame Check Sequence Field

As noted in Chapter 2, the last field in the Ethernet frame is the frame check sequence field, which permits error detection. This is a four-octet field. If the receiver detects an error in a transmitted frame, the receiver simply discards the frame. There is no retransmission of damaged frames.

TEST YOUR UNDERSTANDING

7. a) What is the purpose of the preamble and start of frame delimiter fields? b) Why are Ethernet addresses called MAC addresses or physical addresses? c) What are the steps in converting 48-bit MAC addresses into hex notation? d) The length field gives the length of what? e) What are the two components of the Ethernet data field? f) What is the purpose of the LLC subheader? g) What type of packet is usually carried in the data field? h) What is the maximum length of the data field? i) Who adds the PAD field— the sender or the receiver? j) Is there a minimum length for the data field? k) If the data field is 40 octets long, how long a PAD field must the sender add? l) If the data field is 400 octets long, how long a PAD field must the sender add? m) What is the purpose of the frame check sequence field? n) What happens if the receiver detects an error in a frame? o) Convert 11000010 to hex.

BASIC DATA LINK LAYER SWITCH OPERATION

In this section, we will discuss the basic data link layer operation of Ethernet switches. This is also governed by the 802.3 MAC layer standard. In the section after this one, we will discuss other aspects of Ethernet switching that a firm may or may not use.

Frame Forwarding with Multiple Ethernet Switches

Figure 4-9 shows an Ethernet LAN with three switches. Larger Ethernet LANs have dozens of switches, but the operation of individual switches is the same whether there are only a few switches or many. Each individual switch between the source and destination host reads the address of an incoming frame, looks up an output port in its switching table, and sends the frame out through that port.

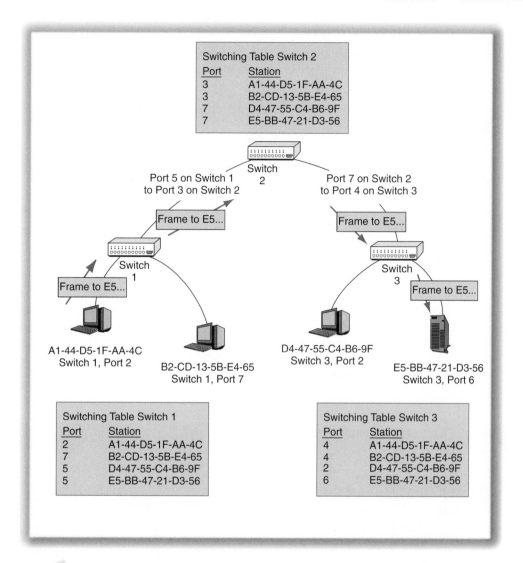

Figure 4-9   Multiswitch Ethernet LAN

In a multiswitch LAN, the switch may be forwarding the frame to another switch on the data link, instead of directly to the destination station. However, switches do not know whether they are forwarding the frame to the destination station or to another switch. They merely send the frame out through the indicated port.

For example, suppose that Station A1-44-D5-1F-AA-4C on Switch 1 transmits a frame destined for Station E5-BB-47-21-D3-56, which attaches to Switch 3. The frame will have to pass through Switch 2 along the way.

Switch 1 will look at its switching table and note that Station E5-BB-47-21-D3-56 is out Port 5. The switch will send the frame out through that port. The link out Port 5 will carry the frame to Switch 2 instead of directly to the destination station.

TEST YOUR UNDERSTANDING

8. a) In a single-switch LAN, the switch reads the address of an incoming frame, looks up an output port in the switching table, and sends the frame out through that port. Do individual switches work differently in multiswitch LANs? Explain. b) What happens on Switch 2? c) What happens on Switch 3?

Hierarchical Switch Topology

Hierarchical Switch Organization

Note that the switches in Figure 4-10 form a **hierarchy**, in which each switch has only one parent switch above it. In fact, the Ethernet standard *requires* a **hierarchical topology** for its switches. Otherwise, loops would exist, causing frames to circulate endlessly from one switch to another around the loop or causing other problems. Figure 4-10 shows a larger switched Ethernet LAN organized in a hierarchy.

Ethernet *requires* a hierarchical switch hierarchy.

Single Possible Path between End Stations

In a hierarchy, there is only a single possible path between any two end stations. (To see this, select any two stations at the bottom of the hierarchy and trace a path between them. You will see that only one path is possible.)

In a hierarchy, there is only a single possible path between any two end stations.

Figure 4-10   Hierarchical Ethernet LAN

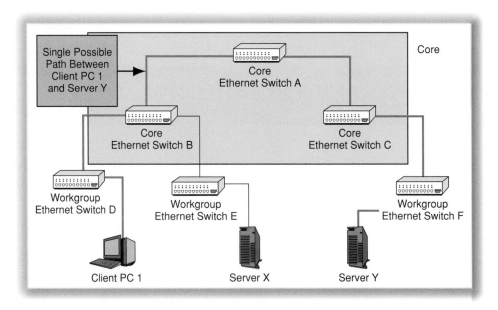

Workgroup versus Core Switches

In a hierarchy of Ethernet switches, there are workgroup switches and core switches. Figure 4-10 illustrates these two types of switches. We saw these two types of switches in Chapter 1, but we will recap the main points:

➤ **Workgroup switches.** Switches that connect computers to the network via access lines are called *workgroup switches* (switches D, E, and F in Figure 4-10).

➤ **Core switches.** Switches farther up the hierarchy (switches A, B, and C in Figure 4-10) that carry traffic via trunk lines between pairs of switches, switches and routers, and pairs of routers are called *core switches*. The collection of all core switches plus the trunk lines that connect them is the network's **core**.

Workgroup switches handle only the traffic of the stations they serve. However, core switches must be able to carry the conversations of dozens, hundreds, or thousands of stations. Consequently, core switches need to have much higher capacity than workgroup switches. Their cost also is much higher.

The dominant port speed for workgroup switches today is 100 Mbps. In contrast, the dominant port speed for core switches today is 1 Gbps, and some core switches already use port speeds of 10 Gbps.

TEST YOUR UNDERSTANDING

9. a) How are switches in an Ethernet LAN organized? b) Because of this organization, how many possible paths can there be between any two stations? c) In Figure 4-10, what is the single possible path between Client PC 1 and Server Y? d) Between Client PC 1 and Server X?

10. a) Distinguish between workgroup switches and core switches in terms of which devices they connect. b) How do they compare in terms of switching capacity? Explain. c) How do they compare in terms of port speeds? Explain.

Only One Possible Path: Low Switching Cost

We have just seen that a hierarchy allows only one possible path between any two hosts. If there is only a single possible path between any two stations, it follows that in every switch along the path, the destination address in a frame will appear only once in the switching table—for the specific outgoing port needed to send the frame on its way.

This allows a simple table lookup operation that is very fast and therefore costs little, per frame handled. This is what makes Ethernet switches inexpensive. As noted in the introduction, simple switching operation and therefore low cost has led to Ethernet's dominance in LAN technology.

> The fact that there is only a single possible path between any two end stations in an Ethernet hierarchy makes Ethernet switch forwarding simple and therefore inexpensive. This low cost has led to Ethernet's dominance in LAN technology.

In Chapter 8, we will see that routers have to do much more work when a packet arrives, because there are multiple alternative routes between any two hosts. Each of these alternative routes appears as a row in the routing table. Therefore, a router must first identify all possible routes (rows) and then select the best one—instead of simply finding a single match. This additional work per forwarding decision makes routers very expensive for the traffic load they handle.

TEST YOUR UNDERSTANDING

11. a) What is the benefit of having a single possible path? Explain in detail. b) Why has Ethernet become the dominant LAN technology?

ADVANCED ETHERNET SWITCH OPERATION

Now that we have discussed basic Ethernet switch operation involved in frame forwarding, we will begin looking at additional aspects of Ethernet switch operation that are important in larger Ethernet networks.

802.1w: The Rapid Spanning Tree Protocol (RSTP)

Single Points of Failure

Having only a single possible path between any two stations allows rapid frame forwarding and, therefore, low switch cost. Unfortunately, having only a single possible path between any two computers also makes Ethernet vulnerable to **single points of failure**, in which the failure of a single component (a switch or a trunk line between switches) can cause widespread disruption.

Having only a single possible path between end stations in a switched Ethernet network reduces cost, but it creates single points of failure, meaning that a single failure can cause widespread disruption.

To understand this, suppose that Switch 2 in Figure 4-11 fails. Then the stations connected to Switch 1 will not be able to communicate with stations connected to Switch 2 or Switch 3. For a second example, suppose that the link between Switch 1 and Switch 2 fails. In this case, too, the network also will be broken into two parts.

Although the two parts of the network might continue to function independently after a failure, many firms put most or all of their servers in a centralized server room. In such firms, clients on the other side of the broken network would lose most of their ability to continue working. For example, in the figure, Client A1-44-D5-1F-AA-4C, which connects to Switch 1, cannot reach Server E5-BB-47-21-D3-56, which connects to Switch 3. External connections also tend to be confined to a single network point for security reasons. Computers on the wrong side of the divide after a breakdown would lose external access.

802.1w (the Rapid Spanning Tree Protocol)

Fortunately, the 802.1 Working Group (not to be confused with the 802.3 Working Group) created a way to provide **backup links**. As Figure 4-11 shows, a company can install a backup link—in this case, between Switch 1 and Switch 3. The backup link provides redundancy so that frames can take alternative paths if there is a failure.

The backup line will create a loop, which is forbidden in Ethernet. Fortunately, under the **802.1w** standard, which is called the **Rapid Spanning Tree Protocol (STP)**,[9]

---

[9]There was an earlier standard, the Spanning Tree Protocol (802.1D), which is now deprecated because of its slow operation.

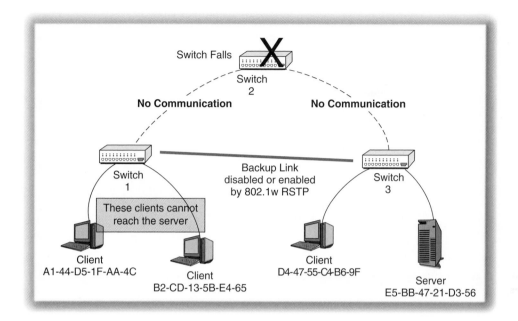

Figure 4-11   Single Point of Failure and 802.1D

switches can detect a loop and disable it by turning off switch ports selectively to disable some links. If the switches are configured properly, the switches will specifically disable the backup link between Switches 1 and 3. Later, if Switch 2 fails, the RSTP protocol will reestablish the backup link between Switches 1 and 3. This will repair most parts of the network temporarily, until the failure point can be restored.

Unfortunately, the use of RSTP is anything but automatic. Although RSTP will always break loops, getting it to restore particular backup links when specific problems occur requires a great deal of planning and extensive switch configuration.

TEST YOUR UNDERSTANDING

12. a) Why is having a single possible path between any two stations in an Ethernet network dangerous? b) What is a single point of failure? c) What standard allows redundancy in Ethernet networks? d) Is it easy or difficult to create backup links effectively in 802.1w RSTP?

Virtual LANs and Ethernet Switches

VLANs

In a normal Ethernet network, any client can send frames to any server, and any server can reach any client. However, many Ethernet switches can now create virtual **LANs (VLANs)**. As Figure 4-12 shows, VLANs are groups of clients and servers that can

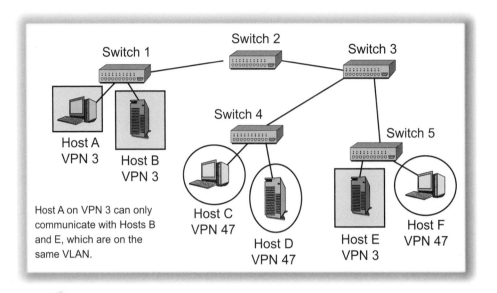

Figure 4-12   Virtual LAN (VLAN) with Ethernet Switches

communicate with each other, but not with clients and servers on other VLANs. In the figure, clients and servers on VLAN1 (indicated by ellipses) cannot communicate with clients and servers on VLAN2 (indicated by rectangles).

Congestion Reduction

VLANs are used for two main reasons. For instance, certain servers tend to **broadcast** frames to all clients. (One reason for the server to do this is to advertise its availability to its clients every 30 seconds or so.) In a large network, this can create a great deal of congestion. With VLANs, however, the server will not flood the entire network with traffic; the frames will go only to the clients on the server's VLAN.

Security

A second reason for using VLANs is security. If clients on one VLAN cannot reach servers on other VLANs, they cannot attack these servers. In addition, if a client becomes infected with a virus, it can only pass the virus on to other clients and servers on its own VLAN.

The 802.1Q VLAN Standards

Until recently, there was no standard for VLANs, so if you used VLANs, you had to buy all of your Ethernet switches from the same vendor. However, as Figure 4-13 shows, the **802.1Q** standard extends the Ethernet MAC layer frame to include two optional **tag fields**.

The first tag field **(Tag Protocol ID)** has the two-octet hexadecimal value 81-00. This field simply indicates that this is a tagged frame.

The second tag field (the **Tag Control Information** field) contains a 12-bit VLAN ID that the sender sets to 0 if the firm does not use VLANs. If the firm does use VLANs, the firm will give each VLAN a different VLAN ID. When a station on a VLAN transmits,

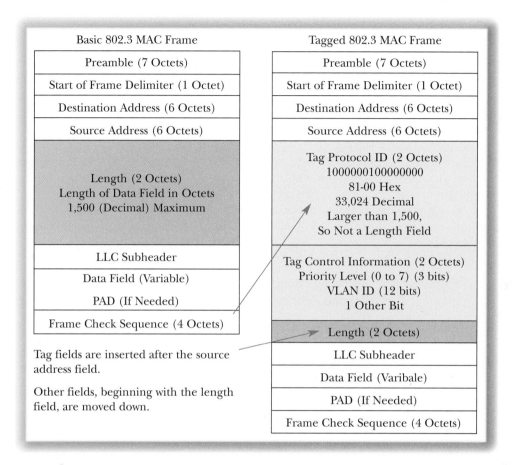

Figure 4-13   Tagged Ethernet Frame (Governed by 802.1Q)

the station adds the VLAN ID of its own VLAN to the Tag Control Information field. The switches will read the VLAN ID to determine how to forward the frame. (The destination Ethernet address is set to forty-eight 1s in broadcasts, so a switch can use only the VLAN ID to determine how to forward the frame.) With 12-bit VLAN IDs, there can be 4095 ($2^{12} - 1$) different VLANs on an Ethernet network.

802.1Q is the standard for frame tagging.

With VLANs, switches do not use their switching tables that contain MAC address–port pairs. Rather, they use a VLAN switching table that associates VLAN ID numbers with one or more ports. Switches from different vendors can all build their VLAN switching tables using standardized VLAN ID numbers. This will allow them to interoperate.

TEST YOUR UNDERSTANDING

13. a) What is a VLAN? b) What two benefits do VLANs address bring? c) How do VLANs bring security? d) What two fields does the 802.1Q standard add to Ethernet frames? e) What does the Tag Protocol ID tell a receiving switch or NIC? f) What information does the tag control information field tell the switch or receiver?

Handling Momentary Traffic Peaks

Momentary Traffic Peaks

If traffic volume is comfortably below a network's traffic capacity, traffic should get through promptly. However, sometimes there are **momentary traffic peaks** that briefly exceed the network's capacity, as Figure 4-14 illustrates.

During momentary traffic peaks, the switch will not be able to handle all of the frames that arrive. Some frames will be delayed. In networking, delay is called **latency**.

Figure 4-14 Handling Momentary Traffic Peaks with Overprovisioning and Priority

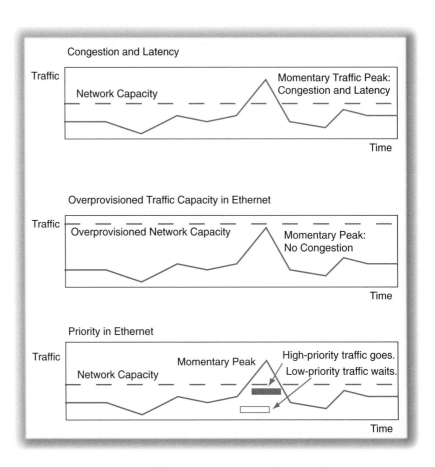

During momentary traffic peaks, delayed frames will be held in a memory area called the switch's *buffer*. If traffic peaks last beyond a certain time, the buffer will become full. Subsequent frames that cannot be delivered will be dropped entirely.

Momentary traffic peaks create congestion that leads to latency and may even cause frame loss.

Although these peaks normally last only a fraction of a second to a few seconds, they can be highly disruptive for some applications, especially voice and video. These are called **latency-intolerant** applications. In contrast, the users of **latency-tolerant** applications, such as e-mail, will not even notice brief delays.

Overprovisioning Ethernet

Most organizations have discovered that the least expensive way to get around peak traffic congestion today is simply to **overprovision** the Ethernet LAN—that is, to install much more capacity in switches and trunk lines than will be needed most of the time. If 100BASE-TX would be sufficient most of the time, for example, they install a 1000BASE-T line. When there are brief traffic bursts, these bursts will very rarely exceed capacity. Although this method wastes capacity most of the time, it works without adding to the cost of switch management.

Priority in Ethernet

Another way to address momentary traffic peaks is to give **priority** to certain traffic, meaning that high-priority traffic will go first. Commercial air travel is a good analogy. Passengers with children, disabled passengers, and first-class passengers are given priority in boarding flights.

Figure 4-13 showed that 802.1Q Ethernet frame tagging can give priority to individual frames. The Tag Control Information field contains not only a 12-bit VLAN ID, but also a 3-bit **priority level** to give a frame one of eight priority levels from 000 (low) to 111 (high). The definition of these eight priority levels is in the **802.1p** standard.[10]

In switching, priority is based on the tolerance of traffic for latency. When brief traffic peaks occur, latency-intolerant traffic, such as voice and video, will be given high priority and so will be switched first. Of course, latency-tolerant applications will be delayed, but users probably will not even notice brief delays during momentary traffic peaks. Priority is also used to guarantee that network control messages get through, which may be crucial during periods of high congestion.

Most switches today support priority. However, priority can be difficult to manage. Its management costs have made priority substantially more expensive than **overprovisioning** Ethernet. In addition, prioritizing traffic can lead to pitched political battles within a firm over whose traffic should have the highest priority.

---

[10]In addition to the 12 VLAN ID bits and the 3 priority bits, the 16-bit Tag Control Information field has a 1-bit canonical format bit. This bit is set to 1 for all networks except 802.5 Token-Ring Networks and FDDI, for which it is set to 0. Both Token-Ring Networks and FDDI are now extremely rare. In canonical format, the rightmost bit in each byte is sent first. In 802.5 and FDDI, the leftmost bit in each byte is sent first. The canonical format bit is rarely used.

Momentary versus Chronic Lack of Capacity

Note that we have been discussing momentary traffic peaks in a network that has sufficient capacity nearly all of the time. This is very different from **chronic lack of capacity**, in which the network lacks adequate capacity much of the time. In such cases, the only good solution is to upgrade capacity. Otherwise, either congestion always will be high or some applications will receive almost no capacity.

TEST YOUR UNDERSTANDING

14. a) What are momentary traffic peaks? b) What problems do they create? c) In what two ways can Ethernet address momentary traffic peaks? d) What is the advantage of each? e) Distinguish between momentary traffic peaks and chronic lack of capacity. f) What can firms do if there is a chronic lack of capacity?

Hubs and CSMA/CD

Today, Switches Dominate in Ethernet
   Earlier Ethernet networks used hubs
   When a bit came in one port, the hub broadcast the bit out all other ports

Media Access Control (MAC)
   With hubs, if two stations transmitted at the same time, their signals would collide
   Media access control (MAC) must be used to prevent that.
   Controls when a station may transmit

CSMA/CD
   The Ethernet hub MAC protocol
   CSMA (carrier sense multiple access)
      If a station wants to transmit, it may do so if no station is already transmitting
      But it must wait a random amount of time if another station is already sending.
      After that random amount of time, the station begins CSMA again
   CD (collision detection)
      If there is a collision because two stations send at the same time, all stations
      stop transmitting, wait a random period of time, and then apply CSMA again

Latency
   When one station transmits, others must wait
   This creates latency
   Latency became bad in large Ethernet hub networks
   Switches solved this problem by avoiding the need to wait
   Multiple conversations can take place simultaneously

Figure 4-15   Hub versus Switch Operation (Study Figure)

Note to students: Neither hubs nor CSMA/CD are important in LANs today. Almost all organizations have replaced their Ethernet hubs with Ethernet switches. So why learn about hubs or the CSMA/CD mechanism they use? The answer is perverse, yet important to you personally. For some strange reason, when recruiters hear that you have taken a networking course, they ask you to explain CSMA/CD. They even ask about it if you are not applying for a networking job. Questions about CSMA/CD and the OSI layers are still the most widely asked questions regarding networking in job interviews. Honest.

BROADCASTING

Before switches became economical to use, Ethernet LANs generally used simpler devices called hubs. As Figure 4-16 shows, switches send an incoming frame out through a single port. In contrast, **hubs** broadcast each arriving bit out through all ports except the port that received the signal.

MEDIA ACCESS CONTROL (MAC)

With switches, several stations can transmit at the same time. With a hub, however, if two stations transmit at the same time, their signals will have a **collision**. Their signals will add together, and the combined signal will be unreadable. Consequently, NICs connected to hubs must not transmit at the same time. Controlling when stations transmit is **media access control (MAC)**.

CSMA/CD

The media access control method used with Ethernet hubs is **carrier sense multiple access with collision detection (CSMA/CD)**. Although this sounds complex, it is very simple.

> ➤ Under carrier sense multiple access (CSMA), if a station wants to transmit, it may do so if no station is already transmitting; but must wait a random amount

Figure 4-16   Hub versus Switch Operation

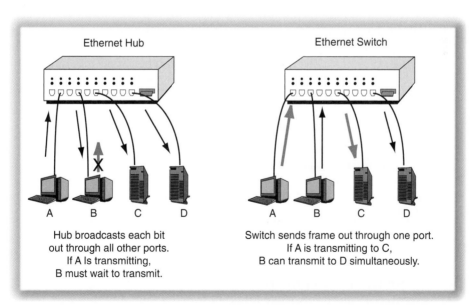

Ethernet Hub

A   B   C   D

Hub broadcasts each bit
out through all other ports.
If A Is transmitting,
B must wait to transmit.

Ethernet Switch

A   B   C   D

Switch sends frame out through one port.
If A is transmitting to C,
B can transmit to D simultaneously.

(continued)

of time if another station is already sending. After that random amount of time, the station begins CSMA again.

➤ Under collision detection (CD), if there is a collision because two stations send at the same time, all stations stop transmitting, wait a random period of time, and then apply CSMA again.

LATENCY

The problem with hubs is that they do not scale. If even one station is transmitting, the other NICs must wait. If there is a network with many hubs, only one station on the entire hub network can transmit at any moment. If there are many stations on the LAN, there will be a great deal of delay (latency) when a NIC wishes to send. The need to eliminate this latency was the main reason why firms changed from hubs to switches.

TEST YOUR UNDERSTANDING

15. a) What Ethernet technology broadcasts each arriving bit out through all other ports? b) What is media access control, and why must NICs that work with hubs use it? c) When NICs work with hubs, what media access control method do they use? d) How does CSMA/CD work? e) Why are hubs undesirable?

PURCHASING SWITCHES

We will end this chapter with a discussion of the issues you will have to deal with when purchasing Ethernet switches. Purchasing an Ethernet switch is a complex task.

Number and Speeds of Ports

The most basic issue is how many ports you will need to have and what their individual speeds need to be. For instance, you might need a workgroup switch with 12 or 24 100BASE-TX ports. To give another example, you might need a core switch with four gigabit Ethernet SC optical fiber ports and two 10GBASE-SX SC optical fiber ports. Fortunately, you can buy switches with a fixed number of ports in almost any port and speed configuration you wish. In addition, many switches are modular, meaning that you can buy the basic chassis and insert cards with the ports you require.

TEST YOUR UNDERSTANDING

16. a) What is the first issue to consider in switch purchases? b) Do you have many choices in port numbers and speeds?

Store-and-Forward versus Cut-Through Switching

One purchasing consideration that was once very important is the ability of a switch to use either store-and-forward or cut-through switching. We saw earlier that an Ethernet frame contains multiple fields and that the data field alone can be as large as 1,500 octets long.

Store-and-Forward Ethernet Switches
As Figure 4-18 illustrates, some Ethernet switches wait until they have received the entire frame before sending it out. This is **store-and-forward** switching. This technology

Number and Speeds of Ports
>   Buyers must decide on the number of ports needed and the speed of each
>   Buyers often can buy a prebuilt switch with this configuration

Store-and-Forward versus Cut-Through Switching (Figure 4-18)
>   Store-and-forward Ethernet switches read whole frame before passing the frame on
>   Cut-through Ethernet switches read only some fields before passing the frame on
>   Cut-through switches have less latency, but this is rarely important

Manageability
>   SNMP Manager controls many managed switches (see Figure 4-19)
>   Polling enables managers to collect data and diagnose problems
>   Switches can be fixed remotely by changing their configurations
>   Manager provides the network administrator with summary performance data
>   Managed switches are substantially more expensive than unmanaged switches
>   However, in large networks, the savings in labor costs and rapid response are
>   worth it

Figure 4-17   Switch Purchasing Considerations (Study Figure)

Figure 4-18   Store-and-Forward versus Cut-Through Switching

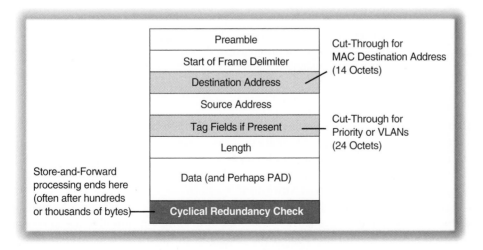

allows switches to check each frame for errors and to discard incorrect frames to reduce traffic. Frames often are hundreds or thousands of octets long, so store-and-forward switching adds a slight delay to frame transmission at each switch.

Cut-Through Ethernet Switches

In contrast, **cut-through** Ethernet switches examine only a few bits in a frame before sending the bits of the frame back out. This allows switches to begin sending out the bits they first received, despite the fact that they have not yet received all the bits of the frame.

> ➤ Obviously, as shown in Figure 4-18, switches must at least read the destination address in order to know which port to use to send the frame back out. This requires reading the preamble, start of frame delimiter, and destination address—a total of only 14 octets.

> ➤ Handling VLANs and priority also requires the reading of tag fields if they are used.

By examining only a few dozen octets at most, then, cut-through switching can reduce latency at each switch, compared with store-and-forward switching, which typically has to examine hundreds or thousands of octets.

Perspective

Although vendors once touted cut-through operation as a major advantage, the greater amount of latency added by store-and-forward switching tends to be negligible today. In addition, most switches today can do both cut-through and store-and-forward operations. Many offer cut-through operation as the default, but sample full frames occasionally and change to store-and-forward operation if the error rate becomes too high. Cut-through versus store-and-forward operation is no longer a significant issue in switch purchasing. However, you need to know these terms to read the vendor literature, and network managers tend to ask about these concepts in job interviews.

TEST YOUR UNDERSTANDING

> 18. a) Which is likely to have less latency—a cut-through switch or a store-and-forward switch? Explain. b) What is the advantage of the other mode of operation? c) Is determining which mode of operation a switch uses a major purchasing issue today?

Manageability

If there is an Ethernet switch problem, discovering which switch is malfunctioning can be very difficult. Fixing the problem, furthermore, may require traveling to the switch to change its configuration. This can be very expensive.

Managed Switches and the Manager

As Figure 4-19 shows, the bank mitigates these problems by using only **managed switches**. As the name suggests, these switches have sufficient intelligence to be managed from a central computer called the **manager**. In most cases, management communication uses the Simple Network Management Protocol discussed in Chapter 1.

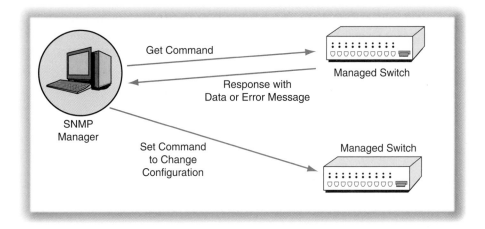

Figure 4-19   Managed Switches

Polling and Problem Diagnosis

Every few seconds, the SNMP manager polls each managed switch and asks for a copy of the switch's configuration parameters. If a problem occurs, the manager can discover quickly which switches are not responding and so can narrow down the source of the problem. In many cases, the status data collected frequently from the switches can pinpoint the cause of a problem.

Fixing Switches Remotely

In some cases, the network administrator can use the manager to fix switch problems remotely by sending commands to the switch. For instance, the manager can command the switch to do a self-test diagnostic. To give another example, the manager can tell the switch to turn off a port suspected of causing problems.

Performance Summary Data

At the broadest level, the manager can present the status data to the network administrator in summarized form, giving the administrator a good indication of how well the network is functioning and of whether changes will be needed to cope with expected traffic growth.

The Cost of Manageability

Manageability is not cheap. Managed switches are much more expensive than nonmanaged switches. However, in firms with many switches, central management reduces management labor, which is considerable. This labor cost reduction more than offsets its higher purchase cost. The main benefit of network management, then, is to reduce overall costs.

Managed switches are more expensive than nonmanaged switches, but they reduce management labor in large networks enough to more than offset managed switch purchase costs. Managed switches reduce overall costs.

TEST YOUR UNDERSTANDING

19. a) What are managed switches? b) What benefits do they bring? c) Do managed switches increase or decrease total costs?

Advanced Purchasing Considerations: Physical and Electrical Features

PHYSICAL SIZE

Almost all switches are 48-cm wide. This allows them to fit into standard 48-cm wide telecommunications racks long used in telephony and today used for data switches and routers. In equipment racks, one U is 4.4 cm in height. Most switches, although not all, are multiples of U. For instance, a 2U switch is 8.8 cm tall.

PORT FLEXIBILITY

There are four basic types of switch organization, each giving a different degree of flexibility over how many ports you may have.

➤ **Fixed-port switches**, as their name suggests, give no port flexibility. The ports you buy them with are the ports you will have to live with for the life of the switch. They are one or two U tall. Most workgroup switches are fixed-port switches.

➤ **Stackable switches**, like fixed-port switches, also are 1 or 2U tall and have a fixed number of ports. However, as the name suggests, they can be stacked on top of one another. A special interconnect bus connects them at speeds that are higher than port-to-port Ethernet connections would permit. With stackable switches, companies can add ports in increments of as few as 12.

➤ **Modular switches** also are 1 or 2U tall, but do not have a fixed number of ports. They have one or more slots for modules containing one to four ports.

➤ **Chassis switches** are several U tall. The box has expansion slots into which a firm can place modular expansion boards. The expansion boards contain six to twelve ports. The box itself contains a high-speed backplane bus that links the ports on all of the expansion cards together. Most core switches are chassis switches.

UPLINK PORTS

Ethernet NICs transmit on Pins 1 and 2 and listen on Pins 3 and 6. Normal Ethernet RJ-45 switch ports, in turn, transmit on Pins 3 and 6 and listen on Pins 1 and 2. If you connect two normal RJ-45 ports on different switches via a UTP cord, the ports will not hear each other. To address this problem, most switches have at least one **uplink port**, which transmits on Pins 1 and 2 and listens on Pins 3 and 6. You can use a standard UTP cable to connect a UTP uplink port on one switch to any normal port on its parent switch.[11]

---

[11]If you have a broadband router at your home, it probably has one port named WAN or Internet. This is an uplink port.

Physical Size
    Switches fit into standard 48-cm wide equipment racks
    Switch heights usually are multiples of 1U (4.4 cm)

Port Flexibility
    Fixed-port switches
        No flexibility: the number of ports is fixed
        1 or 2U tall
        Most workgroup switches are fixed-port switches
    Stackable switches
        Fixed number of ports
        1 or 2U tall
        High-speed interconnect bus connects stacked switches
        Ports can be added in increments as few as 12
    Modular switches
        1 or 2U tall
        Contain one or a few slots
        Each slot module contains 1 to 4 ports
    Chassis switches
        Several U tall
        Contain several expansion slots
        Each expansion board contains 6 to 12 ports
        Most core switches are chassis switches

Uplink Ports
    Normal Ethernet RJ-45 switch ports transmit on Pins 3 and 6 and listen on Pins 1 and 2
    If you connect two normal ports on different switches via UTP cords, they will not be able to communicate
    Most switches have an uplink port, which transmits on Pins 1 and 2. You can use an ordinary UTP cord to connect a UTP uplink port on one switch to any normal port on a parent switch

Electrical Power
    Switches require electrical power
    In addition, switches can provide electrical power to devices connected by UTP
    Under Power over Ethernet (POE), switches can supply power to devices connected by UTP
    Under the original 802.3af POE standard
        Provide up to 13 watts to attached devices
        Sufficient for simple wireless access points
        Sufficient for VoIP phones
    Now, the 802.3at POE plus is under development
        30 or 60 watts
        Backwardly compatible with 802.3af
        Sufficient for multiband wireless access points (see Chapter 5)
        Sufficient for other small devices
        Not sufficient for PCs
    New switches can be purchased with POE and POE plus
        Can also add equipment to an existing switch
    Providing power can raise heat in wiring/switching rooms and switch rooms

Figure 4-20   Physical and Electrical Features (Study Figure)

(*continued*)

In a growing number of switches, every port automatically acts as a normal port or an uplink port by detecting what type of port it is connected to. This eliminates the worry about which ports are regular ports and which are uplink ports.

POWER OVER ETHERNET (POE)

The telephone company wires that come into your home bring a small amount of power. You can plug a basic telephone into a telephone wall jack without having to plug it into a power outlet. USB cables also provide a small amount of electrical power to the devices they connect. Similarly, the **power over Ethernet (POE)** standard can bring power to RJ-45 wall jacks. POE is important to corporations because it can greatly simplify electrical wiring for installing voice over IP (VoIP) telephones, wireless access points, and surveillance cameras. Instead of power having to be provided to each installation via electrical wall jacks, the device can simply be plugged into the Ethernet wall jack.

The original POE standard, 802.3af, brings this capability to Ethernet networking. POE brings up to 13 watts of power to the wall jack. Over time, however, small devices have outgrown the 802.3af power limit. The 803.2 Working Group is now putting the finishing touches on an enhanced version called **POE plus**. This standard, 802.3at, will be backward-compatible with 802.3at, but will also provide up to about 30 watts over two wire pairs and perhaps up to about 60 watts over all four wire pairs. (The 60-watt capability is still being debated.) POE plus will be sufficient for advanced wireless access points that operate in both the 2.4 GHz and 5 GHz bands simultaneously (as discussed in Chapter 5), for high-resolution and motorized surveillance cameras, and even for thin clients (which are not full PCs). This will not be enough wattage for desktop computers, and it probably will not be enough for most notebook computers.

Companies that wish to supply power through their RJ-45 wall jacks will have to install either new switches compatible with the POE standards or modification kits that can add POE plus to existing switches. Companies will also have to deal with the additional heat generated by POE and POE plus.

TEST YOUR UNDERSTANDING

20. a) How tall are most Ethernet switches? Why is this so? b) How wide are most Ethernet switches? Why is this so? c) Distinguish between fixed-port switches, stackable switches, modular switches, and chassis switches. d) How tall are most Ethernet switches? e) Why may uplink ports be needed on Ethernet switches? f) How do uplink ports work? g) What is POE? h) Why is POE attractive to corporations? i) How much power do POE and POE plus provide? j) For what types of devices is POE plus sufficient? k) Is POE sufficient for desktop computers and most notebook computers?

ETHERNET SECURITY

Until recently, few organizations worried about the security of their wired Ethernet networks, presumably because only someone within the site could get access to the network and security should be strong within the site. Unfortunately, experience has shown that attackers can easily get into sites, especially if a site has public areas.

Port Access Control (802.1X)

One threat to Ethernet is that any attacker can plug his or her notebook PC into any Ethernet wall jack and have unfettered access to the network. To thwart this attack, companies can implement **802.1X**, which is a standard for **port control** on the workgroup switches that give users access to the network. Quite simply, the switch port will

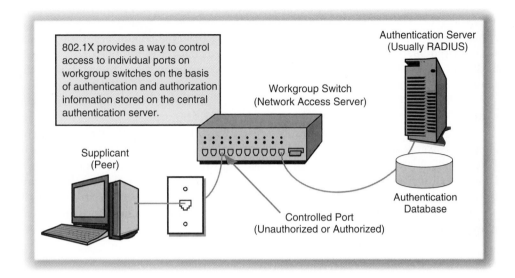

Figure 4-21   802.1X Ethernet Port-Based Access Control

not allow the computer attached to the port to send traffic other than authentication traffic until the computer has authenticated itself.

Figure 4-21 illustrates 802.1X. The workgroup switch is called the **network access server (NAS)**. It gets this name because the workgroup switch provides authentication service to the supplicant computer. The 802.1X standard normally also uses a central authentication server to do the actual supplicant credentials checking. Typically, this is a **RADIUS** server.

When the supplicant computer transmits its authentication credentials (password, etc.), the NAS passes these credentials on to the authentication server. The authentication server checks these credentials against its authentication database. If the authentication server authenticates the credentials, it sends back a confirmation to the workgroup switch. The workgroup switch then allows the supplicant PC to send frames to other devices in the network.

Using a central authentication server provides four benefits:

➤ First, it minimizes the processing power needed in the workgroup switch. Given the large number of NASs, this can produce a major cost saving, compared with having the workgroup switches do the authentication.

➤ Second, having all credentials on the authentication server gives consistency in authentication; there is no danger in having different workgroup switches with incompatible authentication databases.

➤ Third, management cost is reduced because credentials need to be changed on the central authentication server only when user authentication information is changed whenever a user joins the firm, leaves the firm, or needs other credential changes.

➤ Fourth, the credentials of individuals who are fired or suspended can be invalidated in seconds.

Media Access Control (MAC) Security (802.1AE)

A more subtle threat to an Ethernet network comes from management protocols such as the Rapid Spanning Tree Protocol (802.1w). If an attacker can impersonate a switch and send management frames to real switches, he or she may be able to disrupt the network. For instance, the attacker can send a rapid series of RSTP frames telling the switches that they need to reconverge. This frequent reconvergence may leave the switches without the resources they need to handle real frames.

MAC security (802.1AE) attempts to thwart attacks by using RSTP or other management protocols by creating security between each pair of switches, as shown in Figure 4-22. A switch will not accept management messages from another switch unless the sending switch has already authenticated itself to the receiving switch. In addition, 802.1AE has other cryptographic protections, which we saw in Chapter 1 and will see in more depth in Chapter 9. This includes encryption for confidentiality for all management messages.

Security Standards from the 802.1 Working Group

Note from their names that 802.1X and 802.1AE were created by the 802.1 Working Group, not by the 802.3 Working Group that creates Ethernet standards. The 802.1 Working Group produces standards that cut across all 802 network technologies. This includes security standards.

Figure 4-22　Media Access Control (MAC) Security (802.1AE)

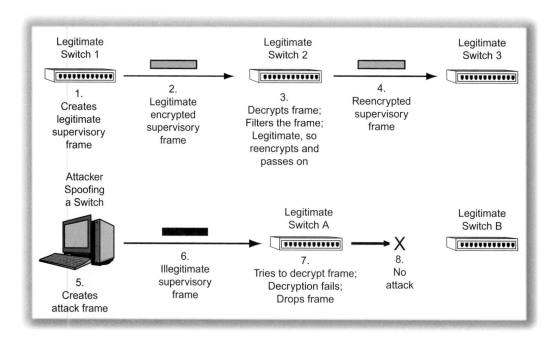

TEST YOUR UNDERSTANDING

21. a) What threat does 802.1X address? b) How does the standard address the threat? c) In 802.1X, what device is the NAS? d) What are the benefits of using a central authentication server instead of having the individual NASs do all authentication work? e) What threat does 802.1AE address? f) How does the standard attempt to thwart such attacks? g) Did the working group that creates Ethernet standards create the 802.1X and 802.1AE standards? Explain.

ETHERNET IN ROUTED LANS

For small LANs, Ethernet is sufficient for most firms. However, for very large LANs, like university campus LANs, it may be better to have a routed network. We have seen that the Rapid Spanning Tree Protocol is complex to implement in large networks. So are VLANs, priority, and a number of other capabilities of Ethernet management. Quite simply, Ethernet does not scale well to very large networks.

Figure 4-23 shows a large routed LAN that adds routing to Ethernet switching. Routers divide the Ethernet LAN into a number of smaller **subnets**. Each subnet is an ordinary Ethernet LAN. This approach joins the simplicity and low cost of basic Ethernet operation with the ability of routed networks to manage large networks well.

TEST YOUR UNDERSTANDING

22. a) Why do many companies use routed LANs instead of pure Ethernet LANs? b) In routed LANs, what are subnets?

Figure 4-23   Routed LAN with Ethernet Subnets

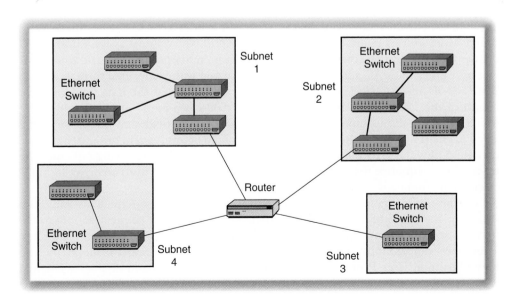

CONCLUSION

Synopsis

This chapter looked in some depth at Ethernet local area networking. Ethernet is the dominant technology for corporate LANs today. Ethernet's only serious competitor is wireless LANs, which we will see in the next chapter. We will learn that wireless LANs are not direct competitors to wired Ethernet LANs, but rather usually work in conjunction with wired Ethernet LANs.

The IEEE 802 LAN/MAN Standards Committee creates many LAN standards. The Committee's 802.3 Working Group specifically creates Ethernet standards. Like all networking standards, Ethernet standards exist at both the physical and data link layers. Therefore, they are OSI standards.

The 802.3 Working Group has created many physical layer Ethernet standards and is still creating better physical layer standards. Speeds range from 10 Mbps to 10 Gbps and are still moving higher. These standards use both 4-pair UTP and optical fiber.

The dominant Ethernet standard for access lines from stations to switches is 100BASE-TX, while the dominant standard for switch-to-switch trunk lines is gigabit Ethernet using optical fiber. The newest and fastest standards (10GBASE-x and beyond) are being created first for metropolitan area networks (MANs), but are moving into LANs as well. Chapter 7 discusses Ethernet MANs.

Although transmission links have maximum lengths, switches regenerate signals. Regeneration allows firms to send frames across many switches connected by trunk lines, with little degradation.

The 802 Committee subdivided the data link layer into two layers. The media access control layer is specific to a particular technology, such as Ethernet or 802.11 wireless LANs. The logical link control layer deals with matters common to all LAN technologies. Ethernet has only a single MAC standard—the 802.3 Media Access Control standard. This standard specifies frame organization and switch operation.

The Ethernet frame has multiple fields. The preamble and start of frame delimiter fields synchronize the receiver's clock with the sender's clock. The destination and source MAC address fields are each 48 bits long, and NIC vendors assign Ethernet addresses to NICs at the factory. Because of human memory limitations (and to simplify writing), Ethernet MAC addresses usually are written in hexadecimal format, such as B2-CC-67-0D-5E-BA. The length field specifies the length of the data field (not the length of the frame as a whole). The data field has two parts: the LLC subheader, which describes the type of packet contained in the data field; and the packet itself. The PAD field is added if the data field is less than 46 octets long, in order to make the data field plus the PAD field exactly 46 octets in length. The receiving NIC uses the frame check sequence field to check for errors. If the receiver finds an error, it simply discards the frame.

Firms must organize their Ethernet switches in a hierarchy. This simplifies switching, making Ethernet switches inexpensive. Switches that connect stations to the network are workgroup switches. Switches higher in the hierarchy are core switches. There must not be loops among switches, because this would break the hierarchy. The Rapid Spanning Tree Protocol (802.1w) automatically detects and disables accidental loops. RSTP can also provide backup links in case of link or switch failures.

Most Ethernet switches can divide an Ethernet LAN into a number of VLANs that are groups of clients and servers which can talk to each other, but cannot talk to clients and servers on different VLANs. Using VLANs reduces congestion when servers broadcast. It also provides security. To standardize VLANs (and priority), two tag fields are added to the Ethernet frame, right after the source address. The Tag Protocol ID field merely indicates that the frame is tagged. The Tag Control Information field has a 12-bit VLAN number to indicate to which VLAN a particular frame belongs.

Even networks that have sufficient capacity most of the time will experience momentary traffic peaks that exceed their capacity. Overloaded switches may have to drop frames, and congestion will cause latency (delay). One approach to handling momentary traffic peaks is to overprovision the network—that is, to install much larger Ethernet lines and switches than are needed most of the time. A more efficient way to manage resources is to give latency-intolerant applications, such as voice, high priority so that they will go first during periods of congestion, minimizing their latency. Priority management uses the 3-bit priority level in the Tag Control Information field to indicate the priority levels of specific Ethernet frames. Unfortunately, priority is management-intensive.

(From the box Hubs and CSMA/CD.) Some older LANs still use Ethernet hubs instead of Ethernet switches. Hubs broadcast incoming bits, so only one station may transmit at a time. Consequently, NICs that use hubs must use CSMA/CD media access control, which only allows NICs to transmit if no other NIC is transmitting and which handles retransmission if there is a collision. NICs that use switches turn off CSMA/CD, giving them full-duplex (simultaneous two-way) transmission.

Purchasing switches is very complex. The most basic issue is the number and speeds of the ports needed. Core switches should have nonblocking or nearly nonblocking capacity, meaning that even if each port is receiving at its maximum speed, the switching matrix will have the capacity needed to switch the input traffic.

Store-and-forward switches forward frames only after receiving the entire frame. In contrast, cut-through switches start sending the frame back out after receiving only a few octets. Cut-through switches reduce latency at each switch, but this is rarely important in practice.

Managed switches are more expensive than other switches, but companies can manage them remotely. Using managed switches saves money overall by reducing management labor.

(From the box Advanced Purchasing Considerations: Physical and Electrical Features.) Switches come in various sizes, with varying basic numbers of ports and varying expandability. Some even provide electrical power to the stations they serve.

Two security standards have been created by the 802.1 Working Group. These security standards are used in all 802 technologies, including 802.3. The 802.1X standard requires a computer that plugs into an RJ-45 wall jack to authenticate itself before it is allowed to use the network. The 802.1AE standard, in turn, protects management frames so that an attacker impersonating a switch cannot succeed when sending false management frames.

Finally, for large LANs, Ethernet runs into serious management problems. Consequently, many companies use routers to turn their LANs into routed LANs. This divides the LAN into several smaller segments that use Ethernet technologies.

End-of-Chapter Questions

THOUGHT QUESTIONS

1. NICs can tell whether an arriving frame is tagged or not simply by looking at it. How can they do so? (Hint: They look at the value in the two octets following the address fields.)

2. If the sender adds a PAD field to an Ethernet frame, the combined data field and PAD will be 46 octets long. How can the receiving NIC tell which part is the data field?

DESIGN QUESTIONS

1. Two switches are 47 meters apart. They need to communicate at 600 Mbps. What Ethernet physical layer standard should you use?

2. Two switches are 200 meters apart. They need to be able to communicate at 1.7 Gbps. What Ethernet physical layer standard should you use?

3. Site Q attaches to Site R, which attaches to Site S. Site Q is 130 meters west of site R. Site R is 180 meters east of Site S. Site Q needs to be able to communicate with Site R at 45 Mbps. Site R needs to be able to communicate with Site S at 2 Gbps. Site Q needs to be able to communicate with Site S at 300 Mbps.

   a) Draw a picture of the situation.
   b) What traffic must the trunk line between Site Q and Site R be able to carry?
   c) What Ethernet standard should be used for the trunk line between Site Q and Site R?
   d) What traffic must the trunk line between Site R and Site S be able to carry?
   e) What Ethernet standard should be used for the trunk line between Site R and Site S?

4. You will create a design for a network connecting four buildings in an industrial park. Hand in a picture showing your network. There will be a core switch in each building.

   ➤ Building A is the headquarters building.
   ➤ Building B is 85 meters south and 90 meters east of the headquarters building. A line will run directly from Building A to Building B.
   ➤ Building C is 150 meters south of the headquarters building. A line will run directly from Building A to Building C.
   ➤ Building D is 60 meters west of Building C. A line will run directly from Building C to Building D.
   ➤ Computers in Building A need to communicate with computers in Building B at 600 Mbps.
   ➤ Computers in Building A need to be able to communicate with computers in Building C at 3 Gbps.
   ➤ Computers in Building A must communicate with computers in Building D at 500 Mbps.
   ➤ Computers in Building C must communicate with computers in Building D at 750 Mbps.

   a) Draw a picture of the situation.
   b) Specify the Ethernet standard you will use to connect Building A to Building B.
   c) Specify the Ethernet standard you will use to connect Building A to Building C.
   d) Specify the Ethernet standard you will use to connect Building C to Building D.

HANDS-ON EXERCISES

Binary and Hexadecimal Conversions

If you have Microsoft Windows, the Calculator accessory shown in Chapter 1a can convert between binary and hexadecimal notation. Go to the *Start* button, then to *Programs* or *All Programs*, then to *Accessories*, and then click on *Calculator*. The Windows Calculator will pop up.

Binary to Hexadecimal

To convert eight binary bits to hexadecimal (hex), first choose *View* and click on *Scientific* to make the Calculator a more advanced scientific calculator. Click on the *Bin* (binary) radio button, and type in the 8-bit binary sequence you wish to convert. Then click on the *Hex* (hexadecimal) radio button. The hex value for that segment will appear.

Hexadecimal to Binary

To convert hex to binary, go to *View* and choose *Scientific*. Click on *Hex* to indicate that you are entering a hexadecimal number. Type the number. Now click on *Bin* to convert this number to binary.

One additional subtlety is that Calculator drops initial 0s. So if you convert 1 hex, you get 1. You must add three initial 0s to make this a 4-bit segment: 0001.

a) Convert 1100 to hexadecimal.
b) Express the following MAC address in binary: B2-CC-67-0D-5E-BA, leaving a space after every eight bits.
c) Express the following MAC address in hex: 11000010 11001100 01100111 00001101 01011110 10111010.

PERSPECTIVE QUESTIONS

1. What was the most surprising thing you learned in this chapter?

2. What was the most difficult material for you in this chapter?

PROJECT

Do a report on the current state of the effort to produce 40 Gbps and 100 Gbps.

GETTING CURRENT

Go to the book website's New Information and Errors pages for this chapter to get new information since this book went to press and to note corrections for any errors in the text.

Token-Ring Networks and Early Ethernet Technologies

INTRODUCTION

Certification exams still require you to know about the 802.5 and FDDI token-ring network technologies, despite the fact that the last hardware for these networks was produced around 1998 and they have almost never been seen in organizations since the turn of the century. If you are going to take a certification exam, you have to learn about token-ring networks. Think of it as a ragging exercise.

Before you begin this chapter, read the box "Hubs and CSMA/CD" in Chapter 4. The token passing access control method used in token-ring networks serves a similar function to that served by CSMA/CD.

The chapter also looks at the two earliest Ethernet standards: 10BASE2 and 10BASE5. Again, this topic is included only because some certifications still require knowledge of that material. Both Ethernet standards are long gone, although the author still has some of the cabling running through his office. (The network staff promises to remove it someday.)

TOKEN-RING TECHNOLOGY

The term **token-ring network** involves two things:

➤ One is the topology of the network, which is a ring topology rather than the hierarchical topology that Ethernet must use whether switches or hubs are employed.

➤ Second, as a box in Chapter 4 discussed, in Ethernet hub networks, only one station may transmit at a time across the entire network. This limitation is also true in token-ring networks. Ethernet hub networks control the times that stations may transmit while using CSMA/CD. Token-ring networks use a process called token passing to control such times.

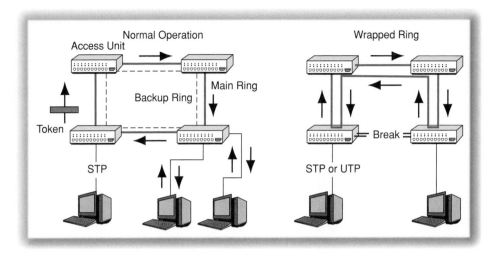

Figure 4a-1   Ring Network

Ring Networks

Figure 4a-1 shows the **ring topology** used in 802.5 and FDDI. The network is laid out in a ring (loop), with transmission lines connecting a number of access units. Stations attach to these access units.

As the figure shows, both technologies actually use a double ring. In normal operation, either both rings are used for transmission or one is left in standby mode. Frames travel in one direction around the active ring or rings.

The figure also shows that if a break occurs between access units, the ring is wrapped in a way that enables the two rings to become a single, longer loop. Frames take longer to get around the ring, but all access units continue to communicate with one another.

This **ring wrapping** makes ring networks exceptionally reliable. There is no single point of failure for transmission lines as there is in Ethernet. Even if an access unit fails, only the stations attached to it are out of service.

However, ring wrapping technology to implement ring wrapping electronically rather than manually is quite expensive. Even adding manual ring wrapping increases the expense. In addition, a large amount of wiring must be laid if a ring topology is used.

TEST YOUR UNDERSTANDING

1. a) How many rings are used in a ring topology? b) Why are ring topologies reliable? c) What is the disadvantage of ring topologies?

Token Passing

As the box on hubs and CSMA/CD in Chapter 4 discussed, when hubs are used in Ethernet, only one station can transit at a time. As the number of stations on an Ethernet hub network increases, delay (latency) grows. Beyond about 100 stations, this latency becomes problematic.

In ring networks, a different access control method called **token passing** is used. When no station is transmitting, a special frame called a **token** circulates around the ring continuously. When a station wishes to transmit, it captures the token when the token reaches the station's position. The station may then transmit. After the transmission ends, the station releases the token again.

Unfortunately, token passing is expensive to implement. The basic idea is straightforward, but token passing has many subtleties. It is said that the devil is in the details. In token passing, there are a surprisingly large number of details. Token passing is much more expensive to implement than CSMA/CD.

TEST YOUR UNDERSTANDING

2. a) Why is token passing needed? b) What is the advantage of token passing, compared with CSMA/CD? c) What is the disadvantage of token passing, compared with CSMA/CD?

EARLY ETHERNET AND 802.5 TOKEN-RING NETWORKS

Early Ethernet: CSMA/CD–Bus Networks

Figure 4a-2 illustrates (in a simplified way) the **bus topology** used by early versions of Ethernet (10BASE5 and 10BASE2). All stations on a segment connected to the segment. When one station transmitted, the signal traveled in both directions, passing all other stations. Essentially, the station broadcast its frame to all other stations. These early Ethernet systems used CSMA/CD for access control, so they were called CSMA/CD–bus networks.

Figure 4a-2 Ethernet Bus Topology

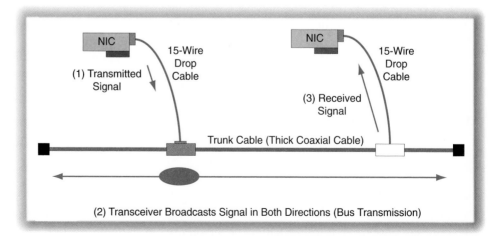

Figure 4a-2 specifically shows 10BASE5, which was the first major Ethernet standard. The trunk line used coaxial cable similar to the cable that connects your VCR to your TV and that brings television into your house if you have cable or satellite dish television service. However, the coax for this trunk line was much thicker than the coax currently used for television. (Installers called it a frozen yellow garden hose.) The drop cable, in turn, consisted of 15 untwisted wires.

Later, the 802.3 Working Group created the 10BASE2 standard, which used thinner (and less expensive) coaxial cable. In addition, 10BASE2 did not use drop cables. Computers were connected one to another in series, like a daisy chain.

TEST YOUR UNDERSTANDING

3. a) What topology did early Ethernet networks use? b) What access control method did early Ethernet networks use? c) What was the main transmission medium for 10BASE5 and 10BASE2? d) Describe 10BASE5 technology. e) Describe 10BASE2 technology. f) What was the attraction of 10BASE2, compared with 10BASE5?

802.5 Token-Ring Networks Appear

Ethernet technology was created before the 802 LAN/MAN Standards Committee existed. When the committee was created, most people thought that it would simply accept Ethernet (or a slightly modified version of Ethernet) as its LAN standard.

However, when the committee formed, IBM introduced and pushed its technology for a token-ring network. The 802 LAN/MAN standards committee failed to select between them. Instead, it released standards for both. Ethernet standards, as noted in Chapter 4, were created by the 802.3 Working Group (and still are). In contrast, the **802.5 Working Group** was created to standardize Token-Ring Network (TRN) technology. (Yes, "Token-Ring Network" is always capitalized when referring to 802.5.)

Given the complexity of token-ring technology, by the time the standard was specified and products began selling, Ethernet was already beginning to be widely installed. In addition, when 802.5 products were finally sold, the complexity of Token-Ring Networks meant that they were twice as expensive as Ethernet products. This was important because NICs in the 1980s often cost $100 to $500.

TEST YOUR UNDERSTANDING

4. Why were 802.5 Token-Ring Networks not successful when they came out?

Ethernet Wins

For a time, many companies felt that it made sense to install 802.5 LANs. This was especially true for companies that used large IBM mainframes, which worked very well with 802.5 networks.

Another reason for going with 802.5 technology was speed. Early versions of Ethernet could only operate at 10 Mbps, while TRNs worked at a whopping 16 Mbps (after a very short time at 4 Mbps). These speed differences seem trivial today, but they were considered to be very fast then.

However, 802.5 market penetration was always small. When Ethernet moved to 100 Mbps speeds and later to switches (which do not need access control mechanisms), 802.5 Token-Ring Networks lost their advantages.[1]

In addition, Ethernet's market share lead generated economies of scale in manufacturing. This increased the price gap between Ethernet and 802.5 networks.

Vendors stopped shipping 802.5 Token-Ring Networks in the late 1990s. In today's network world of 100 Mbps or more to the network, almost every last company that used TRN has replaced it with Ethernet.

TEST YOUR UNDERSTANDING

5. a) At what speed did 802.5 Token-Ring Networks operate? b) What Ethernet developments killed 802.5 networks? c) With switching, do you need either CSMA/CD or token passing?

Shielded Twisted Pair Wiring

Chapter 3 noted that Category 7 wiring is shielded twisted pair (STP) wiring. There is a foil shield around each pair, and there is a metal mesh shield just inside the jacket. This almost completely eliminates interference. In fact, the 802.5 standard specified STP wiring, although optical fiber could also be used. The STP wiring for 802.5 was quite thick, heavy, and expensive.

TEST YOUR UNDERSTANDING

6. What type of wiring were 802.5 networks designed to use?

FIBER DISTRIBUTED DATA INTERFACE (FDDI)

By the mid-1980s, individual companies often had many 10 Mbps Ethernet networks and 16 Mbps Token-Ring Networks. These companies needed a way to link them all together into a larger site LAN. This backbone network, shown in Figure 4a-3, had to be much faster than the individual networks in order to be able to carry cross traffic.

In response to this need, ANSI, the American National Standards Institute (rather than the IEEE 802 LAN/MAN Standards Committee), created the X3T9.5 **Fiber Distributed Data Interface (FDDI)** standard. As its name suggests, FDDI was defined, from the beginning, to use optical fiber. Fiber enabled it to operate at 100 Mbps.

FDDI also is a token-ring network. It is similar to 802.5, primarily because it borrowed heavily from the earlier 802.5 specification. However, instead of beginning as a LAN standard, FDDI actually began as a metropolitan area network standard. Even with multimode fiber, there could be 2 km between access units (which FDDI calls concentrators). The entire loop could be 100 km in circumference with normal operation. Although FDDI was rarely used as a metropolitan area network, its ability to span longer distances made it ideal as a backbone to connect Ethernet hub networks and 802.5 networks, which both had similar distance limitations.

---

[1]Although the 802.5 Working Group eventually defined 100 Mbps operation and switches, this development came too late to save Token-Ring Networks. Even before these developments were finalized, IBM ceased its participation in the 802.5 Working Group.

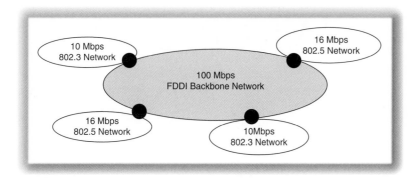

Figure 4a-3   Fiber Distributed Data Interface (FDDI) Backbone
Network

Although FDDI was ideal when it came out, Ethernet speeds rapidly grew from 10 Mbps to 100 Mbps. At that point, the 100 Mbps speed limitation of FDDI became insufficient for its backbone role. Within a short period of time, companies that had installed FDDI backbones began to remove them.

By the turn of the 21st century, the production of FDDI components had stopped. With the demise of FDDI, even the concept of backbone networks connecting smaller networks of limited size disappeared. With Ethernet, there usually is a single hierarchy of Ethernet switches connecting all stations.

Ironically, the growth of Ethernet speeds from 10 Mbps to 100 Mbps that doomed FDDI was due to the 802.3 Working Group's "borrowing" of FDDI technological components and adapting them to a switched Ethernet topology.

TEST YOUR UNDERSTANDING

7. a) Compare 802.5 TRNs and FDDI in terms of speed and ring circumference. b) In local area networks, what role did FDDI take? c) What killed FDDI?

RETURN OF THE RING

Although ring networks passed out of favor for LANs, the reliability of the dual ring topology reappeared in wide area networking, especially metropolitan area networking. The SONET/SDH technology discussed in Module C is a dual-ring network. However, it is used to connect switches, so there is no need for token passing access control.

Wireless LANs (WLANs)

INTRODUCTION

Ethernet technology dominates in wired corporate LANs. Unfortunately, an Ethernet host must plug into a wall jack or directly into an Ethernet switch. This is a problem for the growing number of notebook computers, personal digital assistants, and other mobile devices used in organizations. To serve mobile users, companies are using a new type of LAN technology, **wireless LAN (WLAN)** technology, which uses radio[1] for physical layer transmission.

Wireless LANs (WLANs) use radio for physical layer transmission.

802.11 WLAN Standards

The most important WLAN standardhs today are the **802.11** standards, which are created by the **IEEE 802.11 Working Group**. Recall that Ethernet standards are created by a different working group, the 802.3 Working Group.

---

[1]In addition to radio, it is possible to use infrared light for transmission. (Your television remote control works by infrared transmission.) However, infrared transmission is too slow for corporate WLANs.

> **Wireless LAN Technology**
>   The dominant WLAN technology today
>   Standardized by the 802.11 Working Group
>
> **Wireless Computers Connect to Access Points (Figure 5-2 and Figure 5-3)**
>
> **Supplement Wired LANs**
>   Access points connect to the corporate LAN
>   So that wireless hosts can reach servers on the Ethernet LAN
>   So that wireless hosts can reach Internet access routers on the Ethernet LAN
>
> **Large 802.11 WLANs**
>   Organizations can provide coverage throughout a building or a university campus
>   By the judicious installation of many access points
>
> **Speeds and Distances**
>   Speed up to 300 Mbps, but usually 10 to 100 Mbps
>   Distances of 30 to 100 meters

Figure 5-1   802.11 Wireless LAN (WLAN) Standards (Study Figure)

Rather than being a competitor for wired Ethernet LANs, **802.11 WLANs** today primarily *supplement* wired LANs, but do not replace them. Figure 5-2 shows that mobile users typically connect by radio to devices called **wireless access points**, or simply, **access points**. These wireless access points link the mobile user to the firm's wired Ethernet LAN. Figure 5-3 shows a wireless access point and the wireless NICs that computers need to work with access points.

Why is there normally a connection to the firm's main wired LAN? Quite simply, the servers that mobile host devices need, as well as the firm's Internet access router, usually are on the wired LAN. Wireless hosts need the wired LAN to reach the resources they need.

A single 802.11 wireless access point can serve multiple hosts up to 30 to 100 meters away. Typical throughput today is only a few megabits per second for individual hosts, although throughput is over 100 Mbps per second in the fastest 802.11 networks.

In a home, you are only likely to have a single access point. Businesses need far larger coverage areas. By placing wireless access points judiciously throughout a building, a company can construct a large 802.11 WLAN "cloud" that can serve mobile users anywhere in the building.

We will spend most of this chapter looking at 802.11 WLANs, starting with technology and then moving on to security issues and the management of large 802.11 WLANs.

Other Local Wireless Technologies

We will spend some time toward the end of the chapter looking at other WLAN technologies. We will look most closely at Bluetooth. Bluetooth is not a full corporate

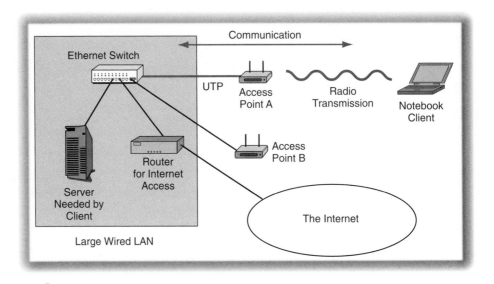

Figure 5-2 802.11 Wireless LAN (WLAN) Operation

Figure 5-3 802.11 Wireless Access Point and Wireless NICs

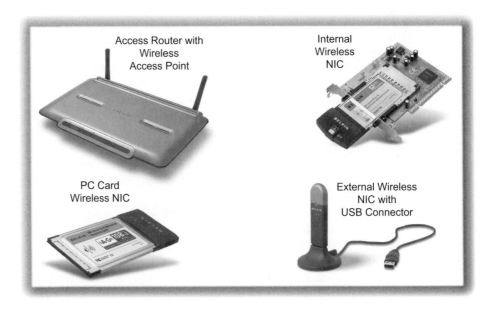

Frequency
     Radio waves are measured in terms of frequency
     Measured in hertz (Hz)—the number of complete cycles per second

Most Common Frequency Range for WLANs
     High megahertz to low gigahertz range

Propagation Problems
     Rapid inverse-square law attenuation
     Attenuation through absorption
     Multipath interference
     Electromagnetic interference
     Shadow zones (dead spots)

Propagation Problems that Increase at Higher Frequencies
     Greater attenuation through absorption
     Deader shadow zones

Figure 5-4   Recap of Radio Propagation Concepts from Chapter 3 (Study Figure)

WLAN technology. Instead, it is designed to provide wireless connections to devices that are physically very close to each other, such as a computer and a wireless mouse. It is only a cable replacement technology.

General Radio Propagation

Before we look at specific wireless LAN standards, we should review what you learned about radio transmission in Chapter 3.

Radio Frequencies

As we saw in Chapter 3, while optical fiber transmission is measured by wavelength, radio transmission is measured in terms of *frequency*. One cycle per second is one hertz (Hz). Higher frequencies are expressed in metric notation. Radio transmission normally operates in the high-megahertz to the low-gigahertz range.

Radio Propagation Effects

In Chapter 3, we also looked at radio propagation. We saw that while wire and fiber transmission are predictable, radio propagation is highly unpredictable. As you have probably have noticed with your mobile phone, sometimes moving even a few meters can change radio propagation dramatically. Among the propagation problems we saw in Chapter 3 were

➤ rapid inverse-square law attenuation,

➤ attenuation through absorption,

➤ multipath interference,

➤ electromagnetic interference, and

➤ shadow zones (dead spots).

Chapter 3 also noted that shadow zones and absorption attenuation problems increase at higher frequencies. Most WLAN transmission takes place at very high frequencies—around either 2.4 GHz or 5 GHz. Propagation effects, consequently, are very significant. Even the doubling of frequency between 2.4 GHz and 5 GHz can result in much higher attenuation and much darker shadow zones.

RADIO BANDS, BANDWIDTH, AND SPREAD SPECTRUM TRANSMISSION

Bands and Bandwidth

Now we can begin looking at new information about radio transmission, beginning with the frequency spectrum, service bands, and channels.

The Frequency Spectrum and Service Bands

The **frequency spectrum** consists of all possible frequencies from zero hertz to infinity, as Figure 5-5 shows.

Service Bands

The frequency spectrum is divided into **service bands** that are dedicated to specific services. For instance, in the United States, the AM radio service band lies between 535 kHz and 1,705 kHz. The FM radio service band, in turn, lies between 88 MHz and 108 MHz. The 2.4 GHz unlicensed band that we will see later for wireless LANs extends from 2.4000 GHz to 2.4835 GHz.

Figure 5-5   The Frequency Spectrum, Service Bands, and Channels

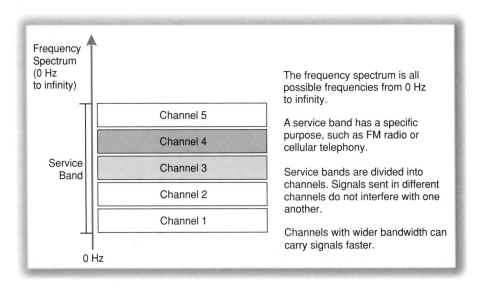

Channels

Service bands are subdivided into smaller frequency ranges called **channels**. A different signal can be sent in each channel because signals in different channels do not interfere with one another. This is why you can receive different television channels successfully.

Signal and Channel Bandwidth

Figure 5-7 shows that signals do not operate at a single wavelength. Rather, signals spread over a range of frequencies. This range is called the signal's **bandwidth**. Signal bandwidth is measured by subtracting the lowest frequency from the highest frequency.

A channel also has a bandwidth. For instance, if the lowest frequency of an FM channel is 89.0 MHz and the highest frequency is 89.2 MHz, then the **channel bandwidth** is 0.2 MHz (200 kHz). AM radio channels are 10 kHz wide, FM channels have bandwidths of 200 kHz, and television channels are 6 MHz wide.

Why are there such large differences in channel bandwidth across service bands? The answer lies in the relationship between possible transmission speed in a channel and channel bandwidth. Shannon found that the maximum possible transmission speed (C) in bits per second when sending data through a channel is directly proportional to the channel's bandwidth (B) in hertz, as shown in the **Shannon Equation** (Equation 5-2).[2]

$$C = B\,[\text{Log}_2\,(1 + S/N)] \qquad \text{(Equation 5-2)}$$

The maximum possible speed is directly proportional to bandwidth, so if you double the bandwidth, you can potentially transmit up to twice as fast. However, C is the *maximum* possible speed for a given bandwidth and signal-to-noise ratio. *Real transmission throughput* will always be less.

To transmit at a given speed, you need a channel wide enough to handle that speed. For example, video signals produce many more bits per second than audio signals, so television uses much wider channels than AM radio (6 MHz versus 10 kHz in AM radio transmission).

The signal-to-noise (S/N) ratio discussed in Chapter 3 is also important, but it is very difficult to modify in practice. By the way, the equation requires the signal-to-noise ratio to be expressed as a ratio (P_2/P_1). If you are given an S/N value in decibels, you must convert it to a simple ratio.

Channels with large bandwidths are called **broadband** channels. They can carry data very quickly. In contrast, channels with small bandwidths, called **narrowband** channels, can only carry data slowly.

Although the terms *broadband* and *narrowband* technically refer only to the width of a channel, broadband has come to mean "fast," while narrowband has come to mean "slow." In terms of transmission speed, the dividing line between broadband and narrowband speeds traditionally has been 200 kbps. As we will see in Chapter 6, a much higher cutoff is being discussed by some regulators for residential Internet access.

The dividing line between broadband and narrowband speeds is 200 kbps.

---

[2]Claude Shannon, "A Mathematical Theory of Communication," *Bell System Technical Journal,* July 1938, pp. 379–423, and October 28, 1938, pp. 623–56.

Signal Bandwidth
- Chapter 3 discussed a wave operating at a single frequency
- However, most signals are spread over a range of frequencies (See Figure 5-7)
- The range between the highest and lowest frequencies is the signal's bandwidth
- The maximum possible transmission speed increases with bandwidth

Channel Bandwidth
- Channel bandwidth is the highest frequency in a channel minus the lowest frequency
- An 88.0 MHz to 88.2 MHz channel has a bandwidth of 0.2 MHz (200 kHz)
- Higher-speed signals need wider channel bandwidths

Shannon Equation
- $C = B\,[\mathrm{Log}_2\,(1+S/N)]$
  - C = Maximum possible transmission speed in the channel (bps)
  - B = Bandwidth (Hz)
  - S/N = Signal-to-noise ratio measured as the power ratio, not as decibels
- Note that doubling the bandwidth doubles the maximum possible transmission speed
- Multiplying the bandwidth by X multiplies the maximum possible speed by X
- Wide bandwidth is the key to fast transmission
- Increasing S/N helps slightly, but usually cannot be done to any significant extent

Broadband and Narrowband Channels
- Broadband means wide channel bandwidth and therefore high speed
- Narrowband means narrow channel bandwidth and therefore low speed
- Narrowband is below 200 kbps
- Broadband is above 200 kbps

The Golden Zone
- Most organizational radio technologies operate in the golden zone in the high megahertz to low gigahertz range
- Golden zone frequencies are high enough for there to be large total bandwidth
  - At higher frequencies, there is more available bandwidth
- Golden zone frequencies are low enough to allow fairly good propagation characteristics
  - At lower frequencies, signals propagate better
- Growing demand creates intense competition for frequencies in the Golden Zone

Channel Bandwidth and Spectrum Scarcity
- Why not make all channels broadband?
- There is a limited amount of spectrum at desirable frequencies
- Making each channel broader than needed would mean having fewer channels or widening the service band
- Service band design requires tradeoffs between speed requirements, channel bandwidth, and service band size

Figure 5-6 Channel Bandwidth and Transmission Speed (Study Figure)

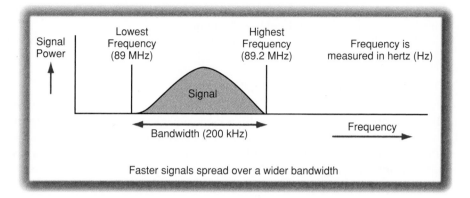

Figure 5-7 Signal Bandwidth

The Golden Zone

Commercial mobile services operate in the high megahertz to low gigahertz range (approximately 800 MHz to 6 GHz). This is the **golden zone**. At lower frequencies, the spectrum is limited and has been almost entirely assigned. At higher frequencies, radio waves attenuate more rapidly with distance and cannot flow through or around objects as they do at lower frequencies. Consequently, the sender and receiver must have a **clear line of sight** (unobstructed direct path) between them. Even at the high end of the golden zone, absorption and shadow zone propagation problems are large. The golden zone is limited, and demand for channels and service bands in the golden zone is increasing rapidly. Consequently, there is strong competition for bandwidth in the golden zone.

The golden zone for commercial mobile services is the high megahertz to low gigahertz range.

TEST YOUR UNDERSTANDING

1. Distinguish among the a) the frequency spectrum, b) service bands, and c) channels. d) In radio, how can you send multiple signals without the signals interfering with one another?

2. a) Does a signal usually travel at a single frequency, or does it spread over a range of frequencies? b) What is channel bandwidth? c) If the lowest frequency in a channel is 1.22 MHz and the highest frequency is 1.25 MHz, what is the channel bandwidth? d) Why is large channel bandwidth desirable? e) What do we call a system whose channels have large bandwidth?

3. a) Write the Shannon Equation. List what each letter is in the equation. b) What information does C give you? c) What happens to the maximum possible propagation speed in a channel if the bandwidth is tripled while the signal-to-noise ratio remains the same? d) Given their relative bandwidths, about how many times as much data is sent per second in television than in AM radio? e) In the Shannon Equation, should S/N be

entered as a simple power ratio (P_2/P_1), in decibels, or either way? f) What is the dividing line between narrowband and broadband speeds?

4. a) What is the golden zone in commercial mobile radio transmission? b) What is a clear line-of-sight limitation?

Licensed and Unlicensed Radio Bands

If two radio hosts transmit at the same frequency, their signals will interfere with each other. In the terminology of Chapter 3, this is electromagnetic interference. To prevent such chaos, governments regulate how radio transmission is used. The International Telecommunications Union, which is a branch of the United Nations, creates worldwide rules for how the radio spectrum is to be used. Individual countries enforce these rules and are also given some discretion over how to implement controls.

Licensed Radio Bands

In **licensed radio bands**, hosts must have a license to operate. They also need a license if the hosts are moved. Television bands are licensed bands, as are AM and FM radio bands. Government agencies control who may have licenses. By doing so, the government limits interference to an acceptable level. In some licensed bands, the rules allow *mobile hosts* to move about while only central antennas are regulated. This is the case for mobile telephones.

Unlicensed Radio Bands

However, for companies that have wireless access points and mobile computers, even the requirement to license central antennas (in this case, access points) is an impossible

Figure 5-8 Licensed and Unlicensed Radio Bands (Study Figure)

Licensed Radio Bands
If two nearby radio hosts transmit in the same channel, their signals will interfere
Most radio bands are licensed bands, in which hosts need a license to transmit
Governments limit licenses to reduce interference
Television bands, AM radio bands, etc. are licensed
In cellular telephone bands, which are licensed, only the central transceivers are licensed, not the mobile phones

Unlicensed Radio Bands
Some bands are set aside as unlicensed bands
Hosts do not need to be licensed to be turned on or moved
802.11 operates in unlicensed radio bands
This allows access points and hosts to be moved freely
However, there is no way to stop interference from other nearby users
Your only recourse is to negotiate with others
At the same time, you may not cause unreasonable interference—for instance, by transmitting at too high power

burden. Consequently, the government has created a few **unlicensed radio bands**. In these bands, any wireless host can be turned on or moved around without the need for government approval.

The problem with unlicensed radio bands is that users of unlicensed radio bands must tolerate interference from others. If your neighbor sets up a wireless LAN next door to yours, you have no recourse but to negotiate with him or her over such matters as which channels each of you will use. At the same time, the law prevents you from creating unreasonable interference—for instance, by using illegally high transmission power.

TEST YOUR UNDERSTANDING

5. a) Do WLANs today use licensed or unlicensed bands? b) What is the advantage of using unlicensed bands? c) What is the disadvantage?

802.11 in the 2.4 GHz and 5 GHz Unlicensed Bands

It would be impossible for a company to have licenses for all of its access points and wireless hosts, so 802.11 operates in unlicensed radio bands. More specifically, WLANs today use two unlicensed bands. One is the 2.4 GHz band. The other is the 5 GHz band.

The 2.4 GHz Unlicensed Band

The 2.4 GHz unlicensed band is the same in most countries in the world, stretching from 2.40 GHz to 2.4835 GHz. This commonality allows companies to sell generic 2.4 GHz

Figure 5-9   802.11 in the 2.4 GHz and 5 GHz Unlicensed Bands (Study Figure)

The 2.4 GHz Unlicensed Band
    Defined the same in almost all countries (2.400 GHz to 2.485 GHz)
    Commonality reduces radio costs
    Propagation characteristics are good
    For 20 MHz 802.11 channels, only three nonoverlapping channels are possible
        Channels 1, 6, and 11
    This creates mutual channel interference between nearby access points transmitting in the same 20 MHz channel
    Difficult or impossible to put nearby access points on different channels (See Figure 5-1)
    Also, potential problems from microwave ovens, cordless telephones, etc.

The 5 GHz Unlicensed Band
    Radios are expensive because frequencies in different countries are different
    Shorter propagation distance because of higher frequencies
    Deader shadow zones because of higher frequencies
    More bandwidth, so between 11 and 24 non-overlapping channels
    Allows different access points to operate on non-overlapping channels
    Some access points can operate on two channels to provide faster service

radios, driving down the price of radios. In addition, radio propagation is better in the 2.4 GHz unlicensed band than in the higher-frequency 5 GHz band.

Unfortunately, the 2.4 GHz band is very limited. It has only 83.5 MHz of bandwidth. Each 802.11 channel is 20 MHz wide. Furthermore, due to the way channels are allocated, there are only three possible nonoverlapping 20 MHz 802.11 channels, which are centered at Channels 1, 6, and 11.[3] If nearby access points operate in the same channel, their signals will interfere with each other unless the access points are far apart. This interference is called **mutual channel interference**.

If you have only three access points that can all hear each other, there is no problem with having only three channels. You simply run each on a different channel, and there will be no interference. However, when you have four access points that can all hear each other, there is no way to avoid having some mutual channel interference. You can minimize mutual channel interference somewhat by giving the shared channel to the two access points that are farthest apart, but this will reduce interference only somewhat.

In addition, the frequencies used in the 2.4 GHz band overlap the frequencies used in microwave ovens, cordless telephones, and Bluetooth equipment. This results in occasional interference that is difficult to diagnose.

The 5 GHz Unlicensed Band

The 802.11 standard can also operate in the 5 GHz band. There have been two problems with this band. The first is the cost of radios. In the 2.4 GHz band, high sales have allowed manufacturers to ride the learning curve down to lower costs. In addition, while the 2.4 GHz band is standardized throughout most of the world, different

Figure 5-10 Mutual Interference in the 2.4 GHz Unlicensed Band

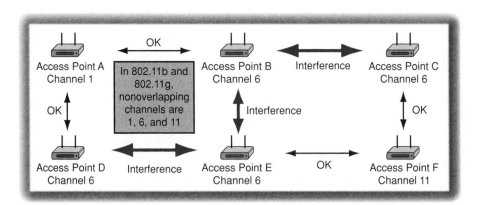

---

[3]Channel numbers were defined for the 2.4 GHz band when channels were narrower. A 20 MHz 802.11 channel overlaps several defined channels. Channels 1, 6, and 11 operate in the 2402 MHz to 2422 MHz, 2427 MHz to 2447 MHz, and 2452 MHz to 2472 MHz frequency ranges, respectively. Note that there are 5 MHz unused "guard bands" between the channels to prevent interchannel interference.

countries use different parts of the 5 GHz band. This also makes 5 GHz radios more expensive due to the need to use only those channels permitted within a given country.

Second, because of the 5 GHz band's higher frequencies, signals do not travel as far, and shadow zones are deader. This means that access points have to be placed closer together. It also means that siting access points to avoid dead spots is more difficult.

The big advantage of the 5 GHz band is that it is much wider than the 2.4 GHz band. In contrast to the 2.4 GHz band's mere three channels, the 5 GHz band provides between 11 and 24 nonoverlapping 20 MHz channels, depending on the frequencies allocated to unlicensed operation in a company.

Having many channels eliminates the mutual channel interference problem because it is easy to assign noninterfering channels to access points even in multifloor buildings (which introduce interference in three dimensions).

In addition, some access points can operate simultaneously on two different channels. This doubles the amount of bandwidth available to devices.

TEST YOUR UNDERSTANDING

   6. a) In what two unlicensed bands does 802.11 operate? b) How wide are 802.11 channels, usually? c) Which licensed band is defined the same way in most countries around the world? d) Does the 2.4 GHz band or the 5 GHz band allow longer propagation distances for a given level of power? e) How many nonoverlapping channels does the 2.4 GHz band support? f) Why is the number of nonoverlapping channels that can be used important? g) How many nonoverlapping channels does the 5 GHz channel support?

NORMAL AND SPREAD SPECTRUM TRANSMISSION

Why Spread Spectrum Transmission?

At the frequencies used by WLANs, there are numerous and severe propagation problems. In these unlicensed bands, regulators mandate the use of a form of transmission called spread spectrum transmission. Spread spectrum transmission is transmission that uses far wider channels than transmission speed requires.

Spread spectrum transmission is transmission that uses far wider channels than transmission speed requires.

Regulators mandate the use of spread spectrum transmission primarily to minimize propagation problems—especially multipath interference. (If the direct and reflected signals cancel out at some frequencies within the range, they will be double at other frequencies.)

In commercial transmission, security is *not* a reason for doing spread spectrum transmission. The military uses spread spectrum transmission for security, but it does so by keeping certain parameters of its spread spectrum transmission secret. Commercial spread spectrum transmission methods must make these parameters publicly known in order for two parties to communicate easily.

In wireless LANs, spread spectrum transmission is used to reduce propagation problems and to reduce mutual interference between nearby hosts transmitting in the same channel, not to provide security.

Spread Spectrum Transmission
  You are required by law to use spread spectrum transmission in unlicensed bands
  Spread spectrum transmission reduces propagation problems
    Especially multipath interference
  Spread spectrum transmission is NOT used for security in WLANs

Normal Transmission versus Spread Spectrum Transmission (See Figure 5-12)
  Normal transmission uses only the channel bandwidth required by your signaling speed
  Spread spectrum transmission uses channels much wider than signaling speed requires

Several Spread Spectrum Transmission Methods (See Figure 5-13)
  Frequency Hopping Spread Spectrum (FHSS) can be used up to 2 Mbps
  Direct Sequence Spread Spectrum (DSSS) is used at 11 Mbps
  Orthogonal Frequency Division Multiplexing (OFDM) is used at 54 Mbps
    Dominates today

Figure 5-11   Spread Spectrum Transmission (Study Figure)

How wide are spread spectrum channels? In 802.11 wireless LANs, the channel bandwidth is at least 20 MHz and may be twice as wide.

TEST YOUR UNDERSTANDING

7. a) In unlicensed bands, what type of transmission method is required by regulators?
   b) What is the benefit of spread spectrum transmission for business communication?
   c) Is spread spectrum transmission done for security reasons in commercial WLANs?

Spread Spectrum Transmission Methods

Normal versus Spread Spectrum Transmission

As noted earlier in our discussion of the Shannon Equation, if you need to transmit at a given speed, you must have a channel whose bandwidth is sufficiently wide.

To allow as many channels as possible, channel bandwidths in *normal radio transmission* are limited to the speed requirements of the user's signal, as Figure 5-12 illustrates. For a service that operates at 10 kbps, regulators would permit only enough channel bandwidth to handle this speed.

In contrast to normal radio transmission, which uses channels just wide enough for transmission speed requirements, **spread spectrum transmission** takes the original signal, called a **baseband signal**, and spreads the signal energy over a much broader channel than is required.

Frequency Hopping Spread Spectrum (FHSS)

The simplest form of spread spectrum transmission is **frequency hopping spread spectrum (FHSS)**. As Figure 5-13 illustrates, the signal in FHSS uses only the bandwidth

Note: Height of Box Indicates Bandwidth of Channel

Channel bandwidth
required for signal speed

Normal Radio: Bandwidth is
no wider than required
for the signal's speed

Spread Spectrum
Transmission:
Channel bandwidth is
much wider than required
for the signal's speed

Commercial spread spectrum transmission reduces certain
propagation effects (multipath interference and narrowband EMI)

Does not provide security as in military spread spectrum systems

Figure 5-12   Normal Radio Transmission and Spread Spectrum Transmission

Figure 5-13   Spread Spectrum Transmission Methods

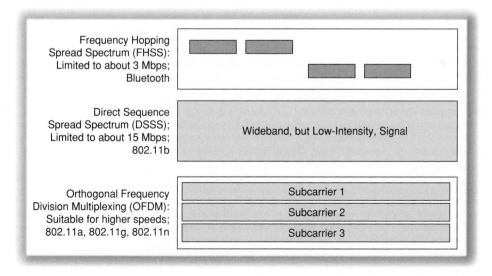

Frequency Hopping
Spread Spectrum (FHSS):
Limited to about 3 Mbps;
Bluetooth

Direct Sequence
Spread Spectrum (DSSS);
Limited to about 15 Mbps;
802.11b

Wideband, but Low-Intensity, Signal

Orthogonal Frequency
Division Multiplexing (OFDM):
Suitable for higher speeds;
802.11a, 802.11g, 802.11n

Subcarrier 1

Subcarrier 2

Subcarrier 3

required by the signal, but hops frequently within the spread spectrum channel. If the signal runs into strong EMI or multipath interference in one part of the broad channel, that part of the message will be lost and must be retransmitted, but parts of the message in other parts of the channel will get through.

FHSS is useful only for relatively low speeds—up to 2 Mbps.[4] Consequently, as we will see later, the 802.11 Working Group for wireless LANs only specified frequency hopping spread spectrum transmission for speeds at or below about 2 Mbps. Few 802.11 LANs operating at these low speeds were purchased. Bluetooth, which is discussed later in this chapter, currently uses FHSS, which is sufficient for Bluetooth's low speed of 3 Mbps today.

Direct Sequence Spread Spectrum (DSSS)

Another spread spectrum technique shown in Figure 5-13 is **direct sequence spread spectrum (DSSS)** transmission, in which a signal is spread over the entire bandwidth of a channel. Interference and multipath interference will affect only small parts of the signal, allowing most of the signal to get through for correct reception. To give an analogy, if an ocean wave hits an obstacle such as a pier, it will still hit the shore at almost full strength.

Although DSSS is more complex than FHSS, DSSS can support speeds up to about 15 Mbps and is used in the 11 Mbps 802.11b wireless LANs that we will see later in this chapter.[5] However, 802.11b WLANs are rapidly being replaced by faster WLAN technologies that do not use DSSS.

Orthogonal Frequency Division Multiplexing (OFDM)

As transmission speeds move above about 15 Mbps, another form of spread spectrum transmission dominates. This is **orthogonal frequency division multiplexing (OFDM)**, which Figure 5-13 also illustrates.

In OFDM, each broadband channel is divided into many smaller subchannels called **subcarriers**. Parts of each frame are transmitted in each subcarrier.[6] OFDM sends data redundantly across the subcarriers, so if there is impairment in one or even a few subcarriers, all of the data usually will still get through.

OFDM is complex and therefore expensive. However, sending data over a single very large channel reliably with DSSS is difficult. In contrast, OFDM can be used at very high speeds because it is easier to send many slow signals reliably in many small subcarriers than it is to send one signal rapidly over a very wide-bandwidth channel. Both of the 54 Mbps 802.11 wireless LAN standards (802.11a and 802.11g) discussed in the next subsection use OFDM.[7]

---

[4]This speed limitation exists because the transmitter must be retuned with each frequency hop. This takes a small, but significant, amount of time, slowing the transmission rate.

[5]The 802.11b standard also has some very-low-speed fallback modes that use FHSS.

[6]In the 802.11a and 802.11g wireless LAN standards discussed later, each 20 MHz channel is divided into 52 subcarriers, each 312.5 kHz wide. Of the 52 subcarriers, 48 are used to send data and 4 are used to control the transmission.

[7]The ADSL services discussed in Chapter 6 generally also use OFDM, although in ADSL service, OFDM is called discrete multitone (DMT) service.

TEST YOUR UNDERSTANDING

8. a) In normal radio operation, how does channel bandwidth usually relate to the bandwidth required to transmit a data stream of a given speed? b) How does this change in spread spectrum transmission?

9. a) Describe FHSS. b) What is its limitation? c) Describe DSSS. d) For what WLAN standard is it used? e) What spread spectrum transmission method is used for 54 Mbps 802.11g WLANs? f) Describe it.

802.11 WLAN OPERATION

As noted at the beginning of this chapter, wireless LANs replace signals in wires with radio waves. WLANs allow mobile workers to stay connected to the network as they move through a building. In some cases, wireless LANs are less expensive to install than wired LANs, but this certainly is not always the case.

Extending the Wired LAN

As noted at the start of the chapter, and as Figure 5-14 shows, an 802.11 wireless LAN typically is used to connect a small number of mobile devices to a large wired LAN—

Figure 5-14   Typical 802.11 Wireless LAN Operation with Wireless Access Points

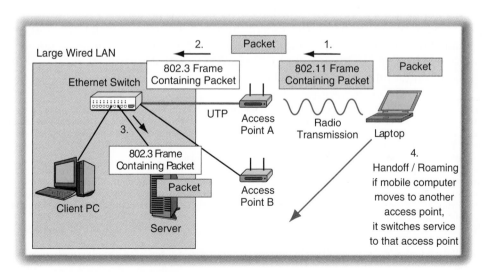

typically, an Ethernet LAN—because the servers and Internet access routers that mobile hosts need to use usually are on the wired LAN.[8]

Hosts

Each mobile host must have a **wireless NIC**. Laptops almost always come with built-in wireless NICs. It is also possible to add upgrade wireless NICs to notebook computers by using PC Card NICs, which simply snap a notebook slot. Desktop computers normally use internal NICs. Both types of hosts also can use an external USB wireless NIC, which sits outside of a PC and plugs into a USB port.

Wireless Access Points

When a wireless host wishes to send a frame to a server, it transmits the frame to a wireless access point.

Technically, an access point is a *bridge* between wireless hosts and the wired LAN. **Bridges** connect two different types of 802 LANs—in the case of 802.11 access points, an 802.11 wireless LAN and an 802.3 wired LAN.[9] Bridges convert between both physical layer signals and data link layer frame formats.

Bridges connect two different types of LANs.

As Figure 5-14 shows, when a wireless NIC transmits to a server on the wired LAN, it places the packet into an 802.11 frame.[10] The wireless access point removes the packet from the 802.11 frame and places the packet in an 802.3 frame. The access point sends this 802.3 frame to the server, via the wired Ethernet LAN. When the server replies, the wireless access point receives the 802.3 frame, removes the packet from the frame, and forwards the packet to the wireless host in an 802.11 frame.

The wireless access point also controls hosts. It assigns transmission power levels to hosts within its range and performs a number of other supervisory chores.

Handoff/Roaming

When a mobile host travels too far from a wireless access point, the signal will be too weak to reach the access point. However, if there is a closer access point, the host can be **handed off** to that access point for service. In WLANs, the ability to use handoffs is also called **roaming**.[11] This aspect of 802.11 WLANs was standardized as 802.11F in 2003, but vendor interoperability has been limited.

---

[8]There is a rarely used 802.11 *ad hoc mode*, in which no wireless access point is used. In ad hoc mode, computers communicate directly with other computers. (In contrast, when an access point is used, this is called 802.11 infrastructure mode.) In addition, 802.11 can create point-to-point transmission over longer distances than 802.11 normally supports. This approach, which normally is used to connect nearby buildings, uses dish antennas and higher power levels authorized for this purpose.

[9]Do bridges sound like routers? Routers can connect any two single networks (LANs or WANs), while bridges can only connect different LANs—specifically, 802 LANs. Also, routers can forward packets across complex internets, while bridges only forward packets and do this forwarding simply and with low functionality. Fortunately, these limitations make bridges much less expensive than routers. For 802.11 access points, bridging is sufficient and inexpensive.

[10]Note that 802.11 frames are much more complex than 802.3 Ethernet frames. Much of this complexity is needed to counter wireless propagation problems.

[11]In cellular telephony, which we will see in the next chapter, the terms *handoff* and *roaming* mean different things.

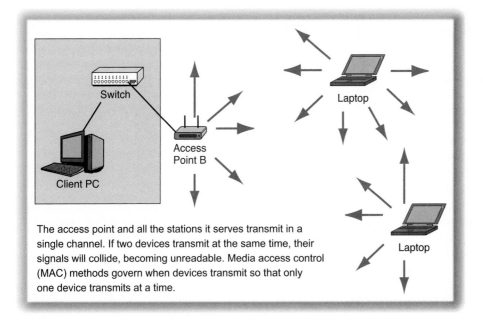

The access point and all the stations it serves transmit in a single channel. If two devices transmit at the same time, their signals will collide, becoming unreadable. Media access control (MAC) methods govern when devices transmit so that only one device transmits at a time.

Figure 5-15   Hosts and Access Points Transmit in a Single Channel

Sharing a Single Channel

As Figure 5-15 shows, the access point and all of the wireless hosts it serves transmit and receive in a single channel. When a host or the access point transmits, all other devices must wait. (If two devices transmit in the same channel at the same time, their signals will interfere with each other.) As the number of hosts served by an access point increases, individual throughput falls. The box "Controlling 802.11 Transmission" discusses how **media access control** methods govern when hosts and access points may transmit so that collisions can be avoided.

The access point and all of the wireless hosts it serves transmit and receive in a single channel. When a host or the access point transmits, all other devices must wait.

TEST YOUR UNDERSTANDING

10. a) List the elements in a typical 802.11 LAN today. b) Why is a wired LAN usually still needed if you have a wireless LAN? c) Is a wireless access point a bridge or a router? d) Why must the access point remove an arriving packet from the frame in which the packet arrives and place the packet in a different frame when it sends the packet back out? e) What is a handoff in 802.11? f) What is the relationship between handoffs and roaming in WLANs? g) When there is an access point and several wireless hosts, why may only one device transmit at a time?

Controlling 802.11 Transmission

MEDIA ACCESS CONTROL

As noted in the body of the text, the access point and the hosts it serves all transmit in the same channel. If two 802.11 devices (hosts or wireless access points) transmit at the same time, their signals will be jumbled together and will be unreadable.

The 802.11 standard has two mechanisms for **media access control**—ensuring that hosts and the access point do not transmit simultaneously. The first, CSMA/CA+ACK, is mandatory and is normally used. The second, RTS/CTS, is optional except in one special case.

TEST YOUR UNDERSTANDING

11. (From the box, "Controlling 802.11 Transmission") a) What is the purpose of media access control? b) Does media access control limit the actions of wireless hosts, the access point, or both?

CSMA/CA+ACK MEDIA ACCESS CONTROL

Problems in Hearing Collisions

If you read the box, "Hubs and CSMA/CD," in Chapter 4, then you know that Ethernet hubs use carrier sense multiple access with collision detection for media access control. Unfortunately, wireless LANs cannot use collision detection. Although all wireless hosts can hear the access point, hosts cannot necessarily hear one another. They may be in dead spots relative to one other, or they may be too far apart. Although CSMA—controlling media access by listening to (sensing) the carrier is still possible, collision detection is not.

CSMA/CA

Instead of using CSMA/CD, 802.11 LANs use **carrier sense multiple access with collision avoidance (CSMA/CA)**. Note that collision *detection* is

replaced by collision *avoidance*. Figure 5-16 illustrates CSMA/CA.

CSMA, as discussed in Chapter 4, requires that a host refrain from transmitting if it hears traffic. This is a very simple rule.

➤ If there is no traffic, collision avoidance comes into play.

➤ If the host does not hear traffic, it considers the last time it heard traffic. If the time since the last transmission exceeds a critical value, the host may transmit immediately.

➤ However, if the time is less than the critical value, the host sets a random timer and waits. If there still is no traffic after the random wait, the host may send.

If the last two points seem odd, remember that the goal is to avoid collisions as much as possible. Two hosts are most likely to transmit at the same time if they both have been waiting for another host to finish transmitting. Without the random delay, both will transmit at the same time, causing a collision.

ACK

More specifically, 802.11 uses **CSMA/CA+ACK**. Collisions and other types of signal loss are still possible with CSMA/CA. When a wireless access point receives a frame from a host, or when a host receives a frame from an access point, the receiver *immediately* sends an acknowledgment frame, an **ACK**. A frame that is not acknowldged is retransmitted. Note that there is no wait when transmitting an ACK. This ensures that ACKs get through while other hosts are waiting.

Note also that retransmission makes CSMA/CA+ACK a reliable protocol. We saw in Chapter 2 that very few protocols are reliable because reliability usually costs more than it brings in benefits. The

CSMA/CA (Carrier Sense Multiple Access with Collision Avoidance)
    Sender listens for traffic
        1. If there is traffic, waits
        2. If there is no traffic:
            2a. If there has been no traffic for less than the critical time value, waits a random amount of time, then returns to Step 1
            2b. If there has been no traffic for more than the critical value for time, sends without waiting
            This avoids collision that would result if hosts could transmit as soon as one host finishes transmitting

ACK (Acknowledgment)
    Receiver immediately sends back an acknowledgment
        If sender does not receive the acknowledgment, retransmits using CSMA
    CSMA/CA plus ACK is a reliable protocol

Figure 5-16   CSMA/CA+ACK in 802.11 Wireless LANs

low error rates in wired media simply do not justify implementing reliability in Ethernet and other wired LAN protocols. However, wireless transmission has many errors, so a reliable protocol is required for reasonably good operation.[12]

Thanks to CSMA/CA+ACK, 802.11 is a reliable protocol.

Inefficient Operation

CSMA/CA+ACK works well, but it is inefficient. Waiting before transmission wastes valuable time. Sending ACKs also is time-consuming. Overall, an 802.11 LAN can only deliver throughput (actual speed) of about half the rated speed of its standard—that is, the speed published in the standard.

This throughput, furthermore, is aggregate throughput shared by the wireless access point and all of the hosts sharing the channel. Individual host throughput will be substantially lower.

TEST YOUR UNDERSTANDING

12. (From the box, "Controlling 802.11 Transmission") a) Describe CSMA/CA+ACK. Do not go into detail about how long a host must wait to transmit if there is no traffic. b) Is CSMA/CA+ACK transmission reliable or unreliable? Explain. c) Why is CSMA/CA+ACK inefficient?

---

[12]In addition, 802.11 uses forward error correction. It adds many redundant (extra) bits to each frame. If there is a small error, the receiver can use these redundant bits to fix the frame. If the receiver can make the repair, it does so and then sends back an ACK. This process makes wireless NICs more expensive than Ethernet NICs, but wireless transmission errors are so common that it makes economic sense to correct errors at the receiver in order to minimize retransmissions.

(*continued*)

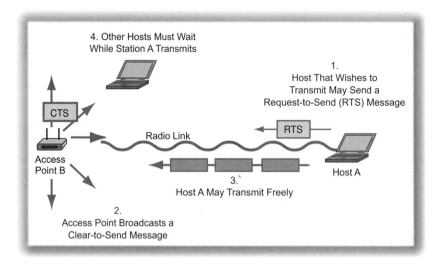

Figure 5-17   Request to Send/Clear to Send (RTS/CTS)

Request to Send/Clear to Send (RTS/CTS)

Although CSMA/CA+ACK is mandatory, there is another control mechanism called **request to send/clear to send (RTS/CTS)**. Figure 5-17 illustrates RTS/CTS. As noted earlier, the RTS/CTS protocol is optional except in one special case we will see later. Avoiding RTS/CTS whenever possible is wise because RTS/CTS is much less efficient, and therefore slower, than CSMA/CA+ACK.

When a host wishes to send and is able to send because of CSMA/CA, the host may send a **request-to-send (RTS)** message to the wireless access point. This message asks the access point for permission to send messages.

If the access point responds by broadcasting a **clear-to-send (CTS)** message, then other hosts must wait. The host sending the RTS may then transmit, ignoring CSMA/CA.

Although RTS/CTS is widely used, keep in mind that it is only an option, while CSMA/CA is mandatory for at least initial communication. Also, tests have shown that RTS/CTS reduces throughput when it is used.

The one special situation in which RTS/CTS is mandatory rather than optional is when 802.11b hosts operating at 11 Mbps and 802.11g hosts operating at 54 Mbps share an 802.11g wireless access point. In this case, hosts must use request to send/clear to send.

TEST YOUR UNDERSTANDING

13. (From the box, "Controlling 802.11 Transmission") a) Describe RTS/CTS. b) Is CSMA/CA+ACK usually required or optional? c) Is RTS/CTS usually required or optional? d) Which is more efficient, RTS/CTS or CSMA/CA+ACK?

Characteristic	802.11	802.11a	802.11b	802.11g	802.11g with 802.11b*	802.11n
Spread Spectrum Method, etc.	FHSS	OFDM	DSSS	OFDM	OFDM	OFDM + MIMO
Channel Bandwidth	20 MHz	20 MHz	20 MHz	20 MHz	20 MHz	20 MHz or 40 MHz
Unlicensed Band	2.4 GHz	5 GHz	2.4 GHz	2.4 GHz	2.4 GHz	2.4 GHz and 5 GHz
Rated Speed	2 Mbps	54 Mbps	11 Mbps	54 Mbps	Not Specified	100 Mbps to 300 Mbps
Actual Throughput, 3 m	1 Mbps	25 Mbps	6 Mbps	25 Mbps	12 Mbps	Closer to rated speed than earlier standards
Actual Throughput, 30 m	Unknown	12 Mbps	6 Mbps	20 Mbps	11 Mbps	High at longer distances
Is throughput shared by all stations using an access point?	Yes	Yes	Yes	Yes	Yes	Yes
Number of Nonoverlapping Channels (varies by country)**	3	12 to 24 in most countries	3	3	3	3 in 2.4 GHz and 12 to 24 in most countries in 5 GHz
Remarks	Dead and gone	Little market acceptance	Bloomed briefly	Today's dominant 802.11 standard	Get rid of old 802.11b equipment	Offers both longer speeds and greater distances

*Speed for an 802.11g host if there is even a single 802.11b associated with the access point.

**The number of nonoverlapping channels is important because nearby access points should operate on different channels.

Source for throughput data: Broadcom.com.

Figure 5-18   Specific 802.11 Wireless LAN Standards

802.11 TRANSMISSION STANDARDS

As Figure 5-18 shows, the 802.11 Working Group has created several WLAN transmission standards. As we will see later, however, 802.11g products and draft 802.11n products dominate in today's marketplace.

Spread Spectrum Transmission and MIMO

The figure shows the technology used in each standard. For spread spectrum transmission, this is either FHSS, DSSS, or OFDM. In addition, the 802.11n standard, which is in the late stages of development, uses a speed-enhancing technology called MIMO, which we will see later in the chapter.

How Fast Are 802.11 Networks?

Rated Speeds and Throughput

Figure 5-18 shows rated speed and throughput for 802.11 transmission standards.

➤ The figure shows that the *rated speeds* of 802.11 WLANs today lie between 11 Mbps and 54 Mbps. (This will soon be more than doubled.) However, these rated speeds are misleading.

➤ As Figure 5-18 also shows, actual *throughput* usually is considerably lower than rated speeds and falls off rapidly with distance.

➤ Furthermore, this 802.11 throughput is *aggregate throughput* shared by all hosts that wish to transmit at the same time. For instance, if the aggregate shared throughput is 5 Mbps, hosts using a wireless access point serving 10–20 hosts might see individual throughput of only one or two megabits per second, despite the fact that only a few hosts are likely to be transmitting at any given moment.

Transmission Modes

Each 802.11 transmission standard has multiple modes that operate at different speeds. Figure 1-24 lists the *highest-speed* mode for each 802.11 standard. When a host is near an access point, it normally can use the highest-speed mode. However, as a computer moves away from an access point, the highest-speed mode begins producing errors. The computer and access point will both switch to a lower-speed mode to reduce transmission errors. As a computer moves even farther from an access point, further drops in transmission mode will occur.

TEST YOUR UNDERSTANDING

14. a) Distinguish between rated speed, aggregate throughput, and individual throughput in 802.11 WLANs. b) Why does transmission speed drop as a computer moves farther from an access point?

Early 802.11 Standards

802.11

In 1997, the 802.11 Working Group produced its first standard, which was called, simply, **802.11**. It operated in the 2.4 GHz unlicensed band. The 802.11 standard used FHSS and only offered a rated speed of 2 Mbps. Given this low speed, few companies bought 802.11 products except to experiment with wireless transmission.

802.11a

Two years later, the 802.11 Working Group created two more standards onh the same day. The first was **802.11a**, which offered a high rated speed (54 Mbps). However, 802.11a operated in the 5 GHz unlicensed band instead of the 2.4 GHz band. Radio equipment was more expensive in the higher unlicensed band. In addition, as noted earlier, while the 2.4 GHz band is standardized around the world, different countries define the limits of the channel rather differently. This further added to the cost of radios. Also, to achieve its high speeds, 802.11a used the complex OFDM spread spectrum technology. Despite the standard's high speed, 802.11a products sold poorly.

802.11b

The 802.11b standard, in contrast, was an instant market success. Although 802.11b only had a rated speed of 11 Mbps and at best delivered half of that, equipment was reasonably priced because 802.11b operated in the 2.4 GHz band and because 802.11b used the simple DSSS spread spectrum technology rather than 802.11a's OFDM technology. In addition, few companies had a need for higher speeds, because there were few users to share the meager bandwidth of 802.11b access points. In addition, when users did begin to outgrow 802.11b, the faster 802.11g technology emerged, which was backwardly compatible with installed 802.11b equipment. For the 802.11g working group, it was literally true that the third time was the charm.

TEST YOUR UNDERSTANDING

15. a) What are the rated speeds of 802.11, 802.11a, and 802.11b? b) In what band do 802.11, 802.11a, and 802.11b operate? c) Why was 802.11a not popular when it first appeared? d) What was the first widely used 802.11 LAN standard? e) What was the main advantage of 802.11b over 802.11a?

802.11g: Today's Dominant Technology

54 Mbps and OFDM

Many analysts expected companies to move to 802.11a once 11 Mbps became too slow for growing wireless LANs. However, in 2001, the 802.11 Working Group ratified the **802.11g** standard. This was another 54 Mbps standard using OFDM. Although the complexity of OFDM originally helped to stunt the growth of 802.11a, OFDM today is not much more expensive than the DSSS spread spectrum method used in 802.11b.

5 GHz Operation

However, while 802.11a operates in the 5 GHz band, 802.11g operates in the 2.4 GHz unlicensed band. This allows for less expensive equipment and provides better propagation characteristics than 5 GHz operation, while offering the same rated speed. Consequently, 802.11g provides faster throughput at all distances than 802.11a.

Backward Compatibility with 802.11b

Of equal importance, 802.11g is **backward-compatible** with 802.11b, which operates in the same 2.4 GHz band. This means that if you have an 802.11g access point, you can serve both 802.11b and 802.11g wireless hosts. (Or, if you buy a new 802.11g wireless NIC, it will work with existing 802.11b access points at 11 Mbps until the company upgrades its access points to 802.11g.) Because it is backward-compatible with older equipment, 802.11g is a seamless upgrade. There is no need to throw away all existing equipment, as there would be in a shift to 802.11a.

Market Dominance

The higher speed of 802.11g caused many companies to increase their spending on WLAN equipment. Very soon, 802.11g dominated sales. Soon after that, it became the most widely installed 802.11 technology.

The most widely installed 802.11 technology is 802.11g.

Many firms are phasing out their 802.11b equipment aggressively. As Figure 5-18 shows, if even a single 802.11b device associates with an 802.11g access point, throughput plummets for all devices associated with the access point.[13]

TEST YOUR UNDERSTANDING

16. a) What 802.11 standard has the largest market share today? b) How is 802.11g better than 802.11b? c) Can you use 802.11b hosts with an 802.11g access point? d) Is there a speed penalty for using 802.11b hosts with an 802.11g access point? e) What disadvantages does 802.11a have compared with 802.11g? f) Is 802.11a backward-compatible with 802.11b or with 802.11g?

802.11n and MIMO

The 802.11 Working Group has recently drafted a new transmission standard, **802.11n**. Figure 5-18 shows that 802.11n will be able to provide a rated speed of 100 Mbps to 300 Mbps. In addition, because of data link layer improvements, 802.11 should be able to provide throughputs much closer to their rated speeds than earlier 802.11 standards.

Wider Channels

How will 802.11n provide these higher speeds? First, instead of being limited to 20 MHz 802.11 channels, as today's versions of 802.11 are, 802.11n will be able to use either 20 MHz or 40 MHz channels. Using a 40 MHz channel alone can roughly double the maximum speed.

However, using wider channels is not always possible. If an 802.11n host detects that one of its two channels is being used by nearby devices, it drops back to using a single channel, cutting its speed in half. In corporate environments, these drop-backs are very frequent, given the high density of access points and the hosts that they serve.

MIMO

More spectacularly, 802.11n will use a new transmission method called **multiple input/multiple output (MIMO)**, which sends two or more radio signals in the same channel between two or more different antennas on access points and wireless NICs. Figure 5-19 illustrates MIMO transmission.

The two or more signals, being in the same channel, will interfere with each other. However, the two signals sent by different antennas will arrive at the two receiving antennas at slightly different times. Using special detection and separation methods based on differences in arrival times for direct signals and reflections, the receiver can separate the different signals in the same channel and read them individually.

Even with only two radio signals and two antennas each on the sender and receiver, MIMO can substantially increase throughput. Using more antennas and radio signals can increase throughput even more. The MIMO standard will permit 2, 3, or 4 antennas on each device.

Of equal importance, MIMO substantially increases propagation distance. It will provide better coverage at current 802.11 distances. It will also allow access points to be placed farther apart, reducing the number of access points needed to serve a building and therefore reducing costs.

---

[13]This occurs because when an 802.11b device associates with an 802.11g access point, transmission control is required to change from CSMA/CA+ACK to RTS/CTS. This greatly reduces throughput for 802.11g stations.

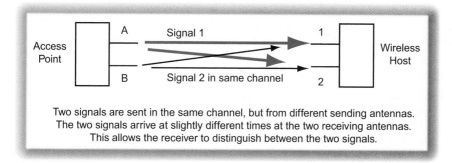

Two signals are sent in the same channel, but from different sending antennas.
The two signals arrive at slightly different times at the two receiving antennas.
This allows the receiver to distinguish between the two signals.

Figure 5-19   Multiple Input/Multiple Output (MIMO) Transmission

The 802.11n standard brings both higher speeds and longer transmission distances.

Backward Compatibility with 802.11g

Due to the large installed base of 802.11g devices in business, the 802.11 Working Group ensured that 802.11n access points and network interface cards were created to be backwardly compatible with 802.11g equipment.

Draft 802.11n Equipment

Although the 802.11n standard has not been ratified at the time of this writing, many **"draft 802.11n"** products are already being sold. Some of these products will be upgradable later to be compatible with the final 802.11n standard. Others will not be upgradable or will only be upgradable at significant cost. Purchasing 802.11n equipment before the final standard is ratified will create both financial risks and the risks of locking the firm into a single vendor. However, this risk will be reduced when the Wi-Fi Alliance implements its plans to certify compatibility between brands claiming draft 802.11n compatibility. Companies today should probably avoid products marked as 802.11n unless they really need the additional speed and reach of 802.11n.

2.4 GHz and 5 GHz Operation

Initial draft 802.11n products are operating in the 2.4 GHz band. This band is crowded but allows cheaper radios to be used.

However, 802.11n also specifies operation in the 5 GHz band, and it is in this band that 802.11n may really shine. The 5 GHz band offers between 11 and 24 twenty megahertz channels, depending on the country. This means that there should be no problem in finding two channels to bond into a 40 MHz channel.

In addition, with so many available channels, access points may operate on several channels simultaneously. This increases the speed available to each user served by the access point.

Voice over IP (VoIP)

If an access point operates on two (or more) channels, one channel can be dedicated to voice over IP, which needs very fast and steady speed. This channel can be kept relatively unpopulated so that VoIP users get good service.

In addition, we are likely to see **802.11e** capabilities built into future products. The 802.11e standard allows managers to specify a very high level of quality of service, which is crucial for VoIP as well as other multimedia applications.

TEST YOUR UNDERSTANDING

17. a) What are the two benefits of MIMO? b) What will be 802.11n's rated speed? c) In what two ways does 802.11n increase throughput? d) How does MIMO work? e) What is the risk in buying draft 802.11n equipment? f) For what unlicensed radio band or bands will the 802.11n standard work? g) What are the two attractions of 5 GHz 802.11n operation? h) What is the advantage of dedicating channels to VoIP service? i) What is the goal of the 802.11e standard?

Advanced Operation

Two other 802.11 developments are beginning to appear in products in very limited ways. In the future, they may become much more important.

Mesh Networking

As we saw at the beginning of this chapter, the main purpose of access points today is to connect wireless hosts to the main wired LAN. However, it is possible to have all-wireless networks. As Figure 5-20 shows, it is theoretically possible for wireless access points to organize themselves into a mesh, routing frames from one to another to deliver frames to the right wireless hosts. Nonstandard 802.11n mesh networking products exist, and the 802.11s standard for mesh networks is under development.

However, significant problems face mesh networking. Meshes must be self-organizing. If hosts and access points enter and leave the mesh frequently, the amount of processing power consumed in maintaining the access points' routing tables could make mesh networking prohibitively expensive.

In addition, it will be difficult to avoid access points near the geographical center of the mesh from being overloaded by the need to route many packets. (Think of sitting in the middle seat at a table during a holiday dinner and constantly having to pass food back and forth.) If mesh networking works, but does not work well, it will have little value.

Figure 5-20 Wireless Mesh Network

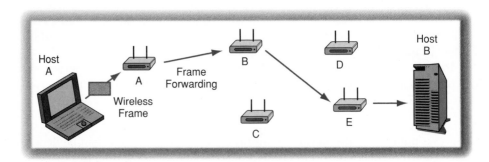

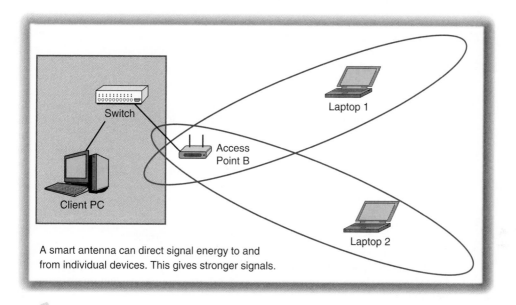

A smart antenna can direct signal energy to and
from individual devices. This gives stronger signals.

Figure 5-21   Smart Antenna

Smart Antennas

Another development will be smart antennas. By having multiple antennas and changing the phase of waves coming from different antennas, an access point can theoretically direct signals toward individual hosts instead of broadcasting them, as Figure 5-21
indicates. This gives more power and, therefore, a stronger signal to each wireless
host. In fact, access points with nonstandard smart antennas are already being sold.

TEST YOUR UNDERSTANDING

18. a) Describe how mesh networking would work in 802.11 WLANs. b) What two prob-
    lems mentioned in the text would 802.11 mesh networking designers have to overcome?
    c) What benefit will smart antennas bring?

802.11 WLAN SECURITY

WLAN Security Threats

For most companies, the biggest problem with 802.11 wireless LANs has been security.
There are four major security threats to 802.11 WLANs:

➤   Most seriously, **drive-by hackers** park just outside a company's premises and eaves-
    drop on the firm's data transmissions. They can also mount denial-of-service attacks;
    send viruses, worms, and spam into the network; and do other mischief. Using

Drive-By Hackers
> Sit outside the corporate premises and read network traffic
> Can send malicious traffic into the network
> Easily done with readily available downloadable software

War Drivers
> Merely discover unprotected access points—become drive-by hackers only if they break in

Rogue Access Points
> Unauthorized access points that are set up by a department or an individual
> Often have very poor security, making drive-by hacking easier
>> Often operate at high power, attracting many hosts to their low-security service

Evil Twin Access Points (Figure 5-23)
> Drive-by hacker sets up an access point outside walls of firm
> Internal hosts associate with the evil twin access point
> Evil twin intercepts authentication credentials, associates with real access point
> Decrypts message traffic, reads it, re-encrypts it, passes it on to a legitimate access point
> Does the same in the other direction

Figure 5-22   WLAN Security Threats (Study Figure)

readily downloaded attack software, drive-by hackers can succeed easily against many WLANs.

➤ Less importantly, **war drivers** are people who, as the name suggests, drive around a city looking for working access points that are unprotected. Only if they try to break in do they become drive-by hackers.

➤ Often, departments and individual employees set up unauthorized access points. These are called **rogue access points**. They often have poor security and provide entry points for drive-by hackers who could not otherwise break into the network. They also frequently operate at high power, attracting many hosts to their poor-security service.

➤ Finally, drive-by hackers can configure their computers to act as access points. These **evil twin** access points operate at high power, enticing internal hosts to associate with them rather than with legitimate internal access points. As Figure 5-23 shows, the evil twin then relays traffic between the duped wireless host and a legitimate access point. It can read the encrypted traffic because it learns the key used in encryption. Evil twin access points are also found in wireless hot spots.

TEST YOUR UNDERSTANDING

19. a) Distinguish between war drivers and drive-by hackers. b) What is an evil twin access point? c) What is a rogue access point? d) Distinguish between evil twin access points and rogue access points.

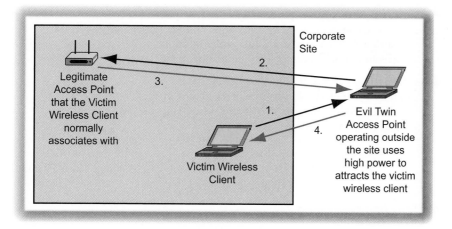

Figure 5-23   Evil Twin Access Point

WEP Security

Problems with 802.11 security began when the first version of the standard was released in 1997. The standard only included a very rudimentary security method called **wired equivalent privacy (WEP)**.[14]

The biggest problem with WEP was that everyone sharing an access point had to use the same key, and WEP had no mechanism for changing this key. In addition, a firm with many access points and users normally gave all access points the same WEP key. When everybody in the firm is told the WEP key, many employees think it is not really secret and often are willing to share the key with unauthorized users. In addition, there is no fast and inexpensive way to change all of the access point keys when an employee is fired.

Shared keys have low security, and WEP had other design problems. By 2001, software that would crack WEP keys quickly was readily available. Initially, cracking a WEP key with one of these programs took one or two hours. Today, it often takes only about 10 minutes. Many companies began putting the brakes on WLAN implementation, and many pulled out their existing access points.[15]

TEST YOUR UNDERSTANDING

20. a) When 802.11 was created, what security protocol did it offer? b) How long does it take to crack WEP today?

---

[14]The WEP specification was only 10 pages long. In contrast, the specification for 802.11i security (discussed later) is 200 pages long.
[15]Some companies began taking other steps, like hiding the SSID (Service Set Identifier) of the access point. Users need to know this SSID to use an access point even if WEP is not used. Another common step was to only accept computers whose wireless NICs had registered MAC addresses. (All 802 LANs have 48-bit MAC addresses.) Unfortunately, these measures take a great deal of work, and they are easily cracked by readily available hacking software. They make sense only if you are only concerned about unsophisticated but nosy neighbors at home.

Provide Security between the Wireless Station and the Wireless Access Point
  Client (and perhaps access point) authentication
  Encryption of messages for confidentiality

Wired Equivalent Privacy (WEP)
  Initial rudimentary security provided with 802.11 in 1997
  Everyone shared the same secret encryption key, and this key could not be changed automatically
  Because secret key was shared, it does not seem to be secret
    Users often give out freely
  Key initially could be cracked in 1–2 hours; now can be cracked in 3–10 minutes using readily available software

Wireless Protected Access (WPA)
  The Wi-Fi Alliance
    Normally certifies interoperability of 802.11 equipment
    Created WPA as a stop-gap security standard in 2002 until 802.11i was finished
  Designed for upgrading old equipment
    WPA uses a subset of 802.11i that can run on older wireless NICs and access points
    WPA added simpler security algorithms for functions that could not run on older machines
  Equipment that cannot be upgraded to WPA should be discarded

802.11i (WPA2)
  Uses AES-CCMP with 128-bit keys for confidentiality and key management
  Gold standard in 802.11 security
  But companies have large installed bases of WPA-configured equipment

Figure 5-24 802.11 Security Standards (Study Figure)

WPA (Wireless Protected Access)

The **Wi-Fi Alliance** is an industry trade group that certifies 802.11 products for interoperability. Normally, the Wi-Fi Alliance leaves standards creation to the 802.11 Working Group. However, when WEP's fatal flaws were discovered in 2000 and 2001, the market for wireless products began to falter. Furthermore, the 802.11i security standard being developed by the 802.11 Working Group was going to take years to develop.

As a stop-gap measure until 802.11i could be developed, the Wi-Fi Alliance created an interim security standard, **Wireless Protected Access (WPA)**. It announced this standard in 2002 and began certifying products for compliance in early 2003.

The alliance decided to keep protection fairly simple so that many older NIC and access point products, which had limited memory and little processing power, could be upgraded to WPA via software only. Older products would not be upgradeable to full 802.11i security, which requires more memory and processing power in wireless NICs and access points. The ability to upgrade many existing products meant that WPA protected the installed base of WLAN equipment for corporations. More specifically, WPA consists of the parts of 802.11i that would fit on most existing products so that installed products could be updated to WPA. For other parts of 802.11, the alliance chose weaker processes that were easier to implement.

WPA provides far stronger security than WEP does. Firms that use WPA should discard legacy wireless NICs and access points that cannot be upgraded to WPA.

TEST YOUR UNDERSTANDING

21. a) Who created WPA? b) What is WPA's attraction compared with 802.11i? c) What is WPA's disadvantage compared with 802.11i? d) What should companies do if they have access points or NICs that cannot be upgraded to WPA?

802.11i (WPA2)

In 2004, the 802.11 Working Group finally ratified the **802.11i standard**. Most important, the 802.11i standard uses extremely strong AES-CCMP encryption for confidentiality. **AES-CCMP** has 128-bit keys and a key management method for changing keys that was too processing intensive to be included in WPA. Confusingly, the Wi-Fi Alliance refers to the 802.11i standard as **WPA2**.

Today, 802.11i is the gold standard in WLAN security. Many corporations are now restricting their purchases of products that do not support 802.11i. They also require the discarding of all existing wireless LAN products that do not have WPA and cannot be upgraded to WPA.

However, most companies have already implemented WPA on their access points and wireless hosts. Reconfiguring all of these devices to work with 802.11i would be expensive, and until there are known cracks for WPA, companies will be reluctant to make this investment.

TEST YOUR UNDERSTANDING

22. a) What is the strongest security protocol for 802.11 today? b) What does the Wi-Fi Alliance call 802.11i? c) What encryption method does 802.11i use? d) What is deterring companies from converting from WPA to 802.11i?

802.1X Mode Operation with Secure EAP

WPA and 802.11i have two modes of operation. For large firms, the only mode that makes sense is **802.1X mode**. (The Wi-Fi Alliance calls it **enterprise mode**.) In Chapter 4, we saw that 802.1X was created for Ethernet wired networks. Its goal is to prevent attackers from simply walking into a building and plugging a computer into any wall jack or directly into a switch.

Each workgroup switch acts as a network access server (NAS). The computer wishing access to the Ethernet network connects to the NAS via UTP. Actual authentication is done by a central authentication server that normally uses the RADIUS protocol.

For Ethernet access, there is no need to have security between the computer seeking access and the workgroup switch that controls access. It is difficult for another person to tap the wired access line between the computer and the switch, and there are easier ways to break into a network. Consequently, as Figure 5-26 shows, there is no

802.1X Mode (See Figure 5-26)
      Uses a central authentication server for consistency and speed of change
      Authentication server also provides key management
      Wi-Fi Alliance calls this *enterprise mode*
      802.1X standard protects communication with an extensible authentication protocol
            Several EAP versions exist with different security protections
            Firm implementing 802.1X must choose one
            Protected EAP (PEAP) is popular because Microsoft favors it

Pre-Shared Key (PSK) Mode: Stations Share a Key with the Access Point
      For networks with a single access point
      Access point does all authentication and key management
      All users must know an initial pre-shared key (PSK)
      Each, however, is later given a unique key
      If the pre-shared key is weak, it is easily cracked
      Pass phrases that generate key must be at least 20 characters long
      Wi-Fi Alliance calls this *personal mode*

Figure 5-25   802.11 Security in 802.1X and PSK Modes (Study Figure)

Figure 5-26   802.1X with Ethernet and 802.11 WLANs

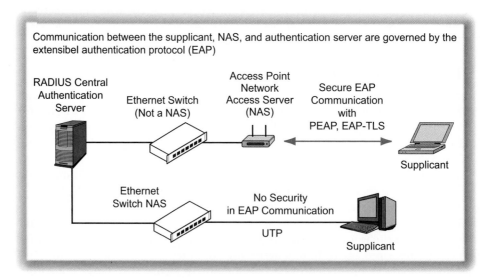

Communication between the supplicant, NAS, and authentication server are governed by the extensibel authentication protocol (EAP)

RADIUS Central Authentication Server

Ethernet Switch (Not a NAS)

Access Point Network Access Server (NAS)

Secure EAP Communication with PEAP, EAP-TLS

Supplicant

Ethernet Switch NAS

No Security in EAP Communication

UTP

Supplicant

security between the wired hosts and the switch that is the 802.1X network access server.

With access points, however, transmissions between a wireless host and the access point are easy to intercept and mimic. The path between the host and the access point needs to be secure.

Although it is not important to understand the details (unless you take a security course), authentication in 802.1X uses the **extensible authentication protocol (EAP)**. To secure communication in wireless access, EAP extensions were developed to add security.

Several extensions to EAP have been developed and are used to secure authentication between access point and the host. The most widely used extension is **Protected EAP (PEAP)**, which is Microsoft's preferred extension. Other popular extensions are **EAP-TLS** and **EAP-TTLS**. It is important to configure both the wireless computer and the access point with the same EAP extension.

TEST YOUR UNDERSTANDING

23. a) In what mode of 802.11i and WPA operation is a central authentication server used? b) What does the Wi-Fi Alliance call 802.1X mode? c) Why does 802.1X not need security between the NAS and the computer in Ethernet? d) Why does 802.1X need security between the NAS and the computer in 802.11 wireless access? e) What standard governs authentication exchanges in 802.1X? f) Why are EAP extensions necessary in 802.11 networks using 802.1X? g) What are three security extensions to EAP? h) Which EAP security extension is favored by Microsoft?

Pre-Shared Key (PSK) Mode

For homes and small businesses, which cannot afford a separate authentication server, 802.11i and WPA offer a simpler operating mode, **pre-shared key (PSK)** mode. PSK does all authentication and key management. The Wi-Fi Alliance calls this **personal mode**.

In this mode, normally used when there is only a single access point, the access points and hosts begin with a shared 64-bit key. Everybody allowed to use the access point is told the shared key. This sounds like WEP, but after authentication, the wireless access point gives each authenticated user a new unique key to use while on the Internet. It also changes this key frequently.

Although PSK mode can be very strong, if the organization implements a weak key as its pre-shared key, the key can be cracked even faster than a WEP key. To create the pre-shared key, the company usually creates a long **pass phrase** (which is much longer than a password) that the access point and host use to generate the key. This pass phrase must be at least 20 characters long to provide adequate security. In fact, if short pass phrases are used, 802.11i and WPA in PSK mode are more easily cracked than WEP.

Using Virtual Private Networks (VPNs)

Chapter 7 discusses **virtual private networks (VPNs)**. VPNs implement security all the way from the host to the server (or to an intermediate termination device). This security is independent of the transmission path. (Think of it as an add-on layer of security.) Consequently, if a mobile host communicates with a server using VPN security, evil twin access points will not be able to decrypt the traffic. Nor will drive-by hackers be able to read messages even if they crack 802.11 security. Unfortunately, VPNs are

Virtual Private Networks (VPNs)
> Provides end-to-end protection from the client all the way to the server on the wired LAN
> If 802.11 security is defeated by an evil twin or by cracking 802.11 keys, communication between the client and the server will still be protected
> Somewhat expensive to implement
> Of greatest importance in high-threat environments, like public hot spots

Virtual LANs (VLANs)
> Client stations are initially assigned to a virtual LAN that only has access to an authentication server
> Only after they successfully authenticate themselves are they allowed into the broader LAN
> Especially useful in universities, where requiring strong 802.11 security on clients is impractical

Figure 5-27   Added Wireless Protection: VPNs and VLANs (Study Figure)

somewhat expensive to implement. However, they may be worth the cost, especially in high-threat areas such as public hot spots.

Virtual LANs

In Chapter 4, we saw that Ethernet networks can implement **virtual LANs (VLANs)**. Recall that hosts in a VLAN can only communicate with other systems on the same VLAN. Some organizations put all switch ports attached to wireless access points on a single VLAN. When a wireless host connects to an access point, it does not have access to other parts of the corporate network. Only after the hosts authenticate themselves to a corporate authentication server are they permitted to enter the broader network.

TEST YOUR UNDERSTANDING

24. a) How does PSK mode differ from 802.1X mode? b) What is a potential weakness of PSK mode? c) How long should pass phrases be with PSK? d) How do VPNs enhance wireless security? e) How can VLANs enhance wireless security?

802.11 WIRELESS LAN MANAGEMENT

Until recently, the term *WLAN management* was almost an oxymoron. Large WLANs were like major airports without air traffic control towers. To a considerable extent, this condition is still true today, but there has been some progress.

Access Point Placement

To provide WLAN service throughout a building, it is crucial to determine where to place access points. Otherwise, there will be many dead spots, interference between many access points, or both.

Access Points Placement in a Building
>   Must be done carefully for good coverage and to minimize interference between access points
>   Lay out 30-meter to 50-meter radius circles on blueprints
>   Adjust for obvious potential problems such as brick walls
>   In multistory buildings, must consider interference in three dimensions
>   Install access points and do site surveys to determine signal quality
>   Adjust placement and signal strength as needed

Remote Access Point Management
>   The manual labor to manage many access points: can be very high
>   Centralized management alternatives (See Figure 5-29)
>>   Smart access points
>>   Dumb access points, with intelligence in WLAN switches
>   Desired functionality
>>   Notify the WLAN administrators of failures immediately
>>   Support remote access point adjustment
>>   Should provide continuous transmission quality monitoring
>>   Allow software updates to be pushed out to all access points or WLAN switches
>>   Work automatically whenever possible

Figure 5-28   Wireless LAN Management (Study Figure)

Initial Planning

The first step is to determine how far signals should travel. In many firms, a good radius is 30–50 meters. If the radius is too great, many hosts will be far from their access points. Hosts far from the access point must drop down to lower transmission speeds, and their frames will take longer to send. This will reduce the access point's effective capacity. If the radius is too small, however, the firm will need many more access points to cover the space to be served.

Once an appropriate radius is selected (say, 30 meters) the company gets out its building blueprints and begins to lay out 30-meter circles with as little overlap as possible. Where there are thick walls or other obstructions, shorter propagation distances must be assumed. In addition, in a multistory building, this planning must be done in three dimensions. Also, planners must install channels to access point positions to minimize interference.

Installation and Initial Site Surveys

Next, the access points are installed provisionally in the indicated places. However, the implementation work has just begun. When each access point is installed, an **initial site survey** must be done of the area around the access point to discover whether there are any dead spots or other problems. This requires a signal analyzer, which can be a personal digital assistant (PDA) or a notebook computer with signal strength analysis software.

When areas with poor signal strength are found, surrounding access points must be moved appropriately, or their signal strengths must be adjusted until all areas have good signal strength.

Ongoing Site Surveys

Although the initial site survey should result in good service, conditions will change constantly. More people may be given desks in a given access point's range, signal obstructions may be put up for business purposes, and other changes must occur. Site surveys must be done periodically to ensure good service. They may also be done in response to specific reports of problems.

TEST YOUR UNDERSTANDING

25. a) Describe the process by which access point locations are selected. b) After access points are installed provisionally, what must be done next?

Remote Management: Smart Access Points and Wireless Switches

Large organizations have hundreds or even thousands of 802.11 wireless access points. Traveling to each one for manual configuration and troubleshooting would be extremely expensive. To keep management labor costs under control, organizations need to be able to manage access points remotely, from a central management console. Figure 5-29 illustrates two approaches to centralized wireless access point management.

Figure 5-29   Wireless Access Point Management Alternatives

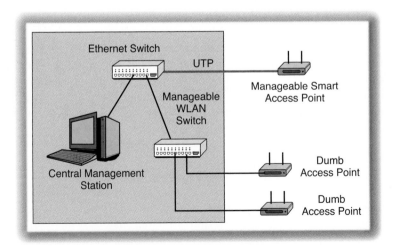

Smart Access Points

The simplest approach architecturally is to add intelligence to every access point. The central management console can then communicate directly with each of these **smart access points**[16] via the firm's Ethernet wired LAN. However, adding management capacity raises the price of access points considerably. Using smart access points is an expensive strategy.

Wireless LAN Switches

A second approach illustrated in Figure 5-29 is to use **WLAN switches**. As the figure shows, multiple access points connect to each wireless LAN switch. The management intelligence is placed in the WLAN switch rather than in the access points themselves. Vendors who sell WLAN switches claim that this approach reduces total cost because only inexpensive **dumb access points** are needed. Of course, smart access point vendors dispute these cost comparisons.

Wireless LAN Management Functionality

Although technological approaches to centralized WLAN management vary, vendors generally agree on the types of functionality these systems should provide.

➤ These systems should notify the WLAN administrators of failures immediately so that malfunctioning access points can be fixed or replaced rapidly.

➤ They should provide continuous transmission quality monitoring at all access points. In effect, they should provide continuous site surveys.

➤ They should help provide security by detecting rogue access points, evil twin access points, or legitimate access points that have improperly configured security.

➤ They should allow remote adjustment—for instance, telling nearby access points to increase their power to compensate for an access point failure. Such adjustments are also needed over time as furniture is moved (creating different shadow zones) or as the number of users in an area changes.

➤ They should allow software updates to be pushed out to all access points or WLAN switches, bypassing the need to install updates manually.

➤ The management software should be able to work automatically, taking as many actions as possible without human intervention.

TEST YOUR UNDERSTANDING

26. a) Why is centralized access point management desirable? b) What are the two technologies for remote access point management? c) What functions should remote access point management systems provide?

OTHER LOCAL WIRELESS TECHNOLOGIES

Although 802.11 WLANs are dominating corporate attention today, other local wireless technologies are important or may soon become important.

---

[16] Smart access points are also called fat access points.

For Personal Area Networks (PANs)
    Devices on a person's body and nearby (mobile phone, PDA, notebook
    computer, etc.)
    Devices around a desk (computer, mouse, keyboard, printer)

Cable Replacement Technology
    For example, with a Bluetooth PDA, print wirelessly to a nearby Bluetooth-
    enabled printer
    Does not use access points
        Uses direct device-to-device communication
    Disadvantages Compared with 802.11
    Short distance (10 meters)
    Low speed (3 Mbps today, with a slower reverse channel)

Advantages Compared with 802.11
    Low battery power drain, so long battery life between recharges
    Application profiles (printing, etc.)
        Somewhat rudimentary
        Devices typically automate only a few

Expanding Bluetooth Radio Options
    The idea: Run Bluetooth application profile standards over other radio standards
    Bluetooth over USB (described later): 480 Mbps to 10 meters
    Bluetooth over 802.11:802.11 speeds and distances!

Figure 5-30   Bluetooth Personal Area Networks (PANs) (Study Figure)

Bluetooth Personal Area Networks (PANs)

Personal Area Networks (PANs) for Cable Replacement

Although 802.11 is good for fairly large wireless LANs, another wireless networking standard, **Bluetooth**,[17] was created for wireless **personal area networks (PANs)**, which are intended to connect devices used by a single person. Bluetooth basically offers *cable replacement*—a way to get rid of cables between devices. It is not designed for full WLANs.

Using Bluetooth, a notebook computer can print wirelessly to a printer and synchronize its files wirelessly with those on a desktop computer. To give another example, a mobile phone can print to the same wireless printer and place a call through the firm's telephone system instead of paying to make a cellular call. Bluetooth does not use access points. Rather, it uses direct device-to-device communication.

Disadvantages Compared with 802.11

Limited Distance   While 802.11i normally has propagation distances up to 100 meters, 30–50 meters is a more realistic range in practice. Bluetooth, however, is normally

---

[17]Bluetooth is named after King Harald Bluetooth, a Scandinavian king of the tenth century. As you might guess, Bluetooth was developed in Sweden, although it is now under the control of an international consortium.

limited to 10 meters or less.[18] For cable replacement around a desk or among the devices carried by a person, there is no need for longer distances. For a home or office WLAN, this distance is insufficient.

Low Speed   Bluetooth was not designed to handle heavy transmission loads. It currently offers a speed of only 3 Mbps with a slower reverse channel. This is sufficient for printing and most other Bluetooth applications, but it is not fast enough for WLANs.

Advantages Compared with 802.11

Long Battery Life   Although Bluetooth offers only low speeds and short distances, these limitations mean that radio transmission power is low, so battery life is quite long. This is very important for small portable devices.

Application Profiles   Bluetooth has one very important capability that 802.11 does not—**application profiles**, which are application-layer standards designed to allow devices to work together automatically, with little or no user intervention. For instance, you may be able to take a Bluetooth-enabled notebook computer to a Bluetooth-enabled printer and print as soon as the two devices recognize each other. The 802.11 standard has nothing like this feature. Unfortunately, the Bluetooth application profiles introduced to date have been rudimentary. In addition, most devices implement only a few of these application profiles, so there is no guarantee that two Bluetooth devices that you wish to connect will be able to work together.

Expanding Bluetooth Radio Options

Bluetooth's popularity has been good, but its radio technology has limited it to low speeds and short distances, In the near future, Bluetooth may break those limitations by expanding beyond its original radio technology. As noted in the next subsection, Bluetooth's application profiles and higher-level components may soon run over ultrawideband radio technology, which will permit Bluetooth to operate at 480 Mbps over traditional short Bluetooth distances (10 meters). Even more importantly, Bluetooth over the next five years may operate over 802.11 radio technology at the physical and data link layers. This will permit it to operate at tens of megabits per second over 100 meters or more. These changes may radically change the uses of Bluetooth.

TEST YOUR UNDERSTANDING

27. a) Contrast how 802.11 and Bluetooth are likely to be used in organizations. b) What is a PAN? c) Why are Bluetooth application profiles attractive? d) Why do they not always fulfill their promise? e) What are the speeds of Bluetooth transmission today? f) What is the normal maximum distance for Bluetooth propagation? g) What benefit do low speeds and short distances bring? h) How may using different radio technologies extend Bluetooth's speed and distance limitations.

---

[18]This is with the standard 2.5 milliwatts (mW) of power. There is an option for 100 mW power, which can raise the maximum propagation distance to 100 meters. This option is rarely built into Bluetooth products.

Ultrawideband (UWB)
> Uses channels several gigahertz wide (spans multiple frequency bands)
> Low power per hertz to avoid interference still gives very high speeds
> But limited to short distance
> Wireless USB provides 480 Mbps up to 3 meters, 110 Mbps up to 10 meters

ZigBee for Almost-Always-Off Sensor Networks at Low Speeds
> Very long battery life
> 250 kbps maximum

RFIDs: Like UPC Tags, But Readable Remotely

Software-Defined Radio
> Can implement multiple wireless protocols
> No need to have separate radio circuits for each protocol
> Reduces the cost of multi-protocol devices

Figure 5-31 Emerging Local Wireless Technologies (Study Figure)

Emerging Local Wireless Technologies

Ultrawideband (UWB) Transmission

We saw earlier in this chapter that maximum speed in a channel is governed primarily by a channel's bandwidth. However, spread spectrum transmission uses much wider channel bandwidths than it needs for its transmission speeds. It does this because regulators require spread spectrum transmission in the unlicensed 2.4 GHz and 5 GHz bands in order to improve transmission.

An expanded form of the spread spectrum transmission is **ultrawideband (UWB)** transmission. While 802.11 spread spectrum transmission uses channels that are 20 MHz to 40 MHz wide, UWB uses channels that may be a gigahertz or wider. These UWB channels actually span several entire service bands.

However, the amount of energy transmitted per hertz of bandwidth is very small in UWB. This prevents it from interfering with other transmissions within its huge range of frequencies. However, small energy per hertz times huge bandwidths is capable of transmitting data at high speeds, although only over short distances. This makes UWB capable of delivering video (even in high definition) and other bandwidth-hungry applications.

Ultrawideband transmission is already beginning to appear in actual products. The first application has been **wireless USB**. Normal wireless USB ports and cords can deliver speeds of 480 Mbps over short distances. Wireless USB, in turn, can transmit at 480 Mbps up to 3 meters and 110 Mbps up to 10 meters. Computers and other devices with built-in USB are already on the market. As noted in the last subsection, Bluetooth will also move to USB over short distances. The high-speed UWB version of Bluetooth may be defined before this book is published.

ZigBee

The oddly named **ZigBee** technology is designed for wireless monitoring and control systems in businesses and homes. While Bluetooth is an always-on technology, ZigBee is an almost-always-off technology for sensors that rarely send signals. In addition, when sensors and other ZigBee devices do transmit signals, they only need to send them at very slow speeds (up to 250 kbps, but usually slower). Such low performance requirements actually have a benefit: They lead to extremely long battery lives—months or even years.

Radio-Frequency IDs (RFID)

Today, bar-coded products must be run carefully over a laser scanner. In contrast, new **radio frequency ID (RFID)** tags can be used in place of UPC tags and only have to be brought *near* an RFID scanner to be read. An RFID reader sends a probing radio signal, and the RFID tag responds with the information it contains.

There are several RFID technologies. Active RFID tags have batteries and can be read meters away. They can be used to record such things as when a firm's delivery trucks arrive and leave. Passive RFID tags derive their power from the radio signal sent by the reader. Passive RFID tags can only absorb a little power from the signal, so when they respond, their signals will not travel more than a few centimeters to a bit more than a meter. Passive ID tags are placed on pallets and boxes, but they are still much too expensive ($0.20 to $0.30 apiece) to be placed on most individual products.

Software-Defined Radio

Another trend to watch is **software-defined radio**, which will enable a wireless device to switch between 802.11, Bluetooth, ZigBee, and other wireless standards by switching software programs instead of having the functionality built entirely into hardware. In contrast, today's hardware-only approach to implementing multiple technologies is expensive because it requires multiple circuits to be built into devices.

TEST YOUR UNDERSTANDING

28. a) Compare the speeds of 802.11, Bluetooth, UWB, and ZigBee. b) Compare the distance limits of 802.11, Bluetooth, and UWB. c) What technology could replace universal product code (bar code) tags on products? d) What is the major promise of software-defined radio?

CONCLUSION

Synopsis

The mantra of networking has always been "anything, anywhere, any time." With the advent of wireless data transmission, this promise is finally being extended to mobile users. This chapter focused on 802.11 WLANs and, to a lesser extent, on Bluetooth. However, other local wireless technologies are appearing, including RFIDs, ultrawideband (UWB) transmission, ZigBee, and software-defined radio.

Wireless networks predominantly use radio transmission. Radio waves are described by frequency (hertz). Real radio signals contain a mix of frequencies. The range of frequencies between the lowest frequency and the highest frequency of a signal is the signal's bandwidth.

The frequency spectrum consists of all frequencies from 0 Hz to infinity. The frequency spectrum is divided into service bands for particular types of services. Service bands are further divided into channels. Different signals can be sent at the same time if they are sent in different channels.

The 802.11 standards use unlicensed radio bands. This means that you can set up access points and move them without worrying about licenses. However, it also means that your network may get interference from a neighbor's network operating in the same unlicensed band.

Most 802.11n standards operate in the 2.4 GHz unlicensed band. This includes the dominant 802.11g standard. Propagation in the 2.4 GHz band is very good, and the band is defined the same way in almost all countries. However, there are only three non-overlapping channels in this band. This makes it difficult to install access points that do not interfere with each other. The 5 GHz band is the mirror image of the 2.4 GHz band. Signals do not travel as far at 5 GHz, and different countries define the band somewhat differently. However, the 5 GHz band offers 11 to 24 channels, which prevents access point interference and even allows an access point to offer service on two channels.

The maximum possible transmission speed within a channel is directly proportional to the channel's bandwidth. Normally, channel bandwidth is set only wide enough to meet transmission speed requirements. However, spread spectrum transmission uses a much higher bandwidth than the signal requires. Spread spectrum transmission is done to reduce propagation problems, which often occur only at certain frequencies. It also reduces mutual interference between nearby devices transmitting in the same channel. Most wireless 802.11 WLAN standards today use orthogonal frequency division multiplexing (OFDM) spread spectrum transmission, which divides the broadband channel into many smaller subcarriers, each of which carries part of the signal.

The 802.11 Working Group sets most wireless LAN (WLAN) standards. Normally, 802.11 wireless LANs serve users through access points, which connect wireless hosts to resources on the company's wired LAN. The most widespread WLAN technology today is 802.11g, which has a rated speed of 54 Mbps. Actual throughput (speed delivered to users) is about half of the rated speed near the wireless access point and falls off with distance. In addition, all hosts using an access point share this throughput, so individual throughput is even lower.

The 802.11n technology, which is currently a draft standard, will use MIMO technology and will have the option of using wider channels. Compared with older standards, 802.11n will produce both much faster communication and longer propagation distances. In addition, 802.11n is designed to operate in both the 2.4 GHz band and the 5 GHz band. These factors are likely to make 802.11n dominant in the future. For now, however, many devices labeled as "draft 802.11n" may only be upgradable to the final standard at considerable cost, if they can be upgraded at all.

The box entitled "Controlling 802.11 Transmission" demonstrated that the access point and the wireless hosts it serves must share a single channel. Only one device may transmit at any time. This means that as the number of wireless hosts increases, individual throughput goes down. To control when hosts may transmit, the access point and hosts implement either the CSMA/CA+ACK or RTS/CTS media access control (MAC) method. Apart from one special case, CSMA/CA+ACK is mandatory while RTS/CTS is optional.

Wireless LANs create security threats because radio signals spread over considerable distances. Drive-by hackers can listen in on traffic or send messages into the network. Evil twin access points can "get between" the wireless host and a legitimate access point. Rogue access points are set up by employees without permission and often without security.

The first 802.11 LANs used weak wired equivalent privacy (WEP) security, and many users did not even turn on this anemic protection. Even with WEP enabled, drive-by hackers could easily eavesdrop on conversations and send attack packets into a network. The 802.11 Working Group finally created a robust security standard, 802.11i; but the development took several years, and older access points and wireless NICs cannot be upgraded to 802.11i. The interim WPA standard has aspects of 802.11i that can be applied to older wireless NICs and access points. Both 802.11i and WPA can operate in 802.1X (enterprise) mode, which uses a central authentication server and extensible authentication protocol communication. These two standards also can work in pre-shared key (PSK) mode, which uses a shared key based on a pass phrase. If the pass phrase is too short (fewer than 20 characters), security will be weaker than it was with WEP. To further increase security, companies can use additional VPN or VLAN security.

Wireless LANs need extensive management. The first concern is where to place access points in a building to provide good service. In addition, there now are technologies to manage all of a site's access points from a central location. Some vendors do remote management through expensive smart access points. Others use traditional "dumb" access points, and the management intelligence is placed in wireless LAN switches.

Bluetooth today is a low-speed, short-distance personal area network (PAN) technology designed to replace wired connections between devices within a few meters of each other. In the future, ultrawideband (UWB) transmission should increase Bluetooth speeds considerably. UWB is already used in 480 Mbps wireless USB.

End-of-Chapter Questions

THOUGHT QUESTIONS

1. Telephone channels have a bandwidth of about 3.1 kHz, as we will see in the next chapter.

    a) If a telephone channel's signal-to-noise ratio is 1000:1, how fast can a telephone channel carry data? (Check figure: Telephone modems operate at about 30 kbps, so your answer should be roughly this speed.)

    b) How fast could a telephone channel carry data if the SNR were increased to 10,000:1?

    c) With an SNR of 1000:1, how fast could a telephone channel carry data if the bandwidth were increased to 4 kHz? Show your work or no credit.

2. A friend wants to install a wireless home network. The friend wants to use either 802.11n or 802.11g. What advice would you give your friend?

3. Suppose that 802.11a access points have a range of 50 meters and that 802.11g access points have a range of 100 meters. If a firm needs 30 access points with 802.11g, how many will it need with 802.11a?

4. What advice would you give a company about WLAN security? This answer should be about a full page, single spaced. It requires you to integrate what you learned in this chapter. This is not a short answer.

5. a) Do you think that 802.11g and Bluetooth might interfere with each other if they are used in the same office? Explain your reasoning.

   b) Do you think that 802.11a and Bluetooth might interfere with each other if they are used in the same office? Explain your reasoning.

DESIGN QUESTION

Consider a square one-story building. It will have an access point in each corner and one in the center of the square. *All access points can hear one another.*

a) Assign access point channels to the five access points if you are using 802.11g. Try not to have any access points that can hear each other use the same channel. Available channels are 1, 6, and 11.

b) Were you able to eliminate interference between access points?

c) Repeat the first two parts of the question, this time using 802.11a. Available channels are 36, 40, 44, 48, 53, 56, 60, 64, 149, 153, 157, and 161, but many NICs and access points only support channels below 100.

TROUBLESHOOTING QUESTION

When you set up an 802.11g wireless access point in your small business, your aggregate throughput is only about 6 Mbps. List at least two possible reasons for this low throughput. Describe how you would test each. Describe what you would do if each proved to be the problem.

INTERNET EXERCISE

1. Home wireless access points today are inexpensive, but enterprise access points cost much more. Go to cnet.com and compare prices for Cisco's residential Linksys access points and Cisco's Aironet enterprise access points. List model numbers and other product details. Speculate on what may be causing the price difference.

PERSPECTIVE QUESTIONS

1. What was the most surprising thing you learned in this chapter?

2. What was the most difficult material for you in this chapter?

PROJECT

Write a short report on a new wireless technology—UWB, ZigBee, mesh networking, software-defined radio, or smart antennas.

GETTING CURRENT

Go to the book website's New Information and Errors pages for this chapter to get new information since this book went to press and to view corrections of any errors in the text.

Telecommunications

Learning Objectives

By the end of this chapter, you should be able to discuss the following:

- Telecommunications.
- The technology of the public switched telephone network (PSTN), including customer premises equipment, the access system, the transport core, and signaling.
- Circuit switching.
- The digital nature of the PSTN, except for the analog local loop to residential customer premises.
- (In a box) PCM conversion of customer signals at the end office.
- Cellular telephony.
- Voice over IP (VoIP).
- The last kilometer (access technologies), including telephone modems, cable modems, DSL, 3G cellular services, WiMAX, and fiber to the home (FTTH).

INTRODUCTION

Telecommunications and the PSTN

Traditionally, voice and data were carried by separate networks. We have seen since Chapter 1 that data is normally transmitted through packet switching. We will see later in this chapter that voice (and video) transmission have traditionally used a very different technology called circuit switching. Due to these traditional technological differences, corporations customarily have built and maintained separate networks for data and voice.

In this book, we will be concerned primarily with data transmission. However, corporate networking staffs also have to deal with voice and video communication, which are traditionally known collectively as **telecommunications.**

Voice and video communication are known collectively as telecommunications.

The Public Switched Telephone Network (PSTN)

The worldwide telecommunications network is officially called the **public switched telephone network (PSTN).** Every time you place a telephone call anywhere in the world, your call travels over the PSTN. Although the PSTN carries video as well as voice traffic, it is still called the public switched *telephone* network.

Telecommunications Carriers and Wide Area Networks

Why are we looking at telecommunications at this point in the book? The answer is that we need a transition between switched local area networks, which we saw in Chapters 4 and 5, and switched wide area networks, which Chapter 7 discusses. Telephony begins taking us outside of the corporate walls.

More fundamentally, we will see in Chapter 7 that wide area data networks normally use the existing telecommunications transmission system for data transport—adding switching and other functionality in order to be able to handle data. The carriers that provide telephone service to your home and company either provide data service or support those carriers which do.

TEST YOUR UNDERSTANDING

1. a) What is the name of the worldwide telecommunications network? b) Does the PSTN handle only telephone traffic? Explain. c) Why are we looking at the PSTN in this chapter?

The Four Elements of the PSTN

Figure 6-1 shows the four technical elements of today's worldwide public switched telephone network: customer premises equipment, access, transport, and signaling.

Figure 6-1   Elements of the Public Switched Telephone Network (PSTN)

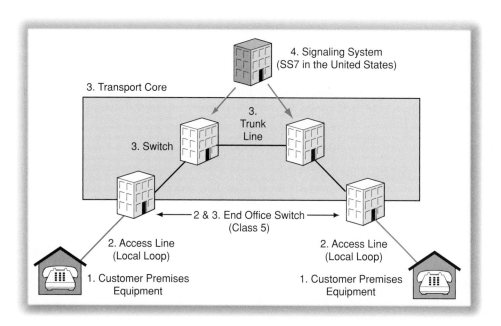

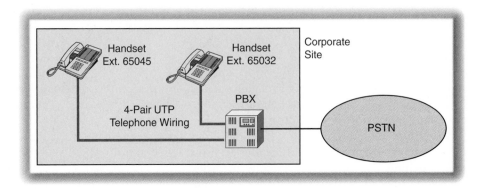

Figure 6-2   Customer Premises Equipment at a Business Site

Customer Premises Equipment

Equipment owned by the customer is called **customer premises equipment.** *Customer premises* is always written in the plural.

Figure 6-2 shows that there are three elements to customer premises equipment at a business site—the telephone handsets, wiring, and a device called a private brafnch exchange (PBX).

The PBX is like an internal telephone switch. The PBX routes internal calls between handsets at the site, and it routes calls between the firm and public switched telephone networks.

Business telephone wiring in buildings uses 4-pair UTP. In fact, 4-pair UTP was created for business telephone usage and was used for business telephony long before it was used to carry data. Given 4-pair UTP's ubiquity in firms, it simply made sense to use 4-pair UTP for LAN data transmission.

Wiring is the most expensive component in customer premises equipment. Fortunately, once wiring is laid—at the cost of several hundred dollars per wall jack—maintenance cost usually is modest.

Access

The customer needs an **access line** to reach the PSTN's central transport core. Collectively, access lines are known as the **local loop.** The **access system** includes both access lines and **termination equipment** in the **end office** at the edge of the transport core.

Although access systems are relatively simple technologically, they represent a huge capital investment. The hundreds of millions of access lines that run to customer premises cost much more collectively than the trunk lines that run between the internal switches in the transport core. Similarly, there are many more end office switches in the PSTN than there are internal switches. The huge capital investment in today's access system is an impediment both to change and to the entry of new access competitors.

Transport

Transport means transmission—taking voice signals from one subscriber's access line and delivering them to another customer's access line. Internally, the **transport core** consists of trunk lines and switches. The end office switch is the transition point between the access system and the transport core, and it is a member of both.

While changes in access line technology have been slow, changes in the transport core have been very rapid (at least by telephony's standards). Changing the transport core represents a smaller investment than changing the access system, and changes in the transport core can save carriers a great deal of money.

Signaling

Finally, **signaling** means the controlling of calling, including setting up a path for a conversation through the transport core, maintaining and terminating the conversation path, collecting billing information, and handling other supervisory functions.

In the PSTN, *transport* is the transmission of voice communication. In contrast, *signaling* is the process of supervising voice communication sessions.

TEST YOUR UNDERSTANDING

2. What are the four technical elements in the PSTN?
3. a) What is customer premises equipment? b) What is the purpose of the PBX? c) What type of wiring does business telephony use for building wiring? d) What is the most expensive part of customer premises equipment to purchase and install?
4. a) What is the local loop? b) What is the function of the transport core? c) What are the two elements of the transport core? d) Which is changing more rapidly—the access system or the transport core? e) Explain why.
5. a) What is signaling? b) In telephony, distinguish between transport and signaling.

Who Owns the PSTN?

The public switched telephone network normally works so well that most people do not realize that the worldwide telephone system really is a linked collection of smaller telephone networks owned by different carriers. Generally speaking, these carriers fall into three tiers:

➤ Local carriers provide access lines and handle transport within a city or other small area.

➤ Long-distance domestic (within a country) carriers transport traffic between different local areas.

➤ International carriers transport traffic between countries.

To complicate matters, many companies fit into multiple categories, offering two or even three layers of service. We are even beginning to see carriers operating in multiple countries.

In Chapter 1, we saw a similar situation—the organization of the Internet. Although there are many ISPs, they all follow the same technical standards (TCP/IP), and they all connect together at network access points. Similarly, all telephone companies follow international standards, and they are all interconnected. As Figure 6-3 shows, these interconnection points are called **points of presence (POPs)** in the United States.

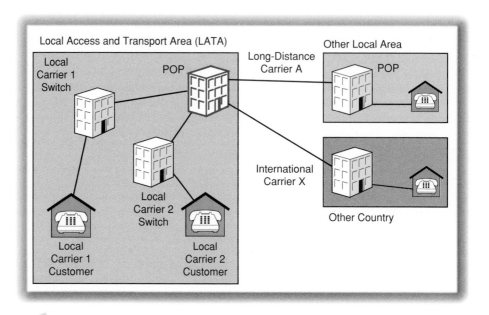

Figure 6-3   Points of Presence (POPs)

TEST YOUR UNDERSTANDING

6. a) What are the three tiers of carriers in the PSTN? b) What are connection points between carriers called in the PSTN?

CIRCUIT SWITCHING

Circuits

In contrast to the packet-switched networks that we saw in previous chapters, the telephone system has traditionally offered **circuit switching,** in which capacity for a voice conversation is reserved on every switch and trunk line end-to-end between the two subscribers. (See Figure 6-4.) Although you may find it difficult to get a dial tone during natural disasters or on Mother's Day, once a circuit (a two-way connection with reserved capacity) is set up, there is no slowing of speech or delay when you talk.

A circuit is a two-way connection with reserved capacity.

Voice versus Data Traffic

Circuit switching works well for voice. As Figure 6-5 illustrates, voice traffic is fairly constant. In a conversation, one side or the other is talking most of the time. Usually, about 30 percent of the capacity of each full-duplex (two-way) telephone circuit is actually used. Relatively little reserved capacity is wasted.

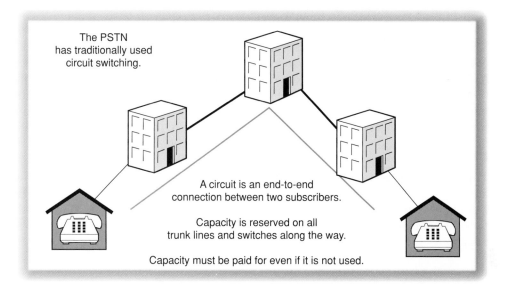

The PSTN has traditionally used circuit switching.

A circuit is an end-to-end connection between two subscribers.

Capacity is reserved on all trunk lines and switches along the way.

Capacity must be paid for even if it is not used.

Figure 6-4   Circuit Switching

Figure 6-5   Voice and Data Traffic

Full-Duplex (Two-Way) Circuit

Voice Traffic:
Fairly Constant Use;
Circuit Switching Is
Fairly Efficient

Full-Duplex (Two-Way) Circuit

Data Traffic:
Short Bursts,
Long Silences;
Circuit Switching Is
Inefficient

Voice traffic uses about 30% of circuit capacity on average.
Data traffic uses about 5% of circuit capacity on average.

In contrast, data traffic is **bursty,** with short, high-speed bursts separated by long silences. For instance, when you are using a website, your request message is very brief. The response message takes a bit longer to transmit, but, particularly on a broadband connection, transmitting the response message takes only a few seconds. After receiving a webpage, you are likely to look at it for 30–60 seconds on average. During this time, no data is transmitted in either direction. Other data applications are similarly bursty. Reserved capacity in circuit switching is extremely wasteful for data transmission, which typically uses only 5 percent of capacity.

Dial-Up Circuits

The PSTN provides two types of circuits. From your personal experience, you probably are only familiar with **dial-up circuits.** When you place a call, the PSTN sets up a circuit. When you finish this call, your circuit is ended, and reserved capacity is released for other circuits. You also know that modem-based data transmission over a dial-up telephone circuit is very slow.

Leased Line Circuits

Figure 6-6 compares the dial-up circuit with the other type of circuit offered by telephone carriers. This is the **leased line circuit,** also called the **private line circuit.**

Always-On

In contrast to dial-up circuits, a leased line circuit is permanent and **always on.** Once a leased line is provisioned (set up) by the telephone company, it is always available for transmission.

High Data Speed

Leased line circuits also carry data much faster than dial-up circuits. Even the slowest leased line circuits carry data at 56 kbps or 64 kbps. The fastest carry data at several gigabits per second.

Multiplexing Several Voice Calls

As just noted, you have traditionally used dial-up circuits at home when you placed telephone calls. In contrast, businesses primarily use leased line circuits for phone

Figure 6-6   Dial-Up Circuits versus Leased Line Circuit

	Dial-Up Circuits	Leased Line Circuits
Operation	Dial-up. Separate circuit for each call.	Permanent circuit, always on
Speed for Carrying Data	Up to 33.6 kbps	56 kbps to gigabit speeds
Number of Voice Calls	One	Several due to multiplexing

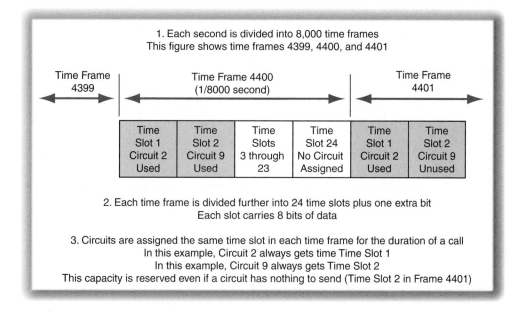

1. Each second is divided into 8,000 time frames
This figure shows time frames 4399, 4400, and 4401

Time Frame 4399	Time Frame 4400 (1/8000 second)	Time Frame 4401

Time Slot 1 Circuit 2 Used	Time Slot 2 Circuit 9 Used	Time Slots 3 through 23	Time Slot 24 No Circuit Assigned	Time Slot 1 Circuit 2 Used	Time Slot 2 Circuit 9 Unused

2. Each time frame is divided further into 24 time slots plus one extra bit
Each slot carries 8 bits of data

3. Circuits are assigned the same time slot in each time frame for the duration of a call
In this example, Circuit 2 always gets time Time Slot 1
In this example, Circuit 9 always gets Time Slot 2
This capacity is reserved even if a circuit has nothing to send (Time Slot 2 in Frame 4401)

Figure 6-7   Time Division Multiplexing (TDM) in T1 Lines

calls. These circuits typically connect a corporate PBX to the nearest end office switch of the public switched telephone network. In doing so, they **multiplex** (mix together on the same line) multiple voice circuits. For instance, the most popular leased line, the T1 line, multiplexes 24 voice circuits. In Chapter 7, we will see that leased lines carry data traffic as well as voice traffic. We saw in Chapter 1 that packet switching uses multiplexing. In contrast, leased line circuits use a less efficient multiplexing method called time division multiplexing. Module B discusses time division multiplexing. In Chapter 7, we will see how these lines can be used to carry data.

Time Division Multiplexing (TDM)

Figure 6-7 shows how the telephone network multiplexes several telephone calls and how it ensures capacity for individual telephone calls. This process is called **time division multiplexing (TDM)**. The figure specifically shows how TDM is done in the T1 lines, about which we will learn more in the next chapter.

The entire process consists of dividing each second into successively finer pieces.

➤ First, each second is divided into 8,000 frames of equal length. Looked at another way, there are 8,000 frames per second.

➤ Second, each frame is further divided into 24 time slots. Each slot carries a single byte (octet) of voice data. This gives 192 bits per frame.

➤ In addition, each frame has an additional bit for control purposes, giving 193 bits per frame. Multiplying this by 8,000 frames per second gives a speed for a T1 line of 1.544 Mbps.

Each voice channel is assigned to a specific time slot in each frame. Each telephone channel, then, gets 8 bits in each frame and 8,000 frames per second, for a total of 64,000 bits per second.

TEST YOUR UNDERSTANDING

7. a) What is circuit switching? b) Why does circuit switching make sense for voice communication? c) What does it mean that data transmission is bursty? d) Why is burstiness bad for circuit switching?

8. a) What are the differences between dial-up and leased line circuits? b) What is multiplexing, in the context of telephone calls and leased lines? c) What method do leased lines use to ensure that every call never gets less than a fixed amount of capacity? d) In TDM, distinguish between frames and slots. e) How are slots assigned in TDM? f) How many calls can a T1 circuit provide? Explain. g) Explain why TDM leads to 64 kbps per channel. h) Explain why TDM leads to T1's 1.544 Mbps speed.

THE ACCESS SYSTEM

The PSTN's access system is the only part of the PSTN that corporations work with directly. Consequently, we will look at it in the most detail.

The Local Loop

As noted earlier, the local loop, although fairly simple, represents an enormous capital investment and so is very difficult to change. Figure 6-8 shows the three main technologies that dominate the local loop. In addition to these three, radio-based local loops and fiber to the home (FTTH) for residential customers may also be important in the future.

Figure 6-8   Local Loop Technologies

Technology	Use	Status
1-Pair Voice-Grade UTP	Residences	Already installed, so no installation cost
2-Pair Data-Grade UTP	Businesses for high-speed access lines	Must be pulled to the customer premises. (This is expensive.)
Optical Fiber	Businesses for high-speed access lines	Must be pulled to the customer premises. (This is expensive.)

Note: *Within* buildings, corporate telephony uses 4-pair UTP.

Single-Pair Voice-Grade UTP

Traditionally, the telephone system brought a single pair of voice-grade UTP to each subscriber home and office. **One-pair voice-grade** copper has much lower transmission quality than the 4-pair UTP used in LANs.

2-Pair Data-Grade UTP

Even for the slowest leased lines, telephone carriers have to run higher-quality access lines, namely **2-pair data-grade** access lines. Note that two pairs are used—one for communication in each direction. In addition, the wiring is of higher quality than voice-grade UTP. This allows it to carry signals much faster.

Although 2-pair data-grade wiring is very good, the telephone carrier has to pull two new pairs of data-grade wiring UTP to each customer who needs it. The labor to do this is very expensive, and this labor cost translates into high monthly prices. In fact, customers must sign leases for certain periods to allow the telephone carrier to recoup its investment. Also, it may take several weeks to **provision** (install and set up) a 2-pair data-grade UTP access line.

Optical Fiber

For leased lines running faster than about 2 Mbps, the telephone carrier has to pull an even more expensive two-strand optical fiber cord to the customer premises. The cost of running fiber to the customer premises is even higher than the cost of running 2-pair data-grade wiring.

TEST YOUR UNDERSTANDING

9. a) Compare the 4-pair UTP wiring used in corporate buildings and the UTP wiring in the residential local loop. b) Distinguish between the UTP wiring in the residential local loop and the UTP wiring used for lower-speed leased lines (under about 2 Mbps). c) What technology do the highest-speed leased lines use? d) What is provisioning?

The End Office Switch

The access line runs from the customer premises to the nearest switch of the telephone company. As Figure 6-1 showed, this is called an **end office switch.**[1]

TEST YOUR UNDERSTANDING

10. What is an end office switch?

Analog–Digital Conversion for Analog Local Loops

Analog Voice Signals

In Chapter 3, we saw digital signals. However, traditional telephones produce a different type of signal, an analog signal. Figure 6-9 shows that when a person speaks into a telephone mouthpiece, his or her sound waves (which are pressure waves) cause a

---

[1]End office switches are sometimes called Class 5 switches, because the telephone network uses a five-class switch hierarchy, and end office switches are the lowest switches in the hierarchy.

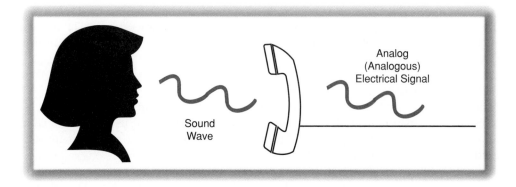

Figure 6-9   Analog Telephone Transmission

diaphragm inside the mouthpiece to vibrate. This generates an analogous electrical disturbance that propagates down the local loop to the nearest switching office. This **analog signal** rises and falls in intensity smoothly, with no clock cycles and no limited numbers of states as in digital signaling.

> An analog signal rises and falls in intensity smoothly, with no clock cycles and no limited numbers of states as in digital signaling.

Mostly Digital
As Figure 6-10 shows, the PSTN transport core, which was originally completely analog, is almost entirely digital today. Almost all of its switches are digital, as are almost all of its trunk lines. Large businesses get digital access lines even for their local loop connections.

Codecs: Analog-to-Digital and Digital-to-Analog Conversion
On the local loop that connects residential customers to the nearest end office, the customer's telephone sends and receives analog signals, so the end office switch needs equipment to convert the analog local loop signals to digital and to translate the digital signals of the end office switch to analog.

Figure 6-11 shows that this termination equipment is called a **codec**. Incoming signals from the subscriber go through an **analog-to-digital conversion (ADC)** process, which is called *coding*. In turn, the codec converts digital signals from the switch into analog signals for subscribers. This is the **digital-to-analog conversion (DAC)** process, which is called *decoding* (hence the name *codec*).

Leased Lines
Leased lines do not need codecs because they carry digital customer signals. Although this eliminates the need for analog–digital conversion, selecting leased lines

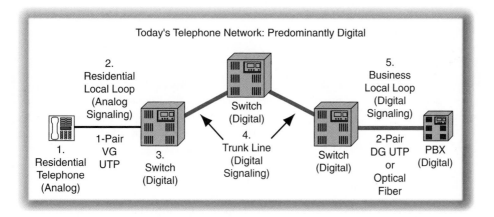

Figure 6-10   The PSTN: Mostly Digital, with Analog Local Loops

is complex because they come in a wide range of speeds, as we will see in the next chapter.

TEST YOUR UNDERSTANDING

11. a) Distinguish between analog and digital signals. b) What parts of the telephone system are largely digital today? c) What parts of the telephone system are largely analog today? d) What is the role of the codec in the end office switch? e) When a customer signal arrives on the access line to the end office switch, does the codec perform ADC or DAC? Explain.

Figure 6-11   Codec at the End Office Switch

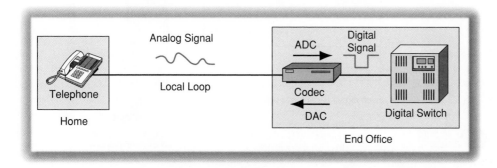

Codec Operation

This box looks in more detail at the analog-to-digital and digital-to-analog conversion processes discussed in the chapter text.

ANALOG-TO-DIGITAL CONVERSION

Analog-to-digital conversion is the more complex of the two processes. Earlier, we noted that TDM creates a data stream of 64 kbps per conversation (in each direction). We will now see why 64 kbps is necessary to support a conversation.

ADC STEP 1: BANDPASS FILTERING

Microwave radio was once used heavily for trunk lines. As Figure 6-12 shows, microwave transmission uses **frequency division multiplexing (FDM),** in which the microwave bandwidth is subdivided into channels, each carrying a single circuit.

How wide should a microwave channel be? Figure 6-13 shows the frequency spectrum for the human voice. Human hearing can range up to 20 kHz, although for most people the maximum is substantially lower. Most voice energy, furthermore, comes at frequencies below 4 kHz, so using 20 kHz channels would do fairly little to improve sound quality.

Instead, microwave channel bandwidths were set to 4 kHz. This allowed five times as many voice signals to be carried by a microwave system than a 20 kHz-channel system would have permitted.

To limit voice bandwidth to 4 kHz, termination equipment in the access system passes subscriber incoming signals through a codec only *after* first passing the signals through a **bandpass filter,** which passes only signals between 300 Hz and about 3.4 kHz. This gives guard bands around

Figure 6-12   Frequency Division Multiplexing (FDM) in Microwave Transmission

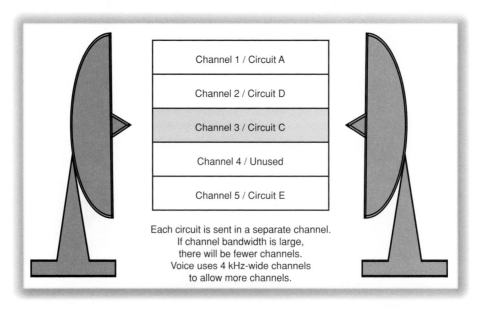

Channel 1 / Circuit A

Channel 2 / Circuit D

Channel 3 / Circuit C

Channel 4 / Unused

Channel 5 / Circuit E

Each circuit is sent in a separate channel.
If channel bandwidth is large,
there will be fewer channels.
Voice uses 4 kHz-wide channels
to allow more channels.

the signal—one below 300 Hz and one between 3.4 kHz and 4 kHz. Although microwave trunk lines are rarely used today, this bandpass filtering continues.

ADC STEP 2: SAMPLING

To digitize voice, the codec ADC **samples** (reads the intensity of) the bandpass-filtered voice signal 8,000 times per second, as Figure 6-13 illustrates.

Figure 6-13   Analog-to-Digital Conversion (ADC): Bandpass Filtering and Pulse Code Modulation (PCM)

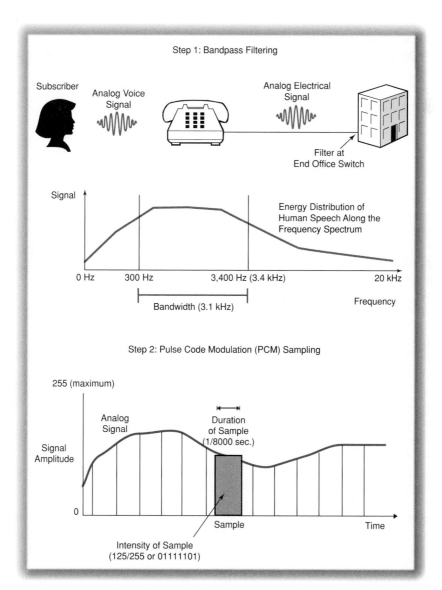

(*continued*)

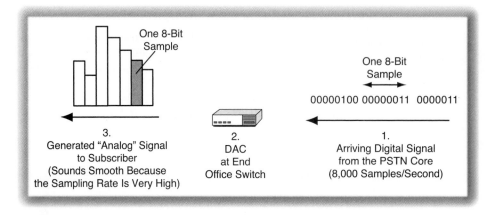

Figure 6-14   Digital-to-Analog Conversion (DAC)

Nyquist showed that if you sample at twice the highest frequency in the signal, you can reproduce the signal with no loss of information. This is why we need 8,000 samples per second (twice 4,000 Hz after bandpass filtering).

During each sampling period, the codec measures the intensity of the signal. The ADC represents the intensity of each sample by a number between 0 and 255. For instance, a signal of half the maximum intensity would be represented by 127. With 256 possible values, a single octet of binary data is needed to store each sample's value.

If you multiply 8 bits per sample by 8,000 samples per second, you get 64,000 bits per second. This specific analog-to-digital conversion technique is called **pulse code modulation (PCM).** Because PCM ADCs produce a data stream of 64 kbps for voice, most telephone lines and equipment are built around 64 kbps channels.[2] Earlier in this chapter, we saw how 64 kbps data streams are implemented in T1 lines.

64 kbps versus 56 kbps

In many cases, the telephone carrier will "steal" 8 kbps from each channel for supervisory signaling, leaving 56 kbps for transmission. This is why the telephone system is built around units of 56 kbps or 64 kbps.

DIGITAL-TO-ANALOG CONVERSION (DAC)

ADCs are used for transmissions from the customer premises to the end office switch. In contrast, digital-to-analog converters (DACs) are for converting transmissions from the digital telephone network's core to signals on the analog local loop. (See Figure 6-11.)

Figure 6-14 shows that as the DAC reads each sample, it puts a signal on the local loop that has the intensity indicated for that sample. It keeps the intensity the same for 1/8000 of a second. If the time period per intensity level is very brief, the amplitude changes will sound smooth to the human ear.

---

[2]The full PCM process is even more complex, but additional details add only marginally to sound quality.

Is the resultant signal really analog? Precisely speaking, it still is digital. However, to the human ear, the sampling and playback rates are so high that the choppiness of the signal shown in Figure 6-14 is not apparent at all. The final signal *sounds* analog to users, so it is considered to be an analog signal.

TEST YOUR UNDERSTANDING

12. a) Explain why bandpass filtering is done. b) Explain how the ADC generates 64 kbps of data for voice calls when it uses PCM. c) Why do we need DACs? d) How do DACs work?

CELLULAR TELEPHONY

Cellular Service

Nearly everybody today is familiar with cellular telephony. In most industrialized countries, half or more of all households now have a cellular telephone.[3] Many people now have *only* a cellular telephone and no landline.

Today, cellular telephony is used primarily for voice communication. Many users also do text messaging (**texting**), in which they type and send short messages to other cellular subscribers. Texting is particularly attractive during peak periods, when voice calling rates are highest. Many people text during the day, but call after peak hours. Texting has developed a rich shorthand to save thumb movements. For example, *you* has devolved into *U.*

It is also common for cellular subscribers to take pictures on built-in cameras and transmit these pictures to other users. With new "3G" technologies, cellular telephony can even be used for moderately high-speed data transmission when a mobile phone connects to a computer with a cellular data modem.

Cells

Cells and Cellsites
Figure 6-15 shows that cellular telephony divides a metropolitan service area into smaller geographical areas called **cells.**

The user has a cellular telephone (also called a **mobile phone, mobile,** or **cell phone**). Near the middle of each cell is a **cellsite,** which contains a **transceiver** (transmitter/receiver) to receive mobile phone signals and to send signals out to the mobiles. The cellsite also supervises each mobile operation (setting its power level, initiating calls, terminating calls, and so forth).

Mobile Telephone Switching Office (MTSO)
All of the cellsites in a cellular system connect to a **mobile telephone switching office (MTSO),** which connects cellular customers to one another and to wired telephone users.

---

[3]Although cellular telephony was first developed in the United States, the United States has slightly lower market penetration than most other countries. One reason is that normal telephony is inexpensive in the United States, so moving to cellular service is an expensive choice. Another reason is that when someone calls a cellular phone in the United States, the cellular owner receiving the call pays; in most other countries, the caller pays. These two factors increase the relative price of using a cellular phone compared with using a landline phone in the United States. A third factor is that U.S. cellular carriers give inadequate coverage, even in large metropolitan areas. In most other countries, dropped calls and dead spots are rare.

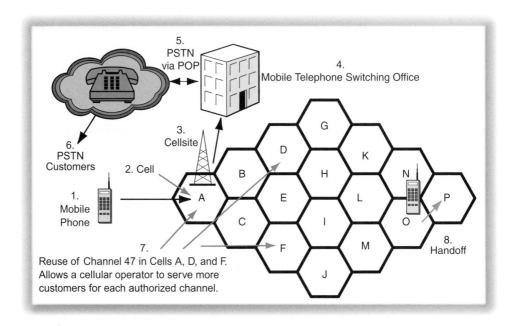

Figure 6-15   Cellular Technology

The MTSO also controls what happens at each of the cellsites. It determines what to do when people move from one cell to another, including deciding which cellsite should handle the transmission when the caller wishes to place a call. (Several cellsites may hear the initial request at different loudness levels; if so, the MTSO selects a service cellsite on the basis of signal loudness—not necessarily on the basis of physical proximity.)

TEST YOUR UNDERSTANDING

13. a) In cellular technology, what is a cell? b) What is a cellsite? c) What are the two functions of the MTSO?

Why Cells?

Why not use just one central transmitter/receiver in the middle of a metropolitan area instead of dividing the area into cells and dealing with the complexity of cellsites?

Channel Reuse

The answer is **channel reuse.** The number of channels permitted by regulators is limited, and subscriber demand is heavy. Cellular telephony uses each channel multiple times, in different cells in the network. This multiplies the effective channel capacity, allowing more subscribers to be served with the limited number of channels available.

Cellular technology is used because it provides channel reuse—the ability to use the same channel in different cells. This allows cellular systems to support more subscribers.

Traditionally, No Channel Reuse in Adjacent Cells

With traditional cellular technologies, such as the GSM technology (to be discussed later), you cannot reuse the same channel in adjacent cells, because there will be interference. For instance, in Figure 6-15, suppose that you use Channel 47 in Cell A. You cannot use it in Cells B or C. This reduces channel reuse. In general, the number of times you can reuse a channel is only about the number of cells, divided by seven. In other words, if you have 20 cells, you can reuse each channel only about 3 (20/7) times.

Channel Reuse in Adjacent Channels with CDMA

Some cellular systems in the United States use a new cellular technology, **code division multiple access (CDMA).** CDMA is a form of spread spectrum transmission. However, in contrast to the types of spread spectrum transmission used in 802.11 wireless LANs, which allow only one station to transmit at a time in a channel, CDMA allows multiple stations to transmit at the same time in the same channel. Furthermore, these channels are very wide, so several stations can transmit at the same time.

In addition, CDMA permits stations in adjacent cells to use the same channel, without serious interference. In other words, if you have 20 cells, with CDMA you can reuse each channel 20 times. This allows you to serve far more customers with CDMA than you could with older forms of cellular telephony.

If CDMA is so good, why do only some systems use it? The answer is that it is new. The first CDMA cellular systems were not built until 1993, and even then, they represented a technological and economic risk. However, CDMA has now proven itself, and all future cellular telephone standards will use CDMA.

Cells and Wireless LAN Access Points

In a sense, enterprise wireless LANs with many access points are like cellular technologies. They allow users to employ the limited number of frequencies available in WLANs many times within a building.

TEST YOUR UNDERSTANDING

14. a) Why does cellular telephony use cells? b) What is the benefit of channel reuse? c) If I use Channel 3 in a cell, can I reuse that same channel in an adjacent cell with traditional cellular technology? d) Can I reuse Channel 3 in adjacent cells if the cellular system uses CDMA transmission?

Handoffs versus Roaming

Handoffs

If a subscriber moves from one cell to another within a system, the MTSO will implement a **handoff** from one cellsite to another. For instance, Figure 6-15 shows a handoff from Cell O to Cell P. The mobile phone will change its sending and receiving channels during the handoff, but this occurs too rapidly for users to notice.

	802.11 WLANs	Cellular Telephony
Relationship	Handoff and roaming mean the *same thing*	Handoff and roaming mean *different things*
Handoffs (means the same in both)	Wireless host travels between access points in an organization	Mobile phone travels between cellsites in the same cellular system
Roaming (means different things)	Wireless host travels between access points in an organization	Mobile phone travels to a *different* cellular system

Figure 6-16   Handoff and Roaming in 802.11 Wireless Networking and Cellular Telephony (Study Figure)

Roaming

In contrast, if a subscriber leaves a metropolitan cellular system and goes to another city or country, this is called **roaming.** Roaming requires the destination cellular system to be technologically compatible with the subscriber's mobile. It also requires administration permission from the destination cellular system. Roaming is as much a business and administrative problem as it is a technical problem.

> In cellular telephony, handoffs occur when a subscriber moves between cells in a local cellular system. Roaming occurs when a subscriber moves between cellular systems in different cities or countries.

Handoff and Roaming in 802.11 WLANs

Recall from Chapter 5 that *handoff* and roaming mean the same thing in 802.11 WLANs. They both mean moving from one access point to another within the same WLAN. In other words, the terms *handoff* and *roaming* are used differently in cellular telephony from the way they are used in WLANs.[4]

TEST YOUR UNDERSTANDING

15. a) Distinguish between handoffs and roaming in cellular telephony. b) Distinguish between handoffs and roaming in 802.11 wireless LANs.

VOICE OVER IP (VoIP)

Basics

One of the newest areas in telephony is **voice over IP (VoIP),** which is the transmission of telephone signals over IP routed networks (including the Internet) instead of over circuit-switched networks. VoIP offers the promise of reducing telephone costs by moving from traditional circuit switching to more efficient packet switching.

---

[4]Wouldn't it be nice if there were a networking terminology court that could punish this sort of thing?

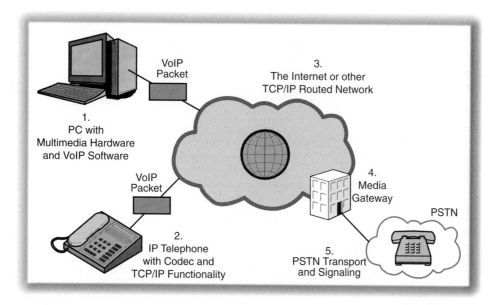

Figure 6-17   Voice over IP (VoIP)

Clients

Figure 6-17 illustrates VoIP operation. The figure shows two clients. One is a client PC with multimedia hardware (a microphone and speakers) and VoIP software. The other is a **VoIP telephone,** which has the electronics to encode voice for digital transmission and to handle packets over an IP internet. With VoIP, these two clients' users can talk with each other.

Media Gateway

The figure also shows a media gateway. The media gateway connects a VoIP system to the ordinary public switched telephone network. Without a media gateway, VoIP users could talk only to one another.

TEST YOUR UNDERSTANDING

16. a) What is VoIP? b) What is the promise of VoIP? c) What two devices can be used by VoIP callers? d) What is the purpose of a media gateway? e) Why is having a media gateway in a VoIP system important?

VoIP Signaling

As discussed earlier in this chapter, there is a fundamental distinction in telecommunications between signaling and transport. Signaling consists of the communication needed to set up circuits, tear down circuits, handle billing information, and do other supervisory chores. Transport is the actual carriage of voice.

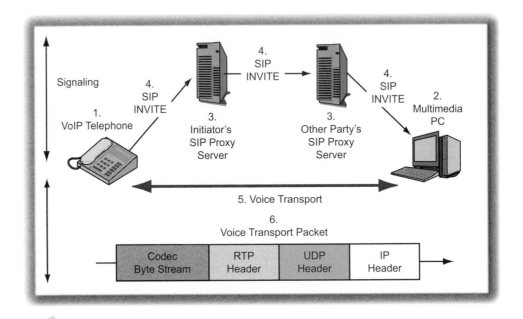

Figure 6-18   VoIP Signaling and Transport

There are two major VoIP signaling standards. The first was the ISO **H.323** standard, which was effective, but very complex. More recently, the IETF created the **Session Initiation Protocol (SIP)** standard. Most older VoIP systems use H.323 to control signaling. However, the use of SIP is growing rapidly, and most VoIP systems today use SIP for signaling.

Figure 6-18 illustrates the SIP protocol. Each subscriber has an SIP proxy server. The calling VoIP telephone sends a SIP INVITE message to its IP proxy server. This message gives the IP address of the receiver. The caller's SIP proxy server then sends the SIP INVITE message to the called party's SIP proxy service. The called party's proxy server sends the INVITE message to the called party's VoIP telephone or multimedia PC.

TEST YOUR UNDERSTANDING

17. a) What are the two major protocols for VoIP signaling? b) Which of these protocols is growing rapidly? c) Describe how SIP initiates a communication session.

VoIP Transport

After SIP or H.323 creates a connection, the two VoIP clients begin communicating directly. This is the beginning of transport, which is the transmission of voice between callers. VoIP, as its name suggests, operates over routed IP networks. Therefore, digitized voice has to be carried from the sender to the receiver in packets.

Codec	Transmission Rate
G.711	64 kbps
G.721	32 kbps
G.722	48, 56, 64 kbps
G.722.1	24, 32 kbps
G.723	5.33, 6.4 kbps
G.723.1A	5.3, 6.3 kbps
G.726	16, 24, 32, 40 kbps
G.728	16 kbps
G.729AB	8 kbps

Figure 6-19   VoIP Codecs

Codecs

VoIP telephones and multimedia PCs need codecs to convert analog voice signals into digital voice data streams. Earlier, if you read the box "Codec Operation," you saw that the codecs in end office switches convert voice into 64 kbps data streams. VoIP systems can use many different codecs. As Figure 6-19 shows, some codecs convert voice streams into bit streams as small as 5 kbps. However, the codecs that do the most compression also lose the most voice quality. Selecting codec in a VoIP network means making a trade-off between voice quality and cost reduction.

VoIP Transport Packets

As noted in Chapter 1, long application messages have to be fragmented into smaller pieces that can be carried in individual packets. Each packet carries a small part of the application message. In VoIP, packets carry a small snippet of digital voice bytes created by the codec.

The Application Layer: Codec Segments

Figure 6-18 shows a VoIP transport packet. At the application layer, there is the snippet of voice codec octets.

UDP

TCP allows reliable application message delivery. However, the retransmission of lost or damaged TCP segments can take a second or two—far too long for voice conversations. Voice needs to be transmitted in real time. Consequently, VoIP transport uses UDP at the transport layer. UDP reduces the processing load on the VoIP telephones, and it also limits the high network traffic that VoIP generates.

The Real Time Protocol (RTP)
Between UDP and the application layer, VoIP adds an additional header, a Real Time Protocol (RTP) header to make up for two deficiencies of UDP.

➤ First, UDP does not guarantee that packets will be delivered in order. RTP adds a sequence number so that the application layer can put packets in the proper sequence.

➤ Second, VoIP is highly sensitive to jitter, which is variable latency in packet delivery. Jitter literally makes the voice sound jittery. RTP contains a time stamp for when its package of octets should be played relative to the octets in the previous packet. This allows the receiver to provide smooth playback.

IP Television (IPTV)
In addition to VoIP, companies are beginning to provide **IP television (IPTV).** Like voice, video traditionally has been an analog service and must be converted to digital to travel over corporate data networks.

TEST YOUR UNDERSTANDING

18. a) What is the purpose of a VoIP codec? b) Some codecs compress voice more. What do they give up in doing so? c) In a VoIP transport packet, what is the application message? d) Does a VoIP transport packet use UDP or TCP? Explain why. e) What two problems with UDP does RTP fix? f) List the headers and messages in a VoIP transport packet, beginning with the first packet header to arrive at the receiver.

Corporate VoIP Alternatives
When companies plan for VoIP, they have two decisions to make. One is how to provide VoIP between sites. The other is how to provide VoIP within sites.

VoIP between Sites
Most companies begin experimenting with VoIP between sites. The reason for this is simple economics. Long-distance telephone calling is expensive, and international telephone calling is much more expensive.

One approach is to place a PBX at each site and get leased lines to connect sites together. Most companies already connect their PBXs at multiple sites with leased lines in order to reduce the cost of telephone calls without VoIP. The company simply has to add a VoIP module to each PBX. PBX–PBX VoIP is attractive because it does not affect anything inside the sites. The VoIP module in the PBX translates between external and internal telephone signaling.

In addition, many carriers are beginning to offer VoIP services directly to firms. These **carrier VoIP services** typically are not limited to transmissions between the firm's sites. Companies can place any long-distance or international call at attractive prices. Carrier VoIP services are possible because many carriers are replacing their traditional circuit-switched transport cores with IP packet-switched cores designed for VoIP.

VoIP within Sites
More controversial is using VoIP within sites. In LANs, transmission costs per bit are low, so no major cost savings result from using VoIP within sites. In fact, the cost of

installing VoIP within sites would probably raise internal telephony costs because of the cost of installing VoIP telephones and the need to upgrade the LAN for additional traffic and, hopefully, QoS improvements.

The main benefit of internal VoIP for corporations is the ability to create applications that integrate voice and data. For instance, when a customer calls, information about the customer can be brought up on a salesperson's computer screen before the salesperson answers the phone. With a LAN already installed, the firm has to purchase IP telephones or upgrade PCs with multimedia equipment and VoIP software.

The biggest technical issue is how to handle mobile users with wireless LAN connections within the site. VoIP needs very good quality of service, and wireless LANs create too much latency (delay) in transmissions. They also have two much variability in latency between adjacent packets. This makes the voice sound jittery, so variable latency is called jitter. The 802.11e standard, which adds QoS to WLANs, may address this problem sufficiently.

VoIP Carriers

Although corporations can implement VoIP internally, many carriers are beginning to offer VoIP service to both residences and businesses. These **VoIP carriers** include both traditional telephone carriers and new entrants such as cable television companies, Internet service providers, and companies like Skype that provide service over any ISP.

TEST YOUR UNDERSTANDING

19. a) If a company already has a multisite PBX network installed for site-to-site voice service, what must it add for site-to-site VoIP? b) Do many carriers offer VoIP services to business customers? c) Can a company save more money with long-distance VoIP or with VoIP within the company's sites? Explain. d) What is the main advantage of VoIP within sites? e) Is it easier to implement VoIP on wired LANs or WLANs? Explain. f) What carriers provide VoIP service?

Telecommunications Staff Concerns with VoIP

Although corporations are implementing VoIP widely, they still have some serious concerns about their VoIP systems. In addition to the questions discussed earlier about costs and QoS for mobile users, corporate telecommunications staffs have high expectations for voice quality and availability. They have long been able to deliver voice signals of high quality 99.999 percent of the time. VoIP networks are likely to bring lower quality and significantly lower availability.

TEST YOUR UNDERSTANDING

20. What two concerns do corporate telecommunications staffs have about VoIP?

REBUILDING THE LAST KILOMETER

Telephone professionals call the access line to your home "**the last kilometer**," despite the fact that wires to the home often run more than a kilometer. However, the term is well established.

For more than a hundred years, the technology for the last kilometer was the telephone company's single twisted pair of voice grade wire. In the 1960s, a few businesses

The Last Kilometer
   The access line to your home
   Traditionally, a 1-pair VG UTP line from the telephone company
   In the 1960s, a few businesses started getting 2-pair data-grade UTP and optical fiber
   Given the cost of upgrading the 1-pair VG UTP plant, it seemed eternal

Telephone Service and Cable TV
   1950s brought cable television
      Used coaxial cable with a central wire and a coaxial conductive ring or mesh
      Thick coax trunk lines past homes
      Thin coax drop lines to homes
   Television services soon went beyond delivering over-the-air signals
   A static situation emerged
      Telephone companies controlled telephone service
      Cable companies controlled television delivery service

Data Transmission
   Telephone modems
      Convert digital computer to analog and send these over the access line
      Convert incoming analog signals into digital signals for the computer
   Asymmetric digital subscriber lines (ASDLs) (See Figure 6-21)
      Digital subscriber lines (DSLs) send digital signals over the existing 1-pair VG UTP access line
      DSL modem at the home
      Splitter at each outlet in each home
      End office has a digital subscriber line access module (DSLAM) to connect the voice signal to the telephone network and the data signal to a data network.
      Asymmetric SDL (ASDL) service to homes—asymmetric speed with higher downstream speed to the home and lower upstream from the home
   Cable modem service (See Figure 6-22)
      Customer has a cable modem
      Cable company needs two-way amplifiers
   In general, ADSL service is somewhat slower, but less expensive, than cable modem service

Figure 6-20 "Traditional" Technologies for the Last Kilometer (Study Figure)

got 2-pair data-grade UTP access lines, and even fewer got optical fiber, but most of the access plant continued to consist of 1-Pair VG UTP. Given the cost of replacing this enormous access plant, it seemed like 1-Pair VG UTP would dominate forever.

TEST YOUR UNDERSTANDING

   21. a) What is the last kilometer? b) What access line do telephone companies already have installed to most customer premises?

Telephone Service and Cable TV

In the 1950s, **cable television** companies sprang up in the United States and several other countries, bringing television into the home. Initially, it brought over-the-air TV only to rural areas. Later, it began to penetrate urban areas by offering far more channels than urban subscribers previously could receive over the air. In the 1970s, many books and articles about the "wired nation" argued that two-way cable would soon turn into an information highway. However, available services did not justify the heavy investment to make cable a two-way service.

Instead of using twisted-pair wiring, cable television companies used **coaxial cable,** which has a central wire and a conductive ring or mesh around the central wire on the same axis at the wire. This is the source of "coaxial" in the name. (Coaxial is often shortened to **coax.**) From the cable television's head end, the cable company ran thick **trunk lines** past homes. They then ran thinner drop cables to the home.

For many years, the telephone network and the cable companies had noncompeting monopolies. The telephone company provided telephone service, and cable offered television service.

TEST YOUR UNDERSTANDING

22. a) What transmission medium do cable television companies primarily use? b) Distinguish between the cable television trunk cable and drop cable. c) What was the traditional division of market dominance between telephone companies and cable television companies?

Data Transmission

Telephone Modems

In the 1970s, data transmission began to be important. Telephone companies, under the spur of national regulators, allowed companies to attach telephone modems to their 1-pair VG UTP lines. **Telephone modems** receive digital signals from a computer, convert the digital signals into analog signals, and transmit the analog signals over the subscriber's access line to the end office switch. When signals come back from the computer at the other end of the line, the modem converts the incoming analog signal into digital signals that the computer can use.

Asymmetric Digital Subscriber Line (ADSL) Service

Around 1990, telephone companies began to bring higher-speed data transmission from households. Figure 6-21 shows that they did this by sending digital data over existing telephone access lines. This was important, because it meant that the telephone company did not have to install entirely new wiring.

The 1-pair VG lines that already ran to references then became **digital subscriber lines (DSLs).** Transmission lines themselves do not care whether you send digital signals or analog signals. The only important requirement is that you have analog devices at the two ends or digital devices at the two ends of the transmission line.

For DSL service, the subscriber needs a **DSL modem.** The subscriber's computer sends Ethernet or USB digital signals to the DSL modem, which, in turn, modifies the digital signals and sends them over the subscriber line to the **digital subscriber line access multiplexer (DSLAM)** at the end office switch. The DSLAM then passes the data signals to a data network.

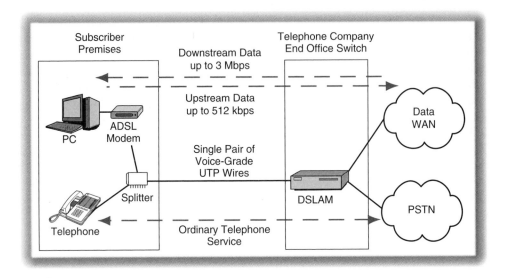

Figure 6-21   Asymmetric Digital Subscriber Line (ADSL)

In addition to giving subscribers more speed than a telephone modem could provide, DSL service does not tie up the subscriber's telephone line when data are sent or received. Ordinary voice signals are sent up to 4 MHz. Data are sent and received at higher frequencies so it will not interfere with the voice signal. As a consequence of this division of frequencies, ADSL users need to have a splitter installed in every telephone outlet. The **splitter** separates the voice and data signals.

The service provided to subscribers actually is called **asymmetrical digital subscriber line (ADSL)** service because speed is different in the two directions. Download speed (to the subscriber) is higher than upload speed (from the subscriber). This is good for WWW access because downloading pages takes many more bits per second than transmitting HTTP request messages.

Cable Modem Service

Cable television companies, not to be left out of the market for broadband (high-speed) access lines, began to offer data transmission over their cable plants. For television, the repeaters that boost signals periodically along the cable run only had to boost television signals traveling downstream. Data transmission required cable companies to install **two-way amplifiers,** which could carry data in both directions. Although this was expensive, it allowed cable companies to compete in the burgeoning market for broadband service. As in the case of ADSL, cable television service was asymmetric, offering faster downstream speeds than upstream speeds. Instead of having a DSL modem, the subscriber had to have a cable modem. In general, this cable service was called **cable modem service.** Figure 6-22 illustrates cable modem service.

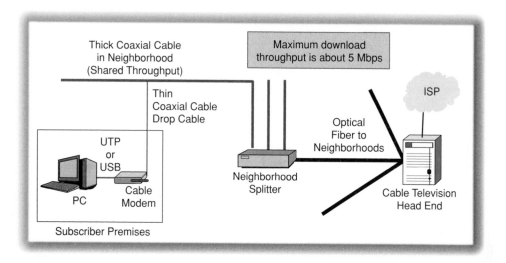

Figure 6-22   Cable Modem Service

The figure also shows that most cable systems today are **hybrid cable/fiber** systems, not pure coaxial cable systems. The line from the head end to the neighborhood is optical fiber. Only the local distribution system in the neighborhood and the drop cables use coaxial cable.

Competition among Wired Data Carriers

Initially, the throughputs of ADSL service and cable modem service were both very slow—delivering about half a megabit per second downstream and roughly half that speed upstream. Fortunately, both ADSL and cable modem systems are several times faster today. In general, ADSL service is somewhat slower and less expensive, while cable modem service is somewhat faster and more expensive.

TEST YOUR UNDERSTANDING

23. a) What does a telephone modem do? b) Why is the telephone company's use of existing access lines important in ADSL service? c) In ADSL service, what does the DSLAM do? d) Do telephone modems tie up your telephone line when you use them to carry data? Do ADSL modems? Explain. e) In DSL, what must be plugged into each telephone outlet? f) Why is residential DSL service called ADSL service?

Wireless Access Services

While traditional telephone and cable companies focused on each other as competitors, additional competition began to come from **wireless access service** carriers. These carriers set themselves up quickly once they procured the necessary radio spectrum. Not having to invest in a wired infrastructure allowed them to begin operation without a massive outlay of capital.

Wireless Access Service

3G Cellular Data Transmission

　　　2G cellular service is for voice, texting, and photographs
　　　　　Can send data via a cellular modem, but only at 10 kbps
　　　3G cellular was created to send data faster
　　　　　Most current services offer low DSL speeds at higher prices
　　　　　2 Mbps to 3 Mbps speeds are arriving, but will be even more expensive
　　　　　Consumer usage is dominating, with downloading music, videos, and games

WiMAX Metropolitan Area Networks

　　　Designed to compete with DSL and cable modem service
　　　Promises to be faster than 3G service at lower cost
　　　　　Beginning with 1 to 4 Mbps, and will be faster
　　　　　Mobile subscribers with omnidirectional antennas will receive speeds at the lower end
　　　　　Fixed subscribers in homes will have directional antennas, and speeds will be at the higher end
　　　Created by the WiMAX Forum
　　　WiMAX depends heavily on the IEEE 802.16 standard
　　　A single MAC-layer standard for all service bands between 2 GHz to 11 GHz
　　　　　WiMAX forum is initially developing profiles for the 2.3, 2.5, 3.5, and 5.8 GHz bands
　　　　　The 2.3GHz, 2.5 GHz, and 3.5 GHz service bands are licensed
　　　　　WiMAX providers want licensed bands for higher quality service
　　　Uses advanced technologies
　　　　　Scalable OFDM, MIMO, adaptive antennas systems (AAS) that do beam forming, and cellular organization for its base stations
　　　WiMAX technology provides high-quality service
　　　　　TDM gives each subscriber its fair share of the capacity

Satellite Access Service

　　　Very expensive because of long transmission distance to satellites

Figure 6-23 Wireless Technologies for the Last Kilometer (Study Figure)

3G Cellular Data Transmission

The traditional cellular technology you probably grew up with is called **second-generation (2G) cellular** service.[5] It offers voice, texting, and probably camera picture transmission. You can also get a cellular modem to work with 2G cellular telephones. This allows you to send and receive data, but only at the miserably slow speed of 10 kbps.

---

[5]The first generation began in the early 1980s. It used analog transmission, while subsequent generations have been digital. First-generation cellular systems had only a fraction of the channels that are available today in second-generation and third-generation systems.

In order to move into the data market, cellular vendors began to offer **third-generation (3G) cellular** services that could move data faster. Initially, 3G services were only about as fast as telephone modems. Today, they mostly have the same throughput as low-end ADSL lines. In addition to being somewhat sluggish, cellular service is expensive. These factors have combined to stymie its growth. Faster 3G service is coming, and this will bring throughputs of 2 Mbps to 3 Mbps. However, these services may also be even more expensive.

Although initially viewed as an Internet access technology, 3G is being used more heavily as a consumer service that brings the Internet to a miniscule browser window, provides e-mail service, and downloads music, games, and even video.

WiMAX Wireless Metropolitan Area Networks (WMANs)

The **WiMAX Forum** has developed a wireless metropolitan area network service called **WiMAX.** WiMAX is designed for carriers who want to provide metropolitan area networking. Its signals can easily reach all customers in a metropolitan area.

WiMAX is specifically designed to compete with ADSL and cable, but it promises to give higher speed than most 3G cellular service at lower cost. It began to reach the market in 2006 and is growing rapidly.

WiMAX today generally offers individual throughput of 1 Mbps to 4 Mbps, and higher speeds are coming soon. Mobile stations, which use omnidirectional antennas, receive service at the lower end of this speed range. Fixed residential customers (the primary market for WiMAX as an ADSL and cable competitor) will have directional antennas and so will receive service at the high end of the frequency range.

WiMAX is primarily based on the **IEEE 802.16** standards from the IEEE 802.16 Working Group, but it also draws upon other standards. The 802.16 standards can operate in any licensed or unlicensed service band from 2 GHz to 11 GHz,[6] but the WiMAX Forum has defined three interoperability **profiles** at 2.3, GHz, 2.5 GHz, 3.5 GHz, and 5.8 GHz. The first three are licensed bands, which makes sense because carriers want to offer good service quality. This is possible, however, only when competition can be limited, and that is possible only in regulated bands. To reduce the cost of equipment, WiMAX (like 802.3 Ethernet) only has a single MAC layer standard that will work with radios operating in different frequency bands.

Arriving much later than 802.11 or even 3G telephone service, WiMAX includes all of the most advanced wireless technologies. It offers a more sophisticated form of OFDM transmission (scalable OFDM), MIMO, **adaptive antenna systems (AAS)** that do beam forming, and cellular organization for its base stations.

Because it is being created primarily for commercial carriers, WiMAX technology ensures that each customer gets a fair share of the available throughput by using TDM in downstream links. TDM, as noted earlier in this chapter, guarantees a time slot in each frame for each subscriber.

Satellite Access Service

A minor player in the wireless access segment is **satellite access service.** Due to the fact that satellites are high above the Earth, transmission distances are long, and this

---

[6]The 802.16 standard also defines operation up to 66 GHz, but at such enormously high frequencies, attenuation is high and shadow zones are completely black. There must be a clear line of sight between the base station and the subscriber equipment. Even then, transmission only runs a few hundred meters.

translates into relatively low speeds and definitely high costs. Satellite access service today is a niche market. It can be attractive only in rural areas and in cities that do not have 3G or WiMAX service yet.

TEST YOUR UNDERSTANDING

24. a) Distinguish between 2G and 3G cellular service for data transmission. b) What may limit 3G cellular's acceptance? c) Do WiMAX carriers primarily use licensed or unlicensed bands? Why? d) What are WiMAX profiles, and why are they important? e) What four advanced technologies does WiMAX use? f) How does WiMAX ensure that each customer gets its fair share of transmission speed? g) What is the biggest problem for satellite access service?

The Triple Play

Carriers are beginning to look at what is being called the triple play—offering telephone service, data access, and television in an integrated package. This moves carriers into both the traditional realms of telephone and cable monopolies.

Voice and Data

Two elements of the triple play—data and voice over IP—can be integrated fairly easily, and almost all cable and telephone carriers today offer both.

Figure 6-24   The Market Situation (Study Figure)

The Triple Play
    The goal of access carriers
    Telephony, data, and video
    Video is the hardest
        People want multiple incoming TV signals
        They also want HDTV

Very High Speed Access
    Fiber to the home (FTTH)
    Speeds up to 100 Mbps or more
    The backhaul issue: the entire network must be upgraded in capacity

The International Situation
    United States ranks 16th internationally in broadband speed and availability
    Korea and Japan provide 50 Mbps speeds or faster at prices comparable to U.S. prices (for slower speeds)
    Leadership in speed brings leadership in applications

Very High Speed Access

Video is another matter. Subscribers who take video want multiple television shows coming into their homes simultaneously because most homes have multiple televisions and also want to record additional television signals coming into their homes for later viewing. Furthermore, people willing to pay for television are increasingly demanding high-definition service.

Overall, triple play service that includes video will require much higher data transmission rates. Cable is jumping to download speeds of about 15 Mbps, and ADSL is also reaching toward these speeds. Among cellular providers, 4G systems will bring far higher speeds, and so will WiMAX.

For telephone companies, one alternative is **fiber to the home (FTTH),** in which carriers will bring fiber all the way into the home (or very near it), just as they do for larger businesses. FTTH will bring speeds of 100 Mbps or even more, depending on how close the fiber actually gets to the home.

The Backhaul Issue

Bringing a large amount of bandwidth to every home is very exciting, but the financial impacts go well beyond the last kilometer. The entire carrier backbone plant has to be able to handle the much higher speeds that customer premises generate. This is called, generically, the **backhaul issue.**

TEST YOUR UNDERSTANDING

25. a) What do companies call the "triple play"? b) Which aspect of the triple play requires a huge increase in bandwidth? c) What technology will bring speeds of 100 Mbps or more to subscribers? d) What is the backhaul issue, and why is it important?

The International Situation

The United States tends to think of itself as a technology innovator, especially when it comes to the Internet. In high-speed access, however, the U.S. is a laggard.

The website speedmatters.org compares download and upload speeds for the United States and several other countries in the world. In late 2007, U.S subscribers had a median download speed of 1.9 Mbps. In comparison, Canada had a median download speed of 7 Mbps; France had 17 Mbps; South Korea, 45 Mbps; and Japan, 61 Mbps. The United States ranked 16th in speed and high-speed availability.

Countries with high speeds are leading in the development of next-generation applications. At the speeds enjoyed in South Korea and Japan, for example, video to the home is commonplace.

The higher speeds in most countries other than the United States do not come at a significant price premium. In Japan, for instance, 50 Mbps service costs only $25 to $35 per month. For this amount, U.S. subscribers can only get about 5 Mbps in download speed.

TEST YOUR UNDERSTANDING

26. a) Is the U.S. the leader in broadband speeds? Explain. b) What two countries, according to the figures given, are the leaders in broadband service? c) What is the advantage of being a leader in broadband service?

CONCLUSION

Synopsis

Telecommunications traffic—voice and video—normally travels over the worldwide telephone network, which is officially called the public switched telephone network (PSTN). The PSTN has four technological elements:

➤ Customer premises equipment includes the PBX, telephones, and transmission lines in the customer's buildings.

➤ The access system is the way customers connect to the PSTN. The access system is the most expensive part of the PSTN, and it is the only part of the PSTN that corporations see directly. The subscriber access line is called the local loop.

➤ The transport core carries signals between customer access lines across distances ranging from less than a mile to halfway around the world.

➤ Signaling is the way the PSTN sets up connections, breaks down connections, provides billing information to carriers, and handles other supervisory chores.

Almost all data networks use packet switching, but the PSTN uses circuit switching, in which capacity is reserved at the start of a call on all switches and trunk lines along the path (circuit) of the call. Time division multiplexing (TDM) provides guaranteed capacity for telephone channels. Circuit switching is reasonably efficient for voice conversations, in which one side or the other usually is talking. However, it makes little sense for data transmission, which is bursty, with short traffic bursts separated by long silences. Data transmission must pay for reserved capacity even during these long silences.

For telephone calls, you probably are most familiar with dial-up circuits, which are set up at the start of a call and broken down afterward. However, most corporations use leased lines, which are circuits that are always on, can multiplex many voice calls, and can carry high-speed data.

The PSTN access system consists of single pairs of voice-grade UTP to residential homes and two pairs of data-grade UTP or an optical fiber cord to businesses with leased lines. These wires (or fiber cords) carry customer signals to the PSTN carrier's end office switch. The PSTN is almost entirely digital internally, but residential customers send and receive analog signals, which rise and fall smoothly in intensity over time.

At the end office switch, as the box entitled "Codec Operation" stated, a device called a codec converts residential analog subscriber signals into digital signals that the transport core can carry. This is analog-to-digital conversion (ADC). In ADC, the subscriber's signal is sent through a bandpass filter to filter out all frequencies lower than about 300 Hz and higher than about 3.4 kHz. Pulse code modulation (PCM) samples the bandpass filtered voice signal 8,000 times per second and sends an 8-bit signal for each sample, resulting in a 64 kbps data stream. (This is sometimes 56 kbps because some carriers steal 8 kbps for signaling.) The codec has a digital-to-analog converter (DAC) that reverses the PCM process for telephone systems going to residences.

Cellular networks divide a region into multiple small areas called cells. This arrangement allows the same channel to be used in multiple cells, permitting more subscribers to be served with a limited number of channels. Traditional FDM and TDM cellular systems cannot reuse a channel in adjacent cells, but CDMA systems can, thus allowing more channel reuse. Most current cellular networks are GSM networks, but future cellular technologies will be based on CDMA. Today's cellular systems primarily use second-generation (2G) technologies, which are digital but cannot carry data rapidly.

Today, corporations have separate packet-switched networks for data and circuit-switched networks for voice. However, VoIP (also called IP telephony) promises to integrate all networking via IP transmission over packet-switched networks. Several codecs are available for converting analog voice to digital signals. For transport, these codec data streams are divided into IP packets containing an RTP header, a UDP header, and several bytes of codec data. Signaling technologies include H.323 and the newer and simpler SIP.

Traditionally, the access technology of the public switched telephone network—one-pair voice-grade UTP—almost never changed. However, competition for the last mile (that is, access lines) is becoming fierce, and speeds are increasing rapidly. Until recently, cable television maintained a near monopoly in wired video distribution, while the telephone company dominated telephone service. For data service, the telephone network originally dominated, with slow telephone modems. Then, competition for data access began in earnest, with telephone companies offering high-speed DSL service (including ADSL service for homes) and with cable companies offering cable modem service. Soon, wireless companies got involved. Cellular providers began to offer 3G service for metropolitan area data service (at comparatively low speed and high prices), and the new WiMAX service aims to provide faster speed at a lower price than 3G. Competitors are now focusing on the "triple play"—a combined package of telephony, data access, and video (including multiple high-definition television (HDTV) signals). This will require much faster speeds. One tool for bringing these speeds is fiber to the home.

In assessing new services, do not focus on the U.S. market if you are dealing with multinational firms. In many parts of the world, far faster and less expensive access service and applications are commonplace.

End-of-Chapter Questions

THOUGHT QUESTIONS

1. a) Trace the path between a telephone subscriber in one city and a telephone subscriber in another city. (Hint: Consider Figure 6-3.) b) Trace a path between a cellular subscriber and a wireline (ordinary) telephone subscriber in the same city. (Hint: Also consider Figure 6-15.)

2. (If you read the box "Codec Operation.") Beings of the planet Zamco can hear frequencies up to 30 kHz. However, they can only hear two loudness levels—soft and deafening. The Zamconian telephone system uses PCM codecs. a) How many bits per second will a Zamconian telephone call generate? Do your work in Excel, and copy and paste your results into your answer page. b) How many bytes will it take to store an hour of a Zamconian telephone conversation? (Hint: The answers to a) and b) are 60 bps and 26 MB, respectively.)

3. (If you read the box "Codec Operation.") In this chapter, you saw how PCM generates 64 kbps of data when it digitizes voice. For music CDs (which store information digitally), a PCM-like algorithm was used. However, instead of cutting off sounds above 3.4 kHz, music digitization uses a 20 kHz cutoff in order to capture the higher-pitched sounds of musical instruments. Music digitization also uses 16 bits per sample, instead of the 8 bits per sample used by voice, in order to give more precise volume representation. Furthermore, music is presented in stereo, so there are two 20 kHz channels to digitize.

Audio CDs were designed to store one hour of digitized music. Compute how big audio CDs need to be. Remember to convert your bit rates into bytes per second, and remember that there are 1024 bytes in a kilobyte and 1024 kilobytes in a megabyte. (Hint: The first CD-ROM disks, which were designed on the basis of audio disk technology, had 550 MB of capacity. You should get a number reasonably close to this.) Calculate your answer in a spreadsheet. Copy and paste the spreadsheet into your answer page.

4. Telephone modems convert analog signals to digital signals and digital signals to analog signals. Codecs convert analog signals to digital signals and digital signals to analog signals. How are modems different from codecs?

HANDS-ON EXERCISE

How fast is your Internet connection? See for yourself by going to http://www.pcpitstop.com/internet/default.asp, http://reviews.cnet.com/7004-7254_7-0.html, and http://speedmatters.org. All three provide a test of your Internet download speed. a) How were you connected to the Internet (telephone modem connection, cable modem connection, DSL connection, lab at school)? b) What was your download speed in each of the three services? c) What speeds were cited for different countries at http://speedmatters.org

PERSPECTIVE QUESTIONS

1. What was the most surprising thing you learned in this chapter?

2. What was the most difficult material for you in this chapter?

PROJECTS

1. Look up speed and price alternatives in your area for ADSL, cable modem service, 3G telephony, and WiMAX.

2. Do a report on Verizon's FiOS fiber-to-the-home service.

GETTING CURRENT

Go to the book website's New Information and Errors pages for this chapter to get new information since this book went to press and to find corrections for any errors in the text.

C h a p t e r   7

Wide Area Networks (WANs)

Learning Objectives

By the end of this chapter, you should be able to discuss the following:

- Differences between LANs and WANs, including the high cost of WANs per bit transmitted and, consequently, the dominance of low-speed transmission (56 kbps to a few megabits per second) in WAN service.
- The three purposes of WANs: remote individual access, site-to-site corporate networking, and Internet access.
- Layer 1 leased line carrier services and leased line networks.
- Layer 2 Public Switched Data Networks (PSDNs): Frame Relay, ATM, and metropolitan area Ethernet.
- Layer 3 transmission over the Internet or IP carrier networks using virtual private network (VPN) security using IPsec or SSL/TLS.

INTRODUCTION

Chapter 6 took us beyond the customer premises to telephone services provided by carriers. It also showed us individual Internet access technologies and services. This chapter looks at carrier services for wide area networking, which connects different sites together.

WANs and the Telephone Network

We studied the PSTN before this chapter because most of these **wide area network (WAN)** services are built on top of the telephone network's technology. Sometimes, end-user companies lease circuits from the telephone company to carry their internal data. In other cases, WAN carriers lease telephone circuits, add their own switching or routing, and offer data networking services, including management, to end-user corporations.

Three Reasons to Have a WAN

There are three main purposes for WANs.

> The first is to provide remote access to customers or to individual employees who are working at home or traveling.

Wide Area Networks (WANs)
    Connect different sites

WANs and the Telephone Network
    Use the PSTN transport system for transmission
    Add switching and management to create a WAN

WAN Purposes
    Provide remote access to individuals who are off site
    Link sites within the same corporation
    Provide Internet access

Evolution of WAN Technology
    Layer 1: Leased line service and networks
    Layer 2: Public switched data networks (PSDNs)
    Layer 3: Virtual Private Networks (VPNs) over the Internet and IP carrier networks

Carriers
    Beyond their physical premises, companies must use the services of regulated carriers for transmission
    Companies are limited to whatever services the carriers provide
    Prices for carrier services change abruptly and without technological reasons
    Prices and service availability vary from country to country

High Costs and Low Speeds
    High cost per bit transmitted, compared with LANs
    Consequently, lower speeds (most commonly 256 kbps to about 50 megabits per second)

Figure 7-1   Wide Area Networks (WANs) (Study Figure)

➤ The second is to link two or more sites within the same corporation. Given the large amount of site-to-site communication in most firms, this is the dominant WAN application.
➤ The third is to provide corporate access to the Internet.

In Chapter 6, we looked at data transmission alternatives for providing remote access to customers or to individual employees who are working at home or traveling. In this chapter, we will look at WAN technologies for site-to-site networking and corporate Internet access. We will see that corporations have three major alternatives for these two needs:

➤ Networks of leased lines.
➤ Public switched data networks (PSDNs).
➤ Virtual private networks (VPNs).

Carriers

A company can build its own LANs because they run through the company's own buildings and land. However, you cannot lay wires through your neighbor's garden, and neither can corporations. Transmission beyond the customer premises requires the use of regulated **carriers.**

One shock that companies face when dealing with carriers is pricing. With LAN technology, prices closely follow costs, and prices change gradually as technology matures. However, with carriers, there often is little relationship between prices and costs. For instance, until recently, companies could purchase Frame Relay WAN service, confident that it would be less expensive than leased line networking. Recently, however, many carriers abruptly and dramatically raised their Frame Relay prices and slashed their leased line prices. This created chaos in corporate WAN planning.

Another shock is service limitations. Usually, there are only a few competing carriers that a firm can use, and these carriers often offer only a few service options. There is nothing like the freedom companies have when they create LANs.

Global companies, furthermore, find that pricing and service options vary widely around the world. Options that are widely available in the United States and Europe often are rare in other parts of the world, and prices almost everywhere are higher than they are in the United States.

The Evolution of WANs

WAN technology has gone through three generations.

Layer 1 Service
In the first generation, which began in the 1960s, carriers provided Layer 1 (physical layer) service. They provided physical point-to-point circuits between two sites. These circuits, which are called *leased lines,* are still in widespread use.

Layer 2 Service
In the second generation, which began in the 1970s, but did not hit full stride until the 1990s, carriers provided Layer 2 (data link layer) switched services called *private switched data networks (PSDNs).* Companies simply connected their sites to a PSDN carrier, and the carrier did all of the switching required to move frames from one site to another.

Layer 3 Service
In the third generation, which began in the 1990s and is still fairly early in its life cycle, carriers began to provide Layer 3 (internet layer) WAN service. Actually, end user corporations took the lead in this generation, using the Internet for wide area networking to link their sites together. However, the Internet has security problems and performance problems. Consequently, carriers have begun to offer commercial IP WANs that use the TCP/IP protocols, but that offer quality of service guarantees and have much higher security.

High Costs and Low Speeds

Most LAN users are accustomed to at least 100 Mbps of unshared speed to the desktop. In contrast, long-distance communication is much more expensive per bit transmitted, so companies usually content themselves with slower transmission speeds in

WANs. Most WAN communication links operate at between 256 kbps and 50 Mbps, with most use coming at the lower end of this range.

Most WAN communication links operate at between 256 kbps and 50 Mbps, with most use coming at the lower end of this range.

TEST YOUR UNDERSTANDING

1. a) How are telephony and wide area networking related? b) What are the three main purposes for WANs? c) What are carriers, and why must they be used? d) How are prices and costs related in carrier WAN services? e) Does a company have more service options with LANs or with WANs? f) Are service options and prices similar around the world? g) List the three generations of carrier LAN technology. h) Compare LAN and WAN transmission speeds. i) Why are they different?

LAYER 1 CARRIER SERVICES (LEASED LINES)

In the 1960s, telephone carriers began to develop high-speed transmission lines to connect their internal switches. Seeing market opportunities, these telephone carriers began to use this technology to offer high-speed end-to-end circuits directly to customers. As we saw in Chapter 6, these subscriber-to-subscriber services are called **leased lines.**

Leased Line Networks for Voice and Data

Private Telephone Networks

Figure 7-2 shows that companies have traditionally used leased lines to connect their PBXs at various sites. This allows any telephone at any site to call any other telephone at any other site. Although leased lines are expensive, this arrangement is almost always much cheaper than using normal dial-up service to place long-distance calls between sites.

Leased Line Data Networks

Figure 7-2 also shows an internal corporate data network using leased lines. If the voice and data parts of the figure seem similar, this is because data networking using leased lines is based on the same technology used for leased line telephone networks. The main difference is that data networks use routers at each site rather than PBXs.

TEST YOUR UNDERSTANDING

2. Distinguish between the technologies of leased line voice networks and leased line data networks.

Leased Line Network Topologies

Should many or all pairs of sites be connected to each other, or should there be as few connections as possible? How the organization links its sites to one another is the network's topology. Figure 7-3 shows two topological extremes for building leased line networks.

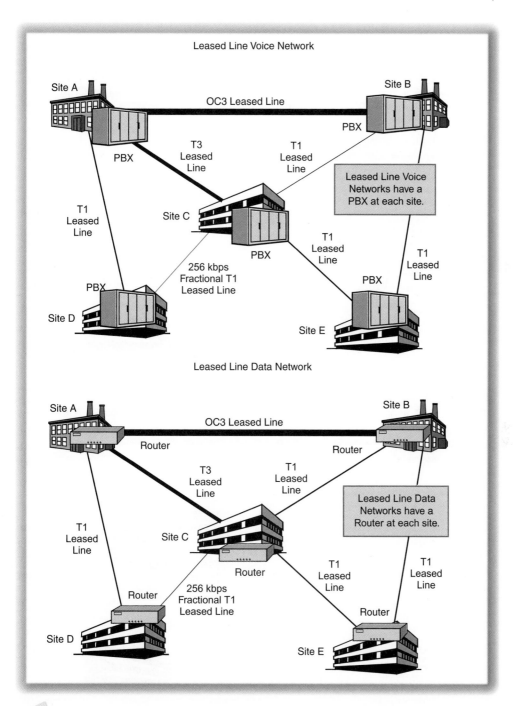

Figure 7-2   Leased Line Networks for Voice and Data

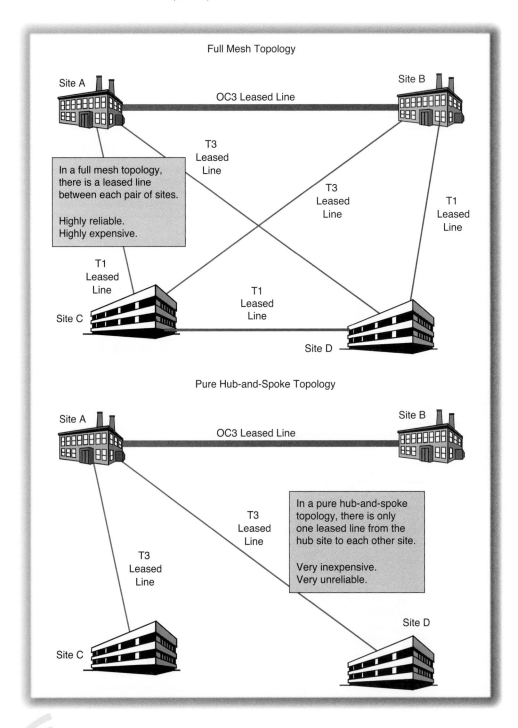

Figure 7-3   Full Mesh and Pure Hub-and-Spoke Topologies for Leased Line Data Networks

Full Mesh Topology Leased Line Networks

The first is a **full mesh topology,** which provides direct connections between every pair of sites. Because there are many redundant paths, if one site or leased line fails, communication can continue unimpeded.

Unfortunately, as the number of sites increases, the cost of a full mesh grows exponentially. For example, if there are N sites, a pure mesh will require $N*(N-1)/2$ leased lines. Likewise, a 5-site pure mesh will require $5*(5-1)/2$ (10) leased lines, a 10-site pure mesh will require 45 leased lines, and a 20-site pure mesh will require 190 leased lines. Full meshes, while reliable, are prohibitively expensive for a company with many sites.

Hub-and-Spoke Leased Line Networks

The second extreme topology for building leased line networks is the **pure hub-and-spoke topology.** This also is illustrated in Figure 7-3 In a pure hub-and-spoke topology, all communication goes through one site. This dramatically reduces the number of leased lines required to connect all sites, compared with a full mesh, and so this kind of topology minimizes cost. However, it also reduces reliability. If a line fails, there are no alternative paths for reaching an affected site. More disastrously, if the hub site fails, the entire network goes down.

Mixed Designs

As you might suspect, full meshes and pure hub-and-spoke topologies represent the extremes of cost and reliability. Most real networks use a mix of these two pure topologies. Real networks must trade off reliability against cost.

TEST YOUR UNDERSTANDING

3. a) What is the advantage of a full mesh leased line network? b) What is the disadvantage of a full mesh leased line network? c) What is the advantage of a pure hub-and-spoke leased line network? d) What is the disadvantage? e) Do most leased line networks use a full mesh or a pure hub-and-spoke topology? Explain.

4. a) A company has three sites: A, B, and C. They form a geographical triangle. A and B need 100 Mbps of transmission capacity between them. B and C need 200 Mbps of transmission capacity between them. A and C need 300 Mbps of transmission capacity between them. Create a hub-and-spoke network with A as the hub. What links will there be, and how fast will they have to be? Explain your reasoning. b) For the same situation, create a full-mesh network. What speeds will the links need to have if you are not concerned with redundancy? Explain you reasoning. c) Building on the last question part, add redundancy so that no single failure will bring down the network.

Leased Line Speeds

Leased line speeds vary from 56 kbps to several gigabits per second. We will now look specifically at the types of leased lines actually offered by telephone carriers.

Figure 7-4 shows that different parts of the world use different standards for leased lines below 50 Mbps. The figure shows lower-speed leased lines in the United States and Europe. There also are differences in other countries.

North American Digital Hierarchy

Line	Speed	Typical Transmission Medium
56 kbps or 64 kbps (rarely offered)	56 kbps or 64 kbps	2-Pair Data-Grade UTP
T1	1.544 Mbps	2-Pair Data-Grade UTP
Fractional T1	128 kbps, 256 kbps, 384 kbps, 512 kbps, 768 kbps	2-Pair Data-Grade UTP
Bonded T1s (multiple T1s acting as a single line)	Small multiples of 1.544 Mbps	2-Pair Data-Grade UTP
T3	44.736 Mbps	Optical Fiber

CEPT Hierarchy

Line	Speed	Typical Transmission Medium
64 kbps	64 kbps	2-Pair Data-Grade UTP
E1	2.048 Mbps	2-Pair Data-Grade UTP
E3	34.368 Mbps	Optical Fiber

SONET/SDH Speeds

Line	Speed (Mbps)	Typical Transmission Medium
OC3/STM1	155.52	Optical Fiber
OC12/STM4	622.08	Optical Fiber
OC48/STM16	2,488.32	Optical Fiber
OC192/STM64	9,953.28	Optical Fiber
OC768/STM256	39,813.12	Optical Fiber

Figure 7-4   Leased Line Speeds

56 kbps and 64 kbps

The lowest-speed lines in these hierarchies operate at 56 kbps or 64 kbps. This is barely higher than telephone modem speeds, and these leased lines are rarely used today.

T1 and E1 Leased Line

At the next level of the hierarchy, the T1 line in the United States operates at 1.544 Mbps. The comparable European E1 line operates at 2.048 Mbps.

Fractional T1/E1 Leased Lines

The gap between 56 kbps/64 kbps and 1.544 Mbps/2.048 Mbps is large, so many U.S. carriers offer **fractional T1** leased lines operating at 128 kbps, 256 kbps, 384 kbps, 512 kbps, or 768 kbps. These fractional leased lines provide intermediate speeds at intermediate prices. Similarly, carriers that offer E1 lines offer fractional lines.

T1/E1 and fractional T1/E1 lines provide speeds in the range of greatest corporate demand for WAN transmission—256 kbps to a few megabits per second. Consequently, T1/E1 and fractional T1/E1 lines are the most widely used leased lines.

T1/E1 and fractional T1/E1 lines are the most widely used leased lines.

Bonded T1s

Sometimes, a firm needs somewhat more than a single T1 line, but does not need the much higher speed of the T3 line (discussed next). Often, a company can **bond** a few T1s to get a few multiples of 1.544 Mbps. This is similar to link aggregation in Ethernet, which we saw in Chapter 4. Bonding is also done with E1 lines, although this is not shown in the figure.

T3 and E3 Leased Lines

The next level of the hierarchy is the T3 line in the United States.[1] It operates at 44.736 Mbps. The comparable E3 line operates at 34.368 Mbps.

SONET/SDH

Beyond T3/E3 lines, the world has nearly standardized on a single technology or, more correctly, on two compatible technologies. These are **SONET (Synchronous Optical Network)** in North America and SDH **(Synchronous Digital Hierarchy)** in Europe. Other parts of the world select one or the other.

Figure 7-4 shows that SONET/SDH speeds are multiples of 51.84 Mbps, which is close to the speed of a T3 line. SONET speeds are given by **OC (optical carrier)** numbers, while SDH speeds are given by **STM (synchronous transfer mode)** numbers.

The slowest offered SONET/SDH speed is 155.52 Mbps. Its speeds range up to several gigabits per second. Note that the SONET speed nearest to 10 Gbps is 9953.28 Mbps. Ethernet uses this speed for WAN usage so that it can transmit data over physical layer SONET lines.

TEST YOUR UNDERSTANDING

5. a) Below what speed are there different leased line standards in different parts of the world? b) At what speeds do the slowest leased lines run? c) What is the exact speed of a T1 line? d) What are the speeds of comparable leased lines in Europe? e) Why are fractional T1 and E1 speeds desirable? f) List common fractional T1 speeds. g) What are the most widely used leased lines? h) What leased line standards are used above 50 Mbps?

---

[1]Although there are T2 and E2 standards, they are not offered commercially.

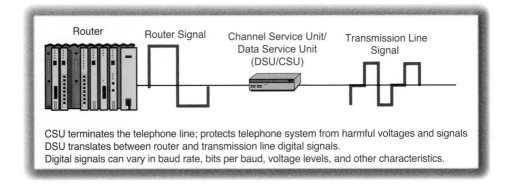

CSU terminates the telephone line; protects telephone system from harmful voltages and signals.
DSU translates between router and transmission line digital signals.
Digital signals can vary in baud rate, bits per baud, voltage levels, and other characteristics.

Figure 7-5 Connecting to a Leased Line

Connecting to Leased Lines

What equipment does the subscriber need to connect to a PSTN leased line? As Figure 7-5 shows, the subscriber needs two pieces of equipment at its site. First, it needs a switch or a router to connect the company to the outside world. Routers are now the most common devices for connecting companies to the outside world.

In addition, the company needs a device called a CSU/DSU.[2] As the acronym suggests, a CSU/DSU has two parts. The **CSU** is designed to protect the telephone network by keeping the subscriber from sending improper signals, such as signals with voltages that are too high. Think of the CSU as a fuse box. In turn, the **DSU** formats the data in the way the leased line requires.

Although the signals that the subscriber sends and the signals that the telephone company will accept over a leased line are both digital, they may have a different number of cycles per second, different voltage ranges, and other differences. The CSU translates between them.

TEST YOUR UNDERSTANDING

6. a) What device must a company have to connect a router to a leased line? b) What is the function is a CSU? c) What is the function of a DSU?

Digital Subscriber Lines (DSLs)

In the last chapter, we learned about digital subscriber lines. Leased lines up to T1 and E1 transmit over 2-pair data-grade UTP, but DSLs send data over 1-pair voice-grade UTP.

One-pair voice-grade UTP is attractive for carrying data because these lines are already in place. In contrast, services that use 2-pair data-grade UTP and optical fiber require new wiring or fiber to be pulled to the customer. This is very expensive. Unfortunately, 1-pair voice-grade UTP lines were not designed to carry data at high speeds, and not all 1-pair voice grade UTP lines can carry DSL signals.

---

[2]The abbreviation CSU/DSU stands for channel service unit/data service unit. However, this is almost never spelled out.

	ADSL	HDSL	HDSL2	SHDSL
Uses existing 1-pair voice-grade UTP telephone access line to customer premises?*	Yes*	Yes*	Yes*	Yes*
Target Market	Residences	Businesses	Businesses	Businesses
Downstream Throughput	A few megabits per second	768 kbps	1.544 Mbps	384 kbps–2.3 Mbps
Upstream Throughput	Slower than downstream	768 kbps	1.544 Mbps	384 kbps–2.3 Mbps
Symmetrical Throughput?	No	Yes	Yes	Yes
QoS Throughput Guarantees?	No	Yes	Yes	Yes

*By definition, all DSLs use 1-pair voice-grade UTP residential access lines.

Figure 7-6   ADSL versus Business-Class Symmetric Digital Subscriber Line (DSL) Service

The last chapter looked at asymmetric digital subscriber lines (ADSL lines), which provide high downstream (from the ISP) speeds but lower upstream (from the PC) speeds. This is fine for residential users, but businesses with site-to-site and Internet access requirements need symmetric high speeds. They also want guaranteed throughput.

HDSL

Fortunately, several business-oriented DSLs are available, as Figure 7-6 indicates. The most popular business DSL is the **high-rate digital subscriber line (HDSL).** This standard allows symmetric transmission at 768 kbps (approximately half of a T1's speed) in both directions. A newer version, **HDSL2,** transmits at 1.544 Mbps in both directions. Like all DSLs, both use a single voice-grade twisted pair. Businesses find HDSL and HDSL2 attractively priced, compared with T1 and fractional T1 lines.

SHDSL

The next step in business DSL is likely to be **SHDSL (super-high-rate DSL),** which can operate symmetrically over a single voice-grade twisted pair and over a speed range of 384 kbps to 2.3 Mbps. In addition to offering a wide range of speeds and a higher top speed than HDSL2, SHDSL also can operate over somewhat longer distances.

Quality-of-Service (QoS) Guarantees

Generally, there are no hard guarantees for ADSL speeds, which are aimed at the tolerant home market. However, throughputs for HDSL, HDSL2, and SHDSL generally come with strong quality-of-service guarantees because they are sold to businesses, which require predictable service. Meeting these guarantees requires more stringent engineering and management by the carrier and so increases carrier costs. This leads to higher prices for HDSL, HDSL2, and SHDSL.

TEST YOUR UNDERSTANDING

7. a) How do the lowest-speed leased lines and DSL lines differ in terms of transmission media? b) Describe HDSL and HDSL2 in terms of speed. c) Describe SHDSL in terms of speed. d) Which DSL services usually offer QoS guarantees?

LAYER 2 CARRIER WAN SERVICES: PUBLIC SWITCHED DATA NETWORK (PSDN)

Leased Lines in Leased Line Data Networks

Earlier, we looked at two topologies for building leased line data networks—mesh and hub-and-spoke topologies. Both approaches use many leased lines, and these leased

Figure 7-7   Public Switched Data Networks (PSDNs) (Study Figure)

Leased Line Data Networks
Use many leased lines, which must span long distances between sites
This is very expensive
Company must design and operate its leased line network

Public Switched Data Networks
PSDN carrier does most of the work
Subscriber needs only a single leased line from each site to the PSDN's nearest point of presence (POP)
PSDN core network is drawn as a cloud to indicate that subscribers do not have to understand it

Costs
Carriers benefit from economies of scale in building and managing the large PSDN network
Consequently, the price to most companies is less than the cost of a network of leased lines

Service Level Agreements (SLAs)
Guarantees for services
Throughput, availability, latency, error rate, and other matters
An SLA might guarantee a latency of no more than 100 milliseconds 99.99 percent of the time

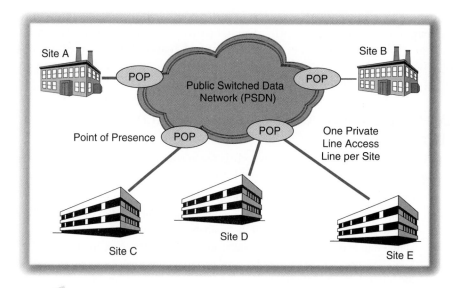

Figure 7-8   Public Switched Data Network (PSDN)

lines must span long distances—all the way between sites. This is very expensive. In addition, each country must design and operate its leased line network.

Public Switched Data Network (PSDN) Access Lines

In contrast, Figure 7-8 illustrates a **public switched data network (PSDN).** With a PSDN, the user needs only one leased line per site. This leased line has to run only from the site to the PSDN's nearest access point, called a **point of presence (POP).**[3]

This means that if you have 10 sites, you need only 10 leased lines. Furthermore, most PSDN carriers have many POPs, so the few leased lines that are needed tend to span only short distances.

The PSDN Cloud

The PSDN's transport core usually is represented graphically as a **cloud.** This symbolizes the fact that although the PSDN has internal switches and trunk lines, the customer does not have to know how things work inside the PSDN cloud. The PSDN carrier handles almost all of the management work that customers have to do when running their own leased line networks. Customers merely have to send data to and receive data from the PSDN cloud, in the correct format. Although PSDN carrier prices reflect their management costs, there are strong **economies of scale**

---

[3]In Chapter 6, we saw that the term *point of presence (POP)* is also used in telephony to mean a place where various carriers interconnect.

in managing very large PSDNs instead of individual corporate leased line networks. It is cheaper to manage the traffic of many firms than of one firm. There also are very large economies of scale in switching and leased line technologies. These economies of scale allow PSDN prices to remain low, compared with the costs of running leased line networks.

OAM&P: Operation, Administration, Maintenance, and Provisioning

All carriers must provide **OAM&P,** which is an acronym for *operation, administration, maintenance, and provisioning.* In Chapter 1, we looked at these concepts.

➤ Operation is the day-to-day provision of service.

➤ Administration includes accounting, billing, and similar tasks. Administration also involves high-level planning and the analysis of performance data.

➤ Maintenance involves fixing problems and doing preventative work.

➤ Provisioning is the supplying of service to a new customer. This may include taking the order, running wire out to the customer, installing customer premises equipment, configuring the customer premises equipment, turning on the service at the POP, and testing the connection. Provisioning can also comprise changing a customer's setup and terminating a customer.

The cost of OAM&P usually is substantially larger than the cost of technology in a PSDN.

Service Level Agreements (SLAs)

Most PSDNs offer **service level agreements (SLAs),** which are quality-of-service guarantees for throughput, availability, latency, error rate, and other service conditions. For instance, an SLA may guarantee a latency of no more than 25 milliseconds 99.99 percent of the time. Although SLAs are very nice to have, they add considerably to the price of a service because PSDN vendors need to allocate more resources to the customer to ensure that SLA guarantees are met.

TEST YOUR UNDERSTANDING

8. a) Describe the physical components of PSDN technology. b) Do customers need leased lines if they use PSDNs? c) Compare leased line costs for leased line networks and PSDNs. d) Which usually is less expensive overall—leased line data networks or PSDN transmission? e) Why is the PSDN transport core drawn as a cloud? f) Why do PSDNs tend to cost less than leased line networks? g) For what is OAM&P an abbreviation? h) What is provisioning? i) What things do SLAs guarantee? j) Why would an SLA guarantee maximum latency rather than minimum latency? Explain.

Virtual Circuit Operation

Figure 7-9 shows the PSDN switches that sit inside the cloud in a mesh topology. In any mesh topology, whether partial or full, there are multiple alternative paths for frames to use to go from a source POP to a destination POP.

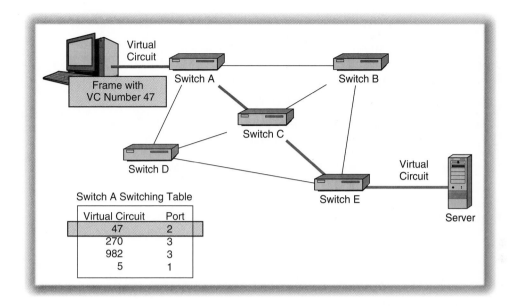

Figure 7-9   Virtual Circuit Operation

Selecting Best Possible Paths through Meshes
Selecting the best possible path for each frame through a PSDN mesh would be complex and, therefore, expensive. In fact, if the best possible path had to be computed for each frame at each switch along its path, PSDN switches would have to do so much work that they would be prohibitively expensive.

Virtual Circuits
Instead, most PSDNs select the best possible path between two sites *before transmission begins.* The actual transmission will flow along this path, called the **virtual circuit.** As Figure 7-9 shows, the switch merely makes a switching decision according to the virtual circuit number in the frame's header. This virtual circuit lookup is much faster than finding the best alternative path every time a frame arrives at the switch. The "heavy work" of selecting the best alternative path is done only once, before the beginning of communication.

PSDN Frame Headers Have Virtual Circuit Numbers Rather than Destination Addresses
Note that PSDNs that use virtual circuits do not have destination addresses in their frame headers. Rather, each frame has a virtual circuit number in its header.

TEST YOUR UNDERSTANDING

9.   a) Why are virtual circuits used? b) With virtual circuits, on what does a switch base its forwarding decision when a frame arrives? c) Do PSDN frames have destination addresses or virtual circuit numbers in their headers?

The Most Popular PSDN Service Today
        56 kbps to 40 Mbps. This fits the range of greatest corporate demand for
        WAN speed
        Usually, less expensive than networks of leased lines

Components of a Frame Relay Network (See Figure 7-11)

In Frame Relay, Virtual Circuit Numbers: DLCIs
        Data Link Control Indicators (pronounced "DULL-sees")
        Normally, 10 bits long

Figure 7-10   Frame Relay (Study Figure)

FRAME RELAY

The Most Popular PSDN

The most popular PSDN service today is Frame Relay. Frame Relay operates at 56 kbps to about 40 Mbps, with most customers operating well below that top speed. This is consistent with the needs of most corporations—256 kbps to a few megabits per second. Furthermore, compared with networks of leased lines, Frame Relay service usually is less expensive.

Components

Figure 7-11 shows the main elements of a Frame Relay network: access devices, access lines, ports at POPs, permanent virtual circuits (PVCs), and management.

Access Devices

Each user site needs an access device to convert between internal and Frame Relay signaling. This is either a router or a dedicated **Frame Relay Access Device (FRAD).**

CSU/DSU

The port on the router or FRAD that terminates the leased access line going to the Frame Relay network must have a CSU/DSU.

Access Lines and Points of Presence (POP)

The customer needs a leased line to use as an **access line** from each customer site to the nearest **point of presence (POP)** of the Frame Relay network. The POP is the entry point to the Frame Relay network. If a carrier has many POPs, then access lines will be relatively short and, therefore, relatively inexpensive.

Port Speed

The POP contains a switch that contains multiple ports of different speeds. At the POP, user transmission speed is limited by port speed. To transmit at a particular

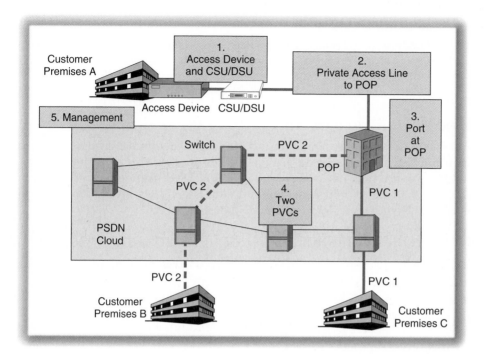

Figure 7-11   Frame Relay Network

speed, the user has to select a port speed suitable for his or her transmission needs. As you would suspect, using faster ports costs more. In fact, port speed usually is the most expensive pricing element in Frame Relay service. Selecting a port speed that is sufficient, but not extravagant, is critical in Frame Relay network design.

Port speed usually is the most expensive pricing element in Frame Relay service.

Virtual Circuits
Between each pair of sites that wish to communicate, there must be a virtual circuit. Collectively, virtual circuit charges usually are the second most expensive element in Frame Relay prices.

Management
Most Frame Relay vendors offer **managed Frame Relay** networks. Although PSDN carriers automatically handle internal management, managed Frame Relay services take on most of the remaining management tasks that customers are still responsible for. They provide traffic reports and actively manage day-to-day traffic to look for and fix problems. They also manage corporate access devices.

TEST YOUR UNDERSTANDING

10. a) List the technical components in a Frame Relay network. b) Briefly explain the purpose of each. c) Which usually is the most expensive component in Frame Relay pricing? d) Which usually is the second most expensive component in Frame Relay pricing? e) What is managed Frame Relay service?

Frame Relay Virtual Circuits

Recall that most PSDNs use virtual circuits. If they do, their frame headers have virtual circuit numbers rather than destination addresses. This virtual circuit number in Frame Relay is the **Data Link Control Identifier,** or **DLCI** (pronounced "DULL-see"). A DLCI is 10 bits long.[4] The switch looks up this DLCI in its virtual circuit switching table and sends the frame out through the indicated port.

TEST YOUR UNDERSTANDING

11. a) What is the name of the Frame Relay virtual circuit number? b) How long is the DLCI, usually? c) How many virtual circuits does this number of bits allow?

ASYNCHRONOUS TRANSFER MODE (ATM)

For PSDN service at speeds greater than Frame Relay can provide, corporations can turn to **asynchronous transfer mode (ATM)** service. ATM services reach gigabits per second. Their speed may extend down to 1 Mbps, but most usage will be much faster.

Not a Competitor for Frame Relay

It might seem that Frame Relay and ATM are competitors. In practice, however, almost all carriers offer both Frame Relay and ATM. Carriers recommend Frame Relay for customers with lower-speed needs and ATM for customers with higher-speed needs. In fact, some vendors have interconnected ATM and Frame Relay networks so that customers can connect low-speed sites with Frame Relay and high-speed sites with ATM.

ATM Frames

Most network protocols have variable-length data fields. This gives flexibility, but switches must do a number of calculations when dealing with variable-length frames. This adds to the work a switch must do and, therefore, to its cost. It also creates a bit of latency at each switch. To reduce switch processing costs and latency, ATM uses short, fixed-length frames.

An ATM frame consists of a 5-octet header[5] and a 48-octet data field (which ATM calls **payloads**). ATM frames do not have trailers.

---

[4]Stations also have true Frame Relay addresses governed by the E.164 standard. These addresses are used to set up virtual circuits. Consequently, if an equipment failure renders a virtual circuit inoperable, a new virtual circuit can be set up using the E.164 addresses.

[5]ATM has a two-part hierarchical virtual circuit number consisting of a virtual path identifier (VPI) and a virtual channel identifier (VCI). A specific VPI might be a path to a particular site. VCIs associated with that VPI might represent paths to specific computers at that site.

For Speeds Greater than Frame Relay Can Provide
 1 Mbps up to several gigabits per second

Not a Competitor for Frame Relay
 Most carriers provide both FR and ATM
 May even interconnect the two services

Short Frames
 Most frames have variable length
 All ATM frames are a very short 53 octets in length
  5 octets of header
  48 octets of data (payload)
  No trailer
  53 octets total
 Short length minimizes latency (delay) at each switch

ATM Has Strong Quality of Service (QoS) Guarantees for Voice Traffic
 Not surprising because ATM was created for the PSTN's transport core
 For pure data transmission, ATM does not provide QoS guarantees

Manageability, Complexity, and Cost
 Very strong management tools for large networks (designed for the PSTN)
 Too complex and expensive for most firms

ATM's Future?
 May flourish after firms outgrow Frame Relay speeds
 However, metropolitan area Ethernet should be a strong competitor
 ATM *is* flourishing in a different market, the PSTN core

Figure 7-12   ATM (Study Figure)

ATM Quality-of-Service Guarantees

ATM supports several different classes of service that receive different guarantees. For voice, ATM can set strict limits on latency and jitter (variable in latency). This makes ATM ideal for voice traffic. Data, however, usually are given no guarantees. In fact, the capacity that has to be reserved to give QoS guarantees for voice means that data traffic gets only leftovers. For pure data transmission, ATM's ability to provide QoS guarantees is not a benefit.

Manageability, Complexity, and Cost

ATM was created to become the transport mechanism for the worldwide PSTN. In fact, most long-distance telecommunications companies have already moved at least partway to having ATM transport cores.

 The requirement to manage the entire worldwide telephone network necessitated the creation of an extremely sophisticated set of ATM management protocols.

This sophistication, of course, results in complexity and high cost. For most firms, ATM is prohibitively expensive.

Market Strengths

As just noted, ATM has become very important in the telephone system's transport core. As corporate demands for WAN speeds increase due to growing needs and falling prices, many Frame Relay users may migrate to ATM. However, for speeds higher than Frame Relay can provide, many firms are considering another PSDN, metropolitan area Ethernet.

TEST YOUR UNDERSTANDING

12. a) Compare Frame Relay and ATM speed ranges. b) Are Frame Relay and ATM competitors? Explain. c) How long are ATM headers and payloads? d) Why does ATM use short frames? Explain your answer. e) Compare what ATM has to offer to voice and data service. f) Why does ATM have strong management tools? g) Why is ATM's sophistication beneficial? h) Why is it problematic? i) Why may ATM usage grow in the future? Why may it not grow in the future?

METROPOLITAN AREA ETHERNET

Metropolitan Area Networking

Ethernet dominates local area networking. Now, Ethernet is now beginning to outgrow the LAN, in the form of **metropolitan area Ethernet (metro Ethernet)**, which spans a single urban area, including its suburbs. Although **metro Ethernet** is still young, it is already beginning to spread rapidly.

E-Line and E-LAN Service

Metro Ethernet is offered in two forms:

➤ Metro Ethernet **e-line** services provide point-to-point connections, as leased lines do.

➤ In turn, **e-LAN** services link multiple sites simultaneously.

To the switches at each site, e-LAN service simply looks like a set of additional trunk lines linking the sites.

Attractions of Metro Ethernet

Low Cost

Although there are several aspects of metropolitan area Ethernet that are attractive, the most important is its low cost. As it does in LANs, Ethernet's simplicity reduces switching costs in MANs. Overall, metro Ethernet is much cheaper than Frame Relay or ATM for comparable speeds.

High Speeds

In addition, metropolitan area Ethernet offers very high speeds. While Frame Relay offers speeds up to a few tens of megabits per second, metro Ethernet offers speeds up to 10 Gbps at only slightly higher cost and will soon offer 40 Gbps. In addition,

Metropolitan Area Network (MAN)
    A carrier network limited to a large urban area and its suburbs
    Metropolitan area Ethernet (metro Ethernet) is available for this niche
    Metro Ethernet is new, but is growing very rapidly

Services
    E-Line Service
        Provides point-to-point connections between sites, as leased lines do
    E-LAN Service
        Links multiple sites simultaneously

Attractions of Metropolitan Area Ethernet
    Low prices
    High speeds
    Familiar technology for networking staff
    Rapid provisioning

Carrier Class Service
    Basic Ethernet standards are insufficient for large WANs (wide area networks)
    Quality of service and management tools must be developed
    The goal: to provide carrier class services that are sufficient for customers
    802.3ad standard
        Ethernet in the first mile
        Standard for transmitting Ethernet signals over PSTN access lines
        1-pair voice-grade UTP, 2-pair data-grade UTP, optical fiber

Figure 7-13   Metropolitan Area Ethernet (Study Figure)

transmission speed can be purchased in small increments, so companies order only the transmission capacity they actually need.

Familiar Technology

A third advantage of metropolitan area Ethernet is that firms can use the standard Ethernet interface they already know well instead of having to master Frame Relay, ATM, or other new interfaces. A site only needs an Ethernet switch, not a router.

    This is not a minor advantage. A help desk employee working at Cisco noted that most calls about border router setup dealt not with routing, but with the details of Frame Relay and other PSDN protocols. If the customer connects to routers via the well-known Ethernet protocol, its staff does not have to master additional PSDN protocols at the data link layer.

Rapid Provisioning

Provisioning is the setting up of service. Once a customer is set up by a metro Ethernet carrier, most carriers can change a customer's setup in a few hours, providing additional speed whenever special circumstances require it.

Carrier Class Service

However, metro Ethernet has not completely developed the quality-of-service and traffic management tools needed to offer true **carrier class service.** Until these tools are finished, corporations will be hesitant to use Ethernet for very large metro networks. The following requirements will have to be met before metro Ethernet can offer carrier class service:

➤ First, metro Ethernet must have much better reliability than LAN versions of Ethernet. This may require the use of a mesh or ring topology.

➤ Second, there must be strong quality-of-service guarantees, not simply priority levels, for voice.

➤ Third, there must be management tools to gain central control of large metropolitan LANs.

So far, only one major aspect of metro Ethernet has been standardized. This is the 802.3ad standard for Ethernet in the first kilometer. This standard describes how to transmit Ethernet signals over several PSTN local loop technologies, including 1-pair voice-grade UTP, 2-pair data-grade UTP, and optical fiber. The 802.3ad standard specifically describes how to do signaling over these lines.

TEST YOUR UNDERSTANDING

13. a) What is metropolitan area Ethernet? b) Distinguish between e-line and e-LAN service. c) Why is metro Ethernet attractive? d) Why are companies hesitant to create large metro Ethernet MANs?

LAYER 3 IP WAN SERVICES

IP over Everything

We learned in Chapter 1 that Cerf and Kahn created the concept of routed networks to provide host-to-host connection across multiple switched networks of different types. Over time, networking professionals have come to focus on TCP/IP routed networks as their basic philosophy of network design. Cerf's motto, "IP over everything," is coming through in large corporations.

Because TCP/IP is the basic skill set for networking professionals, bringing WANs into the familiar realm of IP networking reduces labor costs. In contrast, Frame Relay and other WAN protocols usually are relatively unfamiliar to corporate networking staffs. One Cisco help desk employee noted that most calls for router support dealt with Frame Relay and ATM, not with IP routing per se.

TEST YOUR UNDERSTANDING

14. In terms of ease of use, why is IP WAN transmission attractive compared with Layer 2 WAN services?

The Internet and IP Carrier Networks

Given the importance of IP, companies are beginning to migrate from Layer 1 and Layer 2 WAN service to Layer 3 IP WAN service. There are two basic ways to get Layer 3 IP

IP Is Increasingly Important
  Companies know it and are comfortable with it
  There are two ways to use IP at Layer 3 for WAN transmission:
  Companies can create their own IPsec VPNs
  Companies can use IP carrier networks just as they use PSDNs for Layer 2 WANs
    Reduces corporate labor
    Usually are ISPs
    Sometimes are PSDNs that offer Layer 3 IP carrier service

Advantages Using of the Internet as a WAN
  Low cost per bit transmitted because of economies of scale in the Internet
  Access to other companies, nearly all of which are connected to the Internet
  IP carrier networks can offer QoS SLAs
    IP is only a best-effort protocol
    But companies can engineer their networks for full QoS
    Customers must connect all sites to the same ISP for this to work

Security
  If companies act on their own, they can add virtual private network (VPN) protection to their transmissions
  IP Carrier Network Security
    IP Carrier Networks have some inherent security
      Restrict access to business customers
    However, for real security, virtual private networks (VPNs) are needed
      IP carrier networks provide cryptographic equipment at each site

Figure 7-14   The Internet versus IP Carrier Networks

WAN service. One is simply to use the Internet. The Internet, after all, really is a very large IP WAN. The other is to use an IP carrier network, which essentially is a private version of the Internet. Figure 7-14 compares these alternatives.

Access to Other Companies

The most obvious difference between using the Internet and using an IP carrier network is access to other companies. With the Internet, you can reach almost any company in the world, as well as almost any individual. With IP carrier networks, there usually is no access to other companies at all; IP carrier networks are designed for linking corporate sites together. (Layer 2 PSDNs usually are the same.) Even if you could reach other companies through the same IP carrier network, few of these companies would be your customers, vendors, or other business partners. Your business partners, if they used IP carrier networks at all, probably would use different ones.

Cost

The next issue is cost. The biggest attraction of using the Internet for data transmission is low cost. Thanks to its enormous economies of scale, the Internet's cost per bit transmitted is very low—far lower than Layer 2 PSDN services. IP carrier networks cost significantly more per bit transmitted. However, this higher cost may be justifiable to many firms because of the advantages of IP carrier network, which we will consider next.

Quality of Service (QoS) SLAs

As Chapter 1 emphasized, companies today are very concerned with network quality of service metrics such as speed, latency, and jitter. On the Internet, of course, there are no QoS guarantees. There simply is no way to do traffic engineering end-to-end, across the multiple ISPs that will carry a packet. The Internet, at least for the foreseeable future, will remain a **best effort** network.

In contrast, all IP carrier networks offer QoS service level agreements (SLAs). In fact, this may be their biggest attraction. For site-to-site networking, especially for transaction processing applications, poor quality of service or even inconsistent quality of service can cost the company a great deal of money. As we will see next, security can be added to Internet transmission, but QoS guarantees cannot.

Security: Inherent Security and Virtual Private Networks (VPNs)

The biggest nightmare for most companies considering Internet transmission is breach of security. Conversely, security is the second biggest attraction of IP carrier networks.

First, IP carrier networks are inherently more secure than the Internet. While anyone can get access to the Internet, only commercial companies that subscribe to the same IP carrier network can attack other subscribers. In addition, IP carrier networks segregate the traffic of their customers. Figure 7-15 shows that each customer

Figure 7-15   Route-Based Virtual Private Network (VPN) in an IP Carrier Network

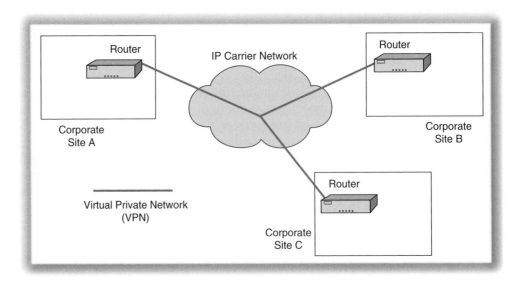

appears to have its own private network completely separate from the traffic of other customers. This further increases the inherent security of IP carrier networks.

Of course, the customer does not literally have a private network; it really has a **virtual private network (VPN),** which appears to the user to be a private network. This type of VPN is called a **route-based VPN** because it provides security by hiding routing information from potential attackers. IP carrier networks also offer **cryptographic VPNs,** which provide cryptographic security for traffic. With cryptographic VPN protection, even if someone outside the customer's company could get access to the company's traffic flowing through the IP carrier network, the attacker could not do damage.

To implement cryptographic VPNs, the IP carrier network places cryptographic edge equipment at each customer site. These edge devices are managed so that the customer can enjoy cryptographic protection without having to do the work of managing the protection.

The Internet, of course, offers no security at all. However, if companies want to take advantage of the low cost of Internet transmission, they can add their own VPN security to Internet transmission, as we will see a little later.

Carriers

Companies are well acquainted with ISPs. What carriers offer IP carrier networks? One group of carriers consists of ISPs themselves. Most offer carrier IP network service for companies that connect all of their sites to a particular ISP.

Another group of companies that offer IP carrier networks comprises the traditional Layer 2 PSDN vendors, including telephone companies. Given the low profitability of Frame Relay and the lack of success of ATM, most PSDN vendors are beginning to push their customers to IP services.

TEST YOUR UNDERSTANDING

15. a) What are the two ways that companies can get IP WAN service? b) Which of these approaches generally is used only for intra-company site-to-site communication? c) What is the big attraction of the Internet for data transmission? d) Are IP carrier networks more expensive or less expensive than transmission over the Internet? e) Which alternative can offer QoS SLAs? f) What security protections do IP carrier networks offer? g) Distinguish between route-based VPNs and cryptographic VPNs. h) For cryptographic VPNs used in IP carrier networks, what organization manages VPN security? i) What companies offer IP carrier network service?

Virtual Private Networks (VPNs)

We have just seen that IP carrier networks provide both route-based and cryptographic virtual private networks. For most networking professionals, the name *virtual private network* means a cryptographic VPN. From this point, we will use the term **virtual private network (VPN)** as a synonym for cryptographic VPN.

Companies do not have to go to IP carrier networks to get VPN protection. They can add their own VPN protection to Internet transmissions. As Figure 7-16 shows, there are two types of VPN connections that can be added to the Internet. The first is a **remote access VPN,** which connects a remote user to a corporate site. The user may be at home or on a business trip. The remote user connects to a **VPN gateway** at the site.

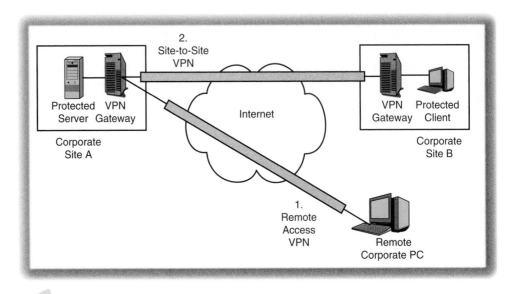

Figure 7-16　Virtual Private Network Connections on the Internet

There also are **site-to-site VPNs.** These VPNs protect all conversations between two sites. The virtual private network connects the VPN gateways at two different sites. Typically, site-to-site VPNs carry far more traffic than remote access VPNs do, because they handle multiple conversations instead of a single conversation.

TEST YOUR UNDERSTANDING

16. a) In the case of an IP carrier network, what is a VPN? b) What are the two types of VPN connections that companies can create for themselves for transmission over the Internet? c) What is a remote access VPN? d) What two devices terminate a remote access VPN connection? e) What is a site-to-site VPN? f) What two devices terminate a site-to-site VPN connection?

IPsec VPNs

There are two standards for virtual VPN security. The most sophisticated VPN technology is a set of standards collectively called **IP security (IPsec).**[6] As its name suggests, IPsec operates at the internet layer. It provides security to all upper layer protocols transparently, protecting everything carried in the IP packet's data field.

Pros and Cons

IPsec offers the strongest security and should eventually dominate remote access VPN transmission, site-to-site VPN transmission, and internal IP transmission. However, IPsec networks are complex to manage and therefore expensive.

---

[6]*IPsec* is pronounced "eye-pea-sek," with equal emphasis on all syllables.

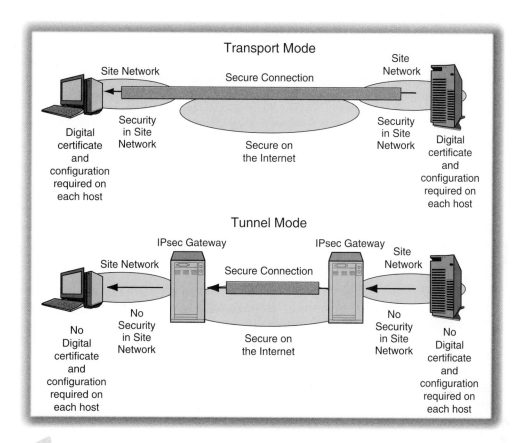

Figure 7-17   IPsec Transport and Tunnel Modes

Transport Mode

Figure 7-17 shows IPsec's two modes of operation. In **transport mode,** the two computers that are communicating implement IPsec. This mode gives strong end-to-end security, but it requires IPsec configuration *and* a digital certificate on all machines. Although the cost per machine for configuration and for the digital certificate is small, the large number of computers in a company makes the aggregate cost of transport mode setup high.

Tunnel Mode

In contrast, in **tunnel mode,** the IPsec connection extends only between **IPsec gateways** at the two sites. This provides no protection within sites, but the use of tunnel mode IPsec gateways offers simple security. The two hosts do not have to implement IPsec security and, in fact, do not even have to know that IPsec is being used between the IPsec gateways. Most importantly, there is no need to install digital certificates on individual hosts. Only the two IPsec gateways need to have IPsec configuration and digital certificates.

TEST YOUR UNDERSTANDING

17. a) At what layer does IPsec operate? b) What layers does IPsec protect? c) Describe IPsec tunnel mode. d) What is the main advantage of tunnel mode? e) What is the main disadvantage of tunnel mode? f) Describe IPsec transport mode. g) What is the main advantage of transport mode? h) What is the main disadvantage of transport mode? i) In which IPsec mode are clients and servers required to have digital certificates? j) Which IPsec mode does not require clients and servers to have digital certificates? k) Is IPsec used for remote access or site-to-site VPNs?

SSL/TLS VPNs

The simplest VPN security standard to implement is **SSL/TLS.** This standard was originally created as **Secure Sockets Layer (SSL)** by Netscape. It was later taken over by the IETF and renamed **Transport Layer Security (TLS).** We will call it SSL/TLS because it is still referred to by both names.

Nontransparent Protection

As Figure 7-18 shows, SSL/TLS provides a secure connection at the transport layer. This protects all applications above it. However, SSL/TLS protects only applications that are **SSL/TLS-aware**—that is, modified to work with SSL/TLS. All browsers and webservers are SSL/TLS-aware. Some e-mail systems also are SSL/TLS-aware. Few other applications are.

Although traditional SSL/TLS is limited to a few applications, many firms need only remote Web access. These firms are likely to use SSL/TLS, which is easy to implement because every browser and webserver application program has SSL/TLS built in.

Authentication Options

In SSL/TLS, one issue is how to authenticate the user—that is, require the user to prove his or her identity. One SSL/TLS option for corporations is to do no authentication for the client; this opens SSL/TLS-based systems to many attacks. Webserver

Figure 7-18   SSL/TLS for Browser–Webserver Communication

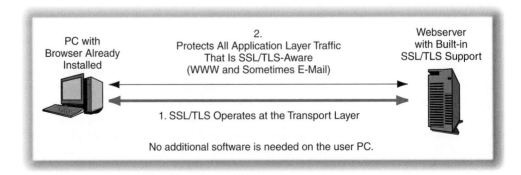

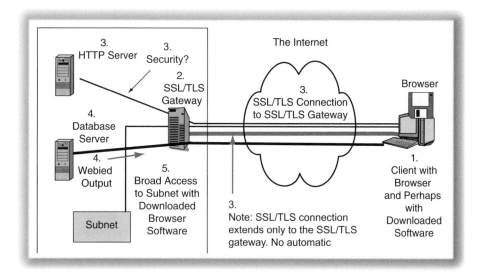

Figure 7-19   SSL/TLS VPN with a Gateway

application programs can supplement this SSL/TLS weakness by adding passwords themselves, but password security is not strong.

The other option is for corporations to use a digital certificate for each client. This provides very strong security; but, as mentioned earlier, implementing client digital certificates is very difficult.

SSL/TLS Gateways

Initially, SSL/TLS protected direct connections between a client and a webserver. However, as Figure 7-19 shows, several vendors have begun to produce **SSL/TLS gateways** to let an authenticated user reach any internal webserver to which he or she should have access. The gateways turn SSL/TLS into a true remote access VPN technology. The user has a single SSL/TLS connection—to the SSL/TLS gateway. The gateway provides access to internal webservers.

Although SSL/TLS normally is limited to HTTP, most SSL/TLS gateway vendors are able to "**webify**" some other applications, such as database applications. Webification involves converting screen images from these nonweb applications into webpages that browsers can read. Webification also involves sending what the user types onto webpage data entry forms into a format the nonweb application can use. This has greatly expanded the ability of remote access users to use nonweb applications, but not all nonweb applications can be webified.

SSL/TLS gateways can also give an external user, after authentication, complete access to a segment of the network containing multiple hosts.

Some gateway vendors even provide add-ins that can be downloaded to browsers. These add-ins are required for some services, such as complete access to a network segment. They may also be desirable, for instance if the user is at a public kiosk and wants to erase all data about his or her session from the computer. Unfortunately, these downloads require administrative access on the computer. On public computers at kiosks and in cybercafés, administrative access is rarely available to users.

TEST YOUR UNDERSTANDING

18. a) How is SSL/TLS limited? b) Why is it attractive? c) Without an SSL/TLS gateway, under what circumstances is SSL/TLS likely to be used? d) When an SSL/TLS gateway is used, how many SSL/TLS connections does the client have? e) What is webification? f) What is webification's benefit? g) Why are downloads for SSL/TLS gateway service not likely to be useful on public PCs?

19. a) Of the two VPN security technologies discussed in this section, which provides transparent security to higher layers? b) Which tends to require the installation of software on many client PCs? c) Which has stronger security? d) Which would you use for an intranet that gives employees remote access to a highly sensitive webserver via the Internet? (This is not a trivial question.) Justify your answer.

MARKET PERSPECTIVE

In this chapter, we have covered several WAN technologies. It is important to understand their relative importance in planning for the future. We will look at them one more time in terms of their future importance.

Leased Line Networks

Leased line networks emerged in the 1960s and dominated corporate WAN technology until the 1990s. Since then, leased line sales have stagnated, and leased lines have primarily been an access technology used to link sites to PSDN POPs. However, sharp drops in vendor pricing around 2002 created a spurt in the market share of leased lines. For the long term, however, leased lines are likely to be used primarily as access lines.

Frame Relay

PSDNs have been around since the 1970s, but they really came into their own in the 1990s when Frame Relay began to grow explosively. Frame Relay was ideal because it allowed companies to get wide area network service without investing extensively in WAN technology or WAN expertise. Routers, dedicated Frame Relay Access Devices, CSU/DSUs, and leased lines were all that companies needed. For an additional fee, Frame Relay vendors would even provide and manage them actively.

Frame Relay prices were very low, but this resulted in low profit margins for carriers. Around 2002, many carriers simultaneously raised Frame Relay prices and dropped leased line prices. Although pricing may change again, Frame Relay is now viewed as a legacy technology rather than as a technology for the future.

Metropolitan Area Ethernet

Many corporations have strong needs for metropolitan area transmission. Although metro Ethernet is still immature, its familiarity, low costs, and high speeds have already brought high market growth.

Leased Line Networks
> Dominated WAN transmission until the 1990s
> But difficult to set up and expensive to run
> Recent spurt in use because of reduced leased line prices and rising Frame Relay prices
> Needed for access lines in PSDNs and VPNs anyway

Frame Relay
> Grew explosively in the 1990s
> Became very widely used
> FR prices have risen recently in an effort by carriers to increase their profit margins
> Widely used and familiar, but now considered a legacy technology

ATM
> Very high speeds, but very high price
> Not thriving in the corporate market

Metro Ethernet
> Price and speed are very attractive
> Growing very rapidly
> Limited to metropolitan area networking
> Still somewhat immature technically

IP Carrier Networks and Internet Transmission with VPNs
> The Internet offers a very low cost per bit transmitted
>> VPNs provide security for Internet transmission
> Companies can also subscribe to IP carrier services
>> IP carrier services also offer QoS
> IP WAN usage is growing rapidly

Figure 7-20   Market Perspective (Study Figure)

Layer 3 Virtual Private Networks (VPNs)

Most corporations now believe that IP transmission over the Internet via VPNs is the wave of the future. IP is a familiar and well-understood technology, and the Internet offers very attractive pricing compared with PSDNs. However, the Internet is not secure, so customers have to add security through virtual private networks (VPNs), which provide cryptographic protections to transmissions. Companies that communicate over the Internet can create remote access VPNs with IPsec or SSL/TLS and site-to-site VPNs with IPsec.

Another problem with the Internet is a lack of quality of service guarantees. To get these, companies are beginning to use IP carrier networks, which are basically private ISPs that are not connected to the Internet (at least, not directly). All communicating partners connect to them. With this level of control over traffic, IP carrier networks can offer strong quality of service guarantees. They also offer VPNs.

Layer 3 networks using VPNs have been growing very rapidly. Many analysts believe that Carrier IP networks and IPsec VPNs will soon dominate WAN technology for long distances and will share dominance of MAN technology with metro Ethernet.

TEST YOUR UNDERSTANDING

20. a) Which of the technologies that we saw in this chapter should be considered legacy technologies—technologies that are not likely to experience rapid growth in the future and that may actually decline? b) Which technologies are growing rapidly?

CONCLUSION

Synopsis

Corporations build wide area networks (WANs) for individual remote access, site-to-site transmission, and Internet access. Among the technologies they use are telephone modems for low speeds, networks of leased lines, public switched data networks (PSDNs), metropolitan area radio transmission (rarely), and virtual private networks (VPNs).

Your personal experience probably has been limited primarily to LAN transmission, for which low cost per bit transmitted allows companies to afford high speeds. You must adjust your thinking for wide area networks (WANs), where long distances make the price per transmitted bit very high, which in turn leads to companies limiting themselves primarily to low speeds—most typically, between 256 kbps and 50 Mbps. Most demand comes at the lower end of this range.

For site-to-site networking, companies have traditionally turned to networks of leased lines. They did this first for telephone services, using leased lines to connect PBXs at different sites. For data networking, they replaced the PBXs with routers. Most leased line networks mix the characteristics of full mesh topologies, which are reliable but expensive, and pure hub-and-spoke topologies, which are inexpensive but have many single points of failure.

Companies that use leased lines typically choose 56 kbps/64 kbps leased lines, T1/E1 leased lines operating at 1.5 Mbps to 2 Mbps, and fractional leased lines below T1/E1 speeds. However, if they have some connections that require much higher speeds, they can choose T3/E3 lines operating at roughly 30 Mbps to 50 Mbps, or SONET/SDH lines operating at 156 Mbps to several gigabits per second. In addition, many firms use HDSL, HDSL2, or SHDSL. These are digital subscriber line services, but they offer high symmetrical speeds and throughput guarantees.

With public switched data networks (PSDNs), the carrier does most of the transmission and management work. Companies merely need access devices (typically routers) at their sites and a single leased line from each site to the PSDN carrier's nearest point of presence. Frame Relay provides speeds of 56 kbps to 40 Mbps. This matches the range of greatest corporate demand, so Frame Relay has dominated the PSDN market.

ATM offers speeds of 1 Mbps to several gigabits per second, with low-megabit speeds being fairly uncommon. For voice traffic, ATM offers stringent latency and jitter control, but it offers no special QoS SLA guarantees for data traffic. In addition, ATM is extremely expensive.

Both Frame Relay and ATM use virtual circuits to simplify the operation of switches and, therefore, minimize switching costs. Switches base forwarding decisions on virtual circuit numbers rather than on destination addresses.

Metropolitan area Ethernet, which extends Ethernet beyond the corporate borders for transmission within an urban area, is very new. It offers the potential to slash transmission costs compared with ATM. However, it may not thrive until quality-of-service standards and general management standards are created that allow Ethernet to work in the large, but highly price-sensitive, world of WAN transmission.

An option for both remote access and site-to-site networking is the virtual private network (VPN), which uses the Internet to lower transmission costs, but adds security to protect sensitive conversations. For remote access, SSL/TLS VPNs can work if the company is primarily using Web-based services, although SSL/TLS gateways have recently extended the number of applications that can use SSL/TLS. Every browser already has the ability to work with SSL/TLS. The IPsec standard offers much better security than SSL/TLS and so should eventually dominate for both remote access and site-to-site transmission. However, IPsec in transport mode requires that digital certificates be provided to all clients, and this is expensive.

End-of-Chapter Questions

THOUGHT QUESTION

Several Internet access systems are asymmetric, with higher downstream speeds than upstream speeds. a) Is this good for client PC access to webservers? Explain. b) Does it matter for client access to e-mail servers? c) Is it good for a file server? Explain. d) Is it good for videoconferencing? Explain.

DESIGN QUESTIONS

1. You have four sites. A (in the Northwest) needs to communicate with B at 100 Mbps. B (in the Northeast) needs to communicate with C at 100 Mbps. C (in the Southeast) needs to communicate with D at 100 Mbps. D (in the Southwest) needs to communicate with A at 200 Mbps. A needs to communicate with C at 100 Mbps. No other traffic is necessary. Design a leased line network with redundancy so that communication will still get through if any single link between sites fails. Specify the speeds of all links, keeping redundancy in mind.
2. A company that does business mostly in the Sydney area is considering replacing its Frame Relay network. What option(s) would you recommend that it consider? Justify your selection(s).

PROJECT

Report on the PSDN service offerings in your area.

GETTING CURRENT

Go to the book website's New Information and Errors pages for this chapter to get new information since this book went to press and to note corrections to any errors in the text.

Case Study: First Bank of Paradise's Wide Area Networks

INTRODUCTION

The First Bank of Paradise (FBP) is a mid-size bank that operates primarily within the state of Hawai'i, although it has one affiliate office on Da Kine Island in the South Pacific.

FBP is a mid-size bank, but it is not a small company. The bank has annual revenues of $4 billion. It has 50 branches and 350 ATMs. It has more than 500 switches, 400 routers, 2,000 desktop and notebook PCs, 200 Windows servers, 30 Unix servers, and 10 obsolete Novell NetWare file servers. Its information systems staff has 150 employees.

ORGANIZATIONAL UNITS

Major Facilities

Figure 7a-1 shows that FBP has three major facilities, all located on the island of O'ahu.

➤ **Headquarters** is a downtown office building that houses the administrative staff.

➤ **Operations** is a building in an industrial area that houses the bank's mainframe operations and other back-office technical functions. It also has most of the bank's IT staff, including its networking staff.

➤ **North Shore** is a backup facility. If Operations fails, North Shore can take over within minutes. North Shore is located in an otherwise agricultural area.

Branches

Although branches are small buildings, they are technologically complex, primarily because the devices there use diverse network protocols. The automated teller machine at a branch uses SNA protocols to talk with the mainframe computer at Operations. The teller terminals use different SNA protocols to talk to the Operations mainframe. File servers require IPX/SPX communication, and branch offices that need Internet access require TCP/IP.

At each branch, there is a Cisco 2600 router to connect the branch to Operations and North Shore (its backup). This is a multiprotocol router capable of handling the many protocols used at the internet and transport layers in branch office communication.

External Organizations

First Bank of Paradise has to deal with several organizations outside the company. Figure 7a-1 shows only one of these—a connection to a credit card authorization

Figure 7a-1   First Bank of Paradise Wide Area Networks

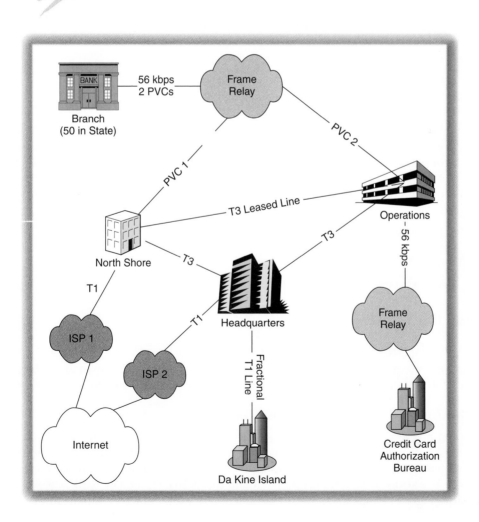

bureau. In fact, FBP deals with more than a dozen outside support vendors, each in a different way. Fortunately, the credit card authorization firm uses TCP/IP, which simplifies matters.

THE FBP WIDE AREA NETWORK (WAN)

Figure 7a-1 shows the complex group of WANs that the bank uses to hold together this geographically dispersed and technologically diverse collection of sites.

T3 Lines

A mesh of T3 lines connects major facilities, as Figure 7a-1 shows. T3 private lines operate at 44.7 Mbps, providing "fat pipes" between these facilities.

Branch Connections

Branches are connected to the major facilities in two ways. Most of the time, they communicate via a Frame Relay network. For each branch, there are two 56 kbps PVCs. One PVC leads to Operations, the other to North Shore.

Da Kine Island Affiliate Branch

For the Da Kine Island affiliate branch, the firm has a 128 kbps fractional T1 digital private line.

Credit Card Service

FBP connects to the credit card processing company using a 56 kbps Frame Relay network connection. This gives adequate speed.

Branch LANs

Branch offices have Ethernet networks. Each branch has a single 48-port 100Base-TX switch connected to the branch's Cisco 2600 border router.

Internet Access

For Internet access, FBP uses two separate ISPs, connecting to each via a T1 private line.

ANTICIPATED CHANGES

Outsourcing

The bank is anticipating two major changes. First, it plans to outsource about 60 percent of its internal operations to a bank processing company in Northridge, California.

Fractional T1 Lines to Branches

Also, in a reversal of past trends, Frame Relay vendors in Hawai'i have been raising their rates in recent years to seek higher profit margins. At the same time, the local telephone company has been dropping its rates on private lines dramatically in response to a strong long-term drop in demand for these circuits. The bank believes that it can bring a 256 kbps fractional T1 connection to each branch economically.

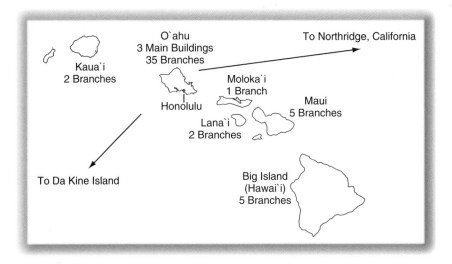

Figure 7a-2   First Bank of Paradise Locations

The bank's main buildings are located on the island of O'ahu. The bank also has 35 branch offices on O'ahu, which is the most populous island. The bank also does business on five "outer islands"—Maui, Kaua'i, Moloka'i, Lana'i, and the Big Island (the Island of Hawai'i). Maui and the Big Island have five branches each. Moloka'i has one branch.

TEST YOUR UNDERSTANDING

1. a) List all examples of redundancy in the FBP network. b) What is the goal of redundancy?

2. a) Why do you think two access points were created instead of one? b) Why are there only two access points to the Internet?

3. Do you think the bank uses the same Frame Relay network to connect its branches as it uses to connect to its credit card processing center?

4. a) Why do you think the bank uses a fractional T1 line to its Da Kine Island branch instead of a full T1 line? b) Instead of a Frame Relay connection?

5. Why do you think the bank uses T3 lines to link its major facilities instead of using ATM?

6. Why do branches need highly capable routers?

End-of-Chapter Questions

DESIGN QUESTIONS

1. What type of connection do you think the bank should have to Northridge, California?
2. Create a rough design for a private line network that would bring a 256 kbps private line to each of the bank's fifty branch offices. Be economical, but ensure that there is redundancy in interisland connections. Assume that connections within an island do not need redundancy because of the high traditional reliability of private lines.

TCP/IP
Internetworking

Learning Objectives

By the end of this chapter, you should be able to discuss the following:

- Basic TCP/IP, IP, TCP, and UDP principles.
- Hierarchical IP addresses, networks and subnets, border and internal routers, and masks.
- Router operation when a packet arrives, including ARP.
- IPv4 fields and IPv6 fields.
- TCP fields, session openings and closings, and port numbers.
- UDP.
- Other important TCP/IP standards, including dynamic routing protocols, ICMP, DHCP, and Layer 3 and 4 switches.

INTRODUCTION

Switched networks are governed by standards at Layer 1 and Layer 2. We began looking at Layer 3 internets (routed networks) for wide area networking in Chapter 6. In this chapter, we will look deeply at internetworking, which operates at Layers 3 and 4—the internet and transport layers.

We will look only at TCP/IP internetworking because TCP/IP dominates the work of network professionals at the internet and transport layers. However, real-world routers cannot be limited to TCP/IP internetworking. They must be multiprotocol routers, which can route not only IP packets, but also IPX packets, SNA packets, AppleTalk packets, and other types of packets. In this chapter, we will look at standards associated with TCP/IP. In Chapter 10, we will look at some aspects of managing TCP/IP internets.

TCP/IP RECAP

The TCP/IP Architecture and the IETF

We first looked at TCP/IP in some depth in Chapter 2. Recall from that chapter that the Internet Engineering Task Force (IETF) sets TCP/IP standards. TCP/IP is an architecture for setting individual standards. Figure 8-1 shows a few of the standards that

5 Application	User Applications			Supervisory Applications		
	HTTP	SMTP	Many Others	DNS	Dynamic Routing Protocols	Many Others
4 Transport	TCP				UDP	
3 Internet	IP			ICMP		ARP
2 Data Link	None: Use OSI Standards					
1 Physical	None: Use OSI Standards					

Note: Shaded protocols are discussed in this chapter.

Figure 8-1   Major TCP/IP Standards

the IETF has created within this architecture. The standards that are shaded in this figure are the ones we will look at in this chapter.

Simple IP at the Internet Layer

Recall also from Chapter 2 that internetworking operates at two layers. The internet layer moves packets from the source host to the destination host across a series of routers. Figure 8-1 shows that the primary standard at the internet layer is the Internet Protocol (IP). Figure 8-2 shows that IP is a simple (connectionless and unreliable) standard. This simplicity minimizes the work that each router has to do along the way, thereby minimizing routing costs.

Reliable Heavyweight TCP at the Transport Layer

In turn, TCP at the transport layer corrects any errors at the internet layer and at lower layers as well. As we saw in Chapter 2, when the transport process on a destination host receives a TCP supervisory or data segment, it sends back an acknowledgment. If the transport process on the source host does not receive an acknowledgment for a TCP segment, it resends the segment. TCP is both connection-oriented and reliable, making it a heavyweight protocol. However, the work of implementing TCP occurs only on the source and destination hosts, not on the many routers between them. This keeps the cost of reliability manageable.

Protocol	Layer	Connection-Oriented/ Connectionless	Reliable/ Unreliable	Lightweight/ Heavyweight
TCP	4 (Transport)	Connection-oriented	Reliable	Heavyweight
UDP	4 (Transport)	Connectionless	Unreliable	Lightweight
IP	3 (Internet)	Connectionless	Unreliable	Lightweight

Figure 8-2   IP, TCP, and UDP

Unreliable Lightweight UDP at the Transport Layer

In Chapters 2 and 6, we saw that TCP/IP offers an alternative to heavyweight TCP at the transport layer. This is the User Datagram Protocol (UDP). Like IP, UDP is a simple (connectionless and unreliable) and lightweight protocol.

TEST YOUR UNDERSTANDING

1. a) Compare TCP and IP along the dimensions in Figure 8-2. b) Compare TCP and UDP according to the dimensions given in Figure 8-2.

IP ROUTING

In this section, we will look at how routers make decisions about forwarding packets—in other words, how a router decides which interface to use to send an arriving packet back out to get it closer to its destination. (In routers, ports are called **interfaces.**) This forwarding process is called **routing.** Router forwarding decisions are much more complex than the Ethernet switching decisions we saw in Chapters 1 and 4. Because of this complexity, routers do more work per arriving packet than switches do per arriving frame. Consequently, routers are more expensive than switches for a given level of traffic. A widely quoted network adage reflects this cost difference: "Switch where you can; route where you must."

> When routers forward incoming packets closer to their destination hosts, this is called *routing.*

Hierarchical IP Addressing

To understand the routing of IP packets, it is necessary to understand IP addresses. In Chapter 1, we saw that IP addresses are 32 bits long. However, IP addresses are not simple 32-bit strings.

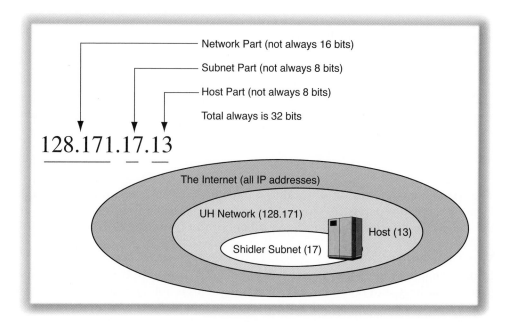

Figure 8-3   Hierarchical IP Address

Hierarchical Addressing

As Figure 8-3 shows, IP addresses are hierarchical. They usually consist of three parts that locate a host in progressively smaller parts of the Internet. These are the network, subnet, and host parts. We will see later in this chapter that hierarchical IP addressing simplifies routing tables.

Network Part

First, every IP address has a **network part,** which identifies the host's network on the Internet. **Internet networks** are owned by single organizations, such as corporations, universities, and ISPs. In the IP address shown in Figure 8-3, the network part is 128.171. It is 16 bits long. This happens to be the network part for the University of Hawai'i network on the Internet. All host IP addresses within this network begin with 128.171. Different organizations have different network parts that range from 8 to 24 bits in length.

Note that *network* in this context does not mean a single network—a single switched LAN or WAN. The University of Hawai'i network itself consists of many single switched networks and routers at multiple locations around the state. In IP addressing, **network** is an organizational concept—a group of hosts, switched networks, and routers owned by a single organization.

In IP addressing, *network* is an organizational concept—a group of hosts, switched networks, and routers owned by a single organization.

Subnet Part

Most large organizations further divide their networks into smaller units called *subnets*. After the network part in an IP address come the bits of the **subnet part.** The subnet part bits specify a particular subnet within the network.

For instance, Figure 8-3 shows that in the IP address 128.171.17.13, the first 16 bits (128.171) correspond to the network part, and the next eight bits (17) correspond to a subnet on this network. (Subnet 17 is the College of Business Administration subnet within the University of Hawai'i network.) All host IP addresses within this subnet begin with 128.171.17.

Host Part

The remaining bits in the 32-bit IP address identify a particular host on the subnet. In Figure 8-3, the **host part** is 13. This corresponds to a particular host, 128.171.17.13, on the College of Business Administration subnet of the University of Hawai'i network.

Variable Part Lengths

In the example presented in Figure 8-3, the network part is 16 bits long, the subnet part is 8 bits long, and the host part is 8 bits long. This is only an example. In general, network parts, subnet parts, and host parts vary in length. For instance, if you see the IP address 60.47.7.23, you may have an 8-bit network part of 60, an 8-bit subnet part of 47, and a 16-bit host part of 7.23. In fact, parts may not even break conveniently at 8-bit boundaries. The only thing you can tell about address bits when looking at an IP address by itself is that the entire address is 32 bits long.

TEST YOUR UNDERSTANDING

2. a) What is routing? b) What are the three parts of an IP address? c) How long is each part? d) What is the total length of an IP address? e) In the IP address, 10.11.13.13, what is the network part?

Routers, Networks, and Subnets

Border Routers Connect Different Networks

As Figure 8-4 illustrates, networks and subnets are very important in router operation. Here we see a simple site internet. The figure shows that a **border router's** main job is to connect different networks. This border router connects the 192.168.x.x network within the firm to the 60.x.x.x network of the firm's Internet service provider.

A border router's main job is to connect different networks.

Internal Routers Connect Different Subnets

The site network also has an **internal router.** An internal router, as Figure 8-4 illustrates, connects different subnets within a firm—in this case, the 192.168.1.x, 192.168.2.x, and 192.168.3.x subnets. Many sites have multiple internal routers to link the site's subnets.

An internal router only connects different subnets within a firm.

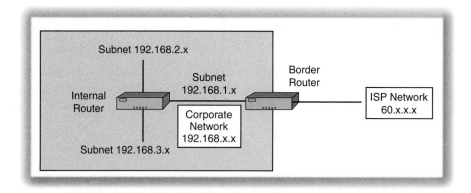

Figure 8-4   Border Router, Internal Router, Networks, and Subnets

TEST YOUR UNDERSTANDING

3. a) Connecting different networks is the main job of what type of router? b) What type of router only connects different subnets?

Network and Subnet Masks

If you know how the University of Hawai'i organizes its IP addresses, you know that the first sixteen bits are always the network part, the next eight are the subnet part, and the final eight are the host part. However, the sizes of the network, subnet, and host parts differ. Routers need a way to tell the sizes of key parts. The tools that allow them to do so are masks.

32-bit Strings

Figure 8-5 illustrates how masks work. A mask is a series of 32 bits, like an IP address. However, a mask always begins with a series of 1s, followed by a series of 0s. In a network mask, the bits in the network part of the mask are 1s, while the remaining bits are 0s. In subnet masks, the bits of both the network and subnet parts are 1s, and the remaining bits are 0s.

A mask is a 32-bit string of 1s and 0s.

The mask has a certain number of initial 1s. The remaining bits are 0s.

In network masks, the initial 1s correspond to the network part.

In subnet masks, the initial 1s correspond to the network and subnet parts.

For example, suppose that the mask is 255.255.0.0. In dotted decimal notation, eight 1s represents 255 and eight 0s represents 0. So this mask has sixteen 1s followed

The Problem

There is no way to tell by looking at an IP address what sizes the network, subnet, and host parts are—only their total of 32 bits

The solution: masks

Series of initial 1s followed by series of final 0s, for a total of 32 bits

Example: 255.255.0.0 is 16 ones followed by 16 zeros

In prefix notation, /16

(Decimal 0 is 8 zeros and Decimal 255 is 8 ones)

Result: IP address where mask bits are 1s and 0s where the mask bits are 0

Mask Operation

Network Mask	Dotted Decimal Notation
Destination IP Address	128.171. 17. 13
Network Mask	255.255. 0. 0
Bits in network part, followed by zeros	128.171. 0. 0

Subnet Mask	Dotted Decimal Notation
Destination IP Address	128.171. 17. 13
Subnet Mask	255.255.255. 0
Bits in network part and subnet parts, followed by zeros	128.171. 17. 0

Figure 8-5   IP Network and Subnet Masks

by sixteen 0s. In prefix notation, the mask is written as /16. Prefix notation gives the number of initial 1s.

Masking IP Addresses

The figure shows what happens when a mask is applied to an IP address, 128.171.17.13. The mask is 255.255.0.0. Where the mask has a 1, the result is the original bits of the IP address. For the remaining bits, which are all 0s, the result is zero. In this case, the result is 128.171.0.0.

Network Masks

Routers use two types of masks. **Network masks** have 1s in the network bits and 0s for remaining bits. If the network mask is 255.255.0.0 and the IP address is 128.171.17.13, then the result is 128.171.0.0. This is the network part, followed by 0s.

Subnet Masks

For **subnet masks,** in contrast, the initial 1s indicate the number of bits in *both* the network and subnet parts. So if 128.171 is the network part and 17 is the subnet part, then the subnet mask will be 255.255.255.0 (/24). If you mask 128.171.17.13 with /24, you get 128.171.17.0.

Why mark the 1s in both parts? Think of a network as a state and a subnet as a city. In the United States, there are two major cities named Portland—one in Maine and the other in Oregon. You cannot just say "Portland" to designate a city uniquely. You must give both the state and city—analogously, the network and subnet parts.

Perspective

Quite simply, network masks give the original bits in the network part, followed by 0s. Subnet masks give the original bits in the network and subnet parts, followed by 0s.

TEST YOUR UNDERSTANDING

3. a) How many bits are there in a mask? b) What do the 1s in a network mask correspond to in IP addresses? c) What do the 1s in a subnet mask correspond to in IP addresses? d) When a network mask is applied to any IP address on the network, what is the result?
4. a) List the bits (1s and 0s) in the mask 255.255.255.0. b) What are the bits (1s and 0s) in the mask /14? c) If /14 is the network mask, how many bits are there in the network part? d) If /14 is the network mask, how many bits are there in the subnet part? e) If /14 is the network mask, how many bits are there in the host part? f) If /14 is the subnet mask, how many bits are there in the network part? g) If /14 is the subnet mask, how many bits are there in the subnet part? h) If /14 is the subnet mask, how many bits are there in the host part?

HOW ROUTERS PROCESS PACKETS

Switching versus Routing

In Chapters 1 and 4, we saw that Ethernet switching is very simple. As Figure 8-6 shows, each row in an Ethernet switching table has a single Ethernet address. This row tells the switch through which port to send the frame back out. This single row can be found quickly, so an Ethernet switch does little work per frame. This makes Ethernet switching inexpensive.

In contrast, firms organize routers in meshes. This gives more reliability because it allows many possible alternative routes between endpoints. Figure 8-6 shows that in a routing table, each row represents an alternative route for a packet. Consequently, to **route** (forward) a packet, a router must first find all rows representing alternative routes that a packet can take. It must then pick the best alternative route from this list. This requires quite a bit of work per packet, making routing more expensive than switching.

TEST YOUR UNDERSTANDING

5. Why are routing tables more complex than Ethernet switching tables? Be articulate.

Routing Table

Figure 8-8 shows a **routing table** with a number of rows and columns. We will see how a router uses these rows and columns to make a routing decision—a decision about what to do with an arriving packet.

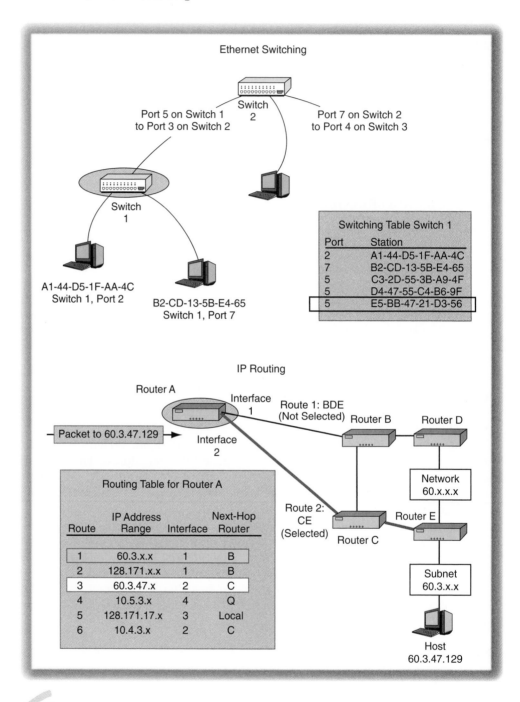

Figure 8-6   Ethernet Switching versus IP Routing

Routing
    Processing an arriving packet and sending it on its way is called *routing*

The Routing Table
    Each router has a routing table that it uses to make routing decisions
    Routing Table Rows
        Each row represents a route for a range of IP addresses—often a network or
        subnet

The Routing Decision for *Each* Packet
    Step 1: Finding All Row Matches
        The router looks at the destination IP address in an arriving packet
        For each row:
            Apply the row's mask to the destination IP address in the packet
            Compare the result with the row's destination value
            If the two match, the row is a match
        The router does this to ALL rows because there may be multiple matches
        This step ends with a set of matching rows
        Example 1: A Destination IP Address That Is NOT in the Range
            Destination IP address of arriving packet 60.43.7.8

Apply the (network) mask	255.255.0.0
Result of masking	60.43.0.0
Destination column value	128.171.0.0
Does destination match the masking result?	No
Conclusion	Not a match

        Example 2: A Destination IP Address That Is in the Range
            Destination IP address of arriving packet 128.171.17.13

Apply the mask	255.255.0.0
Result of masking	128.171.0.0
Destination column value	128.171.0.0
Does destination match the masking result?	Yes
Conclusion	Row is a match

    Step 2: Find the Best-Match Row
        The router examines the matching rows
        It selects the row with the longest match (initial 1s in the row mask)
        If there is a tie on longest match, select among the tie rows on the basis of
        metric values
            For cost metric, choose the row with the lowest metric value
            For speed metric, choose the row with the highest metric value
    Step 3: Send the Packet Back Out
        Send the packet out through the interface (router port) designated in the
        best-match row
        Address the packet to the IP address in the next-hop router column
            If the address says Local, the destination host is out through that interface
            Address the packet to the destination IP address in the packet

Figure 8-7   The Routing Process (Study Figure)

Row	Destination Network or Subnet	Mask (/Prefix)	Metric (Cost)	Interface	Next-Hop Router
1	128.171.0.0	255.255.0.0 (/16)	47	2	G
2	172.30.33.0	255.255.255.0 (/24)	0	1	Local
3	60.168.6.0	255.255.255.0 (/24)	12	2	G
4	123.0.0.0	255.0.0.0 (/8)	33	2	G
5	172.29.8.0	255.255.255.0 (/24)	34	1	F
6	172.40.6.0	255.255.255.0 (/24)	47	3	H
7	128.171.17.0	255.255.255.0 (/24)	55	3	H
8	172.29.8.0	255.255.255.0 (/24)	20	3	H
9	172.12.6.0	255.255.255.0 (/24)	23	1	F
10	172.30.12.0	255.255.255.0 (/24)	9	2	G
11	172.30.12.0	255.255.255.0 (/24)	3	3	H
12	60.168.0.0	255.255.0.0 (/16)	16	2	G
13	0.0.0.0	0.0.0.0 (/0)	5	3	H

Figure 8-8   Routing Table

Rows Are Routes

In the routing table, each row represents a route for all IP addresses within a range of IP addresses—typically a network or subnet. It does not specify the full route, however, but only the next step in the route (either the next-hop router to handle the packet next or the destination host).

> In the routing table, each row represents a route for all IP addresses within a range of IP addresses—typically a network or subnet.

This is important because the routing table does not need a row for *each IP address*, as an Ethernet switching table does. It only needs one row for each *group of IP addresses*. This means that a router needs far fewer rows than an Ethernet switch would need for the same number of addresses.

However, there are many more IP addresses on the Internet than there are Ethernet addresses in an Ethernet network. Even with rows representing groups of IP addresses, core routers in the Internet backbone still have several hundred thousand rows. In addition, while an Ethernet switch needs to find only a single row for each arriving frame, we will see that routers need to look carefully at *all* rows.

Row Number Column

The first column in Figure 8-8 is a route (row) number. Actual routing tables do not have this column. We include it in the figure so we can refer to specific rows in our discussion. Again, each row specifies a route to a destination.

TEST YOUR UNDERSTANDING

6. a) In a routing table, what does a row represent? b) Do switches have a row for each individual Ethernet address? c) Do routers have a row for each individual IP address? d) What is the advantage of the answer to the previous subpart of this question?

Step 1: Finding All Row Matches

We will now see how the router uses its routing table to make routing decisions. The first step for the router is to find which of the rows in the routing table match the destination IP address in an arriving packet. Due to the existence of alternative routes in a router mesh, most packets will match more than one row.

Row Matches

How does the router know which IP addresses should be governed by a row? The answer is that it uses the *Destination Network or Subnet* (*destination*) column and the *Mask* column.

Suppose that all IP addresses in the University of Hawai'i network should be governed by a row. The mask would be the network mask, 255.255.0.0, because the UH has a 16-bit network part. If this mask is applied to any UH address, the result will be 128.171.0.0. This is the value that will be in the destination column. In fact, this is Row 1 in Figure 8-8.

Let's see how routers use these two fields in Figure 8-8. Suppose that a packet arrives with the IP address 60.43.7.8. The router will look first at Row 1 and take the following steps:

➤ In this row, the router applies the mask 255.255.0.0 to the arriving packet's destination IP address, 60.43.7.8. The result is 60.43.0.0.

➤ Next, the router compares the masking result, 60.43.0.0, with the destination value in the row, 128.171.0.0. The two are different, so the row is not a match.

However, suppose that a packet arrives with the IP address 128.171.17.13. Now, the situation is different, and the router does the following:

➤ Again, the router applies the mask 255.255.0.0 in Row 1 to the destination IP address, 128.171.17.13. The result is 128.171.0.0.

➤ Next, the router compares 128.171.0.0 to the destination value in the row, 128.171.0.0. The two are identical. Therefore, the row is a match.

Mask and Compare

This may seem like an odd way to see whether a row matches. A human can simply look at 60.43.7.8 and see that it does not match 128.171.0.0. However, routers do not possess human pattern-matching abilities.

Although routers cannot perform sophisticated pattern recognition, routers (and all computers) have specialized circuitry for masking and comparing—the two operations that row matching requires. Thanks to this specialized circuitry, routers can blaze through hundreds of thousands of rows in a tiny fraction of a second.

The Default Row

The last row in Figure 8-8 has the destination 0.0.0.0 and the mask 0.0.0.0. This row will match *every* IP address because masking any IP address with 0.0.0.0 will give 0.0.0.0—the value in the destination field of Row 13. This row ensures that at least one row will match the destination IP address of every arriving packet. It is called the **default row.** In general, a "default" is something you use if you do not have a more specific choice.

Look at All Rows

Thanks to their mesh topology, router networks have many alternative routes. Consequently, a router cannot stop the first time it finds a row match for each arriving packet, because there may be a better match further on. A router has to look at each and every row in the routing table to find all matches. So far, we have seen what the router does in Row 1 of Figure 8-8. The router then goes on to Row 2 to see whether it is a match, by masking and comparing. After this, it goes on to Row 3, Row 4, Row 5, and so on, all the way to the final row (Row 13 in Figure 8-8).

TEST YOUR UNDERSTANDING

7. a) In Row 3 of Figure 8-8, how will a router test whether the row matches the IP address 60.168.6.7? Show the calculations. Is the row a match? b) Why is the last row called the default row? c) Why must a router look at all rows in a routing table? d) What rows match 172.30.17.6? e) Which rows match 60.168.7.32? Show your calculations for rows that match. f) Which rows in Figure 8-8 match 128.171.17.13? (Don't forget the default row.) Show your calculations for rows that match.

Step 2: Selecting the Best-Match Row

List of Matching Rows

At the end of Step 1, the mask and compare process, the router has a list of matching rows. For a packet with the destination IP address 128.171.17.13, two rows in Figure 8-8 match. The first is Row 1, as we have already seen. The second is Row 7, with a destination of 128.171.17.0 and a mask of 255.255.255.0. From these, the router must select the **best-match row,** the row that represents the best route for an IP address.

First Tie Breaker: Longest Match Rule

How does the router decided whether to follow Row 1 or Row 7? The answer is that it follows the rule of selecting the **longest match.** Row 1 has a mask of 255.255.0.0, which means that it has a 16-bit match. Row 7, in turn, has the prefix /24, meaning that it has a 24-bit match. Row 7 has the longest match, so the router selects it.

Why does following the longest match rule give the best result? The answer is that the closer a route gets a packet to the destination IP address, the better route it probably is. Row 1 gets the packet only to the UH network, 128.171.x.x, while Row 7 gets the packet all the way to the Shidler College of Business subnet of the University of Hawai'i, 128.171.17.x—the subnet that contains host 128.171.17.13.

Second Tie Breaker: The Metric Column for Match Length Ties

What if two rows have the same longest match? For instance, the destination IP address 172.29.8.112 matches both Row 5 and Row 8 in Figure 8-8. Both have a match length of 24 bits—a tie.

In case of a tie for longest match, the tie-breaker rule is to use the **metric column**, which describes the "goodness" of a router. For instance, in Figure 8-8, the metric is cost. Row 5 has a cost of 34, while Row 8 has a cost of 20. Lower cost is better than higher cost, so the router selects Row 8.

In this case, the row with the lowest metric won. However, what would have happened if the metric had been speed instead of cost? More speed is better, so the router would choose Row 5, with the higher speed (34).

TEST YOUR UNDERSTANDING

8. a) Distinguish between Step 1 and Step 2 in the routing process. b) If any row other than the default row matches an IP address, why will the router never choose the default row? c) Which rows in Figure 8-8 match 128.171.17.13? (Don't forget the default row.) Show your calculations. d) Which of these is the best-match row? Justify your answer. e) What rows match 172.40.17.6? Show your work. f) Which of these is the best-match row? Justify your answer. g) Which rows match 172.30.12.47? Show your work. h) Which of these is the best-match row? Justify your answer. i) How would your previous answer change if the metric had been reliability?

Step 3: Sending the Packet Back Out

In Step 1, the router found all rows that matched the destination IP address of the arriving packet. In Step 2, it found the best-match row. Finally, in Step 3, the router sends the packet back out.

Interface

As noted earlier, router ports are called *interfaces*. The fifth column in Figure 8-8 is *interface number*. If a router selects a row as the best match, the router sends the packet out through the interface designated in that row. If Row 1 is selected, the router will send the packet out through Interface 2.

Next-Hop Router

In a switch, a port connects directly to another switch or to a computer. However, a router interface connects to an entire subnet or network. Therefore, it is not enough to select an interface through which to send the packet out. It is also necessary to specify *a particular device* on the subnet.

In most cases, the router will send the packet on to another router, called the **next-hop router (NHR).** The Next-Hop Router column specifies the router that should receive the packet. It will then be up to that router to decide what to do. In Figure 8-8, the next-hop router value is G.[1]

---

[1]Actually, this column should have the IP address of Router G, rather than its name. However, we include the letter designation rather than the IP address for simplicity of understanding.

In some cases, however, the destination host will be on the subnet out through a particular interface. In that case, the router should send the packet to the destination host instead of to another router. In this case, the next-hop router field will say *local*.

TEST YOUR UNDERSTANDING

9. a) Distinguish between Step 2 and Step 3 in routing. b) What are router ports called? c) If the router selects Row 13 as the best-match row, through what interface will the router send the packet out? d) To what device? e) Why is this router called the default router? (The answer is not in the text.) f) If the router selects Row 2 as the best match row for packet 172.30.33.6, through what interface will the router send the packet out? g) To what device? (Don't say, "The local device.")

Cheating (Decision Caching)

We have discussed what happens when a packet arrives at a router. However, what will the router do if another packet for the same destination IP address arrives immediately afterward? The answer is that the router *should* go through the entire process again. Even if a thousand packets arrive that are going to the same destination IP address, the router should go through the entire three-step process for each of them.

As you might expect, a router might cheat, or, as it euphemistically named, *cache* (remember) the decision it makes for a destination IP address. It will then use this decision for successive IP packets going to the same destination. **Decision caching** greatly reduces the work that a router will do for each successive packet.

However, caching is not prescribed in the Internet Protocol. In addition, it is dangerous. The Internet changes constantly as routers come and go and as links between routers change. Consequently, a cached decision that is used for too long will result in non-optimal routing or even in routes that will not work, effectively sending packets into a black hole.

TEST YOUR UNDERSTANDING

10. a) What should a router do if it receives several packets going to the same destination IP address? b) How would decision caching speed the routing decision for packets after the first one? c) Why is decision caching dangerous?

Masking When Masks Do Not Break at 8-Bit Boundaries

MASKS THAT BREAK AT 8-BIT BOUNDARIES

All of the masks that we have seen up to this point have broken at 8-bit segment boundaries. For example, at the University of Hawai'i, the network part is 16 bits long, which corresponds to two segments (128.171); the subnet part is 8 bits long (17); and the host part is 8 bits long. All of the masks in Figure 8-8 break also at 8-bit segment boundaries.

Masks that break at 8-bit boundaries are easy for humans to read. In general, you can look at a mask in the table and decide whether it matches a particular IP address. For instance, if the mask is 255.255.0.0 (/16) and the destination column value is 128.171.0.0, then this definitely matches the IP address 128.171.45.230.

However, masks do *not* always break at 8-bit boundaries. For example, suppose that a mask is 11111111 11111000 00000000 00000000 (spaces added for easier reading). In dotted decimal notation, this is 255.136.0.0. Suppose also that the destination is 00000011 10001000 00000000 00000000 (3.264.0.0).

Now suppose that a destination IP address is 3.143.12.12. Does this IP address match the row? There is no way to tell just by looking at the dotted decimal notation versions of the destination, the mask, and the destination IP address. To solve the problem, you go to the raw 32-bit numbers. The following example shows that the masked destination IP address matches the destination value in the row, so the row is a match.

IP address: (3.143.12.12)	00000011	10001111	00001100	00001100
Mask (255.136.0.0)	11111111	10001000	00000000	00000000
Result (3.136.0.0)	00000011	10001000	00000000	00000000
Destination (3.264.0.0)	00000011	10001000	00000000	00000000

TEST YOUR UNDERSTANDING

10. a) An arriving packet has the destination IP address 128.171.180.13. Row 86 has the destination value 128.171.160.0. The mask is 255.255.224.0. Does this row match the destination IP address? Show your work. Preferably, do the problem in a spreadsheet program, and cut and paste your answer. (Hint: Excel has BIN2DEC and DEC2BIN functions that usually are turned on in the program. They can help you with the conversions.)

The Address Resolution Protocol (ARP)

The final step in the routing process for each arriving packet is to send the packet back out through another port, to a next-hop router or the destination host. That seems easy enough, but there is one additional thing that routers must do.

To send a packet to a next-hop router or a destination host, the router's interface must place the packet into a frame and send this frame to the NHR router or destination host. To do this, the interface must know the data link layer address of the destination host. Otherwise, the router's interface will not know what to place in the destination address field of the frame.

The internet layer process may only know the IP address of the destination host. If the router's interface is to deliver the frame containing the packet, the internet layer process must discover the data link layer address of the destination host as well. This process is called *address resolution*.

(*continued*)

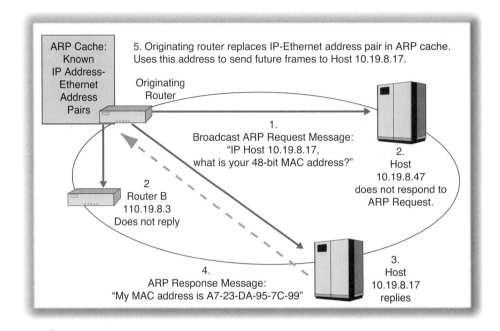

Figure 8-9   Address Resolution Protocol (ARP)

ADDRESS RESOLUTION ON AN ETHERNET LAN WITH ARP

Determining a data link layer address when you know only an IP address is called address resolution. Figure 8-9 shows the **Address Resolution Protocol (ARP)**, which provides address resolution on Ethernet LANs. There are other address resolution protocols for other subnet technologies.

ARP REQUEST MESSAGE

Suppose that the router receives an IP packet with destination address 10.19.8.17. Suppose also that the router determines from its routing table that it

can deliver the packet to a host on one of its Ethernet subnets. Here are the steps that are taken:

➤ First, the router's internet layer process creates an ARP request message that essentially says, "Hey, device with IP address 10.19.8.17, what is your 48-bit MAC layer address?" The router then broadcasts this ARP packet to all hosts on the subnet.[2]

➤ Second, the internet layer process on every host examines the ARP request message. If the target IP address is not that of the host, the host's internet layer process ignores the ARP request

---

[2]Actually, the router passes the packet down to the data link layer process on the subnet's interface. It tells the data link layer process to broadcast its ARP packet. If the subnet standard is Ethernet, the data link layer process places the packet into a frame with the destination Ethernet address FF-FF-FF-FF-FF-FF (forty-eight 1s). This is the Ethernet broadcast address. Switches will send frames with this broadcast address to all stations, and all stations will accept it as they would a frame addressed to their specific Ethernet address.

message. However, host 10.19.8.17 composes an ARP response message that includes its 48-bit MAC layer address (A7-23-DA-95-7C-99). The target host sends this ARP response message back to the router.

➤ Third, the router's internet layer process now knows the subnet MAC address associated with the IP address. It will deliver the packet to that host.

THE ARP CACHE

ARP is a time-consuming process, and the router does not want to do it for each arriving packet. Consequently, the internet layer process on the router saves the IP address–data link layer address information in its **ARP cache** (section of memory). Afterward, whenever an IP packet comes for this IP destination address, the router will send the IP packet down to its NIC, together with the required MAC address. The NIC's MAC process will deliver the IP packet within a frame containing that MAC destination address.

USING ARP FOR NEXT-HOP ROUTERS

We have looked at how routers use ARP when they deliver packets to destination hosts. A router also needs to know the data link layer destination addresses of next-hop routers. Routers use ARP to find the data link layer destination addresses of both destination hosts and other routers.

ARP ENCAPSULATION: FINALLY, ANOTHER INTERNET LAYER PROTOCOL!

In this book so far, we have seen only a single protocol at the internet layer—the Internet Protocol (IP). However, ARP is also a protocol at the internet layer, and ARP messages are called packets. ARP packets are encapsulated directly in frames, just as IP packets are.

TEST YOUR UNDERSTANDING

10. A router wishes to send an IP packet to a host on its subnet. It knows the host's IP address. a) What else must it know? b) Why must the router know it? c) What message will it broadcast? d) What device will respond to this broadcast message? e) Does a router have to go through the ARP process each time it needs to send a packet to a destination host or to a next-hop router? Explain. f) Is ARP used to find the destination data link layer destination addresses of destination hosts, routers, or both? g) At what layer does the ARP protocol operate? h) Draw the final frame when a router transmits an ARP request message over an Ethernet LAN.

THE INTERNET PROTOCOL (IP)

We have focused on IP routing. However, the Internet Protocol has other properties that networking professionals need to understand.

IPv4 Fields

Today, most routers on the Internet and private internets are governed by the **IP version 4 (IPv4)** standard. (There were no versions 0 through 3.) Figure 8-10 shows the IPv4 packet. Its first four bits contain the value 0100 (binary for 4) to indicate that the packet is formatted according to IPv4. Although we have looked already at some of the fields in IPv4, here we will look at fields that we have not seen yet.

IP Time to Live (TTL) Field

In the early days of the ARPANET, which was the precursor to the Internet, packets that were misaddressed would circulate endlessly among packet switches in search of

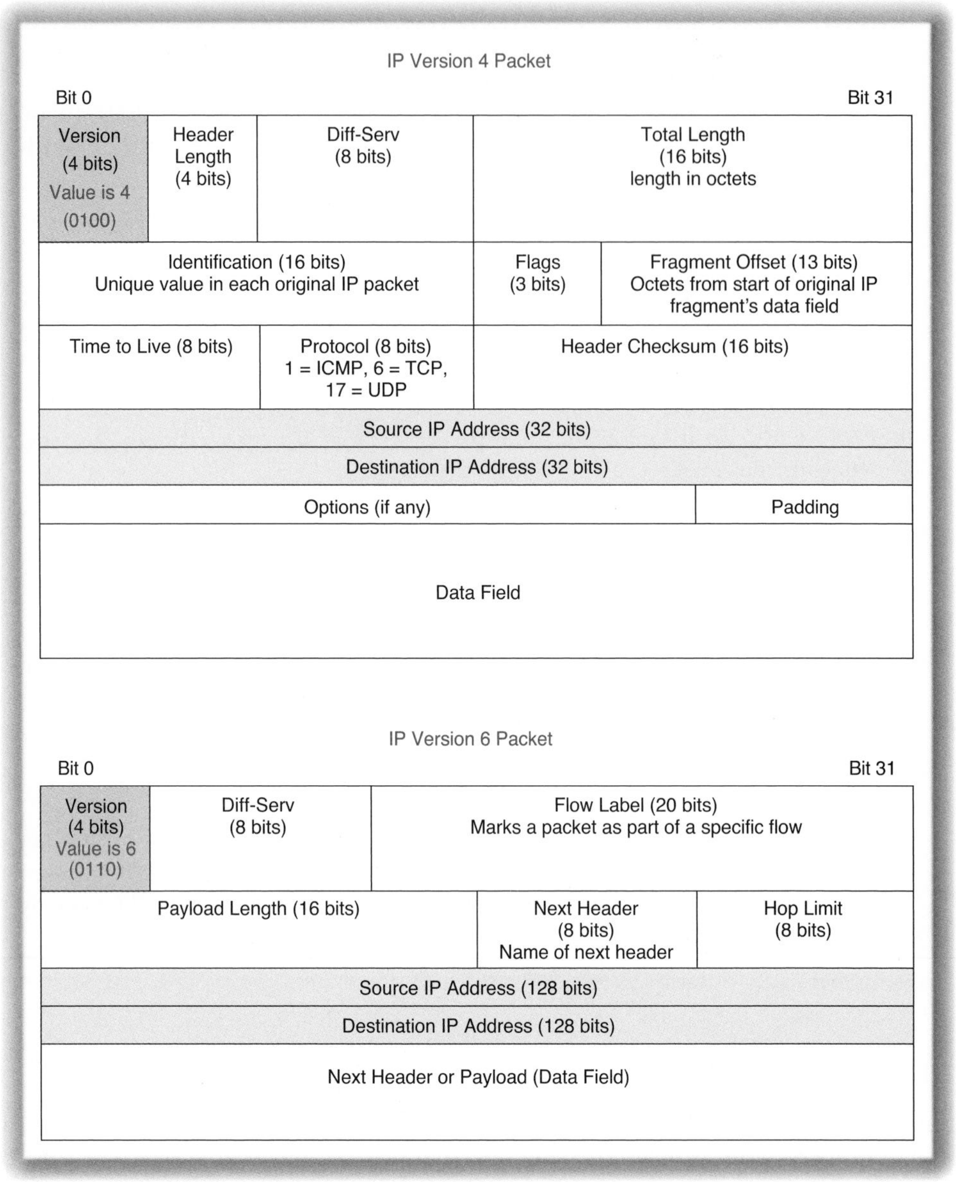

IP Version 4 Packet

Bit 0   Bit 31

Version (4 bits) Value is 4 (0100)	Header Length (4 bits)	Diff-Serv (8 bits)	Total Length (16 bits) length in octets
Identification (16 bits) Unique value in each original IP packet		Flags (3 bits)	Fragment Offset (13 bits) Octets from start of original IP fragment's data field
Time to Live (8 bits)	Protocol (8 bits) 1 = ICMP, 6 = TCP, 17 = UDP	Header Checksum (16 bits)	
Source IP Address (32 bits)			
Destination IP Address (32 bits)			
Options (if any)			Padding
Data Field			

IP Version 6 Packet

Bit 0   Bit 31

Version (4 bits) Value is 6 (0110)	Diff-Serv (8 bits)	Flow Label (20 bits) Marks a packet as part of a specific flow
Payload Length (16 bits)	Next Header (8 bits) Name of next header	Hop Limit (8 bits)
Source IP Address (128 bits)		
Destination IP Address (128 bits)		
Next Header or Payload (Data Field)		

Figure 8-10   IPv4 and IPv6 Packets

their nonexistent destinations. To prevent this, IP added a **time to live (TTL)** field that is given a value by the source host. Different operating systems have different TTL defaults. Most insert TTL values between 64 and 128. Each router along the way decrements the TTL field by 1. A router decrementing the TTL to 0 will discard the packet, although it may send back an ICMP error advisement message to the source host.

IP Protocol Field

The **protocol** field tells the contents of the data field. If the protocol field value is 1, the IP packet carries an ICMP message in its data field. TCP and UDP have protocol values 6 and 17, respectively. After decapsulation, the internet layer process must pass the packet's data field to another process. The protocol field value designates which process should receive the data field.

IP Identification, Flags, and Fragment Offset Fields

If a router wishes to forward a packet to a particular network and the network's maximum packet size is too small for the packet, the router can fragment the packet into two or more smaller packets. Each fragmented packet receives the same **identification** field value that the source host put into the original IP packet's header.

The destination host's internet process reassembles the fragmented packet. It places all packets with the same identification field value together for sorting. It then places them in order of increasing **fragment offset** size. The more fragments bit is set (equal to 1) in all but the last fragment. Not setting it in the last fragment indicates that there are no more fragments.

Fragmentation is uncommon in IP today and is suspicious when it occurs because it is rarely used legitimately and is often used by attackers.

IP Options

Similarly, **options** are uncommon in IP today and also tend to be used primarily by attackers. If an option does not end at a 32-bit boundary, **padding** is added up to the 32-bit boundary.

IP Diff-Serv

The **Diff-Serv** field can be used to label IP packets for priority and other service parameters.

TEST YOUR UNDERSTANDING

11. a) What is the main version of the Internet Protocol in use today? b) What does a router do if it receives a packet with a TTL value of 1? c) What does the protocol field value tell the destination host? d) Under what circumstances would the identification, flags, and fragment offset fields be used in IP? e) Why is IP fragmentation suspicious? f) Why are IP options suspicious?

IPv6 Fields

The IETF has standardized a new version of the Internet Protocol, **IP version 6 (IPv6)**. As Figure 8-10 shows, IPv6 also begins with a version field. Its value is 0110 (binary for 6). This tells the router that the rest of the packet is formatted according to IPv6.

Address Field

The most important change from IPv4 to IPv6 is an increase in the size of IP address fields from 32 bits to 128 bits. The number of possible IP addresses is 2 raised to a power that is the size of the IP address field. For IPv4, this is 2^{32}. For IPv6, this is 2^{128}—an enormous number. IPv6 will support the huge increase in demand for IP addresses that we can expect from mobile devices and from the likely evolution of even simple home appliances into addressable IP hosts.

Slow Adoption

IPv6 has been adopted only in a few geographic regions because its main advantage, permitting far more IP addresses, is not too important yet. However, IPv6 is beginning to gather strength, particularly in Asia and Europe, which were shortchanged in the original allocation of IPv4 addresses.[3] In addition, the explosion of mobile devices accessing the Internet will soon place heavy stress on the IPv4 IP address space. Fortunately, IPv6 packets can be tunneled through IPv4 networks by being placed within IPv4 packets, so the two protocols can (and will) coexist on the Internet for some time to come.

TEST YOUR UNDERSTANDING

12. a) How is IPv6 better than IPv4? b) Why has IPv6 adoption been so slow? c) What forces may drive IPv6's adoption in the future? d) Must IPv6 replace IPv4 all at once? Explain.

THE TRANSMISSION CONTROL PROTOCOL (TCP)

Fields in TCP/IP Segments

Chapter 2 looked at the **Transmission Control Protocol (TCP).** In this section, we will look at this complex protocol in even more depth. When IP was designed, it was made a very simple "best effort" protocol (although its routing tables are complex). The IETF left more complex internetwork transmission control tasks to TCP. Consequently, network professionals need to understand TCP very well. Figure 8-11 shows the organization of TCP messages, which are called *TCP segments.*

TEST YOUR UNDERSTANDING

13. a) Why is TCP complex? b) Why is it important for networking professionals to understand TCP? c) What are TCP messages called?

Sequence Numbers

Each TCP segment has a unique 32-bit **sequence number** that increases with each segment. This allows the receiving transport process to put arriving TCP segments in order if IP delivers them out of order.

Acknowledgment Numbers

In Chapter 2, we saw that TCP uses acknowledgments (ACKs) to achieve reliability. If a transport process receives a TCP segment correctly, it sends back a TCP segment

---

[3]North America has 74 percent of all IPv4 addresses. In fact, Stanford University and MIT have more IPv4 addresses than China, which now has fewer IP addresses than it has Internet users.

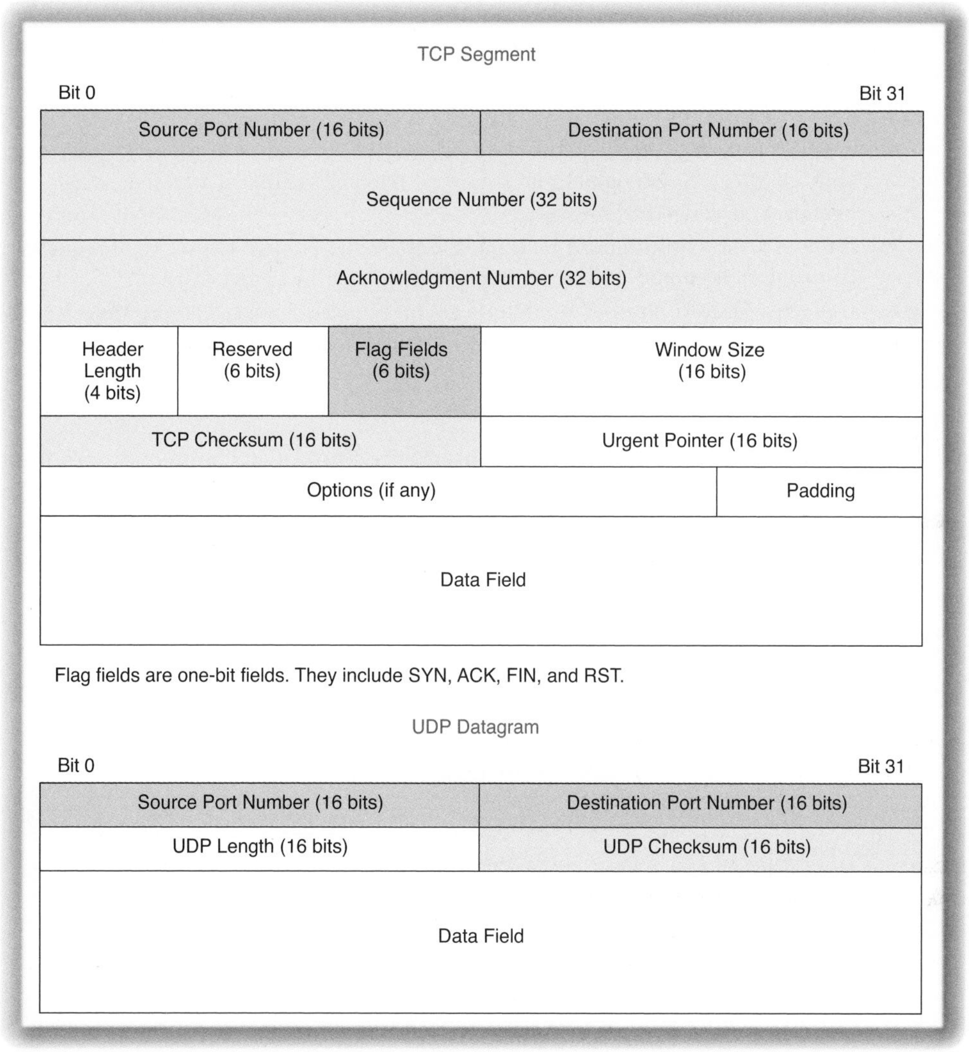

TCP Segment

Bit 0   Bit 31

Source Port Number (16 bits)	Destination Port Number (16 bits)

Sequence Number (32 bits)

Acknowledgment Number (32 bits)

Header Length (4 bits)	Reserved (6 bits)	Flag Fields (6 bits)	Window Size (16 bits)

TCP Checksum (16 bits)	Urgent Pointer (16 bits)

Options (if any)	Padding

Data Field

Flag fields are one-bit fields. They include SYN, ACK, FIN, and RST.

UDP Datagram

Bit 0   Bit 31

Source Port Number (16 bits)	Destination Port Number (16 bits)

UDP Length (16 bits)	UDP Checksum (16 bits)

Data Field

Figure 8-11   TCP Segment and UDP Datagram

acknowledging the reception. If the sending transport process does not receive an acknowledgement, it transmits the TCP segment again.

The **acknowledgment number** field indicates which segment is being acknowledged. One might expect that if a segment has sequence number X, then the acknowledgment number in the segment that acknowledges it would also be X. As Module A notes, the situation is more complex, but the acknowledgment number is at least based on the sequence number of the segment being acknowledged.

Flag Fields

TCP has six single-bit fields. Single-bit fields are called **flag fields,** and if they have the value 1, they are said to be **set.** These fields allow the receiving transport process to identify the kind of segment it is receiving. We will look at four of these flag bits:

➤ If the ACK bit is set, then the segment acknowledges another segment. If the ACK bit is set, the acknowledgment field must be filled in to indicate which message is being acknowledged.

➤ If the SYN (synchronization) bit is set (has the value 1), then the segment requests a connection opening.

➤ If the FIN (finish) bit is set, then the segment requests a normal connection closing.

➤ If the RST (reset) bit is set, then the segment announces an abrupt connection closing.

Openings and Normal Closings

In Chapter 2, we saw that TCP is a connection-oriented protocol. Connection-oriented protocols have formal openings and closings. Figure 8-12 illustrates these openings and closings.

Figure 8-12   TCP Session Openings and Closings

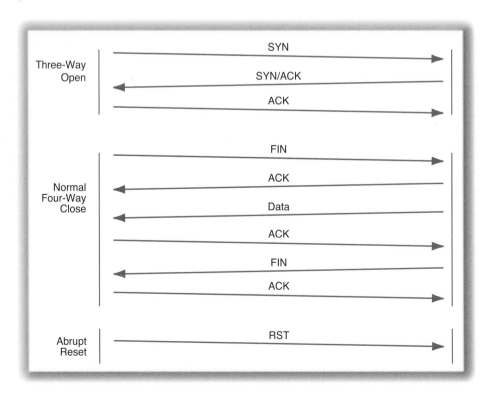

Openings with SYN Segments

Opening a TCP connection requires an interaction called a **three-way handshake.** The interaction happens as follows:

➤ First, one transport process sends a SYN (synchronization) message to open the connection. A synchronization message consists of a header without a body. In a synchronization message, the SYN bit is set.

➤ The other transport process sends back a SYN/ACK. In this message, the SYN and ACK bits are both set. In addition, the acknowledgment number field indicates which segment is being acknowledged (in connection openings, the initial SYN segment).

➤ Finally, the side that initiated the connection opening sends back an ACK segment.

Normal Closes with FIN Segments

Each side normally ends a TCP connection by sending a FIN message in which the FIN bit is set. This closing is called a **four-way close** because four segments must be sent, as Figure 8-12 illustrates. The closing works in the following way:

➤ First, one side initiates the close by sending a FIN (finish) segment. This is a TCP segment whose **FIN** bit is set. The TCP segment is only a header and has no data field.

➤ The other side responds with an ACK segment that acknowledges receiving the FIN segment. This also is a header-only TCP segment.

➤ Later, the other side sends a FIN segment, which the side that initiates the close acknowledges. The connection is now closed.

When the side that initiates the close sends the FIN segment, it signals that it has no more information to send. However, as the figure shows, the other side may continue to send TCP segments containing information, and the side that initiated the close will continue to send back ACK segments. When the other side finishes sending information segments, it sends its own FIN message, which the side that initiated the close acknowledges.

Abrupt Resets

Figure 8-12 shows that TCP also allows a second way to close connections. This is the **abrupt reset.** It is like hanging up during the middle of a telephone conversation. Either side can send a reset message with the **RST bit** set. This is a one-way close. There is no acknowledgment or response from the other side.

TEST YOUR UNDERSTANDING

14. a) Why are sequence numbers good? b) What are 1-bit fields called? c) If someone says that a flag field is set, what does this mean? d) If the ACK bit is set, what other field must have a value? e) What is a SYN segment? f) Describe three-way openings in TCP. g) Distinguish between four-way closes and abrupt resets. h) Do SYN and FIN segments have data fields? i) After a side sends a FIN segment, will it respond to further messages from the other side? Explain.

Port Numbers

As Figure 8-11 shows, both TCP and UDP have **port number** fields. These fields are used differently by clients and servers. In both cases, however, they tell the transport process what application process sent or should receive the data in the data field. This is necessary because computers can run multiple applications at the same time.

Server Port Numbers

For servers, the port number field indicates which application program on the server should receive the message. Major applications have **well-known port numbers** that are usually (but not always) used. These well-known port numbers are from 0 to 1023. Following are some examples:

➤ The well-known TCP port number for HTTP is 80.

➤ For FTP, TCP Port 21 is used for supervisory communication with an FTP application program, and TCP Port 20 is used to send and receive data segments.

➤ Telnet uses TCP Port 23.

➤ TCP Port 25 is for Simple Mail Transfer Protocol (SMTP) messages in e-mail.

➤ UDP has its own well-known port numbers. The well-known UDP port numbers also run from 0 to 1023.

Figure 8-13 shows that every time the client sends a message to a server, the client places the port number of the server application in the destination port number field. In this figure, the server is a webserver, and the port number is 80. When the server responds, it places the port number of the application (80) in the source port number field.

Client Port Numbers

Clients do something very different. Whenever a client connects to an application program on a server, the client creates a random **ephemeral port number,** which it only

Figure 8-13 Use of TCP (and UDP) Port Numbers

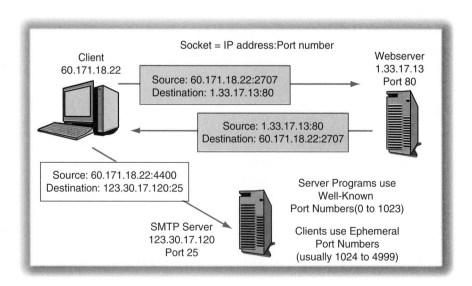

uses for a single TCP session with a single server. According to IETF rules, this port number should be between 49153 and 65535. However, many operating systems ignore these rules and use other ephemeral port numbers. For instance, Microsoft Windows, which dominates the client PC operating system market, uses the ephemeral port number range 1024 to 4999.

Microsoft Windows uses the ephemeral port number range 1024 to 4999.

In Figure 8-13, the ephemeral port number is 2707 for client communication with the webserver. When the client transmits, it places 2707 in the source port field of the TCP or UDP header. The server, in return, places this ephemeral port number in all destination port number fields of TCP segments or UDP datagrams that it sends to the client.

A client may maintain multiple connections to different application programs on different servers. In Figure 8-13, for example, the client has a connection to an SMTP mail server as well as to the webserver. The client will give each connection a different ephemeral port number to separate the segments of the two connections. For the SMTP connection in Figure 8-13, the client has randomly chosen the ephemeral port number to be 4400.

Sockets

The combination of an IP address and a port number designates a specific connection to a specific application on a specific host. This combination is called a **socket.** It is written as an IP address, a colon, and then a port number—for instance, 128.171.17.13:80.

A socket is written as an IP address, a colon, and then a port number. It designates a specific application on a specific host.

TEST YOUR UNDERSTANDING

15. a) What type of port number do servers use? b) What type of port number do clients use? c) What is the port range for well-known port numbers? d) What is the range of Microsoft ephemeral port numbers?
16. A Windows host sends a TCP segment with source port number 25 and destination port number 2404. a) Is the source host a server or a client? Explain. b) If the host is a server, what kind of server is it? c) Is the destination host a server or a client? Explain.
17. a) What is a socket? b) What specifies a particular application on a particular host in TCP/IP? c) How is a socket written? d) When the SMTP server in Figure 8-13 transmits to the client PC, what will the source socket be? e) What will the destination socket be?

THE USER DATAGRAM PROTOCOL (UDP)

We saw in Chapter 2 that the **User Datagram Protocol (UDP)** is a simple (connectionless and unreliable) protocol. We saw in Chapter 6 that VoIP uses UDP to carry voice packets because there is no time to wait for retransmissions. In Chapter 10, we will see that the Simple Network Management Protocol uses UDP for a different reason—to reduce network traffic. UDP does not have openings, closings, or acknowledgments, and so it produces substantially less traffic than TCP.

Because of UDP's simple operation, the syntax of the UDP datagram shown in Figure 8-11 is very simple. After two port number fields, which we just saw in the previous section, there are only two more header fields.

There is a **length** field so that the receiving transport process can know how long the datagram is.

There also is a **UDP checksum** field that allows the receiver to check for errors in this UDP datagram.[4] If an error is found, the UDP datagram is discarded. There is no mechanism for retransmission.

TEST YOUR UNDERSTANDING

18. a) What are the four fields in a UDP header? b) Describe the third. c) Describe the fourth. d) Is UDP reliable? Explain.

OTHER TCP/IP STANDARDS

In this section, we will look briefly at several other important TCP/IP standards that network administrators need to master. We will look at some of these protocols here. Chapter 10 will look more closely at some TCP/IP protocols that are focused on network management.

Dynamic Routing Protocols

How does a router get the information in its routing table? One possibility is to enter routes manually. However, that approach does not scale to large internets. Instead, as Figure 8-15 shows, routers constantly exchange routing table information with one another by using **dynamic routing protocols.**

Figure 8-14   Dynamic Routing Protocols (Study Figure)

Dynamic Routing Protocol	Interior or Exterior Routing Protocol?	Remarks
RIP (Routing Information Protocol)	Interior	Only for small autonomous systems with TCP/IP and low needs for security
OSPF (Open Shortest Path First)	Interior	For large autonomous systems that use only TCP/IP
EIGRP (Enhanced Interior Gateway Routing Protocol)	Interior	Proprietary Cisco Systems protocol. Not limited to TCP/IP routing. Also handles IPX/SPX, SNA, and so forth
BGP (Border Gateway Protocol)	Exterior	Organization cannot choose what exterior routing protocol it will use

---

[4]If the UDP checksum field has the value zero, then error checking is not done at all.

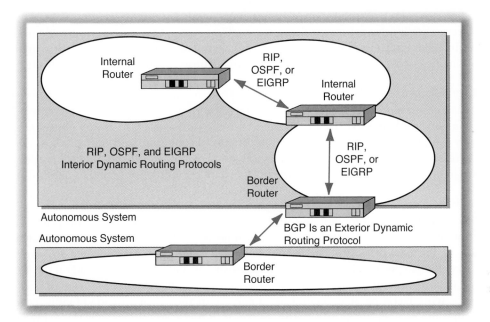

Figure 8-15   Dynamic Routing Protocols

Routing

Note that TCP/IP uses the term **routing** in two different, but related, ways. First, we saw earlier that the process of forwarding arriving packets is called routing. Second, the process of exchanging information for building routing tables is also called **routing.**

In TCP/IP, the term *routing* is used in two ways—for packet forwarding and for the exchange of routing table information through dynamic routing protocols.

Autonomous Systems and Interior Dynamic Routing Protocols

Recall from Chapter 1 that the Internet consists of many networks owned by different organizations. Within an organization's network, which is called an **autonomous system,** the organization owning the network decides which dynamic routing protocol to use among its internal routers, as shown in Figure 8-15. For internal use, the organization is free to choose among available **interior dynamic routing protocols.** There are three popular interior dynamic routing protocols. Each has different strengths and weaknesses.

Routing Information Protocol (RIP)

The simplest interior dynamic routing protocol created by the IETF is the **Routing Information Protocol (RIP).** RIP's simplicity makes it attractive for small internets. Management labor is relatively low. On the negative side, RIP is not very efficient because its metric is merely the number of router hops needed to get to the destination host. However, this is not a serious problem for small internets. The one serious problem with RIP in small internets is poor security. If attackers take over a firm's interior dynamic routing protocol communications, they can maliciously reroute the internet's traffic.

Open Shortest Path First (OSPF)

For larger autonomous systems, or if security is a serious concern, the IETF created the **Open Shortest Path First (OSPF)** dynamic routing protocol. OSPF is very efficient, having a complex metric based on a mixture of cost, throughput, and traffic delays. OSPF also offers strong security. It costs much more to manage than RIP, but unless a corporate internet is very small, OSPF is the only IETF dynamic routing protocol that makes sense.

EIGRP

Cisco Systems is the dominant manufacturer of routers. Cisco has its own proprietary interior dynamic routing protocol for large internets—the **Enhanced Interior Gateway Routing Protocol (EIGRP).** The term **gateway** is another term for *router*. EIGRP's metric is very efficient because it is based on a mixture of interface bandwidth, load on the interface (0 percent to 100 percent of capacity), delay, and reliability (percentage of packets lost). EIGRP is comparable to OSPF, but many companies use it instead of OSPF because it can route SNA and IPX/SPX traffic, as well as IP traffic. On the negative side, EIFRP is a proprietary protocol, and using it forces the company to buy only Cisco routers.

Exterior Dynamic Routing Protocols

When communicating outside the organization's network, the organization is no longer in control. It must use whatever exterior dynamic routing protocol the external network to which it is connected requires. The almost universally used exterior dynamic routing protocol is the **Border Gateway Protocol (BGP).** BGP is designed specifically for the exchange of routing information between autonomous systems.

TEST YOUR UNDERSTANDING

19. a) What is the purpose of dynamic routing protocols? b) In what two ways does TCP/IP use the term *routing*?
20. a) What is an autonomous system? b) Within an autonomous system, can the organization choose its interior routing protocol? c) What are the two TCP/IP interior dynamic routing protocols? d) Which IETF dynamic routing protocol is good for small internets that do not have high security requirements? e) Which IETF dynamic routing protocol is well suited for large internal internets that have high security requirements? f) What is the main benefit of EIGRP, compared with OSPF, as an internal dynamic routing protocol? g) When might you use EIGRP as your interior dynamic routing protocol? h) May a company select the routing protocol its border router uses to communicate with the outside world? i) What is the almost universally used exterior dynamic routing protocol?

Internet Control Message Protocol (ICMP) for Supervisory Messages at the Internet Layer

Supervisory Messages at the Internet Layer

IP is only concerned with packet delivery. For supervisory messages at the internet layer, the IETF created the **Internet Control Message Protocol (ICMP).** IP and ICMP work closely together. As Figure 8-16 shows, IP encapsulates ICMP messages in the IP

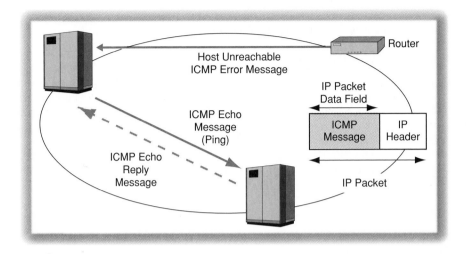

Figure 8-16   Internet Control Message Protocol (ICMP) for Supervisory Messages

data field, delivering them to their target host or router. There are no higher-layer headers or messages.

Error Advisement

IP is an unreliable protocol. It offers no error correction. If the router or the destination host finds an error, it discards the packet. Although there is no retransmission, the router or host that finds the error may send an **ICMP error message** to the source device to inform it that an error has occurred, as Figure 8-16 illustrates. This is **error advisement** (notification) rather than error correction. There is no mechanism within IP or ICMP for the retransmission of lost or damaged packets. ICMP error messages are sent only to help the sending process or its human user to diagnose problems. One important subtlety is that sending error advisement messages is not mandatory. For security reasons, many firms do not allow error advisement messages to leave their internal internets, because hackers can exploit the information contained in them.

Echo (Ping)

Perhaps the most famous ICMP error message type is the **ICMP echo** message. One host or router can send an echo request message to another. If the target device's internet process is able to do so, it will send back an echo reply message.

Sending an echo request is often called **pinging** the target host, because it is similar to a submarine pinging a ship with sonar to see if it is there. In fact, the most common program for pinging hosts is called *ping*.[5] Echo is a good diagnostic tool because

---

[5]The echo reply message also gives the latency for the reply—the number of milliseconds between echo messages and echo reply messages. This is useful in diagnosing problems.

if there are network difficulties, a logical early step in diagnosis is to ping many hosts and routers to see whether they can be reached.

TEST YOUR UNDERSTANDING

21. a) For what general class of messages is ICMP used? b) How are ICMP messages encapsulated? c) An Ethernet frame containing an ICMP message arrives at a host. List the frame's headers, messages, and trailers at all layers. List them in the order in which they will be seen by the receiver. For each header or trailer, specify the standard used to create the header or message (for example, Ethernet 802.3 MAC layer header). (Hint: Remember how ICMP messages are encapsulated.) d) Explain error advisement in ICMP. e) Explain the purpose of ICMP echo messages. f) Sending an ICMP echo message is called____the target host.

DHCP

When you send a packet to a server, you need to know the server's IP address. Servers usually get **static IP addresses** that never change. (Imagine how hard it would be to find your favorite store if it kept changing its street address.)

For client PCs, however, the situation is different. Client PCs get **dynamic IP** addresses that may be different each time the user goes on the Internet. To give customers these dynamic IP addresses, ISPs and most companies use the **Dynamic Host Configuration Protocol (DHCP).**

When a PC goes onto the Internet after booting up, it realizes that it does not have an IP address. The following steps are then taken:

➤ The PC broadcasts a DHCP message asking for a temporary IP address.

➤ A DHCP server sends a DHCP response message giving an IP address and other configuration information, including a subnet mask, the IP address of the default router, and the IP addresses of DNS servers.

Figure 8-17 Dynamic Host Configuration Protocol (DHCP) (Study Figure)

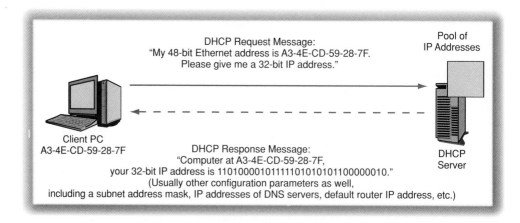

Although PCs could be configured manually, they would have to be manually re-configured every time a company changed its masks, the IP addresses of a DNS server, or other operating details. With DHCP and dynamic configuration, only the DHCP servers have to be updated.

TEST YOUR UNDERSTANDING

22. a) Distinguish between static and dynamic IP addresses. b) Why do servers receive static IP addresses? c) Describe the four-step process for a PC to get an IP address and other configuration information dynamically. d) What configuration information does a DHCP server send to a PC? e) What is the advantage of dynamic configuration?

Layer 3 and Layer 4 Switches

In Chapters 1 and 4, we learned why Ethernet switches are so fast and inexpensive. In this chapter, we saw why routers are so slow and expensive. However, just as nature abhors a vacuum, technology abhors a sharp distinction. New devices called Layer 3 switches fall between routers and traditional Layer 2 (data link layer) switches for single networks, in terms of speed and price.

LAYER 3 SWITCHES

First, we need to get one thing clear. **Layer 3 switches** are true routers. They are *not* switches. They forward IP packets by using routing tables, implement routing protocols to exchange routing table information, and do other things that routers do. In other words, the term "switch" is highly inaccurate—a "gift" of marketers who used the term "switch" because switches traditionally have been faster than routers.

Layer 3 switches are true routers. They are *not* switches.

LAYER 3 SWITCHES ARE FAST AND RELATIVELY INEXPENSIVE

Layer 3 switches are fast primarily because they do almost all processing in hardware. Hardware-based routing is much faster than traditional software-based routing.[6] Largely as a consequence, Layer 3 switches are less expensive to purchase than traditional software routers for a given traffic level. Layer 3 switches cost only about as much as Ethernet switches and are almost as fast.

However, the labor cost of managing any router—including Layer 3 switches—is much higher than the cost of managing Ethernet switches. The total cost of Layer 3 switches therefore lies between the total cost of Ethernet switches and traditional software routers.

---

[6]Programs consist of multiple statements. Often, dozens of program statements may need to be loaded and run to accomplish a simple task. If the task can be done in hardware, in contrast, no time is needed to "load" software statements or write out results to memory. In addition, parallel processing usually allows hardware to perform functions in fewer steps than software implementations would take. On the negative side, it is far more expensive to create hardware to implement required functionality than to write a program to do the work. The technology that makes Layer 3 switches possible is the application-specific integrated circuit (ASIC). This is a production integrated circuit custom-made for a particular application (in this case, routing).

(*continued*)

LAYER 3 SWITCHES HAVE LIMITED FUNCTIONALITY

Because there are limits on what can currently be done in hardware, Layer 3 switches today do not have all of the functionality of full routers. Instead of being full multiprotocol routers, for example, they often handle only IP, or perhaps IP and IPX. In addition, they often have only Ethernet interfaces.

ROLES IN SITE NETWORKS

For many organizations, such as banks, which have multiple internal protocols from different architectures, Layer 3 switches cannot be used at all. However, where the limited functionality of Layer 3 switches is sufficient for an organization, they are ideal.

For instance, Layer 3 switches are often used as internal routers in organizations that have standardized on TCP/IP for internal communication. Figure 8-18 shows that in many site networks, Layer 3 switches are pushing routers to the edges of the site because the ability of routers to support multiple WAN protocols is crucial at borders. In contrast, within the site, the low cost of Layer 3 switches usually makes them dominant in the core above the workgroup switch. For workgroup switches, in turn, labor costs usually are too high for Ethernet switches to be replaced with Layer 3 routers.

LAYER 4 SWITCHES

As noted earlier in this chapter, TCP and UDP headers have port number fields which indicate the application that created the encapsulated application layer message and the application layer program that should receive the encapsulated application message.

Layer 4 switches examine the port number field of each arriving packet's encapsulated TCP segment. This allows them to switch packets according to the application they contain. Specifically, this allows Layer 4 switches to give priority to or even to

Figure 8-18   Layer 3 Switches and Routers in Site Internets

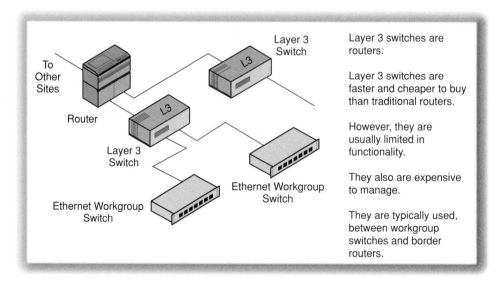

Layer 3 switches are routers.

Layer 3 switches are faster and cheaper to buy than traditional routers.

However, they are usually limited in functionality.

They also are expensive to manage.

They are typically used, between workgroup switches and border routers.

deny forwarding to IP packets from certain applications. For example, TCP segments to and from an SMTP mail server (Port 25) might be given low priority during times of congestion because e-mail is insensitive to moderate latency.

APPLICATION SWITCHES (LAYER 5 OR LAYER 7 SWITCHES)

Finally, some switches are **application switches.** They are also called Layer 5 or Layer 7 switches. Layer 5 is the application layer in the hybrid TCP/IP–OSI architecture, and Layer 7 is the application layer in the OSI architecture. Whatever they are called, application switches make switching decisions according to the content of application messages. For instance, for Web traffic, switching

may be done on the basis of URLs in request messages. Application switches may determine where the application message goes and what priority the message should be given.

TEST YOUR UNDERSTANDING

23. a) Are Layer 3 switches really routers? b) How are they better than traditional software-based routers? c) How are they not as good? Give a full explanation. d) When would you often use Layer 3 switches? e) Where would you not use Layer 3 switches?

24. a) What are Layer 4 switches? b) What field do Layer 4 switches examine? c) Why are Layer 4 switches beneficial?

CONCLUSION

Synopsis

TCP/IP is a family of standards created by the Internet Engineering Task Force (IETF). IP is TCP/IP's main standard at the internet layer. IP is a lightweight (unreliable and connectionless) protocol. At the transport layer, TCP/IP offers two standards: TCP, which is a heavyweight protocol (reliable and connection-oriented); and UDP, which is a lightweight protocol like IP.

IP addresses are hierarchical. Their 32 bits usually are divided into a network part, a subnet part, and a host part. All three parts vary in length. Subnet masks and network masks help devices identify which bits in an IP address make up the network part, the subnet part, the host part, or the total of two parts.

Routers forward packets through an internet. Border routers move packets between the outside world and an internal site network. Internal routers work within sites, moving packets between subnets. In the Internet backbone, core routers handle massive traffic flows. Ports in routers are called interfaces. Different interfaces may connect to different types of networks—for instance, Ethernet or Frame Relay networks. Most routers are multiprotocol routers, which can handle not only TCP/IP internetworking protocols, but also internetworking protocols from IPX/SPX, SNA, and other architectures.

Routers are designed to work in a mesh topology. This topology creates alternative routes through the internet. Alternative routes are good for reliability. However, the router has to consider the best route for each arriving packet, and this is time consuming and therefore expensive.

To make a routing decision (deciding which interface to use to send an incoming packet back out), a router uses a routing table. Each row in the routing table represents a route to a particular network or subnet. All packets to that network or subnet are governed by the one row. Each row (route) has destination, mask, metric, and next-hop router fields.

If the destination IP address in an arriving packet is in a row's range, that range is a match. After finding all matches in the routing table, the router finds the best-match row on the basis of match length and metric values. Once a best-match route (row) is selected, the router sends the packet out to the next-hop router in that row.

If you read the box "The Address Resolution Protocol (ARP)," you saw that the router must encapsulate the packet in a frame in order to send it out. ARP is a protocol for discovering the data link layer destination of the device at the next-hop router's IP address.

IP version 4 has a number of important fields besides the source and destination address fields. The time to live (TTL) field ensures that misaddressed packets do not circulate endlessly around the Internet. The protocol field describes the contents of the data field—ICMP message, TCP segment, UDP datagram, and so forth. IP version 6 will offer many more addresses, thanks to its 128-bit address fields.

The Transmission Control Protocol (TCP) has sequence numbers that enable the receiving transport process to place arriving TCP segments in order. The TCP header has several flag fields that indicate whether the segment is a SYN, FIN, ACK, or RST segment. Connection openings use a three-way handshake with SYN segments. Normal closes involve a four-way message exchange that use FIN segments. Resets close a connection with a single segment (RST) instead of the normal four.

Both TCP and UDP have 16-bit source and destination port number fields that tell the transport process which application process sent or should receive the contents in the segment data field. Major applications have well-known port numbers. For instance, the well-known server port number of HTTP is Port 80. Clients, in contrast, have ephemeral port numbers, which they select randomly for each connection. Microsoft uses the ephemeral port number range 1024 to 4999.

Routers build their routing tables by listening to other routers. Routers frequently exchange messages, giving information stored in their routing tables. These messages are governed by one of several available dynamic routing protocols.

IP itself does not have supervisory messages. For internet layer supervisory messages, hosts and routers use the Internet Control Message Protocol (ICMP). We looked at two types of ICMP messages—error advisement messages and echo messages (ping). ICMP messages are carried in the data fields of IP packets.

In Chapter 1, we saw that DHCP provides a temporary IP address to client PCs when they first boot up. In this chapter, we saw that DHCP also sends other configuration information to the PC—a mask and the IP addresses of local DNS hosts.

If you read the box, "Layer 3 and Layer 4 Switches," you saw that traditionally, there has been a sharp distinction between fast, inexpensive switches and slow, expensive routers. However, Layer 3 switches bridge this gap. They are true routers, but operate in hardware rather than in software as traditional routers do. However, hardware operation can implement only some of the functionality of routers. Layer 3 switches are not multiprotocol routers. In addition, because Layer 3 switches are routers, they are more costly to manage than Layer 2 switches. Layer 4 switches, in turn, make switching decisions on the basis of the content of transport layer messages, usually port number fields. Finally, application switches (Layer 5 or Layer 7 switches) make their switching decisions on the basis of the contents of application layer messages.

End-of-Chapter Questions

THOUGHT QUESTIONS

1. Give a non-network example of hierarchical addressing, and discuss how it reduces the amount of work needed in physical delivery. Do not use any example in the book.

2. A client PC has two simultaneous connections to the same webserver application program on a webserver. (Yes, this is possible, and in fact it is rather common.) What will be different between the TCP segments that the client sends on the two connections?

3. A router that has the routing table shown in Figure 8-8 receives an incoming packet. The source IP address is 10.55.72.234. The destination host is 10.4.6.7. The TTL value is 1. The Protocol field value is 6. What will the router do with this packet?

HARDER THOUGHT QUESTIONS

1. For security reasons, many organizations do not allow error reply messages to leave their internal internets. How, specifically, could hackers use information in echo reply messages to learn about the firm's internal hosts?

2. Continuing from the previous question, how could a hacker use the TTL field to learn about the organization of the firm's routers?

TROUBLESHOOTING QUESTION

You suspect that the failure of a router or of a transmission line connecting routers has left some of your important servers unavailable to clients at your site. How could you narrow down the location of the problem using what you learned in this chapter?

HANDS-ON EXERCISE

To get a second-level domain name, you need to go to an address registrar. Network Solutions is a well-known address registrar. Go to the Network Solutions website, *www.netsol.com*. Pick a second-level domain name for an imaginary business and see whether it is available. If it is not, try other second-level domain names until you find one that is available.

PERSPECTIVE QUESTIONS

1. What was the most surprising thing you learned in this chapter?

2. What was the most difficult material for you in this chapter?

PROJECT

Prepare a short report describing ICMP commands.

GETTING CURRENT

Go to the book website's New Information and Errors pages for this chapter to get new information since this book went to press and to note corrections to any errors in the text.

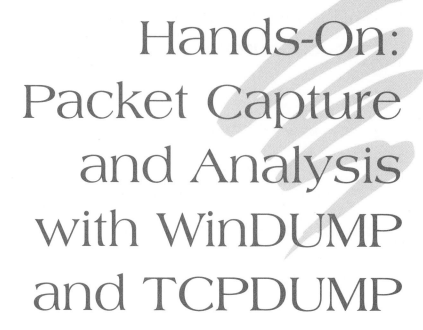

Chapter 8a

Hands-On: Packet Capture and Analysis with WinDUMP and TCPDUMP

Learning Objectives

By the end of this chapter, you should be able to discuss:

■ The purpose of WinDUMP and TCPDUMP.

■ How to obtain and run WinDUMP.

■ How to read "simple" WinDUMP output.

■ How to read hexadecimal WinDUMP output.

WHAT ARE WɪɴDUMP and TCPDUMP?

WinDUMP and TCPDUMP are packet capture and analysis programs. As the names suggest, they capture packets entering and leaving your computer and afterward allow you to look at the contents of selected fields in individual packets.

More specifically, after these programs capture packets, they print one or more lines per packet, as Figure 8a-1 illustrates for an HTTP request–response cycle. The printout is dense and looks forbidding, but with a little practice, it becomes easily readable.

This detailed information can help you troubleshoot problems and detect hacker activity. Using WinDUMP/TCPDUMP is also a great way to solidify your understanding of TCP.

378

Historically, TCPDUMP was created to run on Unix machines, and it has been wildly popular among Unix administrators. TCPDUMP was ported to Windows computers as WinDUMP. Although WinDUMP is not as mature or as full-featured as TCPDUMP, it will run on the Windows clients that most students have.

WORKING WITH WINDUMP

Installing WinDUMP

To get a copy of WinDUMP, go to *http://www.tcpdump.org/*. This website has directions for downloading the program to your computer. Websites do not always remain alive, so if this website is not working, you might have to do an Internet search to find WinDUMP.

WinDUMP does not work by itself. It requires a library of packet capture programs collectively known as WinPCAP. Fortunately, websites that help you download WinDUMP also help you download WinPCAP. During installation, install WinPCAP first, then WinDUMP.

Running WinDUMP

As Figure 8a-1 shows, you begin a WinDUMP session by going to the command line. You do this by clicking Start, then Run, and then typing cmd or command. You can then give WinDUMP commands. In this example the command is the following:

Command prompt>tcpdump www2.pukanui.com -c 40

Note that the command is "tcpdump," not windump. TCPDUMP may have been ported over to WinDUMP, but its Unix commands have been kept.

If you simply type the command "tcpdump" without options, all packets going into and out of the computer interface will be captured. However, the tcpdump program has many options to control what packets are captured and how they are displayed. In the example shown in Figure 8a-1, for instance, only packets to and from the specified host, www2.pukanui.com, will be captured. The packet count, c, is set to 40, meaning that only the first 40 packets will be captured.

To get full information about TCPDUMP, do an Internet search for "tcpdump man page." This will take you to a detailed Unix manual page. Most of what you read there will work with WinDUMP.

Getting Data to Capture

Of course, if no packets go to or from *www2.pukanui.com*, there will be nothing to capture. Without closing your command prompt window, open your browser and go to a website. This will download the home page, giving you at least one HTTP request–response cycle. (If a page has graphics and other elements, each is a separate file, so several request–response cycles will be captured.) In this case, capturing only 40 packets will show you only the connection opening and some of the subsequent packet exchanges.

```
Command prompt>tcpdump www2.pukanui.com

7:50.10.500020 10.0.5.3.62030 > www2.pukanui.com.http: S 800000050:800000050(0) win 4086
<mss1460>

7:50.10.500030 www2.pukanui.com.http > 10.0.5.3.62030 : S 300000030:300000030(0) ack
800000051 win 8760 <mss1460>

7:50.10.500040 10.0.5.3.62030 > www2.pukanui.com.http: . ack 1 win 4086

7:50.10.500050 10.0.5.3.62030 > www2.pukanui.com.http: P 1:100(100)

7:50.10.500060 www2.pukanui.com.http > 10.0.5.3.62030 : . ack 101 win 9000

7:50.10.500070 www2.pukanui.com.http > 10.0.5.3.62030 : . 1:1000(999)

7:50.10.500080 10.0.5.3.62030 > www2.pukanui.com.http: . ack 1001 win 4086

7:50.10.500090 www2.pukanui.com.http > 10.0.5.3.62030 : P 1001:2000(999)

7:50.10.500100 10.0.5.3.62030 > www2.pukanui.com.http: . ack 2001 win 4086

7:50.10.500110 10.0.5.3.62030 > www2.pukanui.com.http: R
```

Figure 8a-1   ASCII WinDUMP Printout

READING WinDUMP OUTPUT

To see how WinDUMP works (and to help you solidify and extend your understanding of IP and especially TCP), we will look at the WinDUMP output in Figure 8a-1.

Opening the TCP Connection

The first three packets in Figure 8a-1 open a TCP connection between a client PC and a server. This is the classic SYN–SYN/ACK–SYN three-way handshake.

Syn

The first packet carries a SYN segment from the client PC running WinDUMP to a web-server. 7:50.10.500020 10.0.5.3.62030 > www2.pukanui.com.http: S 800000050:800000050 (0) win 4086 <mss1460>

➤ The printout begins with a time stamp, 7:50.10.500020. This gives time to the millionth of a second.

➤ Next comes the source host's IP address and port number. The IP address is 10.0.5.3. The source host's port number, 62030, is an ephemeral port; so the source host must be a client PC. It is the computer running WinDUMP.

➤ Next comes the destination host's host name and port number. Unless you tell it otherwise, WinDUMP looks up the host name for the source and destination IP addresses and inserts them in the printout. The client, 10.0.5.3, does not have a host name, so its IP address is used in the packet printout. The destination host is *www2.pukanui.com*. The port number is the well-known port for http (80). Unless you tell it otherwise, WinDUMP substitutes protocol names when it sees well-known port numbers.

➤ The "S" indicates that the SYN flag is set. Flags except for ACK are shown in this position. If no flag is set, a period is shown instead of flags.

➤ Next comes the odd-looking 800000050:800000050(0). When a host begins a TCP session, it randomly generates an initial sequence number. The client has generated an initial sequence number of 800000050 for this session. A SYN message is a pure supervisory message containing no data. The 800000050:800000050(0) shows that the data field has a length of zero (0) because this is a supervisory segment containing no data.

➤ The win 4086 part of the printout shows that the client has told the server to use a window of 4,086 bytes. The server can transmit only 4,086 bytes of data before getting a window extension.

➤ Data within angle brackets describe options. Here the client advertises a maximum segment size (MSS) of 1460. This tells the receiver of the packet (the webserver) to place no more than 1,460 bytes of data in TCP data fields.

Syn/Ack

Now the webserver replies with its SYN/ACK message. 7:50.10.500030 www2. pukanui. com. http > 10.0.5.3.62030 : S 300000030:300000030(0) ack 800000051 win 8760 <mss1460>

➤ The host designations are reversed to indicate that the webserver is sending to the client.

➤ The S flag is again set.

➤ The webserver's initial sequence number is 300000030, and the packet carries no data.

➤ The "ack" indicates that this TCP segment contains an acknowledgment. The acknowledgment number is 800000051. The acknowledgment number is always one byte larger than the last data byte in the segment being acknowledged (800000050). It specifies the next byte of data that is expected after the segment being acknowledged.

➤ The window again indicates how many more bytes may be transmitted before the window size is increased. In this case, the window size is 8760. The client may transmit through byte 800000051 plus 8,760, or 800008811.

ACK The client now sends back an ACK. 7:50.10.500040 10.0.5.3.62030 > www2.pukanui. com.http:. ack 1 win 4086

➤ Notice that there is no flag other than ack, so a period (.) is placed in the flags position.

➤ Also notice that the ack is 1, not 300000031, as you would expect (one more than the last data byte received). The packet really does have 300000031 as the acknowledgment. However, to make the printout easier to read, WinDUMP subtracts the initial sequence number, leaving 1. All subsequent TCP data indications will be based on the bytes sent by a transport process since the beginning of the session.

The HTTP Request Message

Now the client sends an HTTP request message. This consists of a single packet.
7:50.10.500050 10.0.5.3.62030 > www2.pukanui.com.http: P 1:100(100)

> ➤ The flags field shows P, for push. This tells the receiver that a full application message is contained in the message, and that the transport process should pass the application message to the application layer.

> ➤ The 1:100(100) says that the HTTP request message contains data bytes 1 through 100—100 bytes in total.

The webserver responds with an acknowledgment. The acknowledgment number, as always, gives the next data byte the webserver expects to see (101). 7:50.10.500060 www2.pukanui.com.http > 10.0.5.3.62030 :. ack 101 win 9000

The HTTP Response Message

Now the webserver sends back an HTTP response message. This response message is too large to fit into a single packet. Consequently, the webserver sends the HTTP response message in two packets, each of which is acknowledged separately.

7:50.10.500070 www2.pukanui.com.http > 10.0.5.3.62030 :. 1:1000(999)

7:50.10.500080 10.0.5.3.62030 > www2.pukanui.com.http:. ack 1001 win 4086

7:50.10.500090 www2.pukanui.com.http > 10.0.5.3.62030 : P 1001:2000(999)

7:50.10.500100 10.0.5.3.62030 > www2.pukanui.com.http:. ack 2001 win 4086

> ➤ The packets containing the HTTP response message do not have any flag fields set, so a period (.) appears where you would see a flag if one had been included in the segment.

> ➤ Note that the first packet contains bytes 1 through 1000 of the HTTP response message, while the second packet contains remaining bytes, 1001 through 2000.

> ➤ Note also that the second packet containing the HTTP response message contains a P flag field. This is the push flag, which tells the receiving transport process that all data has been delivered, so that the data should be pushed up to the application program. The first response packet does not have a push flag because it only delivers the first part of the HTTP response message. A push is not needed again until all the data is received.

Ending the Connection

7:50.10.500110 10.0.5.3.62030 > www2.pukanui.com.http: R

In Chapter 8, we saw how TCP connections should end with a four-way close in which FINs are sent and acknowledged. In Figure 8a-1, however, the client, for no obvious reason, has sent a reset message (flag R) to the webserver. A reset message abruptly terminates the connection. Neither side transmits again.

SOME POPULAR WinDUMP OPTIONS

Major Options

WinDUMP's tcpdump command has a large number of options. We have already seen two of them.

➤ Giving a hostname or IP address specifies that packets to or from only that address should be captured.

➤ The "-c" option specifies how many packets should be captured.

The following are a few other commonly used options. The tcpdump command has many more.

➤ e: Print the data link layer header fields on each line.

➤ i: Specify an interface (NIC) if a computer has more than one.

➤ n: Do not convert IP addresses into host names (reduces capture processing work).

➤ N: Print only the host part of a hostname. This makes output more condensed and perhaps easier to read.

➤ q: Quiet. Print less information for each packet in ASCII.

➤ s: Snaplen. Specifies how many octets will be shown for each packet. The default is 68.

➤ t: Do not print the time stamp on each line.

➤ v: Verbose output—more detail than the normal output. There also are vv and vvv options for more verbose and incredibly verbose output.

➤ w: Writes the raw packets to a file. A space and then a file name must follow the w option.

➤ r: Reads data from a file instead of capturing the data. Often used when the data was captured in a file with the w option.

➤ The -x option specifies that output should be in hex, while the -X option specifies that output should be in both ASCII and hex. We will see what this means in the next section.

Example

For example, suppose you give the following command.
tcpdump www2.pukanui.com -c 1 -tN

This tells tcpdump to collect data only for packets going to and from *www2. pukanui.com*. It also tells tcpdump to suppress time stamps and to show only the host name. Finally, it tells tcpcump to capture only a single packet. The following shows the output you might see:

10.0.5.3.62030 > www2.http: S 800000050:800000050(0) win 4086 <mss1460>

Expression

In the examples we have been using, we have included the name of a host, *www2.pukanui.com*. This type of option is called an *expression*. It specifies what packets will be captured and analyzed.

➤ Host expressions are the most common. They should be written as "host hostname," but host is the default, so it does not need to be added.

➤ The expression "port 80" tells TCPdump to look only at HTTP traffic. The more finely grained expression "src port 80" captures packets only from HTTP servers.

➤ It is even possible to have expressions that limit traffic to a particular network, as in "net 128.171," to capture traffic going to or from a particular network.

HEXADECIMAL PRINTOUT

ASCII versus Hex

The type of printout we have been seeing is called *ASCII printout,* because it contains alphanumeric characters (keyboard characters) stored in the ASCII format. WinDUMP also offers another way to store and see packet information—hexadecimal format. As discussed in Chapter 4, "hex" represents a group of four bits as a symbol from 0 through F. Hex output typically groups hex symbols in groups of two to represent a byte (octet) or in groups of four, to represent two-byte sequences. WinDUMP does the latter.

Hex Output

Figure 8a-2 shows hexadecimal output for a single packet. To turn on hex output, give the -x (lowercase x) option in a tcpdump command. To get both ASCII and hex output, give the -X (uppercase X) option.

IP Fields

To help you read this very dense output, the IP header is shown in boldface. In the following description, the most widely interpreted sequences are shown in boldface.

➤ **4500.** The 4 indicates that this is an IPv4 packet. The 5 indicates that the length of the header is 5 times 32 bits, or 20 octets. This is the header length of an IP header without options. Options are rare and suspicious, so anything other than a 5 for the header length should be seen as a caution sign. The 00 is the Diff-Serv octet, which normally is not used and so normally is set to 00. Almost all packets should start with 4500.

➤ 00c7. This is the length of the entire IP packet (199 bytes).

➤ ff53 0000. This is the identification field value and other information used to reassemble fragmented packets. Fragmentation is rare.

➤ **8006.** The 80 is the one-byte time to live field value (128 in decimal). The 06 is the one-byte protocol field value. The 06 protocol is TCP. This field is needed to interpret the IP data field, which is not always a TCP segment. Protocol 01 is ICMP, for instance, while Protocol 17 is UDP.

➤ 3d5e. This is the header checksum.

Figure 8a-2   Hexadecimal Output from WinDUMP

```
4500 00c7 ff53 0000 8006 3d5e b87a 3270

b87a c3d0 F230 0050 0023 37d6 1d37 1302

5018 07d0 b329 0000 ...
```

➤ **b87a 3270.** This is the IP source address.

➤ **b87a c3d0.** This is the IP destination address.

TCP Fields

In Figure 8a-2, the TCP fields are underlined. The following are the TCP fields. Again, the most widely interpreted fields are shown in boldface.

➤ **F230.** This is the source port number (62000).

➤ **0050.** This is the destination port number (80 decimal).

➤ **0023 37d6.** This is the sequence number.

➤ **1d37 1302.** This is the acknowledgment number.

➤ **50.** 5 is the header length in 32-bit units; as in IP, 5 is the header length without options. In contrast to IP, options are common in TCP. The 0 is from reserved bits. It should always be 0.

➤ **18.** The 1 indicates that the ack bit is set. The 8 indicates that the push bit is set. These two symbols indicate which bits are set. Experienced WinDUMP users learn the most common combinations.

➤ 07d0. This is the window size (2000 in decimal).

➤ b329. This is the checksum.

➤ This is the urgent pointer. If the urgent (U) bit is set in the flags field, this pointer tells where urgent data begin in the TCP byte sequence.

End-of-Chapter Questions

GENERAL QUESTIONS

1. What does WinDUMP do?

2. Distinguish between WinDUMP and TCPDUMP.

3. Distinguish between ASCII and hex output.

4. What steps should you take to capture and display data?

For Each of the Following, Specify a Command to Do the Work

5. Show all packets.

6. Show the first 100 packets going to or from dakine.pukanui.com.

7. Repeat the preceding command, this time not showing a time stamp or the full host name.

8. Show all packets going to or from HTTP servers.

INTERPRETATION

1. Interpret the following ASCII printout:

   7:50.10.500099 db.pukanui.com.54890 > www2.pukanui.com.http: 1:21(21) ack 52 win 4086

2. The following hex printout shows an IP packet. a) What type of message is in its data field? b) What is the IP destination address?

| 4500 | 00c7 | ff53 | 0000 | 8017 | 3d5e | b87a | 3270 |
| b87a | c3d0 | | | | | | |

3. The following hex printout shows an IP packet containing a TCP segment. a) What is the source port (in decimal)? b) Is the ack bit set? (Tell how you can know.)

4500	00c7	ff53	0000	8006	3d5e	b87a	3270
b87a	c3d0	0060	0050	0023	37d6	0000	0000
5008	07d0	b329	0000	....			

Security

Learning Objectives

By the end of this chapter, you should be able to discuss the following:

- Security threats (worms and viruses, hacking, and denial-of-service attacks).
- Types of attackers.
- Why security is primarily a management issue, not a technical issue.
- Security planning principles.
- Access control, including authentication mechanisms: passwords, digital certificate authentication, and biometrics.
- Firewall protection, including stateful inspection, IDSs, and IPS filtering.
- The protection of dialogues by cryptographic systems; encryption for confidentiality; the phases of cryptographic systems.
- Responding to successful compromises.

INTRODUCTION

A Major Threat

In the 1990s, the Internet blossomed, allowing people to reach their choice of hundreds of millions of servers around the world. Unfortunately, the Internet also gave attackers access to hundreds of millions of users. Security quickly became one of the most important IT management issues.

One thing that sets security apart from other aspects of IT is that the company must battle against intelligent adversaries, not simply against errors and other forms of unreliability. Companies today are engaged in an escalating arms race with attackers, and security threats and defenses are mutating at a frightening rate.

Recap from Chapter 1

In Chapter 1, we looked very briefly at security. That chapter made four points:

➤ In authentication, a verifier requires a supplicant to prove the supplicant's identity before being granted access to resources.

➤ *Cryptography* is the use of mathematics to protect information in storage or during transmission. During transmission, cryptography prevents interceptors from reading your messages, changing your messages and sending them on, or sending new messages in your name.

A Major Threat

Intelligent Adversaries
    Not just human error

Recap from Chapter 1
    Authentication
    Cryptography for messages
    Firewalls
    Host hardening

Figure 9-1 Security (Study Figure)

➤ A firewall examines each packet passing through it. If a packet is a provable attack packet, the firewall drops and logs the packet. Otherwise, it lets the packet through.

➤ Host hardening allows a host to survive an attack if attack packets reach it despite firewalls and other protections.

TEST YOUR UNDERSTANDING

1. a) What is authentication? b) What is cryptography? c) What do firewalls do? d) If a firewall examines a packet that is suspect, but is not a provable attack packet, what does the firewall do? e) Why is host hardening necessary?

SECURITY THREATS

The first principle in security is, "Understand the organization's needs." In security, this requires a solid knowledge of the company's **threat environment**—the sum of all threats facing the corporation. Before discussing how to prevent attacks, we will look at the main types of attacks that corporations face. A successful attack is called a **compromise.** It is also called an **incident** or a **breach.**

Compromises are widespread, varied, and costly.

TEST YOUR UNDERSTANDING

2. a) What is the threat environment? b) What is a compromise? c) Give two other names for compromises.

Viruses and Worms

The most frequent compromises occurring in firms today are virus and worm incidents. Almost all firms suffer a virus or worm incident each year, and most suffer from several. Furthermore, these compromises are occurring despite the fact that nearly all companies have antivirus systems in place.

Understand the Organization's Security Needs
  To do this, understand the threat environment
  Types of attacks a company faces and will face in the future

Successful Attacks Are Called
  Compromises
  Incidents
  Breaches

Figure 9-2   Basic Security Terminology (Study Figure)

Figure 9-3   Malware (Study Figure)

Malware
  A general name for evil software

Viruses
  Pieces of code that attach to other programs
  Virus code executes when infected programs execute
  Infect other programs on the computer
  Spread to other computers by e-mail attachments, IM, peer-to-peer file transfers, etc.
  Antivirus programs are needed to scan arriving files
    Also scan for other malware

Worms
  Stand-alone programs that do not need to attach to other programs
  Can propagate like viruses through e-mail, etc.
    This requires human gullibility, which is unreliable and slow
  Vulnerability-enabled worms jump to victim hosts directly
    Can do this because hosts have vulnerabilities
  Vulnerability-enabled worms can spread with amazing speed
  Vendors develop patches for vulnerabilities, but companies often fail or are slow
  to apply them

Payloads
  After propagation, viruses and worms execute their payloads
  Payloads erase hard disks or send users to pornography sites if they mistype URLs
  Trojan horses are exploitation programs that disguise themselves as system files
  Spyware Trojans collect sensitive data and send the data to an attacker

Viruses and Virus Propagation

Propagation within a Computer   **Viruses** are pieces of executable code that attach themselves to other programs. Within a computer, whenever an infected program runs (executes), the virus attaches itself to other programs.

Propagation between Computers   Between computers, the virus spreads when an infected program is transferred to another computer via a floppy disk, an e-mail attachment, a webpage download, an unprotected disk share, a peer-to-peer file-sharing transfer, an instant message, or some other propagation vector. Once on another machine, if the infected program is executed, the virus spreads to other programs on the machine.

More than 90 percent of viruses today spread via e-mail. Viruses find addresses in the infected computer's e-mail directories. They then send messages with infected attachments to all of these addresses. If a receiver opens the attachment, the infected program executes and the receiver's programs become infected. In addition, instant messaging (IM) and peer-to-peer (P2P) file transfers are also becoming important ways for viruses and worms to spread.

To stop viruses, a company must protect its computers with **antivirus programs** that scan each arriving e-mail message or floppy disk for signatures (patterns) that identify viruses. These antivirus programs also scan for other types of **malware** (evil software).

Worms and Worm Propagation

Another important type of malware is worms. We have just seen that viruses are pieces of code that must attach themselves to other programs. In contrast, **worms** are full programs that operate by themselves. Both viruses and worms can create mass epidemics that infect hundreds or even millions of computers. However, they can spread in different ways.

Worms are capable of propagating, like viruses, through e-mail attachments, IM, and P2P file transfers. These methods require human gullibility to succeed. Although human gullibility is widespread, it is not reliable. More importantly, human gullibility is rather slow. Until someone opens an e-mail attachment, nothing happens.

Unlike viruses, worms have another propagation vector (way to propagate). This is the exploitation of software vulnerabilities rather than human gullibility. Security vulnerabilities are discovered in most programs several times per year. Although software vendors usually issue patches quickly when new vulnerabilities are found, it takes some time to patch all vulnerable computers. Attackers use this window of opportunity to launch **vulnerability-enabled** worms that exploit the vulnerability in the software.

Vulnerability-enabled worms require no human intervention, so they can spread directly from computer to computer with incredible speed. In 2003, the Blaster worm infested 90 percent of all vulnerable hosts on the entire Internet within 10 minutes. The nightmare scenario for security professionals is the prospect of a fast-spreading worm that exploits a vulnerability that damages a large percentage of all computers on the Internet.

Vulnerability-enabled worms require no human intervention, so they can spread with incredible speed.

Payloads

In war, when a bomber aircraft reaches its target, it releases its payload of bombs. Similarly, after they spread, viruses and worms may execute pieces of code called **payloads.** In malicious viruses and worms, these payloads can completely erase hard disks and do other significant damage. In some cases, they can take the victim to a pornography site whenever the victim mistypes a URL. In other cases, they can turn the user's computer into a spam generator or a pornography download site.

Trojan Horses

Often, the payload installs a **Trojan horse** program on the user's computer. Once installed, the Trojan horse continues to exploit the user indefinitely. A Trojan horse does not spread by itself, but rather relies on a virus, worm, hacker, or gullible user to install it on a computer. As its name suggests, a Trojan horse disguises itself as a system file; it typically replaces a legitimate system program, so it is difficult to detect.

Spyware

An especially problematic category of Trojan horses is **spyware**—a name given to programs that are installed on your computer **surreptitiously** (without your knowledge) in order to collect information about you and send this information to the attacker. Some spyware programs collect information about your Web surfing habits and send this information to advertisers. More dangerous are **keystroke loggers,** which record your keystrokes, looking for passwords, social security numbers, and other information that can help the person who receives the keystroke logger's data commit fraud. **Data mining** spyware searches your hard drive for potentially useful information and sends this information to the attacker.

TEST YOUR UNDERSTANDING

3. a) How do viruses propagate within computers? b) How do viruses propagate between computers? c) How can viruses be stopped? d) Distinguish between vulnerability-enabled worms and e-mail worms. e) Which can spread faster—viruses or vulnerability-enabled worms? Explain. f) How can vulnerability-enabled worms be stopped? g) What are payloads? h) What is malware? i) What are Trojan horses? j) How does a Trojan horse get on a computer? k) What is spyware? l) What is a keystroke logger?

Attacks on Individuals

Social Engineering

As technical defenses improve, a growing number of malware attacks today focus on **social engineering,** which is a fancy name for tricking the victim into doing something against his or her interests. Viruses and worms have long tried to do this with e-mail attachments—say, by telling the user that he or she has won a lottery and needs to open the attachment for the details. The range of social engineering attacks has expanded greatly in the last few years.

Social Engineering
> Tricking the victim into doing something against his or her interests

Spam
> Unsolicited commercial e-mail

Fraud

Taking the Reader to a Website with Malware

Credit card number theft
> Performed by carders

Identity theft
> Involves collecting enough data to impersonate the victim in large financial transactions

Phishing
> A sophisticated social engineering attack in which an authentic-looking e-mail or website entices the user to enter his or her username, password, or other sensitive information

Figure 9-4 Attacks on Individuals (Study Figure)

Spam

Perhaps the most annoying type of malware on a day-in, day-out basis is **spam,**[1] which is unsolicited commercial e-mail. Spammers send the same solicitation e-mail message to millions of e-mail addresses in the hope that a few percent of all recipients will respond. Most users today have to delete about 10 or more spam messages for every legitimate message they receive. In addition, many spam messages involve pornography or fraud.

Fraud

Few spam messages are really designed to sell legitimate products. Most are fraudulent attempts to get someone to send money for goods that will not be delivered or that are effectively worthless.

Visiting a Website

One way for spam to create problems is for messages to include a link to a website. If the receiver clicks on the link, he or she will be taken to a website that will complete the fraud or download malware into the victim's computer.

---

[1]Except at the beginnings of sentences, e-mail *spam* is spelled in lowercase. This distinguishes unsolicited commercial e-mail from the Hormel Corporation's meat product, Spam, which should always be capitalized. In addition, Spam is *not* an acronym for "spongy pink animal matter."

Phishing Attacks

An especially effective form of spam is **phishing,**[2] which uses authentic looking e-mail or websites to entice the user to send his or her username, password, or other sensitive information to the attacker. One typical example of phishing is an e-mail message that appears to be from the person's bank. The message asks the person to "confirm" his or her username and password in a return message. Another typical example is an e-mail message with a link to what appears to be the victim's bank website, but that is, in fact, an authentic-looking fake website.

Credit Card Number Theft

In fraudulent spam, the message may convince the user to type a credit card number to purchase goods. The attacker will not deliver the goods. Instead, the **carder** (credit card number thief) will use the credit card number to make unauthorized purchases.

Identity Theft

In other cases, thieves collect enough data to impersonate the victim in large financial transactions. This impersonation is **identity theft.** It allows thieves to take out large loans and do other major damage. Identity theft is more serious than credit card theft.

TEST YOUR UNDERSTANDING

4. a) What is social engineering? b) What is the definition of spam? c) How can spam be used to harm people who open spam messages? d) What is phishing? e) Distinguish between credit card number theft and identity theft. f) What are carders? g) Which is more serious—credit card theft or identity theft?

Human Break-Ins (Hacking)

Viruses and worms spread randomly with fixed attack methods. However, human attackers often wish to break into a specific company's computers manually. Human adversaries can attack a company with a variety of approaches until they find one that succeeds. This makes human break-ins much more likely to succeed than viruses.

What Is Hacking?

We will use the term *hacking* to mean breaking into a computer. More specifically, **hacking** is intentionally using a computer resource without authorization or in excess of authorization. Note that it is still hacking if the attacker is given an account and uses the computer for unauthorized purposes.

> Hacking is intentionally using a computer resource without authorization or in excess of authorization.

The Scanning Phase

When a hacker begins an attack on a firm, he or she usually begins by **scanning** the network. This involves sending **probe packets** into the firm's network. Responses to

---

[2]IT attackers often replace *f* with *ph*. For example, *phone freaking* (dialing long distance numbers illegally) became *phone phreaking* and, later, just *phreaking.*

Human Break-Ins
> Viruses and worms rely on one main attack method
> Humans can keep trying different approaches until they succeed

Hacking
> Hacking is breaking into a computer
> Hacking is intentionally using a computer resource without authorization or in excess of authorization

Scanning Phase
> Send attack probes to map the network and identify possible victim hosts
> The Nmap program is popular for scanning attacks (Figure 9-6)

The Break-In
> Uses an exploit—a tailored attack method that is often a program
> Normally exploits a vulnerability on the victim computer
> Often aided by a hacker tool
> The act of breaking in is called the exploit
> The hacker tool is also called an exploit

After the Break-In
> The hacker downloads a hacker tool kit to automate hacking work
> The hacker becomes invisible by deleting log files
> The hacker creates a backdoor (way to get back into the computer)
>> Backdoor account—account with a known password and full privileges
>> Backdoor program—program to allow reentry; usually Trojanized
> The hacker can then do damage at his or her leisure
>> Download a Trojan horse to continue exploiting the computer after the attacker leaves
>> Manually give operating system commands to do damage

Figure 9-5   Human Break-Ins (Study Figure)

these probe packets tend to reveal information about the firm's general network design and about its individual computers—including the operating systems and the applications these computers are running.

Figure 9-6 shows output from one popular scanning program, Nmap. In the figure, Nmap has identified several open ports on a server—port numbers that will accept connection attempts. The scan has also fingerprinted the operating system of the server because many attacks are operating-system specific.

The Break-In

If the attacker has an **exploit** (tailored attack method) for one of the services associated with these ports, the attacker can now hack the server. Often, the exploit is a program written to automate the steps of the attack method. Normally, exploits take advantage of **vulnerabilities** (security weaknesses) in the software that is being attacked.

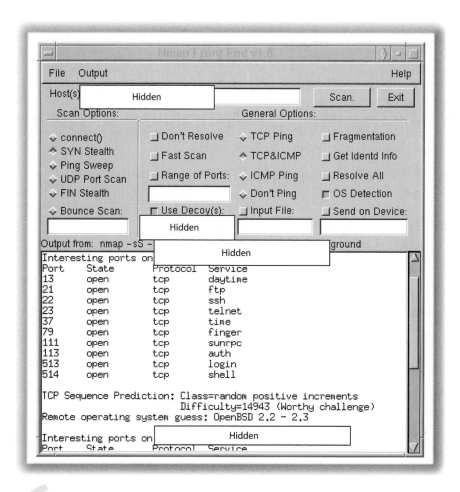

Figure 9-6   Nmap Scanning Output

After the Break-In
After the break-in, the real work begins.

Downloading a Hacker Toolkit   Typically, the first thing a hacker does is to download a **hacker toolkit** to the victim computer. The toolkit is a collection of tools which automate some tasks that the hacker will have to perform after the break-in.

Erasing Log Files   The hacker usually first uses the hacker toolkit to erase the operating system's log files so that the computer's rightful owner cannot trace how the attacker broke in or what the hacker did after the break-in.

Creating a Backdoor   The hacker users the hacker toolkit to create a **backdoor** that will allow the hacker back in later, even if the vulnerability is repaired. The backdoor may simply be a new account with a known password and full privileges. It can also be a Trojan horse program that is difficult to detect.

Downloading Trojan Horses for Continuing Damage   Next, the attacker may download a Trojan horse, which will continue to cause damage after the hacker leaves—for instance, turning the host into a pornography download site or using the compromised host to attack other computers. Keystroke loggers that collect what the user types are also popular Trojan horses. The most dangerous Trojan horses are bots, which we will learn about in the next subsection.

Manual Work   Although hacker toolkits and Trojan horses automate a great deal of what the hacker wishes to do, hackers also work manually. With full access to the computer, the attacker can give ordinary operating system commands to read any file on the computer, change files, delete them, or do anything else that a legitimate user can do.

TEST YOUR UNDERSTANDING

5. a) List the three main phases in human break-ins (hacks). b) What is hacking? c) Why do hackers send probe packets into networks? d) What is an exploit? e) Do most exploits take advantage of vulnerabilities? f) What steps does a hacker usually take after a break-in? g) What software does the hacker download to help him or her do work after compromising a system? h) What does a hacker do to avoid being caught? i) What is a backdoor? j) What is host exploitation software?

Denial-of-Service (DoS) Attacks That Use Bots

Another type of attack, the denial-of-service attack, does not involve breaking into a computer, infecting it with a virus, or infesting it with a worm. Rather, the goal of **denial-of-service (DoS)** attacks is to make a computer unavailable to its users. As Figure 9-7 shows, most DoS attacks involve flooding the victim computer with irrele-

Figure 9-7   Distributed Denial-of-Service (DDoS) Attack Using Bots

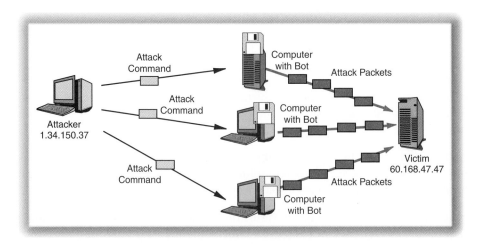

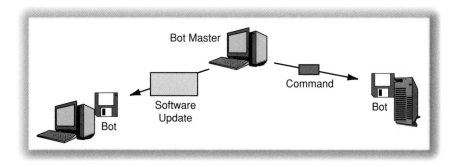

Figure 9-8 Bots

vant packets. The victim computer becomes so busy processing this flood of attack packets that it cannot process legitimate packets. The overloaded host may even fail.

More specifically, the attack shown in the figure is a **distributed DoS (DDoS)** attack. In this type of attack, the attacker first installs programs called bots on hundreds or thousands of PCs or servers. When the user sends these bots an attack command, they all begin to flood the victim with packets.

Bots are not limited to DDoS attacks. **Bots** are general-purpose exploitation programs that can be remotely controlled after installation. The attacker can send commands to the bots and can even upgrade them remotely with new capabilities. Bots are extremely dangerous because they can engage in massive attacks that were previously possible only with relatively dumb and inflexible viruses and worms. Bots bring the flexibility of human thought into the attack, making them very dangerous. (See Figure 9-8)

TEST YOUR UNDERSTANDING

6. a) What is the purpose of a denial-of-service attack? b) What are bots? c) How do distributed DoS attacks work?

Attackers

As Figure 9-9 shows, there are many different types of attackers facing organizations today.

Traditional Attackers

When most people think of attackers, they normally have three pictures in their minds: hackers driven by curiosity, virus writers, and disgruntled employees and ex-employees. Indeed, these used to be the three most important types of attackers.

Hackers   Traditionally, some hackers have been motivated primarily by curiosity and the sense of power they get from breaking into computers. In many cases, they are also motivated by a desire to increase their reputation among their hacker peers by boasting about their exploits. This typically is the image of hackers presented in mainstream movies. However, these are not the typical hackers today.

Traditional Attackers
   Traditional Hackers
      Hackers break into computers
      Driven by curiosity, a desire for power, and peer reputation
   Virus writers
   Script kiddies use scripts written by experienced hackers and virus writers
      They have limited knowledge and abilities
      But large numbers of script kiddies make them dangerous
   Disgruntled employees and ex-employees

Criminal Attackers
   Most attacks are now made by criminals
   Crime generates funds that criminal attackers need to increase attack sophistication

On the Horizon
   Cyberterror attacks by terrorists
   Cyberwar by nations
   Potential for massive attacks

Figure 9-9   Types of Attackers

Virus Writers   **Virus writers,** as the name suggests, create viruses. They also create other types of automated malware. Virus writers appear to enjoy the excitement of seeing their programs spread rapidly. These virus writers tend to be blind to the harm that they do to people.

Script Kiddies   Experienced hackers and virus writers often developed small programs, called scripts, to automate parts of their attacks. Over time, these programs grew more sophisticated. More importantly, they grew easier to use. Many now have graphical user interfaces and the look, feel, and reliability of commercial programs. This has led to the emergence of **script kiddie** attackers, who use these scripts developed by more experienced attackers. Although traditional attackers disparage script kiddies for their lack of skills, there are far more script kiddies than traditional hackers and virus writers, and script kiddies collectively represent a great threat to corporations.

Disgruntled Employees and Ex-Employees   Other traditional attackers have been **disgruntled employees** and **disgruntled ex-employees** who attacked their own firms. Employee attackers tend to do extensive damage when they strike because they typically already have access to systems and a broad knowledge of how the systems work.

Criminal Attackers

In addition, there now are many **criminal attackers** who steal credit card numbers to commit credit card fraud, who extort firms, and who steal trade secrets to sell to competitors. In fact, today, more than half of all Internet attacks are committed by criminals motivated by money, and this fraction is increasing rapidly.

Funded by their crimes, many criminals can afford to hire the best hackers and to enhance their own skills. Consequently, criminal attacks are not just growing in numbers; they also are growing very rapidly in sophistication.

Cyberterrorists and National Governments

On the horizon is the danger of far more massive **cyberterror** attacks by terrorists and even worse **cyberwar** attacks by national governments. These could produce damages in the hundreds of billions of dollars.

TEST YOUR UNDERSTANDING

7. a) Are most attackers today driven by curiosity and a sense of power? b) Why are employees dangerous? c) What type of attacker is the most common today? d) What are cyberterror and cyberwar attacks?

PLANNING

Security Is a Management Issue

People tend to think of security as a technological issue, but security experts agree unanimously that security is primarily a management issue. Unless a firm does excellent planning, implementation, and day-to-day operation, the best security technology will be wasted. As Bruce Schneier, a noted security expert, has often said, "Security is a process, not a product." Unless firms have good security processes in place, the most technologically advanced security products will do little good.

Security is primarily a management issue, not a technology issue.

TEST YOUR UNDERSTANDING

8. Why is security primarily a management issue, not a technology issue?

The Plan-Protect-Respond Cycle

It is useful to think about security processes in terms of the plan-protect-respond cycle. This cycle begins with planning processes so that actions are governed by well-thought-through plans. Next, the plans are implemented to provide ongoing protection. This is the longest and most expensive phase of the cycle. Finally, firms must respond to the inevitable attacks that do get through their defenses from time to time.

Planning Principles

Perhaps more than any other aspect of IT, effective security depends on effective planning. Security planning is a complex process. We will only note four key principles that must be used in planning.

Risk Analysis

In contrast to military security, which often makes massive investments to stop threats, corporate security planners have to ask whether applying a protection against a particular

Security Is a Management Issue, Not a Technical Issue
  Without good management, technology cannot be effective
  A company must have good security processes

Plan-Protect-Respond Cycle
  Organizes thinking about security
  Planning
  Protecting (the largest part of the cycle)
  Responding (when protections break down)

Security Planning Principles
  Risk Analysis
    Risk analysis is the process of balancing threats and protection costs for individual assets
    Cost of protection should not exceed the cost of likely damage
  Comprehensive Security
    An attacker has to find only one weakness
    A firm needs comprehensive security to close all avenues of attack
  Access Control
    Limit access to resources to legitimate users
    Give legitimate users minimum permissions (things they can do)
  Defense in Depth
    Every protection breaks down sometimes
    An attacker should have to break through several lines of defense to succeed
    Providing this protection is called *defense in depth*

Access Control Planning for Individual Resources
  Enumerating and Prioritizing Resources
    Firms must enumerate and prioritize the resources they have to protect
    Otherwise, security planning is impossible
  Companies must then develop an access control plan for each resource
    The plan includes the AAA protections
      Authentication is proving the identity of the person wishing access
      Authorization is determining what the person may do if he or she is authenticated
      Auditing is logging data on user actions for later appraisal

Figure 9-10 Security Planning (Study Figure)

threat is justified economically. For example, if the probable annual loss due to the threat is $100,000 and security measures to thwart the threat will cost $200,000, firms should not spend the money. Instead, they should accept the probable loss. **Risk analysis** is the process of balancing threats and protection costs for individual assets.

Risk analysis is the process of balancing threats and protection costs for individual assets.

Comprehensive Security

Corporate security is an example of asymmetrical warfare in which the attacker has a clear advantage. A company must close off all vectors of attack. If it misses even one and the attacker finds it, the attacker will succeed. The attacker, in contrast, has to find only one security weakness. Although it is difficult to achieve **comprehensive security,** in which all avenues of attack are closed off, it is essential to come as close as possible.

Access Control with Minimum Permissions

Security planners constantly worry about access to resources. There may be several ways for both legitimate users and attackers to access each resource. Consequently, as we will see in the next subsection, companies need to plan for **access control** (determining who may access each resource).

Even if a user has legitimate reasons for access to a resource, he or she should have **minimum permissions** (things they can do to the resource). For instance, a user who has a legitimate reason to read a document might be given read-only access, which would prevent the user from changing the document (resource).

Defense in Depth

Another critical planning principle is defense in depth. Every protection will break down occasionally. If attackers have to break through only one line of defense, they will succeed during these vulnerable periods. However, if an attacker has to break through two, three, or more lines of defense, the breakdown of a single defense technology will not be enough to allow the attacker to succeed. Having successive lines of defense is called **defense in depth.**

TEST YOUR UNDERSTANDING

9. a) What is the plan-protect-respond cycle? b) List the three major planning principles. c) What is risk analysis? d) Why is comprehensive security important? e) What is defense in depth? f) Why is it necessary?

Access Control Plans

Given the high importance of access control planning, we will look at this topic in a little more detail.

Enumerating and Prioritizing Resources

A firm has a wide variety of resources on its client PCs and servers. Some of these resources are extremely crucial, others less so. Consequently, the firm needs an access control plan *for each resource.*

➤ One of the first things a company must do to have adequate security is to enumerate (identify and list) its resources.

➤ The next step is to rank these resources by sensitivity (security risk). Databases of customer information, for instance, would be ranked as very sensitive because the consequences of a security breach in which this information was stolen would be catastrophic.

➤ Finally, the company must develop an access control plan for each resource (or at least for each resource category).

Creating Access Control Plans (AAA) for Individual Assets

In general, access control plans have three key elements: authentication, authorization, and auditing. Collectively, they are known as AAA.

Authentication **Authentication,** as we saw in Chapter 1, is requiring someone wishing to use a resource to prove his or her identity. Grocery store clerks do this when they require you to show them identification when you want to pay by check.

Authorization Just because a person has access to a resource does not mean that the person should be able to do anything he or she wishes with the resource. The person must have specific **authorizations** that define specific actions that the person can take—for instance, deleting files.

Auditing The final element of the access control plan is **auditing**—that is, collecting information about what people do and recording this information in log files for later analysis. Auditing serves roughly the same function that surveillance cameras do. If people know that what they are doing is being recorded, they are less likely to misbehave.

TEST YOUR UNDERSTANDING

10. a) What are the three things that companies must do in security asset planning? b) What does AAA stand for? c) Distinguish between *authentication* and *authorization*. d) What is auditing?

AUTHENTICATION

The most complex element of access control is authentication. Figure 9-11 illustrates what you learned about authentication in earlier chapters. The user trying to prove his or her identity is the **supplicant.** The party requiring the supplicant to prove his or her

Figure 9-11   Authentication with a Central Authentication Server

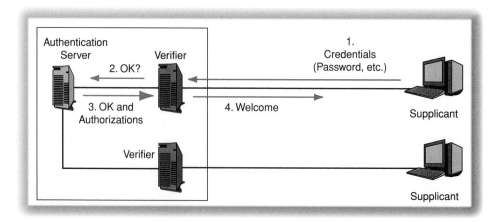

identity is the **verifier.** The supplicant tries to prove his or her identity by providing **credentials** (proofs of identity) to the verifier.

As we saw in Chapters 4 and 5, there often is a third party, the **authentication server,** which stores data to help the verifier check the credentials of the supplicant. Use of a central authentication server helps provide consistent security by ensuring that all verifiers check credentials against the same authentication information.

Use of a central authentication server helps provide consistent security by ensuring that all verifiers check credentials against the same authentication information.

The type of authentication tool that is used with each resource must be appropriate for the sensitivity of that particular resource. Sensitive personnel information should be protected by very strong authentication methods. For relatively nonsensitive data, less expensive and weaker authentication methods may be sufficient.

TEST YOUR UNDERSTANDING

11. a) What is authentication? b) Distinguish between the supplicant and the verifier. c) What are credentials? d) Why are authentication servers used? e) Why must authentication be appropriate for the sensitivity of an asset?

Passwords

The most common authentication credential is the **password,** which is a string of characters that a user types to gain access to the resources associated with a certain **username** (account) on a computer. It also is the weakest form of authentication, and it is appropriate only for the least sensitive assets.

Ease of Use and Low Cost

People find passwords familiar and relatively easy to use. Also, passwords add no additional cost because operating systems and many applications have built-in password authentication.

Word/Name Passwords and Dictionary Attacks

The main problem with passwords is that most users pick very weak passwords. They often pick ordinary **dictionary words** or the **names** of family members, pets, sports teams, or celebrities. Dictionary-word and name passwords often can be **cracked** (guessed) in a few seconds if the attacker can get a copy of the password file (which contains an encrypted list of account names and passwords). The attacker uses a **dictionary attack,** trying all words or names in a standard or customized dictionary. There are only a few thousand dictionary words and names in any language, so dictionary attacks can crack dictionary-word and name passwords almost instantly.

The main problem with passwords is that most users pick very weak passwords.

Dictionary attacks also have **hybrid modes,** in which they look for simple variations on words, such as a word with the first letter capitalized, followed by a single digit

Passwords
> Passwords are strings of characters
> They are typed to authenticate the use of a username (account) on a computer

Benefits
> Ease of use for users (familiar)
> Inexpensive because they are built into operating systems

Often Weak (Easy to Crack)
> Word and name passwords are common
> They can be cracked quickly with dictionary attacks

Passwords Should Be Complex
> Should mix case, digits, and other keyboard characters ($, #, etc.)
> Complex passwords can be cracked only with brute force attacks (trying all possibilities)

Passwords Also Should Be Long
> Should have a minimum of eight characters
> Each added character increases the brute force search time by a factor of about 70

Other Concerns
> If people are forced to use long and complex passwords, they tend to write them down
> People should use different passwords for different sites
> > Otherwise, a compromised password will give access to multiple sites

Figure 9-12   Password Authentication (Study Figure)

(for instance, Dog1). Hybrid word or name passwords are cracked almost as quickly as passwords made of simple words and names.

Complex Passwords and Brute Force Attacks

Dictionary attacks can be thwarted by making passwords more complex. Ideally, the password will be a random string of uppercase letters, lowercase letters, the digits from 0 to 9, and other keyboard symbols, such as & and #. Random or semi-random passwords can be cracked only by **brute force attacks** that try all possible combinations of characters. First, all combinations of a single character are tried, then all combinations of two characters, then all combinations of three characters, and so forth. Brute force attacks take far longer than dictionary attacks.

> Ideally, the password will be a random string of uppercase letters, lowercase letters, the digits from 0 to 9, and other keyboard symbols, such as & and #.

Unfortunately, random or nearly random passwords are difficult for users to remember, so they tend to write them on a sheet of paper that they keep next to their computers. This makes passwords easy to steal so that there is no need to crack them by dictionary or brute force attacks.

Password Length

Increasing **password length** (the number of characters in the password) helps, too. If the password has a combination of uppercase and lowercase letters, digits, and punctuation symbols, each additional character increases, by a factor of about 70, the time needed for a brute force attack to succeed.

Passwords should be at least eight characters long, and even longer passwords are highly desirable.

Passwords should be at least eight characters long, and even longer passwords are highly desirable.

Reusing Passwords at Multiple Sites

Another problem with passwords is that users often use the same password at multiple sites. This is very dangerous because if a password is cracked at one site, the attacker is likely to be able to impersonate the user at other sites.

TEST YOUR UNDERSTANDING

12. a) Distinguish between usernames and passwords. b) Why are passwords widely used? c) What types of passwords are susceptible to dictionary attacks? d) What is a brute force attack? e) What types of passwords can be broken only by brute force attacks? f) Why is password length important? g) How long should passwords be? h) Why are long and complex passwords not likely to be successful? i) Why is it dangerous if users use the same password at multiple sites?

13. Critique each of the following passwords regarding strength and the type of cracking attack that would be used to crack it. a) Viper1 b) R7%t& c) NeVeR.

Digital Certificate Authentication

The gold standard for authentication is digital certificate authentication. For extremely sensitive assets, digital certificate authentication is almost certainly necessary.

Public Keys, Private Keys

In **digital certificate authentication,** each user is given a **public key,** which, as the name suggests, is not kept secret. This public key is paired with a **private key** that only the user should know.

Digital Certificate

The **digital certificate** is a tamper-proof file that gives the name of a subject (person or software process) and the subject's public key.

Operation

For authentication, the supplicant uses his or her *private key* to do a calculation, which only the person who was given the private key should know.

The verifier uses the *public key* contained in the digital certificate of the true party—the person the supplicant claims to be—to test the calculation performed by

Public and Private Keys
Each party has both a public key and a private key
A party makes its public key available to everybody
A party keeps its private key secret

Digital Certificate
Tamper-proof file that gives a party's public key

Operation
1. Supplicant performs a computation with his or her private key, which only he or she should know
2. Verifier tests the calculation with the public key in the digital certificate of the true party—the party the supplicant claims to be
3a. If the test is successful, the supplicant is authenticated as knowing the true party's secret private key
3b. If the test fails, the supplicant is rejected

Perspective
Digital certificate authentication is very strong
However, it is very expensive because companies must set up the infrastructure for distributing public–private key pairs
The firm must do the labor of creating, distributing, and installing private keys

Figure 9-13 Digital Certificate Authentication

the supplicant. If the test is successful, then the supplicant must know the private key of the true party. If the test fails, the supplicant must be an impostor and so is rejected.

Strong Authentication

Digital certificate authentication is extremely strong because private keys are extremely long and perfectly random. There is no known way to calculate private keys from the public key in a digital certificate in a reasonable amount of time.

Expensive to Implement

Unfortunately, digital certificate authentication, also known as **public key authentication,** is expensive to implement. Each server and client PC must have digital certificate authentication software and a private key installed on it. This requires a great deal of expensive labor time. Many companies are reluctant to spend this much money on authentication, despite the strength of digital certificate authentication.

Furthermore, if a company has many digital certificates, it needs a public key infrastructure to manage them. There also must be strong processes in place to ensure that only legitimate people and computers get digital certificates. Overall, managing **public key infrastructures** (the hardware, software, and human processes needed to implement digital certificates) is quite expensive.

TEST YOUR UNDERSTANDING

14. a) In digital certificate authentication, who should know a user's private key? b) Who should know a user's public key? c) What information does a digital certificate provide? d) Describe how digital certificates are used in authentication. e) In digital certificate authentication, what key does the supplicant use to perform a calculation? f) What key does the verifier use to test the calculation performed by the supplicant? g) Is digital certificate authentication strong? h) Why are companies reluctant to implement digital certificate authentication? i) What is a public key infrastructure?

Biometrics

A relatively new form of authentication is **biometrics,** which is the use of body measurements to identify a supplicant. The main promise of biometrics is to eliminate passwords.

The main promise of biometrics is to eliminate passwords.

Figure 9-14   Biometric Authentication (Study Figure)

Biometric Authentication
    Authentication based on body measurements
    Promises to eliminate passwords

Fingerprint Scanning
    Dominates biometrics use today
    Simple and inexpensive
    Substantial error rate (misidentification)
    Often can be fooled fairly easily by impostors

Iris Scanners
    Scan the iris (colored part of the eye)
    Irises are complex, so iris scanning gives strong authentication
    Expensive

Face Recognition
    Camera: allows analysis of facial structure
    Can be done surreptitiously—that is, without the knowledge or consent of the person being scanned
    Very high error rate and easy to fool

Error and Deception Rates
    Error and deception rates are higher than vendors claim
    The effectiveness of biometrics is uncertain

Fingerprint Scanning

The least expensive (and, unfortunately, the weakest) form of biometric authentication is **fingerprint scanning.** In addition to having substantial **error rates** (normal misidentification rates that occur even when the subject is cooperating), many fingerprint scanners can be deceived fairly easily by impostors. This makes fingerprint authentication very weak.

Despite its limitations, fingerprint scanning is by far the most widely used biometric authentication method. Many firms feel that the balance between fingerprint scanning vulnerabilities and the weaknesses of passwords still tips the scale toward fingerprint scanning for noncritical applications.

Iris Scanners

At the other extreme of the strength and cost range are **iris scanners**—cameras that read the very complex pattern of the supplicant's iris (colored part of the eye). Irises are extremely complex, so iris scanners have low error rates, making them suitable for sensitive applications. However, even iris scanners have small error rates and can be deceived, so their use in critical situations needs to be considered very carefully.

Face Recognition

Some airports and other public locations now have cameras with **face recognition** systems that scan passersby to identify terrorists or wanted criminals by the characteristics of their faces. This is controversial because it typically is done **surreptitiously**—that is, without the knowledge or explicit permission of those passersby being scanned.

In surreptitious identification, the person is identified without his or her knowledge.

Error rates have been so high that many of these systems are now being removed. In nearly every case in which a face recognition device identifies someone as a criminal or terrorist, the scanner is producing a false alarm. In addition, criminals and terrorists often can deceive face recognition systems.

Questions about Error Rates and Deception

As just noted, error rates refer to the percentage of mistakes made by a biometric system even when users are not practicing deception to fool the system. When an attacker intentionally uses deception, his or her probability of being falsely authenticated or not authenticated may be much higher than the error rate. How susceptible are biometric error rates and vulnerabilities to deception? This currently is an open question, but it is clear that error rates in practice are much higher than those stated by vendors and that many systems are vulnerable to deception. Biometric authentication must be used carefully and with awareness of its limitations.

TEST YOUR UNDERSTANDING

15. a) What is biometrics? b) What is its main promise? c) Give a pro and a con of fingerprint scanning. d) Give a pro and some cons of iris scanning. e) Give a pro and some cons of face recognition. f) What is surreptitious scanning? g) Distinguish between error rates and deception.

FIREWALLS, IDSs, AND IPSs

We will continue our discussion of the protection phase by looking at firewalls and two close relatives—intrusion detection systems (IDSs) and intrusion prevention systems (IPSs).

Firewalls

In hostile military environments, travelers must pass through one or more checkpoints. At each checkpoint, their credentials will be examined. If the guard finds the credentials insufficient, the guard will stop the arriving person from proceeding and note the violation in a checkpoint log.

Dropping and Logging Provable Attack Packets

Figure 9-19 shows that firewalls operate in similar ways. Whenever a packet arrives, the **firewall** examines the packet. If the firewall identifies a packet as a **provable attack packet,** the firewall discards it. On the other hand, if the packet is not a provable attack packet, the firewall allows it to pass.

The firewall also copies information about the discarded packet into a **firewall log file.** Firewall managers should read their firewall log files every day to understand the types of attacks coming into the resource that the firewall is protecting.

If a firewall identifies a packet as a provable attack packet, the firewall discards it.

Ingress and Egress Filtering

When most people think of firewalls, they think of filtering packets arriving at a network *from the outside.* This is called **ingress filtering.** Figure 9-15 illustrates ingress filtering.

Figure 9-15   Firewall Operation

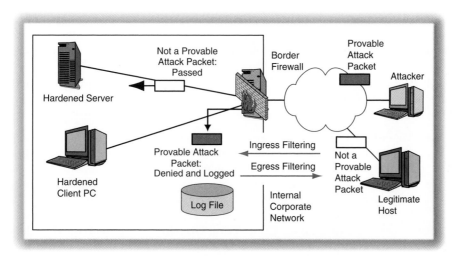

In addition, most firms also do **egress filtering**—that is, they filter packets going from the network *to the outside*. Doing egress filtering makes the corporation a good citizen, ensuring that its computers are not used in attacks against outside firms. Egress filtering also attempts to prevent sensitive corporate information from being sent outside the firm.

TEST YOUR UNDERSTANDING

16. a) What does a firewall do when a packet arrives? b) Why is it important to read firewall logs daily? c) Distinguish between ingress and egress filtering.

Stateful Firewall Filtering

We have used the term *firewall filtering* up until now without explaining it. We did this because firewalls use several different **filtering methods.** Most firewalls today, however, use stateful firewall filtering, which was invented in the early 1990s and has now become the dominant firewall filtering method.

States

During a data conversation between hosts, there often are a number of stages (states) through which the conversation progresses. Often, for example, there are distinct opening, ongoing communication, and closing states. The concept of states is important for security as well. During a connection opening, for instance, there has to be authentication to provide identity and the negotiation of security practices to be used during the connection. During the ongoing communication state, in turn, security requirements usually are simpler because the heavy security work was done in the first state. Stateful inspection firewalls exploit these differences in security requirement between states.

A connection may have several distinct states. For simplicity, however, we will consider a connection that has only two states—connection-opening attempts and ongoing communication.

Packets That Attempt to Open a Connection

Stateful firewalls have specific rules for evaluating packets that attempt to open a connection (such as packets containing TCP segments whose SYN bits are set). There are default rules that are automatically applied unless they are specifically overruled. In addition, it is possible to overrule these default rules by using access control lists.

Default Behavior   As Figure 9-17 shows, **stateful firewall filtering** is concerned with connections, not just individual packets. Its most complex work occurs when a host inside or outside attempts to open a connection.

> ➤ Figure 9-17: Default Stateful Firewall Behavior for a Connection-Opening Attempt
> By default, all connections that are initiated by the outside are prevented. This prevents outside clients from reaching internal servers.

> ➤ By default, all connections that are initiated by an inside host going to an outside host are permitted. This allows clients to connect freely to external servers.

Access Control Lists (ACLs) for Connection Openings   Although the default behavior of stateful firewalls provides both strong protection against outside attacks and easy

Stateful Firewall Filtering
    There are several types of firewall filtering
    Stateful inspection is the dominant filtering method today
    Stateful firewalls often use other filtering mechanisms as secondary mechanisms

States
    Connections often go through several states
    Connection opening, going communication, closing, etc.
    Different security actions are appropriate for different states

Connection Initiation State
    State when packets attempt to open a connection
        Example: packets with TCP segments whose SYN bits are set
    Default connection-initiation behavior (see Figure 9-17)
        By default, all connections that are initiated by the outside are prevented
            This prevents outside clients from reaching internal servers
        By default, all connections that are initiated by an inside host to an outside
        host are permitted
            This allows clients to connect freely to external servers
            The connection is considered to be open
    Access control lists (ACLs) (see Figure 9-18)
        ACLs modify the default behavior for ingress or egress
        Ingress ACL rules allow access to selected internal servers
        Egress ACL rules prevent access to certain external servers

Packets in the Ongoing Communication State
    If the packet does not attempt to open a connection,
        Then if the packet is part of an established connection
            It is passed without further inspection
            (However, these packets can be filtered if desired)
        If the packet is not part of an established connection, it must be an attack
            It is dropped and logged
    This simplicity makes the cost of processing most packets minimal
    Nearly all packets are in the ongoing communication state

Perspective
    Simple operation for most packets leads to inexpensive stateful firewall operation
    However, stateful inspection firewall operation is highly secure

Figure 9-16 Stateful Firewall Filtering (Study Figure)

access to the outside world, this default behavior is not always appropriate. In most firms, some outside clients need to get access to at least a few internal servers. In addition, firms may not want their internal client PCs to connect to all servers on the Internet.

Access control lists (ACLs) are sets of rules that modify the default behavior of stateful firewalls, allowing connections to some internal servers and preventing

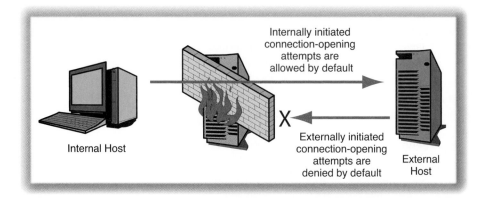

Figure 9-17 Default Stateful Firewall Behavior for a Connection-Opening Attempt

connections to some external servers. ACLs allow security administrators to tune how their stateful filtering mechanisms work.

Figure 9-18 shows a very simple ingress access control list with only three rules. The first two rules provide exceptions to the default behavior of dropping all external connection-opening attempts. These rules permit very specific externally initiated connection openings.

➤ Rule 1 looks at the protocol field in the IP header. The rule requires the protocol value to indicate that the packet data field holds a TCP segment. Furthermore, the destination port number in the TCP segment must be 25, indicating that the packet

Figure 9-18 Ingress Access Control List (ACL) for a Stateful Inspection Firewall

1. If protocol = TCP AND destination port number = 25, PASS and add connection to connection table.
    This rule permits external access to all internal mail servers.

2. If IP address = 10.47.122.79 AND protocol = TCP AND destination port number = 80, PASS and add connection to connection table.
    This rule permits access to a particular webserver (10.47.122.79).

3. Deny All AND LOG.
    If earlier rules do not result in a pass or deny decision, this last rule enforces the default rule of banning all externally initiated connection-opening attempts.

is going to an SMTP mail server—any mail server. If this rule holds, then the firewall will pass the packet and add the connection to the connection table. This connection will include the IP addresses of the source and destination hosts and also the port numbers on the two computers.

➤ Rule 2 is similar, although it applies to HTTP servers instead of SMTP servers. More importantly, it specifies a particular internal server—the server at IP address 10.47.122.79. This is safer than Rule 1, because Rule 1 opens the firewall to every internal mail server while Rule 2 only opens the firewall to connections to a single server.

➤ The last rule is Deny All and Log. This is simply the default rule for incoming packets that try to open a connection. Putting this rule last means that unless a packet is explicitly allowed by an earlier rule, it is dropped and logged.

Handling Packets during Ongoing Communication

If a packet does not attempt to open a connection or is not part of a connection-opening attempt, then either the packet must be part of the ongoing communication state or the packet is spurious. When a packet arrives that does not attempt to open a connection, then the stateful firewall does the following:

➤ If the packet is part of an established connection, it is passed without further inspection. (However, these packets can be further filtered if desired.)

➤ If the packet is not part of an established connection, then it must be spurious. It is dropped and logged.

These rules for ongoing communication are very simple to implement. Consequently, most packets are handled with very little processing power. This makes stateful firewalls very inexpensive.

Perspective

Although the simple operation of stateful inspection makes it inexpensive, stateful filtering provides a great deal of protection against attacks coming from the outside. This combination of low cost and strong security is responsible for the dominance of stateful inspection today.

TEST YOUR UNDERSTANDING

17. a) What is the default behavior for stateful firewalls regarding connection opening attempts? b) Why are ACLs needed for stateful firewalls? c) When a packet that is part of an ongoing connection arrives at a stateful inspection firewall, what does the firewall do? d) When a packet that is not part of an ongoing connection and that does not attempt to open a connection arrives at a stateful inspection firewall, what does the firewall do? e) Why are stateful firewalls attractive? f) What type of firewalls do most corporations use for their main border firewalls?

Intrusion Detection Systems (IDSs)

Figure 9-19 compares firewalls and two other protection devices—intrusion detection systems (IDSs) and intrusion prevention systems (IPSs). Firewalls, as noted earlier, drop *provable* attack packets. If a packet is merely highly *suspect*, the firewall permits it to pass, despite the fact that it may be a real attack packet.

	Firewalls	IDSs	IPSs
Inspect Packets?	Yes	Yes	Yes
Action Taken	Drop and log individual proven attack packets based on individual packet or connection inspections.	Log multipacket attacks based on deep (multilayer) packet inspections of streams of packet flows. Notify an administrator of severe attacks, but do not stop the attacks.	Applies IDS processing methods—deep packet inspection and packet stream inspection. But actually stops some attacks that have high confidence, but are not provably attacks.
Processing Power Required	Modest	Heavy	Heavy
Maturity	Fairly mature	Still immature with too many false positives (false alarms). Tuning can reduce false positives, but this takes a great deal of labor.	New. Used to stop only attacks that can be identified fairly accurately.

Figure 9-19   Firewalls, Intrusion Detection Systems (IDSs), and Intrusion Prevention Systems (IPSs)

Reporting Suspicious Packets

To continue the analogy, **intrusion detection systems (IDSs)** supplement firewalls by identifying *suspicious* packets that may indicate an attack. Although IDSs do not drop these suspicious packets, they do log them, and if the threat of attack looks serious, they notify a security administrator.

To give an analogy, police officers may arrest people only if they have probable cause—a reasonably high standard of proof. They cannot arrest anyone who is merely acting suspiciously. However, although police officers cannot arrest people for merely suspicious behavior, they can note it and investigate them.

Notification Speed and False Positives

Attack notification is important because, unless security administrators identify attacks quickly, the attacker will be able to do extensive damage. Unfortunately, like car alarms, IDSs tend to create too many false alarms. (These are called **false positives** in IDS terminology.) Many firms "tune" their IDS to reduce false positives. For instance, they have the IDS report only attacks with high potential severity. To give another example, if a firm does not have servers running the Solaris operating system, it can turn off alarms over attacks against that operating system. However, many firms disconnect their IDSs after deciding that the benefits are not worth the tuning effort and that the volume of false positives that occur even after tuning is still too large for the IDS to be useful.

TEST YOUR UNDERSTANDING

18. a) Distinguish between the types of packets that firewalls and IDSs seek. b) What type of filtering device is plagued by false alarms?

Intrusion Prevention System (IPS) Filtering

Based on IDS Filtering

Although stateful filtering provides strong security, it cannot stop some highly sophisticated attacks. A relatively new filtering method, **intrusion prevention system (IPS) filtering** is capable of stopping many complex attacks that can bypass even stateful firewall inspection. We discuss firewall IPS filtering after IDSs because most IPS filtering methods derive from IDS inspection methods. More specifically, IPS filtering uses two sophisticated filtering methods.

Deep Packet Inspection

First, IPSs use **deep packet inspection,** which examines internet, transport, and application layer content in an integrated way. This allows an IPS to detect certain types of attacks that other filtering methods cannot, because the clues lie at multiple levels in the packet.

Inspecting Streams of Packets

Second, an IPS examines patterns in *streams of packets,* not just individual packets. To give a simple example, if there are many pings to hosts with consecutive IP addresses, this is almost certainly a scanning attack. Looking at a single ping would not tell you that. To give another example, if an application message is fragmented and sent in multiple packets, there is no way to catch an application layer attack without examining the whole stream of packet fragments and then reconstituting them into the original application message. Again, this process can discover attacks that packet-by-packet inspection or asking whether the packet is part of a connection cannot.

Intensive Processing and ASICs

These two types of inspection are extremely processing-intensive. Earlier firewalls did not have the processing speed to implement them. Now, however, application-specific integrated circuits (ASICs) allow many computations that were previously done in software to be done in hardware. This greatly reduces processing time because hardware processing is much faster than software processing. Firewalls that use ASICs to do IPS filtering now have the processing speed they need.

What's in a Name?

The name *intrusion prevention system* is another unfortunate "gift" from marketers. *All* firewalls are systems to prevent intrusions, so all are IPSs in that sense. In fact, firewalls have long detected and automatically stopped denial-of-service attacks and several other complex attacks. However, the term *intrusion prevention system* today is used only for firewalls that do deep packet inspection and that inspect streams of packets to identify problems.

Choosing What Attacks to Stop

One problem with IPS filtering is that it is based on IDS inspection methods, which have a poor reputation for precision. Consequently, left unchecked, IPS filtering has the potential to stop many legitimate traffic flows. In effect, IPSs could generate a self-inflicted denial-of-service attack. To prevent the stopping of legitimate traffic, most firms allow their IPSs to stop only attacks that can be identified fairly accurately. This need to decide which attacks to stop means that firms must have a very good understanding of IPS filtering in order to use it effectively.

TEST YOUR UNDERSTANDING

19. a) What are the two characteristics of IPS filtering? b) Why must firms that use IPS filtering understand it very well?

Multimethod Firewalls

Although stateful packet inspection is the dominant firewall filtering methodology, most real firewalls supplement stateful inspection with other filtering methodologies. In addition, as noted earlier, firewalls usually stop several types of denial-of-service attacks and other complex attacks. This means that before firewalls filter specific traffic, they first decide what method to use to examine it.

On the other hand, firewalls rarely do antivirus filtering. They either ignore the problem or actively pass webpage downloads, e-mail messages with attachments, and other traffic that should be filtered to an antivirus filtering server.

TEST YOUR UNDERSTANDING

20. a) Do most firewalls provide only a single filtering method? b) Do most firewalls provide content filtering, antivirus filtering, or denial-of-service (DoS) protection?

THE PROTECTION OF DIALOGUES WITH CRYPTOGRAPHIC SYSTEMS

We now continue our discussion of the protection stage in the plan-protect-respond cycle by looking at cryptographic protections for dialogues involving the exchange of many messages.

Cryptographic Systems

In Chapter 7, we saw that many companies use virtual private networks (VPNs), which transmit data over the nonsecure Internet with added security. We looked at two VPN standards: SSL/TLS and IPsec.

These two standards and many others are called cryptographic systems. **Cryptographic systems** provide security to dialogues that involve the exchange of

Cryptographic Systems
> These systems provide security to multimessage dialogues

At the Beginning of Each Communication Session
> The two parties authenticate each other

Message-by-Message Protection
> After this initial authentication, cryptographic systems provide protection to every message
> Encrypt each message for confidentiality so that eavesdroppers cannot read it (see Figure 9-21)
> Adds an electronic signature to each message
>> The electronic signature authenticates the sender
>> It also provides message integrity: The receiver can tell whether a message has been changed in transit
>> Digital signatures use digital certificate authentication
>>> Very strong authentication, but also very expensive
>> HMACs (key-hashed message authentication codes) are less expensive
>>> They are not quite as secure as digital signatures, but are still quite secure
>>> The most widely used electronic signature method

Figure 9-20   Cryptographic Systems (Study Figure)

many messages. At the beginning of each communication session, the two communication partners authenticate each other.

After this initial authentication, cryptographic systems provide protection to every message. First, each message is encrypted for confidentiality. If an attacker intercepts the message in transit, he or she will not be able to read it.

Cryptographic systems also provide message-by-message authentication by adding an electronic signature to each message. These authentication messages also provide message integrity, which is assurance that the message has not been changed in transit. (If the message has been changed, the electronic signature authentication will fail.)

TEST YOUR UNDERSTANDING

21. a) What do cryptographic systems protect? b) What three protections do cryptographic systems provide for individual messages?

Symmetric Key Encryption for Confidentiality

Encryption for Confidentiality

When most people think of cryptographic protection, they think of encryption for confidentiality. **Encryption** is the scrambling of messages so that communication is **confidential** (cannot be read by eavesdroppers). Encryption methods are called

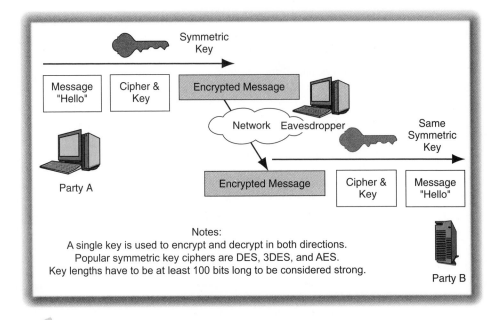

Figure 9-21   Symmetric Key Encryption for Confidentiality

ciphers. The receiver **decrypts** (unscrambles) the message in order to read it. Figure 9-21 illustrates encryption for confidentiality.

Symmetric Key Encryption
Most encryption for confidentiality uses **symmetric key encryption** ciphers, in which the two sides use the same key to encrypt messages to each other and to decrypt incoming messages. As Figure 9-21 shows, symmetric key encryption ciphers use only a single key in two-way exchanges. Popular symmetric key encryption ciphers include DES, 3DES, and AES.[3]

Key Length
Earlier, we looked at brute force password guessing. Symmetric and public/private keys also can be guessed by the attacker's trying all possible keys. This is called **exhaustive search.** The way to defeat exhaustive key searches is to use long keys, which are merely binary strings. For symmetric key ciphers, lengths of 100 bits or greater are

---

[3]Earlier, we saw how public key encryption can be used in authentication. Public key encryption can also be done for confidentiality, but this is rarely done. Nearly all encryption today uses symmetric key encryption because symmetric key encryption is very efficient. In contrast, public key encryption for confidentiality is very slow, placing heavy processing burdens on the sender and receiver. Public key encryption for confidentiality is used only for specific purposes. In contrast, all common applications use symmetric key encryption for message-by-message confidentiality. In public key encryption, RSA key lengths need to be at least 1024 bits to be considered strong. ECC key lengths can be smaller (around 512 bits) and still be strong.

considered to be strong keys. DES has only a 56-bit key length, but 3DES doubles or triples this key length. AES supports multiple strong key lengths up to 256 bits.

TEST YOUR UNDERSTANDING

22. a) What is a cipher? b) What protection does confidentiality provide? c) In two-way dialogues, how many keys are used in symmetric key encryption? d) What is the minimum size for symmetric keys to be considered strong?

Electronic Signatures

Authentication and Message Integrity

In addition to encrypting each packet for confidentiality, cryptographic systems normally add **electronic signatures** to each packet. These are small bit strings that provide message-by-message authentication, much as people use signatures to authenticate written letters. Electronic signatures also provide message integrity, meaning that the receiver will be able to detect whether the packet is changed in transit. Consequently, cryptographic systems provide three protections: message-by-message confidentiality, authentication, and message integrity.

Digital Signatures

There are two types of cryptographic signatures. **Digital signatures** use digital certificate authentication. Digital signatures provide very strong authentication, but they place a very heavy processing load on computers and are expensive to manage.

HMACs

Digital signatures are talked about a great deal in the popular press, but most cryptographic systems use **key-hashed message authentication codes** (HMACs) for authentication.[4] HMACs use a process called hashing, which is very fast. HMACs place a lower processing load on computers than digital signatures do, while providing authentication and message integrity almost as strong as digital signatures. However, for legal transactions between organizations, HMACs have some limitations that may be crucial.

TEST YOUR UNDERSTANDING

23. a) What two protections do electronic signatures provide? b) What are the two popular types of electronic signatures? c) What methods do they use? d) Which type of electronic signature is more widely used? e) What is the relative advantage of HMACs over digital signatures?

OTHER ASPECTS OF PROTECTION

We will end our discussion of the protection phase by looking at a few other aspects of protection. In a sense, these fall outside the realm of network security and into the realm of host security. However, such artificial technical distinctions cannot drive security thinking.

---

[4]No, it's not KHMAC, although it is occasionally rendered kHMAC. "Why not put in the *k* most of the time?" I asked one of the creators of the IETF HMAC standard. His response was, "Well, obviously, it uses a key, so why put that in the acronym?" Then, when asked why "key" is included in the name, he responded, "Because it uses keys."

Hardening Servers and Client PCs
Setting up computers to protect themselves

Server Hardening
Back up so that restoration is possible
Patch vulnerabilities
Use host firewalls

Client PC Hardening
As with servers, patching vulnerabilities, having a firewall, and implementing backup
Also, a good antivirus program that is updated regularly
Client PC users often make errors or sabotage hardening techniques
In corporations, group policy objects (GPOs) can be used to centrally manage security on clients

Vulnerability Testing
Protections are difficult to set up correctly
Vulnerability testing is attacking your system yourself or through a consultant
There must be follow-up to fix vulnerabilities that are discovered

Figure 9-22 Other Aspects of Protection (Study Figure)

Hardening Servers

Even if companies install several layers of firewalls to protect servers, some attack packets will inevitably get through. Consequently, servers have to be **hardened**—that is, set up to protect themselves.

Backup
The most basic computer protection of all is to keep data on servers backed up frequently. Recovery may be difficult or impossible unless the data on affected systems have been backed up very recently.

Vulnerabilities and Patching
As noted earlier in this chapter, security vulnerabilities are found frequently in both operating systems and application programs. When vulnerabilities are found, vendors develop patches (software updates). If companies install these updates, the companies will be immune to worms and hackers who can exploit these vulnerabilities.

Unfortunately, companies tend to be slow to install patches on servers and sometimes do not install them at all. In contrast, attackers are quick to develop exploits (programs that exploit known vulnerabilities). Consequently, hackers often have an easy time breaking into host computers, and worm writers often have massive success.

Host Firewalls

It is difficult to configure border firewalls and even internal firewalls, because these devices must protect many internal servers, each of which has different requirements. Consequently, it makes sense to install firewalls on individual servers. Most servers need only one or two ports open, so firewall protection can be configured simply and adequately.

Hardening Client PCs

The same things that are important in server hardening—installing patches, having a firewall, and implementing backup—are important in client PC hardening as well. In addition, clients need to have good antivirus programs and antispyware programs.

The problem with client security is that it usually is left to end users, who typically lack the knowledge and skills needed to make their computers secure. In addition, many end users actively sabotage security—for instance, by turning off their antivirus programs to speed the transmission of attachments. Users may even install prohibited software, such as music file-sharing software, which attackers can use to compromise the computers.

Larger organizations can "lock down" their desktops by administering them remotely. This is particularly true for Microsoft Windows client computers, which dominate the corporate client base. Corporations that install Microsoft domain controllers can create **group policy objects (GPOs);** these are sets of security policy rules. Domain controllers enforce GPOs on individual client PCs.

Group policy objects (GPOs) are sets of security policy rules that domain controllers enforce on individual client PCs.

TEST YOUR UNDERSTANDING

24. a) What is host hardening? b) What are the three steps in hardening servers? c) What are the steps in hardening client PCs? d) What are group policy objects (GPOs)?

Vulnerability Testing

One problem in creating protections is that it is easy to make mistakes when configuring firewalls and other protections. It is essential to do **vulnerability testing** after configuring protections such as firewall ACLs, in order to catch those mistakes. In vulnerability testing, the company or a consultant attacks protections in the way a determined attacker would and notes which of those attacks that should have been stopped actually succeed.

Vulnerability testing must result in a report that leads to a plan to upgrade protections to remove vulnerabilities. There also must be follow-up to ensure that the vulnerability removal plan actually is carried out.

TEST YOUR UNDERSTANDING

25. a) What is vulnerability testing? b) Why is it important? c) Describe the appropriate steps that the company should take after vulnerability testing is completed.

RESPONSE

The last stage in the plan-protect-respond cycle is *respond*. Inevitably, some attacks will succeed in getting through the company's protection systems. The amount of damage done in these compromises depends heavily on how quickly and how well the organization responds.

Stages

There are four general stages in responding to an attack.

Detecting the Attack

The first stage is detecting the attack. Detection can be done by the firm's IDS or simply by users reporting apparent problems. Obviously, until an attack is detected, the attacker will be able to continue doing damage. Companies need to develop strong procedures for identifying attacks quickly.

Figure 9-23 Incident Response (Study Figure)

Stages
> Detecting the attack
> Stopping the attack
> Repairing the damage
> Punishing the attacker

Major Attacks and CSIRTs
> Major incidents are those that the on-duty staff cannot handle
> Computer security incident response team (CSIRT)
>> Must include members of senior management, the firm's security staff, members of the IT staff, members of functional departments, and the firm's public relations and legal departments

Disasters and Disaster Recovery
> Natural and humanly made disasters
> IT disaster recovery for IT
>> Dedicated backup sites and transferring personnel
>> Having two sites that mutually back up each other
> Business continuity recovery
>> Getting the whole firm back in operation
>> IT is only one player

Rehearsals
> Rehearsals are necessary for speed and accuracy in response
> Time literally is money

Stopping the Attack

The second stage is stopping the attack. The longer an attack has to get into the system, the more damage the hacker can do. Reconfiguring corporate firewall ACLs may be able to end the attack. In other cases, attack-specific actions will have to be taken.

Repairing the Damage

The third stage is repairing the damage. In some cases, this is as simple as running a cleanup program or restoring files from backup tapes. In other cases, it may involve the reformatting of hard disk drives and the complete reinstallation of software and data.

Punishing the Attacker?

The fourth general stage is punishing the attacker, if possible. This is easiest if the attacker is an employee. Remote attackers can be extremely difficult to track down, and even if they are found, prosecution may be difficult or impossible.

If legal prosecution is a goal, it is critical for the company to use proper **forensic procedures** to capture and retain data in ways that fit the rules of evidence in court proceedings. These rules are very complex, and it is important for the firm to use certified forensics professionals. Even if an employee is fired, it is necessary for the company to use good forensic procedures to avoid a potential lawsuit.

Major Incidents and CSIRTs

Minor attacks can be handled by the on-duty IT and security staff. However, during **major incidents,** such as the theft of thousands of credit card numbers from a corporate host, the company must convene the firm's **computer security incident response team (CSIRT),** which is trained to handle major incidents.

The key to creating CSIRTs is to have the right mix of talents and viewpoints. Major attacks affect large parts of the firm, so the CSIRT must include members of senior management, the firm's security staff, members of the IT staff, members of functional departments, and the firm's public relations and legal departments.

Disasters and Business Continuity

When natural disasters, terrorist attacks, or other catastrophes occur, the company's basic operations may be halted. This can be extremely expensive. Companies must have active disaster recovery plans to get their systems working quickly.

IT disaster recovery is the reestablishment of information technology operations. Many large firms have dedicated backup sites that can be put into operation very quickly, after data and employees have been moved to the backup site. Another option, if a firm has multiple server sites, is to do real-time data backup across sites. If one site fails, the other site can take over immediately or at least very rapidly.

More broadly, **business continuity recovery** goes beyond IT disasters to deal with events that affect enough of a firm to pause or stop the functioning of the business. IT security is only one player in business continuity recovery teams.

Rehearsals

"Practice makes perfect" is time-honored advice. It certainly is true for major attacks that must be handled by CSIRTs, and it is equally true for disaster recovery. It is important for the company to establish CSIRT and disaster teams ahead of time and to have them rehearse how they will handle major attacks and disasters. Although practice does not really make perfect, it certainly improves response speed and quality. During the first two or three rehearsals, team members will work together awkwardly, and there will be many mistakes. Rehearsals will also reveal flaws in the company's major attack and disaster response plans. It is far better to go through these problems before the firm is in a real crisis.

TEST YOUR UNDERSTANDING

26. a) What are the four response phases when attacks occur? b) What is the purpose of forensic tools? c) Why are CSIRTs necessary? d) Should the CSIRT be limited to security staff personnel? e) What is disaster recovery? f) Explain how firms use backup sites in disaster recovery. g) Why are CSIRT and disaster response team rehearsals necessary?

CONCLUSION

Synopsis

Attacks

Companies today suffer compromises from many different types of attacks.

➤ Viruses attach themselves to other programs and need human actions to propagate—most commonly by users opening e-mail attachments that are infected programs. Worms are full programs; they can spread by e-mail, but vulnerability-enabled worms can propagate on their own, taking advantage of unpatched vulnerabilities in victim hosts. Some vulnerability-enabled worms can spread through the Internet host population with amazing speed. Many worms and viruses carry damaging payloads. Often, payloads place Trojan horse programs or other types of exploitation software on the victim computer. Malware is the general name for evil software.

➤ Viruses, worms, and Trojan horses are not the only attacks that are aimed at individuals. Spam deluges the victim with unsolicited commercial e-mail, and messages often are fraudulent. Spyware collects information about users and sends this information to an attacker. Adware pops up advertisements constantly. Phishing attacks use an official-looking e-mail message or website to trick users into divulging passwords and other special information. Attacks on individuals, including e-mail virus and worm attacks, often depend on social engineering—tricking the victim into doing something against his or her best interests. Two common goals of attacks on individuals are credit card number theft, in which a credit card number is stolen, and identity theft, in which enough private information is stolen to enable the attacker to impersonate the victim in large financial transactions.

➤ Hacking is the intentional use of a computer resource without authorization or in excess of authorization. Hacking break-ins typically require a prolonged series of actions on the part of the attacker.

➤ Denial-of-service (DoS) attacks overload victim servers so that they cannot serve users.

Attackers

Traditionally, most attackers were curiosity-driven hackers and disgruntled employees and ex-employees. Now, criminals dominate the attack world, and the money their crimes generate enables them to invest in new technology and hire top hackers. On the horizon, cyberterror attacks by terrorists and cyberwar attacks by national governments could do unprecedented levels of damage.

Security Management

Security is primarily a management issue, not a technical issue. Planning involves risk analysis (balancing the costs and benefits of protections), creating comprehensive security (closing all avenues of attack), and using defense in depth (establishing successive lines of defense in case one line of defense fails).

Access Control

Firms need to control access to their assets. The first step, obviously, is to enumerate (identify and list) assets. The second is to rate the sensitivity of each asset in terms of security risks. Then the firm must develop a specific asset control plan for each asset. This asset control plan must be appropriate to the sensitivity of each asset.

Access control plans require authentication (proving a supplicant's identity to a verifier), authorization (specifying what the supplicant can do to various resources), and auditing (recording actions when users work). This trio of actions is called AAA. The most complex aspect of access control is authentication. There are three main technologies for authentication.

➤ Passwords are inexpensive and easy to use, but users typically choose poor passwords that are easy to crack. Passwords should be used only for low-sensitivity resources.

➤ Digital certificate authentication, at the other extreme, gives very strong authentication, but it is complex and expensive to implement.

➤ Biometrics promises to use bodly measurements to authenticate supplicants, replacing other forms of authentication. Concerns with biometrics include error rates and the effectiveness of deliberate deception by supplicants.

Firewalls, IDSs, and IPSs

Firewalls examine packets passing through the firewall. If a firewall finds provable attack packets, it drops them and records information about them in a log file. Ingress filtering examines packets coming into the firm; egress filtering examines packets going out of the firm.

Most firewalls use stateful inspection, which controls which internally initiated and externally initiated connections will be allowed. By default, all connection-opening attempts from inside hosts to outside hosts are allowed, but all connection-opening attempts from outside hosts to inside hosts are prohibited. Access control lists (ACLs) can modify this default behavior.

Once a connection is established, subsequent packets in the connection usually are passed with little or no filtering. However, packets that are not connection-opening attempts and not in established connections are dropped.

A firewall drops provable attack packets, but it does *not* drop packets that are merely suspicious. However, intrusion detection systems (IDSs) are designed to detect suspicious traffic. If an IDS finds suspicious traffic that indicates a serious attack, the IDS will notify the security manager. Although prompt discovery of an attack is crucial, IDSs create many false positives (false alarms). False positives can be reduced, but this takes a great deal of work.

Some new firewalls use intrusion prevention system (IPS) filtering, which is based on IDS filtering methods. IPS filtering can detect complex attacks that stateful inspection and other traditional firewall inspection methods cannot identify. To avoid the false positives problem of IDSs, IPS filtering is used to stop attacks only if there is a high degree of confidence that an attack has been identified.

Cryptographic Systems

Cryptographic systems provide protections to multimessage dialogues. One key protection is encryption for confidentiality, which encrypts messages to prevent attackers from reading any messages that they intercept. Encryption methods are called ciphers. There are two types of ciphers for confidentiality.

➤ In symmetric key encryption, both sides encrypt and decrypt with a single key.
➤ In public key encryption, each side has a private key and a public key.

Symmetric key encryption is the dominant encryption technology for confidentiality because it is very efficient. Symmetric key encryption is used in nearly all applications for confidentiality. Public key encryption for confidentiality is used only for special purposes. Symmetric keys must be at least 100 bits long to be considered strong keys today. Public/private keys need to be much longer—1,024 and 512 bits long for RSA and ECC, respectively.

In addition to providing message-by-message encryption, cryptographic systems also provide message-by-message authentication by adding an electronic signature to each message. This allows the receiver to be sure that each message he or she receives is from the true party. Electronic signatures also provide message integrity.

There are two types of electronic signatures. Digital signatures use digital certificate authentication; digital signatures are extremely strong, but they require a large amount of processing power and time-consuming management. Most electronic signatures are HMACs, which require much less processing power and require little management labor to implement. However, digital signatures are needed for many business-to-business transactions.

Host Hardening

Ongoing protection includes hardening to protect servers and clients from attacks. Servers can be hardened by having vulnerabilities patched, having host firewalls installed, and being backed up regularly. Client PC hardening includes the same protections, plus the installation of an antivirus program and an anti-spyware program. Spam filters and anti-adware filters also are desirable.

Vulnerability Testing

Protections are very difficult to set up and configure, so it is easy to make mistakes that make a firm vulnerable to attack. Consequently, firms should conduct vulnerability tests in which an employee or a consultant attempts to attack the firm (with permission), in order to identify security weaknesses.

Response

Protections occasionally break down. Response to successful compromises must be rapid and effective in order to limit damage. The stages in response to attack typically include identifying the attack, stopping the attack, recovering from the attack, and (sometimes) punishing the attacker. Major incidents require the convening of a computer security incident response team (CSIRT). IT disaster recovery requires getting IT back in operation at another site, while business continuity recovery involves getting the entire firm back in operation. It is important for recovery teams to conduct rehearsals before problems occur.

End-of-Chapter Questions

THOUGHT QUESTIONS

1. a) Suppose that an attack would do $100,000 in damage and has a 15 percent annual probability of success. Spending $9,000 on Measure A would cut the annual probability of success by 75 percent. Do a risk analysis comparing benefits and costs. Show your work clearly. b) Should the company spend the money? c) Should the company spend the money if Measure A costs $20,000 per year? Again, show your work.

2. a) What form of authentication would you recommend for relatively unimportant resources? Justify your answer. b) What form of authentication would you recommend for your most sensitive resources? Justify your answer.

3. Critique each of the following passwords regarding the type of cracking attack that would be used to crack it and the strength of the password. a) swordfish b) Processing1 c) SeAtTLe d) 3R%t e) 4h*6tU9$^l

PERSPECTIVE QUESTIONS

1. What was the most surprising thing for you about the chapter?

2. What was the most difficult part of this chapter for you?

GETTING CURRENT

Go to the book website's New Information and Errors pages for this chapter to get new information since this book went to press and to note corrections to any errors in the text.

Network Management

Learning Objectives

By the end of this chapter, you should be able to discuss the following:

■ Planning: Planning the technological infrastructure, traffic management methods, and network simulation.

■ TCP/IP management: IP subnet planning, Network Address Translation (NAT), Multiprotocol Labor Switching (MPLS), the Domain Name System (DNS), DHCP servers, and the Simple Network Management Protocol (SNMP).

■ Directory servers, including Microsoft's Active Directory.

NETWORK MANAGEMENT

Today we can build huge and complex networks. In fact, our networks are increasing in size and complexity faster than our ability to manage them. Up to this point, we have focused most closely on network technology. However, technology is worthless unless the network is well planned and managed. In this chapter, we will look at some key issues and skills in network management.

The fact that this anchor chapter on network management comes late in the book should not be taken as an indication that management is unimportant. Rather, this chapter comes after discussions of technology because it is impossible to discuss network management in a comprehensive way until the student has a strong understanding of the technologies that network administrators must manage.

Network Management in Chapter 1

In fact, we have been looking at network management from the very first chapter. We first learned that networks must work well, and we looked at several quality-of-service (QoS) metrics for measuring the effectiveness of service delivery to our customer units in the firm and to our business partners outside the firm. We looked at the importance of speed, cost, availability, error rates, latency, and jitter. For speed, we considered the important distinction between rated speed and real throughput. For cost, we discussed the critical concept of total cost of ownership (TCO)—the total cost of a network component over its entire life cycle, not just its development life cycle, and definitely not just hardware and software purchases.

Chapter 1 ended with a section on network management. We learned that network planning must build a road map for the future, and we saw that planning should focus on lock-in decisions, which commit the firm to a particular technology for a considerable length of time. We also noted that companies have to consider their legacy technologies carefully, balancing the functionality gains of newer technologies with their costs.

We also learned about multicriteria decision making, which a company should use to evaluate alternative network technologies. We saw that cost and functionality are crucial, but that other considerations, such as manageability, are also important. Throughout the subsequent chapters, we constantly compared and contrasted different technologies and management options to show the types of factors that must be considered in selection.

We then looked at operational management—the management of systems *after* their development. This is the longest part of any network component's total life cycle. Operational management can be characterized by OAM&P—operations, administration, maintenance, and provisioning. Administration is difficult to discuss in general terms because it consists of high-level activities such as planning, mid-level activities such as data analysis, and low-level activities such as handling the mechanics of purchasing and paying bills. Provisioning—setting up service—is surprisingly expensive even if no new equipment is needed. We noted in Chapter 7, for example, that security options which require the network administrator to touch many client PCs instead of just servers or VPN gateways can be the factors that shift decisions from transport mode to tunnel mode in IPsec.

TEST YOUR UNDERSTANDING

1. a) What is operational management? b) For what does the acronym OAM&P stand? c) What is provisioning? d) What is the total cost of ownership?

Subsequent Chapters

In fact, the book introduced network management issues in almost every chapter up to now. We will list just a few examples:

➤ In Chapter 3, we looked at the relative roles of UTP and optical fiber in the firm and showed that they are not competitors, because they server different roles in most networks. We saw how to pick the right type of fiber for different uses, showing how cost dominates the decision. We also looked at how simple design and implementation actions, namely, limiting UTP runs and not untwisting UTP wires at the end more than a half inch, can marginalize several propagation problems.

➤ In Chapter 4, we saw that Ethernet is no longer the simple plug-in-and-forget technology that it used to be. The chapter discussed how to handle momentary traffic peaks, how to handle the susceptibility of Ethernet to single points of failure, and how to purchase Ethernet switches.

➤ Chapter 5 addressed wireless LAN management heavily. At its end, the chapter focused on such matters as how to lay out access points in a building, how to manage WLANs centrally, and, of course, how to manage WLAN security. Chapter 6 looked at the complex issue of dealing with carriers, including limited options and the lack of clear relationships between prices and costs. We also looked at VoIP and convergence, strategies for sending voice over packet-switched networks, and QoS concerns with availability and sound quality.

➤ Chapter 7 focused specifically on WAN carrier services. We saw that carriers offer many options at Layers 1, 2, and 3. We also learned that some of these options are being adopted rapidly, while others have fallen into legacy status and should no longer be considered. We also noted how rapidly the attractiveness and costs of major technologies can vary. We looked especially at IP carrier networks and metro Ethernet, which are becoming very important. We also focused closely on wireless metropolitan area networking, including 3G cellular telephony and 802.16 WiMAX.

➤ Chapter 8 did not look heavily at TCP/IP management because the topic of tools for managing TCP/IP networks was left to now.

➤ Chapter 9 looked at security from the standpoint of the *management* of security. It introduced the plan-protect-respond cycle and walked us through the major management decisions needed at different phases in the cycle.

PLANNING

Comedian Lilly Tomlin once said, "I always wanted to be somebody. I should have been more specific." This warning is certainly apt for network planning. Unless careful plans are made, networks will grow into very expensive and unmanageable messes of noninteroperable technology that cannot be understood, much less managed.

Planning the Technological Infrastructure

One of the most important things to plan in networking is the technological infrastructure—the firm's arrangement of hardware, software, and transmission lines that allows the network to carry information.

Figure 10-1 Planning the Technological Infrastructure (Study Figure)

What-Is Analysis
    Understand the current network in detail
    Requires a comprehensive inventory of network components and applications

Driving Forces for Change
    Normal growth in application demand
    Disruptive applications
    Organizational changes
    Changes in other aspects of IT (data center consolidation, etc.)

Gaps Analysis
    Identify gaps that exist and that will need to be closed
    Characterize and document each

Options for Closing the Gaps
    Multiple options must be considered
    Select roughly the least expensive option that will fully meet requirements
        Base cost decisions on the total cost of ownership (TCO)
    Select based on scalability
        The ability to grow cost effectively and sufficiently (See Figure 10-2)

What-Is Analysis

Planning for changes in the technological infrastructure must begin with the "what-is network situation"—that is, what network the company has today. This may sound like an easy task, but most firms do not have a thorough understanding of their network components and interactions, much less of their trouble spots. What-is analysis begins with an exhaustive inventory of the network's components and their interrelationships. This sounds simple. It is not.

Driving Forces for Change

Today's technological infrastructure will not be sufficient for the future because many things will change. Companies need to consider the major driving forces that will require changes in the network. Some of these driving forces are:

➤ The normal continuing growth of application traffic demand. In most firms, traffic has been growing at an increasing rate. This will certainly continue in the future.

➤ The introduction of disruptive applications that may create major surges in demand far beyond traditional patterns. Voice over IP is an obvious example. However, if video applications begin to grow—both legal and illegal applications—capacity planning will become extremely difficult.

➤ Organizational change can be a major driving force. If a company is adding a site, not only will the site have to be served, but communication between different parts of the firm will change, depending on what units are moved there. In fact, all corporate reorganizations are likely to impact network planning. At the extreme are nightmare scenarios that exist if the company is bought out or buys out another firm.

➤ Changes in other elements of the IT technological infrastructure can also require extensive network changes. One long-term trend has been the consolidation of data centers from many to few. This can radically change traffic flows within the corporate network.

Gaps Analysis

Comparing driving forces with the what-is network will inevitably reveal gaps between what the firm will need and what the current network can provide. These gaps must be identified, characterized, and documented.

Options for Closing the Gaps

The firm then must develop strategies for closing the gaps. It must consider multiple technologies and multiple topologies (physical connections) for each gap. As we will see later in this chapter, development of a strategy for closing a gap may benefit from network simulation programs.

One key consideration in developing strategies is that options that do not meet requirements are useless, no matter how cheap they are. A second key is that companies must select the lowest-cost technology (or something close to it) that can meet requirements. Spending too much in one area will result in the inability to support other areas. However, comparisons must be based on the Total Cost of Ownership (TCO)—not just initial hardware costs, first-year costs, or total development costs.

One final consideration is that some choices do not scale, meaning that they are not useful beyond a certain traffic volume. As Figure 10-2 shows, a technology may be cost effective when its use is small, but may grow too expensive at higher traffic volumes or may even be unable to support needed capacity. **Scalable** solutions retain their cost advantage and can handle a corporation's future traffic volume.

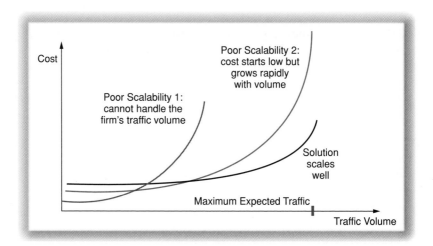

Figure 10-2   Scalability

Scalable solutions retain their cost advantage and can handle a corporation's future traffic volume.

TEST YOUR UNDERSTANDING

2. a) Define and describe what-is analysis. b) List the four driving forces for change. c) For each, give an example not listed in the text. d) What is gaps analysis? e) What is scalability? f) In what two ways can network technologies fail to be scalable?

Traffic Management Methods

One fact of networking life that can never be ignored in planning is that traffic volume varies widely. Peak periods of traffic can overwhelm the network's switches and transmission lines. Given the statistical nature and high variability of traffic, **momentary traffic peaks** lasting a fraction of a second to a few seconds are bound to occur, and a firm must have a plan for managing momentary traffic peaks. (See Figure 10-3.) On the other hand, a chronic or frequent lack of capacity requires other approaches—most commonly, increasing network capacity. Several traditional traffic management methods that deal with traffic and capacity are available to network managers, as Figure 10-3 shows. Network planners must select which approach to use in the corporate network or in different parts of the corporate network.

Momentary Traffic Peaks

In Chapter 4, we saw that congestion problems caused by momentary traffic peaks can be handled in Ethernet in two basic ways: overprovisioning and priority. ATM added a third way to deal with momentary traffic peaks: quality-of-service (QoS) guarantees.

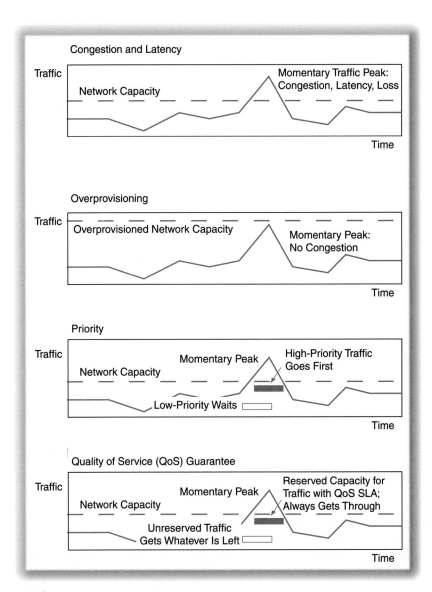

Figure 10-3   Traditional Traffic Management Methods

Overprovisioning Ethernet LANs

On Ethernet LANs, overprovisioning means adding much more switching and transmission line capacity than will be needed most of the time. With overprovisioning, momentary traffic peaks will rarely exceed capacity. This means that no regular ongoing management is required. The downside of overprovisioning is that it wastes capacity.

Today, the desire to minimize labor costs on LANs makes overprovisioning very attractive. On WANs, however, overprovisioning is too expensive to consider.

Priority

Priority, in turn, assigns high priority to latency-intolerant applications, such as voice, while giving low priority to latency-tolerant applications, such as e-mail. Whenever congestion occurs, high-priority traffic is sent through without delay. Low-priority traffic must wait until the momentary congestion clears. Priority allows the company to work with lower capacity than overprovisioning allows, but priority requires more active management labor.

QoS Guarantees

ATM goes a step beyond priority, reserving capacity on each switch and transmission line for certain types of traffic. This enables the firm to create quality-of-service (QoS) service level agreement guarantees for minimum throughput, maximum latency, and even maximum jitter.

QoS requires extremely active management. In addition, traffic with no QoS guarantees only gets whatever capacity is left over after reservations. This may be too little, even for latency-tolerant traffic.

Traffic Shaping

Even with priority and overprovisioning, sufficient capacity must be provided for the total of all applications apart from momentary traffic peaks. Even more active management is needed to control the amount of traffic entering the network in the first place. Restricting traffic entering the network at access points is called **traffic shaping.**

Filtering   Traffic shaping has two options. The first is **filtering** out unwanted traffic at access switches. Some traffic generally has no business on the corporate network, such as the downloading of MP3 files, video files, and software.

Capacity Percentages   The second tool of traffic shaping is to assign specific **percentages of capacity** to certain applications arriving at access switches. Even if file sharing has legitimate uses within a firm, for instance, the firm may wish to restrict the amount of capacity that file sharing can use. Typically, each application or application category is given a maximum percentage of the network's capacity. If that application attempts to use more than its share of capacity, incoming frames containing the application messages will be rejected.

Perspective on Traffic Shaping   Overprovisioning, priority, and QoS guarantees merely attempt to deal with incoming traffic. Traffic shaping actually *reduces* the amount of incoming traffic. Only traffic shaping can dramatically reduce network cost.

Although traffic shaping is very economical in terms of transmission capacity, it is highly labor intensive. It is used today primarily on high-cost WAN links. However, as management software costs fall in price and require less labor to operate, traffic shaping should see increasing use.

Another issue that arises when traffic shaping is used is politics. Telling a department that its traffic will be filtered out or limited in volume is not a good way to make

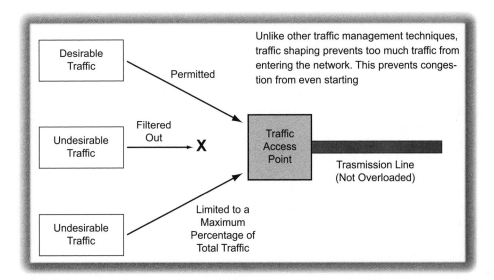

Figure 10-4   Traffic Shaping

friends. Priority and QoS reservations also raise political problems, but in traffic shaping, these problems are particularly bad.

Compression   A final approach is to compress traffic before it enters a network. In the example shown in Figure 10-5, there are two incoming data streams. The first is 800 Mbps. The second is 600 Mbps. This is a total of 1,400 Mbps, or 1.4 Gbps. The capacity of the transmission line is only 1 Gbps. This is not enough.

Figure 10-5   Compression

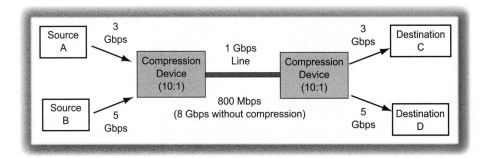

Before the incoming data enters the transmission line, however, the data will be compressed by 10:1, which is often possible for typical data streams. This reduces the compressed data streams to 80 Mbps and 60 Mbps, for a total of only 140 Mbps. This traffic can easily fit on a 1 Gbps line, with ample room for other traffic.

At the other end of the network, another device decompresses the data streams. It sends the 800 Mbps data stream to one destination and the 600 Mbps data stream to another destination.

One requirement for compression is that you must have comparable equipment at the two ends of the network. Given a frequent lack of vendor compatibility, compression tends to lock the company into a single vendor's solution.

A second problem with compression is that encrypted data cannot be compressed. Encryption makes the transmitted message look like a completely random stream of 1s and 0s. Compression is effective only if there is **redundancy** (repeated patterns) in the data. Well-encrypted traffic should completely lack redundancy.

TEST YOUR UNDERSTANDING

3. a) Distinguish between momentary traffic peaks and chronic lack of capacity. b) Distinguish between overprovisioning and priority. c) Distinguish between priority and QoS guarantees. d) What problem can QoS create? e) How is traffic shaping different from traditional approaches to handling traffic overloads? f) In what two ways can traffic shaping reduce traffic? g) Why can compression help in traffic management? h) What is redundancy? i) Why can compression lead to vendor lock-ins? j) What problem does encryption for confidentiality present for compression?

Network Simulation

Designing a new network or a modified network is very difficult, because the designer faces many alternatives and because network components tend to interact in unforeseen ways. Network simulation allows network designers to get a handle on this complexity. In addition, it is far more economical to simulate many alternatives than to build several real systems to study.

Network Simulation Purposes

There are several specific reasons to do network simulation:

➤ Comparing alternatives in order to identify the best one.

➤ Sensitivity analysis involves choosing a most likely situation and seeing how changing ranges of configuration values will affect various performance measures. This identifies variables on which the simulation is most sensitive.

➤ Anticipating problems (for instance, anticipating where bottlenecks will appear.)

➤ Planning for growth requires finding areas where the network will not continue to meet needs as traffic grows by extrapolating traffic. Then, alternatives must be simulated and studied.

Before the Simulation: Collecting Data

The best simulation analysis is worthless if the model does not include realistic data. Analysts use the acronym **GIGO**—"garbage in, garbage out"—to emphasize that if input data is bad, the results cannot be accurate. Networking simulation data is never

Simulation
    What-is versus what-if
    More economical to simulate network alternatives than to build them
    Purposes
        Comparing alternatives to select the best one
        Base case and sensitivity analysis to see what would happen if the values of
        variables were varied over a range
        Anticipating problems, such as bottlenecks
        Planning for growth, to anticipate areas where more capacity is needed

Before the Simulation, Collect Data
    Data must be good
    Otherwise, GIGO (garbage in, garbage out)
    Collect data on the current network
    Forecast growth

The Process
    Based on OPNET IT Guru
    Add nodes to the simulation work area (see Figure 10-7)
        Clients, servers, switches, routers, etc.
    Specify the topology with transmission lines
    Configure the nodes and transmission lines (see Figure 10-8)
    Add applications, which generate traffic data (see Figure 10-9)
    Run the simulation for some simulated period of time
    Examine the output to determine implications
    Validate the simulation (compare with reality, if possible, to see if the simulation
    is correct)
    Conduct what-if analyses (see Figure 10-10)
    Application performance analysis (OPNET ACE)

Figure 10-6   Network Simulation (Study Figure)

perfect, but simulations are best if the firm collects actual data on its traffic— including how traffic varies by time of day and how much it fluctuates from second to second. Of course, simulations usually deal with the future, so traffic needs to be extrapolated to what it is likely to be in the future.

The Process

Once data is collected, the modeler can begin to create the simulation. We will discuss simulation using the popular **OPNET IT Guru** network simulation program.

Adding Nodes

The first step in building a simulation is to place **nodes** (items to be connected by transmission lines) on the simulation work area. These nodes can be clients, servers,

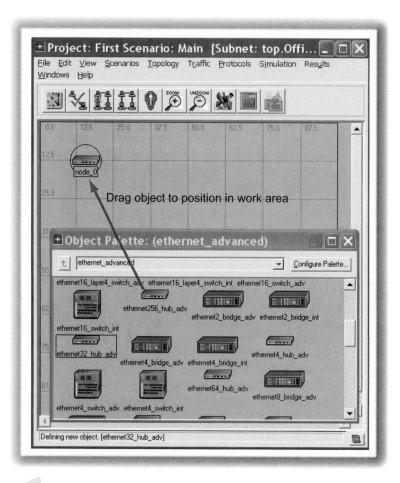

Figure 10-7   OPNET IT Guru Node Template

switches, routers, and other types of devices. As Figure 10-7 shows, IT Guru has templates with icons of common nodes. You can drag and drop these icons onto the simulation work area.

Specifying the Topology

The next step is to specify the **topology**—how the nodes are linked together by transmission lines. IT Guru has a template with icons for transmission lines and networks. You select a transmission icon and then decide what nodes it will link. In this case, a Frame Relay network will be used to connect the host computers. You now have something that looks very much like the network you are modeling.

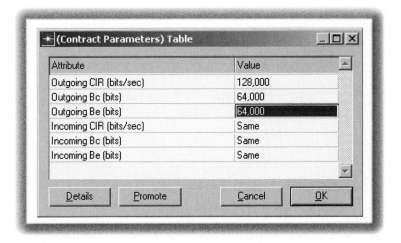

Figure 10-8   Configuring Elements in IT Guru

Configuring Elements

However, the nodes and transmission links need to be configured before your simulation of them is complete. For instance, on a router, you will have to specify the speeds of various interfaces. You also will have to configure specific operating parameters, such as the retransmission time for sending unacknowledged messages. Figure 10-8 shows a Frame Relay transmission link being configured so that its outgoing excess burst size (Be) is set to 64 kbps.

Adding Applications

Your nodes and lines are now ready to work, but you need to specify traffic. IT Guru has you do this by specifying applications, the traffic characteristics of these applications, and on which nodes they will run. This is realistic because traffic is created by applications, not by the network.[1] Figure 10-9 shows the final configured simulation model after applications have been added.

Running the Simulation

Your simulation is now specified. The next step is to run it. You choose a period over which the simulation is to run—for instance, over an entire simulated day or during a simulated busy period. After you do this, IT Guru will run the simulation using sophisticated statistical and queuing theory methods.

---

[1]Somewhat oddly, each application appears as an icon on the simulation working area. It is important not to confuse these software objects with the physical hardware and transmission line objects on the working area.

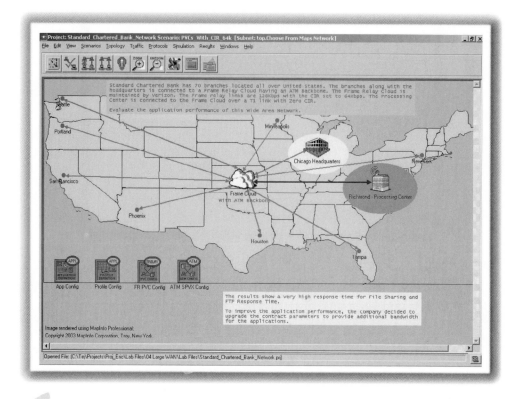

Figure 10-9   Adding Applications

Examining the Output

IT Guru gives you many alternatives for looking at your output. Graphical output is best for searching for trends or anomalies. You often can learn a great deal by looking at the simulation results in detail.

Validating the Simulation

If you are simulating a real network, you can **validate** the model by comparing its performance with that of the real network. If the model gives very different output, you need to revise the model and attempt to validate it again. Of course, for proposed networks, validation generally is impossible.

What-If Analysis

Now comes the real power of simulation—running what-if analyses to see how specific changes would affect the results. You might change a single parameter on a single computer (such as a windows size field in TCP). You might change the speed of a transmission line or consider an upgrade for a router. You might see whether adding a router in a particular place would get rid of a bottleneck. The possibilities are endless. By trying many alternatives, you can develop a network solution that is optimized for the company's business needs.

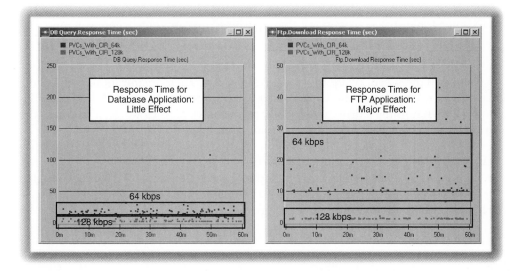

Figure 10-10   What-If Analysis in Network Simulation

Figure 10-10 shows a what-if analysis in which two PVC speeds are tried in order to determine the impact on response times for two applications—FTP and database queries. It shows that PVC committed information rate has a major impact on FTP response time, but not on database response time.

Application Analysis
OPNET offers a product related to IT Guru. This is the **Application Characterization Environment (ACE).** While IT Guru focuses primarily on network-level and internet-level performance, ACE allows the modeler to focus on application performance.

TEST YOUR UNDERSTANDING

4. a) Why is network simulation attractive economically? b) For what purposes is network simulation done? c) What is sensitivity analysis? d) Explain GIGO. e) How can a firm get good data for a simulation?

5. a) What are nodes? b) What is a topology? c) List the steps in building a network simulation. d) What steps should be undertaken after running the simulation? e) What is validation, and why is it important? f) What is what-if analysis? g) Distinguish between IT Guru and ACE.

TCP/IP MANAGEMENT

If a firm uses TCP/IP as its internetworking protocol, it must do a considerable amount of work to build and maintain the necessary infrastructure of TCP/IP. While switched networks are (generally) capable of operating for long periods without intervention by network managers, routed TCP/IP internets require constant tuning and support. This results in a need for considerable TCP/IP expertise and management effort. As we saw in Chapter 1, network managers say, "Switch where you can; route where you must."

IP Subnet Planning

As Chapter 8 discussed, IP addresses are 32 bits long. Each organization is assigned a network part. We saw that the University of Hawai'i's network part (128.171) is 16 bits long. There is nothing a firm can do to alter its network part. However, it was up to the university to decide what to do with the remaining 16 bits.

Subnetting at the University of Hawai'i

The university, like most organizations, chose to subnet its IP address space. It divided the 16 bits over which it has discretion into an 8-bit subnet part and an 8-bit host part. (This is the first column in the example—8/8.)

The 2^N-2 Rule

With N bits, you can represent 2^N possibilities. Therefore, with 8 bits, one can represent 2^8 (256) possibilities. This would suggest that the university can have 256 subnets, each with 256 hosts. However, a network, subnet, or host part cannot be all 0s or all 1s.[2] Therefore, the university can have only 254 (256 − 2) subnets, each with only 254 hosts. Figure 10-11 illustrates these calculations.

Figure 10-11 IP Subnetting

Step	Description				
1	Total size of IP address (bits)	32			
2	Size of network part assigned to firm (bits)	16		8	
3	Remaining bits for firm to assign	16		24	
4	Selected subnet/host part sizes (bits)	8/8	6/10	12/12	8/16
5	Possible number of subnets ($2^N - 2$)	254 ($2^8 - 2$)	62 ($2^6 - 2$)	4094 ($2^{12} - 2$)	254 ($2^8 - 2$)
6	Possible number of hosts per subnet ($2^N - 2$)	254 ($2^8 - 2$)	1022 ($2^{10} - 2$)	4094 ($2^{12} - 2$)	65,534 ($2^{16} - 2$)

---

[2] If you have all ones in an address part, this indicates that broadcasting should be used. All-zero parts are used by computers when they do not know their addresses. As we will see later in this chapter, most client PCs get their IP addresses from DHCP servers. All-zero addresses can be used only in DHCP messages sent from a host to a DHCP server.

In general, if a part is N bits long, it can represent $2^N - 2$ networks, subnets, or hosts. For example, if a subnet part is 9 bits long, there can be $2^9 - 2$, or 510, subnets. Or if a host part is 5 bits long, there can be $2^5 - 2$, or 30, hosts.

In general, if a part is N bits long, it can represent $2^N - 2$ networks, subnets, or hosts.

Balancing Subnet and Host Part Sizes

The larger the subnet part, the more subnets there will be. However, the larger the subnet part is made, the smaller the host part must be. This will mean fewer hosts per subnet. There is always a trade-off. More subnets mean fewer hosts, and more hosts means fewer subnets.

The University of Hawai'i's choice of 8-bit network and subnet parts was acceptable for many years because, then, no college needed more than 254 hosts. The advantage of this network/subnet choice is that its subnet mask (255.255.255.0) was very simple, breaking at 8-bit boundaries. This made it easy to see which hosts were on which subnets. The host at 128.171.17.5, for instance, was the 5th host on the 17th subnet. If the subnet mask did not break at an 8-bit boundary, which subnet a host is on cannot be determined just by looking at the address in dotted decimal notation.

However, many colleges in the university now have more than 254 computers, and the limit of 254 hosts required by the original subnetting decision has become a serious problem. Several colleges now have two subnets connected by routers, which is expensive and awkward.

The university would have been better served had it selected a smaller subnet part, say, 6 bits. As Figure 10-11 shows, this would have allowed 62 college subnets, which probably would have been sufficient. A 6-bit subnet part would give a 10-bit host part, allowing 1,022 hosts per subnet. This would be ample for several years to come.

A Critical Choice

In general, it is critical for corporations to plan their IP subnetting carefully, in order to get the right balance between the sizes of their network and subnet parts.

TEST YOUR UNDERSTANDING

6. a) Why is IP subnet planning important? b) If you have a subnet part of 9 bits, how many subnets can you have? c) Your firm has the 8-bit network part 60. If you need at least 250 subnets, what must your subnet size be? d) How many hosts can you have per subnet? e) Your firm has a 20-bit network part. What subnet part would you select to give at least 10 subnets? f) How many hosts can you have per subnet?

Network Address Translation (NAT)

One security issue that firms face is whether to risk allowing people outside the corporation to learn their internal IP addresses for individual hosts. Attackers who know internal IP addresses can send attack packets to individual hosts. To prevent this, companies can use **network address translation (NAT),** which presents external IP addresses that are different from internal IP addresses used within the firm.

NAT
> Sends false external IP addresses that are different from internal IP addresses

NAT Operation (Figure 10-13)

NAT is Transparent to Internal and External Hosts

Security Reason for Using NAT
> External attackers can put sniffers outside the corporation
> Sniffers can learn IP addresses
> Attackers can send attacks to these addresses
> With NAT, attackers learn only false, external IP addresses

Expanding the Number of Available IP Addresses
> Companies may receive a limited number of IP addresses from their ISPs
> There are roughly 4,000 possible ephemeral port numbers for each IP address
> So for each IP address, there can be 4,000 external connections
> If a firm is given 248 IP addresses, there can be roughly one million external connections
> Even if each internal device averages several simultaneously external connections, there should not be a problem providing as many external IP connections as a firm desires

Private IP addresses
> Can be used only inside firms
> 10.x.x.x
> 192.168.x.x (most popular)
> 172.16.x.x through 172.31.x.s

Protocol Problems with NAT
> IPsec, VoIP, etc.
> Work-arounds must be considered very carefully in product selection

Figure 10-12   Network Address Translation (NAT) (Study Figure)

NAT Operation

Figure 10-13 shows how NAT works. An internal client host, 192.168.5.7, sends a packet to an external server host. The source address in this packet is 192.168.5.7, of course. The source port number is 3333, which is an ephemeral port number that the source host made up for this connection.

When the NAT firewall at the border receives the packet, it makes up a new row in its translation table. It places the internal IP address and port number in the table. It then generates a new external source IP address and external source port number. These are 60.5.9.8 and 4444, respectively.

When packets arrive from the external host, they have 60.5.9.8 in their destination IP address fields and 4444 in their destination port number fields. The NAT looks

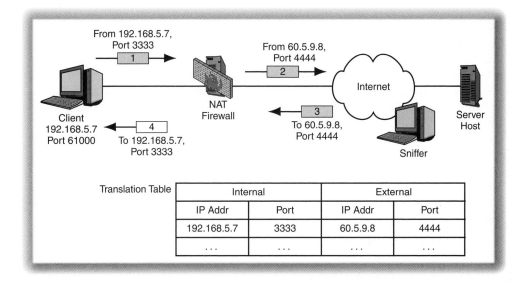

Figure 10-13   Network Address Translation (NAT)

up these values in its translation table, replaces the external values with the internal values, and sends them on to the client PC.

Transparency

NAT is transparent to both internal and external hosts. Hosts do not even know that NAT is happening. Consequently, there is no need to change the ways in which they operate. This makes NAT easy to initiate.

Security

Figure 10.13 shows how NAT brings security. An attacker may be able to install a **sniffer program** beyond the corporation's NAT firewall. The sniffer will be able to read all packets coming out of the firm, but with NAT, an eavesdropper learns only false (external) IP addresses and port numbers. In theory, if an attacker can attack immediately, it can send packets to the external IP addresses and port numbers, and the NAT firewall will pass the packets on to the internal host. However, this immediate reaction is rarely possible. NAT provides a surprising amount of security despite its simple operation.

Expanding the Effective Number of Available IP Addresses

An equally important potential reason for using NAT is to permit a firm to have many more internal IP addresses than it is given by its ISP. Suppose that an ISP gives a firm only 254 IP addresses by giving it a host part with 24 bits. Without NAT, a firm can have only 254 PCs simultaneously using the Internet.

However, NAT is really network address translation/port number translation (NAT/PNT). Even with Microsoft Windows, there are almost 4000 ephemeral port numbers. Even if internal hosts maintained four simultaneous connections to the outside world, each IP address could be used, on average, by 1,000 client PCs to make outside connections if each client PC had a different set of port numbers for its connections. If a firm has 254 IP addresses, it can multiply this already large number of connections by 254. This could support almost a quarter of a million PCs using the Internet simultaneously. While no firm would push the number of connections this far, having 10 to 100 internal hosts for each external IP address is very common.

Using Private IP Addresses

To support NAT, the Internet Assigned Numbers Authority (IANA) has created three sets of **private IP address ranges** that can only be used *within* firms. These are the three ranges:

> ➤ 10.x.x.x
> ➤ 192.168.x.x
> ➤ 172.16.x.x through 172.31.x.x

The 192.168.x.x private IP address range is the most popular because it allows firms to use internal subnet masks of 255.255.255.0. This allows companies to use 255.255.0.0 and 255.255.255.0 network and subnet masks, respectively. These break at convenient 8-bit boundaries. However, the other two private IP address ranges are also widely used.

Protocol Problems with NAT

In terms of security and expanding IP effective address ranges, NAT is a simple and effective tool. However, some protocols cannot work across a NAT firewall or can work only with considerable difficulty. These include the popular IPsec cryptographic system in transport mode and several VoIP protocols. The decision to use NAT must be made only after a careful assessment of protocols.

TEST YOUR UNDERSTANDING

> 7. a) What is NAT? (Do not just spell it out.) b) Describe NAT operation. c) What are the three attractions of NAT? d) How does NAT enhance security? e) How does NAT allow a firm to deal with a shortage of IP addresses given to it by its ISP? f) How are private IP address ranges used? g) What are the three ranges of private IP addresses? h) What problems may firms encounter when using NAT?

Multiprotocol Label Switching (MPLS)

Routing Decisions with IP

As we saw in Chapter 8, when routers make routing decisions for arriving packets, they are supposed to make a full routing decision for every packet, even if the packets are going to the same destination. Although route caching can reduce this work considerably, it is limited and somewhat risky. The need to make full routing decisions is made necessary by the mesh topology of the Internet and its frequent changes in routers and links.

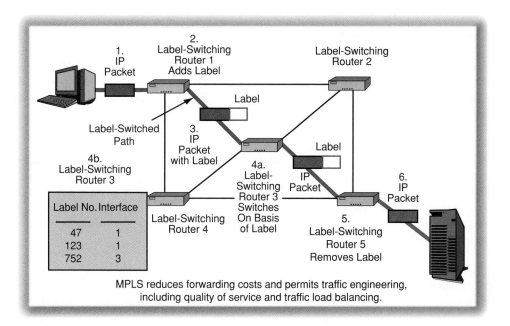

Figure 10-14   Multiprotocol Label Switching (MPLS)

Virtual Circuits and MPLS

Public switched data networks also have a mesh topology. Yet they do not make complex switching decisions for each arriving frame. Instead, they set up virtual circuit paths ahead of time and use these to make easy and fast individual frame switching decisions.

For IP networks, **multiprotocol label switching (MPLS)** does much the same thing. As Figure 10-14 shows, MPLS places a **label header** before the IP header (and after the frame header). This header label's **label number** is like a virtual circuit number.

Label Switching for Cost Reduction

When a labeled packet arrives, a **label-switching router** does not go through the traditional routing calculation processes. Instead, it merely reads the label number from the label header. It then looks into the **label-switching table** to find the interface associated with the label number. (For instance, in the table shown in Figure 10-14, a label with label number 47 indicates that the packet should be sent out through Interface 1.) The router then sends the packet out through the indicated interface. This is much faster than traditional routing calculations, because there is only a single row for each label number. The simplicity of this method dramatically lowers the cost of routing.

All packets between two host IP addresses might be assigned the same label number. This would be similar to creating a virtual circuit at the internet layer. More likely, all traffic between two *sites* might be assigned a single label number because it would all be going to and from the same site.

Quality of Service

Lowering router costs is the main attraction of MPLS, but there are other advantages as well. One is quality of service. Traffic flowing between two sites might be assigned one of two different label numbers. One label number might be for VoIP, which is latency-intolerant, and the other might be for latency-tolerant traffic between the sites. For latency-intolerant traffic, it is even possible to reserve capacity at routers along the selected path. This gives full quality of service.

Traffic Engineering

MPLS can also be used for **traffic engineering** (that is, to determine how traffic will travel through the network). One capability is **load balancing**—to move some traffic from a heavily congested link between two routers to an alternative route that uses different and less-congested links. MPLS does this by setting up multiple label-switched routes ahead of time and by sending traffic based on the congestion along different label-switched routes.

MPLS Boundaries

Corporations can use MPLS internally, and most ISPs now use MPLS for their traffic. However, it is currently impossible to implement MPLS across the entire Internet because of coordination difficulties among the ISPs.

TEST YOUR UNDERSTANDING

8. a) How is MPLS similar to the use of virtual circuits? b) On what does each label-switched router base routing decisions? c) What is MPLS's main attraction? d) What are its other attractions? e) How can MPLS provide quality of service? f) What is traffic engineering? g) Can MPLS provide traffic load balancing?

Domain Name System (DNS)

As we saw in Chapter 1, if a user types in a target host's host name, the user's PC will contact its local Domain Name System (DNS) server. The DNS server will return the IP address for the target host or will contact other DNS servers to get this information. The user's PC can then send IP packets to the target host. In this chapter, we will look at DNS and its management in more detail.

More Detailed Operation

Figure 10-15 illustrates how a DNS provides IP addresses when a host sends a request message containing the IP address. In many cases, however, as we saw in Chapter 1, the local DNS server will know the IP address and send it back. In other cases, the local DNS host will not know the IP address. It must then find the **authoritative DNS server** for the domain in the host name. In the figure, dakine.pukanui.com's authoritative DNS server is authoritative for the pukanui.com domain. That DNS server will send the IP address to the local DNS server, which will pass on the address to the host that sent the DNS request.

What Is a Domain?

Figure 10-16 shows that the **Domain Name System (DNS)** and its servers are not limited to providing IP addresses for host names. More broadly, DNS is a general system for naming domains. A **domain** is any group of resources (routers, single networks,

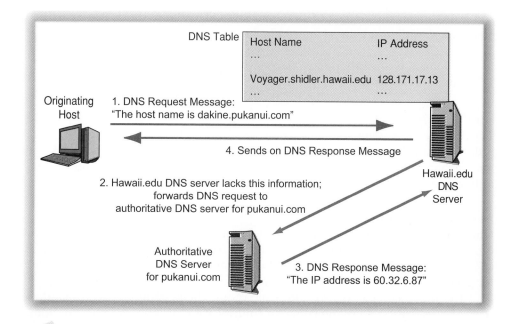

Figure 10-15   Domain Name System (DNS) Lookup

Figure 10-16   Domain Name System (DNS) Hierarchy

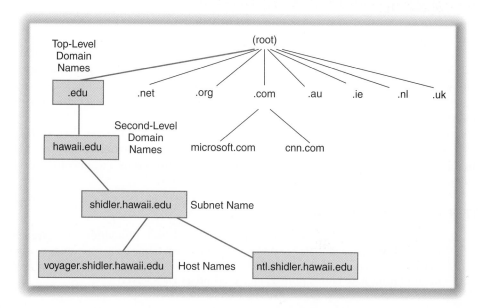

and hosts) under the control of an organization. The figure shows that domains are hierarchical, with host names being at the bottom of the hierarchy.

A domain is any group of resources (routers, single networks, and hosts) under the control of an organization.

Root DNS Servers

The domain name system is organized in a hierarchy. At the top of the DNS hierarchy is the **root,** which consists of all domain names. Under the root are **top-level domains** that categorize the domain by organization type (such as .com, .net, .edu, .biz, and .info) or by country (for example, .uk, .ca, .ie, .au, .jp, and .ch).

Second-Level Domains

Under top-level domains are **second-level domains,** which usually specify a particular organization (microsoft.com, hawaii.edu, cnn.com, etc.). Sometimes, however, specific products, such as movies, get their own second-level domain names. Competition for good second-level domain names is fierce.

Getting a second-level domain name is only the beginning. Each organization that receives a second-level domain name must have a DNS server to host its domain name record. Large organizations have their own internal DNS servers that contain information on all subnet and host names. Individuals and small businesses that use webhosting services depend on the webhosting company to provide this DNS service.

Further Qualifications

Domains can be further qualified. For instance, within hawaii.edu, which is the University of Hawai'i, there is a shidler.hawaii.edu domain. This is the Shidler College of Business. Within *shidler.hawaii.edu* is *voyager.shidler.hawaii.edu,* which is a specific host within the college.

Hierarchy of DNS Servers

To implement this naming hierarchy, the domain name system maintains a hierarchy of DNS servers. As noted earlier, at the root level there are 13 **DNS root servers** that contain information about DNS servers for top-level domains (.com, .edu, .ca., .ie, etc.). Having multiple DNS root servers provides reliability. Each top-level domain itself maintains multiple DNS servers.

Companies with second-level domain names are required to have their own DNS servers and almost always maintain two or more DNS servers for their own second-level domain name. For small businesses whose websites are hosted by a webhosting service, the webhosting service also maintains their DNS function.

TEST YOUR UNDERSTANDING

9. a) Is the Domain Name System used only to send back IP addresses for given host names? b) What is a domain? c) Which level of domain name do corporations most wish to have? d) What are DNS root servers? e) How many DNS root servers are there? f) Why do most firms have both a primary and a secondary DNS server?

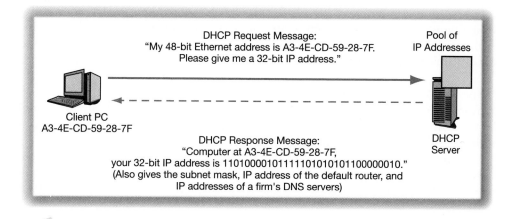

Figure 10-17   Dynamic Host Configuration Protocol (DHCP)

DHCP Servers

Static IP Addresses

Server hosts always use the same IP address. Otherwise, client PC users would not know how to reach them. (Imagine what would happen if your favorite store constantly changed its street address!) Hosts that use the same IP address all the time are said to have **static IP addresses.**

Dynamic IP Addresses and the Dynamic Host Configuration Protocol (DHCP)

In contrast, client PCs connected to the Internet usually do not get the same IP address each time they use the Internet. As shown in Figure 10-17, when a client PC wishes to use the Internet, it first contacts a **Dynamic Host Configuration Protocol (DHCP)** server. This server has a pool of IP addresses that it manages.

The DHCP request message asks the DHCP server for an IP address. The server responds by sending back an IP address to the client PC.[3] The client PC will use this IP address until it signs off its ISP or for several hours. This is only a temporary IP address—or to be technical, a **dynamic IP address.**

Although the DHCP server tries to give your PC the same IP address each time you use the Internet, there is no guarantee that it will do so. Consequently, each time you use the Internet, you may get a different IP address.

---

[3]Actually, the process is somewhat more complex. The client PC first broadcasts a message to all nearby DHCP servers. Each server sends back an offer to provide service. These offers include the length of time the client PC may use the IP address and other parameters. The client PC evaluates these offers and then works with the DHCP server that provides the best offer.

Additional Configuration Information

In addition, the DHCP server gives the client PC additional configuration information, including a subnet mask, the IP address of a default router, and the IP addresses of the firm's Domain Name System (DNS) servers.

Why Dynamic Configuration?

Why use dynamic configuration? Consider what would happen if configuration had to be typed into each client PC manually. This would be extremely expensive, compared with autoconfiguration through DHCP.

In addition, if the company changed the IP addresses of its DNS servers, it would have to manually change this information on all PCs. During the time this change would take to be made, client PC users with obsolete information would be unable to use DNS.

DHCP Management

DHCP provides for flexibility in implementation so that companies can do what is best in their circumstances. Some firms have each subnet manager maintain a DHCP server for his or her subnet. At the University of Hawai'i, a strong tradition of decentralization made this arrangement a logical choice. At many large firms, however, each site has a single DHCP server—even the headquarters site, which has many subnets. This flexibility is possible because DHCP has a configurable **scope** parameter that determines which subnets it will serve.

TEST YOUR UNDERSTANDING

10. a) What hosts are given static IP addresses? b) Why do these hosts need static IP addresses? c) What are dynamic IP addresses? d) How do clients get dynamic IP addresses? e) What additional information does DHCP provide to client PCs? f) Why is dynamic configuration attractive? g) What are DHCP scopes?

Simple Network Management Protocol (SNMP)

We have seen the **Simple Network Management Protocol (SNMP)** in the first chapter and in several subsequent chapters. We will now look at it in a little more detail.

Core SNMP Elements

In Chapter 1, we saw that SNMP has several components.

➤ The network administrator works at a central computer that runs a program called the network management program, or, more simply, the manager.
➤ The SNMP manager is responsible for many managed devices—devices that need to be administered, such as printers, switches, routers, and other devices.
➤ Managed devices have pieces of software (and sometimes hardware) called network management agents, or, more simply, agents. SNMP agents communicate with the SNMP manager on behalf of their managed devices. In other words, the manager does not communicate with the managed device directly, but rather with the device's agent.
➤ The manager stores information it receives in a central management information base (MIB). The term "MIB" is somewhat ambiguous because it can refer either to the database itself or to the design (schema) of the database.

Core Elements (from Chapter 1)
> Manager program
> Managed device
> Agents (communicate with the manager on behalf of the managed device)
> Management information base (MIB)
>> Stores the retrieved information
>> "MIB" can refer to either the database on the manager or on the database schema

Messages
> Commands
>> Get
>> Set
> Responses
> Traps (alarms sent by agents)
> SNMP uses UDP at the transport layer to minimize the burden on the network

RMON Probes
> Remote monitoring probes
> A special type of agent
> Collects data for a part of the network
> Supplies this information to the manager

Objects (see Figure 10-19)
> Information about which information is stored
> Number of rows in the routing table
> Number of discards caused by lack of resources (indicates a need for an upgrade)

Set Commands
> Dangerous if used by attackers
> Many firms disable set to thwart such attacks
> However, they give up the ability to manage remote resources without travel
> SNMPv1: community string shared by the manager and all devices
> SNMPv3: each manager-agent pair has a different password

User Functionality
> Reports, diagnostics tools, etc. are very important
> They are not built into the standard
> They are added by SNMP manager vendors
> Critical in selection

Figure 10-18 Simple Network Management Protocol (SNMP) (Study Figure)

To work together, these devices send messages to one another.

➤ The manager can send commands to the agent, telling the agent what to do. Agents send responses confirming that the command was fulfilled or explaining why it could not be fulfilled.

➤ SNMP Get commands ask for a specific piece of information. In practice, the manager constantly polls all of its managed devices, collecting many pieces of data from each in every round of polling.

➤ SNMP Set commands tell the agent to change the way the device operates, say, by going into self-test mode.

➤ If an agent detects a problem, it can send a trap (alarm) message to the manager without waiting passively to be asked.

➤ These commands are sent via UDP to reduce the traffic burden on the network. This method leads to occasional errors, but occasional errors mean only that a few pieces of information are a few seconds or minutes out of date.

RMON Probes

In addition, there is a specialized type of agent called an **RMON probe** (remote monitoring probe). This may be a stand-alone device or software running on a switch or router. An RMON probe collects data on network traffic passing through its location, instead of information about the RMON probe itself. The manager can poll the RMON probe to get summarized information about the distribution of packet sizes, the numbers of various types of errors, the number of packets processed, the 10 most active hosts, and other statistical summaries that may help to pinpoint problems. This generates far less network management traffic than does polling many devices individually.

Objects

The SNMP MIB schema is organized as a hierarchy of objects (properties of managed devices). Figure 10-19 shows the basic model for organizing SNMP objects. First, there is the system. This might be a computer, switch, router, or another device. In addition, there are TCP, UDP, IP, and ICMP objects and objects for individual interfaces. Each of these objects has categories under it, and those subcategories have further subcategories.

For example, if a router does not appear to be working properly, the manager can issue *Get* commands to collect appropriate router objects. A first step might be to check whether the router is in forwarding mode. If it is not, it will not route packets. If checking does not clarify the problem, the manager can collect more information, including error statistics and general traffic statistics of various types.

Set Commands and SNMP Security

Most firms are very reluctant to use *Set* commands because of security dangers. If setting is permitted and attackers learn how to send *Set* commands to managed devices, the results could be catastrophic. Fortunately, SNMP security has improved over time.

➤ The original version of SNMP, SNMPv1, had almost no authentication at all, making this danger a distinct possibility. The manager and all managed devices merely had to be configured with the same **community name.** With hundreds or thousands of devices sharing the same community name, if attackers can easily learn the community name, they can implement massive attacks.

➤ SNMPv3 has added passwords for each manager–agent pair, and these passwords are encrypted during transmission. In addition, each message is authenticated by the shared password. This requires a great deal of work to set up.

System Objects
 System name
 System description
 System contact person
 System uptime (since last reboot)

IP Objects
 Forwarding (for routers). Yes if forwarding (routing), No if not
 Subnet mask
 Default time to live
 Traffic statistics
 Number of discards because of resource limitations
 Number of discards because could not find route
 Number of rows in routing table
 Rows discarded because of lack of space
 Individual row data

TCP Objects
 Maximum/minimum retransmission time
 Maximum number of TCP connections allowed
 Opens/failed connections/resets
 Segments sent
 Segments retransmitted
 Errors in incoming segments
 No open port errors
 Data on individual connections (sockets, states)

UDP Objects
 Error: no application on requested port
 Traffic statistics

ICMP Objects
 Number of errors of various types

Interface Objects (One per Interface)
 Type (e.g., 69 is 100Base-FX; 71 is 802.11)
 Status: up/down/testing
 Speed
 MTU (maximum transmission unit—the maximum packet size)
 Traffic statistics: octets, unicast/broadcast/multicast packets
 Errors: discards, unknown protocols, etc.

Figure 10-19 SNMP Object Model

Most products today permit two SNMPv3 passwords for each manager–device pair. One is for *Get* commands. The other is for the more dangerous *Set* commands.

Of course, poor implementation can defeat SNMPv3 security. If a lazy administrator uses the same password for all manager–agent pairs, then this is no better security than community strings.

User Interface Functionality

Collecting data in the MIB is worthless if there is no good way to get it out. All SNMP manager products have the ability to produce reports, diagnose problems, and do many other things that network administrators must do to manage networks.

This functionality is not built into the standard. The standard just handles the mechanics of collecting and storing object data from managed devices. User interface functionality is a critical factor in selecting SNMP products.

TEST YOUR UNDERSTANDING

11. a) List the main elements in a network management system. b) Does the manager communicate directly with the managed device? Explain. c) Explain the difference between managed devices and objects. d) Is the MIB a schema or the actual database? (This is a trick question.) e) Why must user interface functionality for the SNMP manager be considered carefully in selecting SNMP manager products?

12. List one object in each of the following areas: the system, IP, TCP, UDP, ICMP, and an interface.

13. a) In SNMP, which device creates commands? b) Responses? c) Traps? d) Explain the two types of commands. e) What is a trap? f) Why are firms often reluctant to use *Set* commands? g) Describe SNMPv1's poor authentication method. h) Describe SNMPv3's good authentication method.

DIRECTORY SERVERS

Directory Server Basics

What Are Directory Servers?

Many firms now have **directory servers,** which centralize information about a firm. For instance, for individual people, the directory server may have a name, a telephone number, and an e-mail address. It might also store the person's login password and permissions on various servers.

In addition, the directory server may store information about individual hosts. For instance, it might contain a set of security rules for a group of PCs or even for a single PC. In Microsoft's Active Directory product, for instance, a firm can create a **group policy object (GPO),** which may state that a user cannot add new programs at all. It can even lock down the way the screen appears.

In previous chapters, we saw that an authentication has a set of information for authenticating users. Several verifiers may rely on the authentication server for authentication services. If a company has multiple authentication servers, they may each get their authentication information from a central directory server.

Hierarchical Organization

As Figure 10-20 shows, information in a directory server is arranged hierarchically, much as entries in a DNS server are organized.

The figure shows the directory structure for the mythical University of Cannes. The top level is the **organization.** Under the top level, there are schools (**organizational units,** in directory server terminology). In each school, there are faculty, staff, and router categories. Under the faculty category, there are the usernames of faculty

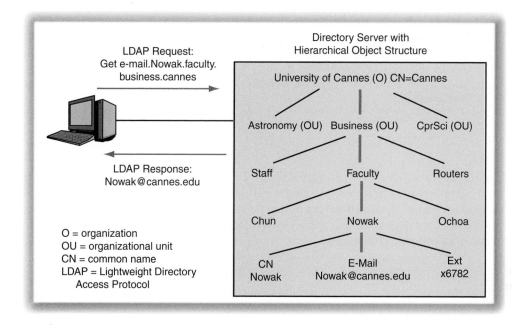

Figure 10-20   Directory Server Organization and LDAP

members. At the bottom of the hierarchy are the properties of individual faculty members, including the faculty member's common name, e-mail address, and telephone extension.

Lightweight Directory Access Protocol (LDAP)

Most directory servers today permit query commands governed by the **Lightweight Directory Access Protocol (LDAP).** The figure shows that LDAP commands specify the path to a property, with individual nodes along the way, separated by dots. This is why the request for Nowak's e-mail address is specified as the following:

GET e-mail.nowak.faculty.business.cannes

Directory Servers and the Networking Staff

Organizations store a great deal of information about themselves in directory servers, including much networking information. Creating a directory server data organization plan (schema) requires extensive planning about what information an organization needs to store and how this information should be arranged hierarchically.

Although creating and managing a directory server goes well beyond networking, the networking staff is often given the task of leading directory server planning projects and managing the directory server on a daily basis.

TEST YOUR UNDERSTANDING

14. a) What kinds of information are stored in a directory server? b) How is information in directory servers organized? c) What is the purpose of LDAP? d) If Astronomy has a similar directory organization to Business (in Figure 10-20), give the specification for the telephone extension of Claire Williams (username cwilliams), who is an Astronomy staff member.

Microsoft's Active Directory (AD)

Active Directory Domains

Microsoft's directory server product is **Active Directory (AD).** Network administrators must become very familiar with AD. Figure 10-21 shows that a firm must divide its computers into logical **Active Directory domains,** which are simply called domains. These AD domains are organized in a hierarchy. The Microsoft concept of domains is similar to the DNS concept of domains.

Domain Controllers

A domain must have one or more **domain controllers,** which are servers that run Active Directory and maintain an AD database for the domain. If there are multiple domain controllers within a domain, then their AD data is *fully* replicated between

Figure 10-21   Active Directory Domains and Domain Controllers

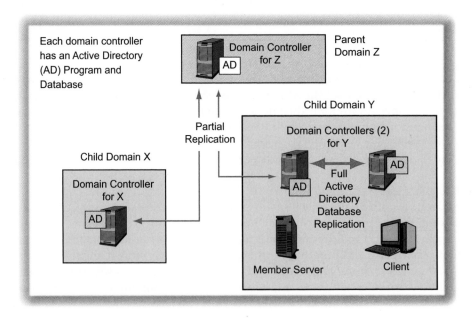

them so that each has the same data. If one domain controller fails, the other takes over automatically. Not all servers in a domain must run AD. Servers that do not are called **member servers.**

Domains in an Active Directory Tree

A logical hierarchy of AD domains is called an **Active Directory tree.** Information in AD databases typically is *partially* replicated across domain controllers at different levels.

Complex Structures

We have looked at a tree organization of AD domains, at domain controllers, and at replication. However, companies may have forests (groups of AD trees), and replication can be handled with almost infinite variations and trust relationships.

TEST YOUR UNDERSTANDING

    15. a) What is Active Directory? b) What is an AD domain? c) What are domain controllers? d) Can a domain have more than one domain controller? e) What are servers called that do not run AD? f) Describe replication among domain controllers in the same domain. g) What is an AD tree? h) Describe replication among domain controllers in parent and child domains. i) What is an AD forest?

CONCLUSION

Synopsis

This chapter deals with three major network management issues: planning, TCP/IP management, and directory servers.

Planning

The first job of network management is planning. The first thing to plan is the technological infrastructure. A firm must conduct what-is analysis to understand its current technological infrastructure. The next step is to understand the driving forces for change. Comparing the current situation with what is likely to be needed will reveal a number of gaps. The company must consider options for closing the gaps. Options must fully meet requirements, but should be as inexpensive as possible as long as requirements are met.

One consideration in network planning is what traffic management methods to use. Earlier in the book, we learned about overprovisioning, priority, and QoS guarantees for some traffic. In this chapter, we also saw traffic shaping, which controls traffic by not allowing some kinds of traffic to enter the network at all and that limits other kinds of traffic to a certain percentage of the total traffic. We also saw how compression can reduce the amount of traffic entering the network. This also helps avoid congestion and its problems.

Rather than physically trying things until something works, companies use network simulation to consider alternatives until they find promising ones. To do a network simulation, it is important to first gather extensive data about the network. Then, a model is built of the network. After this model is validated, simulations are run on alternatives. Although no model is perfect, simulation usually can drastically reduce uncertainty in planning.

TCP/IP Management

The TCP/IP standards that dominate internetworking require a great deal of management attenuation, both initially and on an ongoing basis. The first step is to develop an IP subnet schema for the firm. This creates a basic tradeoff between the number of subnets and the number of hosts per subnet. The firm also has to decide whether or not to use network address translation (NAT). NAT has several benefits, including added security and an increase in the effective number of public IP addresses that a firm has; but NAT causes problems for certain protocols.

To make IP traffic flows more efficient, firms can turn to Multiprotocol Label Switching, which adds a tag to each packet. The tag directs routers along the way to treat packets with specific label numbers in specific ways. This process dramatically reduces the costs of handling IP packets when they arrive at routers. It also permits traffic engineering; for example, there can be two MPLS label numbers between two sites. One can be for high-priority traffic, another for low-priority traffic. Routers along the way will let the high-priority traffic go first.

In this chapter, we looked more closely at the Domain Name System (DNS). We saw that DNS is a hierarchical system of named domains (collections of resources under the control of an organization). Corporations want second-level domain names, such as prenhall.com. After they get one, they must maintain two or more DNS servers for their second-level domain. We also saw that if a local DNS server does not know the IP address for a host name, it contacts the authoritative DNS server for the domain in the IP address.

We looked at static IP addresses and the dynamic configuration of client PCs with DHCP servers. We saw that this is less expensive than manually configuring each PC and reconfiguring each PC when configuration parameters change. However, this process gives each PC a dynamic IP address that may be different each time the PC boots up.

We have seen the Simple Network Management Protocol (SNMP) repeatedly since we first saw it in Chapter 1. This chapter looked at SNMP operation in a bit more detail, focusing on the concept of objects and the types of objects specified in MIB schemas. We also saw RMON probes.

Although TCP/IP works very well, TCP/IP requires a great deal of planning and ongoing work, and this in turn requires extensive knowledge about how TCP/IP's supervisory protocols operate, as well as how the architecture's core protocols (IP, TCP, and UDP) work.

Directory Servers

Increasingly, companies are centralizing information about their people, computers, and other resources in directory servers. Directory servers store information in a hierarchical organization, so careful planning is needed because it is very difficult to change the schema after it is created. Typically, data in directory servers is accessed via the Lightweight Directory Access Protocol (LDAP).

Smaller firms have a single directory server, but larger firms with complex organizational frameworks often have multiple directory servers. We learned how Microsoft's

Active Directory (AD) program enables a company to create a hierarchy of domains, each with one or more domain controllers with an AD program and database. If a domain has multiple domain controllers, they fully replicate the data in their AD database. Domain controllers in parent and child domains may partially replicate their AD databases.

End-of-Chapter Questions

THOUGHT QUESTIONS

1. Suppose that both DNS servers and DHCP servers send your client PC an IP address. Distinguish between these two addresses.

2. Assume that an average SNMP response message is 100 bytes long. Assume that a manager sends 40 SNMP *Get* commands each second. a) What percentage of a 100 Mbps LAN link's capacity would the resulting response traffic represent? b) What percentage of a 128-kbps WAN link would the response messages represent? c) What can you conclude from your answers to parts (a) and (b) of this question?

3. A firm is assigned the network part 128.171. It selects an 8-bit subnet part. a) Draw the bits for the four octets of the IP address of the first host on the first subnet. (Hint: Use Windows Calculator.) b) Convert this answer into dotted decimal notation. c) Draw the bits for the second host on the third subnet. (In binary, 2 is 10, while 3 is 11.) d) Convert this into dotted decimal notation. e) Draw the bits for the last host on the third subnet. f) Convert this answer into dotted decimal notation.

TROUBLESHOOTING QUESTION

In your browser, you enter the URL of a website you use daily. After some delay, you receive a DNS error message that the host does not exist. What may have happened? Explain your reasoning. **Again, do NOT just come up with one or two possible explanations.**

PERSPECTIVE QUESTIONS

1. What was the most surprising thing to you about the material in this chapter?

2. What was the most difficult thing for you in the chapter?

PROJECT

Do a report on Active Directory.

GETTING CURRENT

Go to the book website's New Information and Errors pages for this chapter to get new information since this book went to press and to note corrections to any errors in the text.

Network Management Utilities and Router Configuration

Objectives

By the end of this chapter, you should be able to do the following:

- Know the tools to use to diagnose problems in PC connections to the outside world and know how to use simpler tools.
- Understand transmission analysis tools.
- Discuss network mapping.
- Have an initial understanding of how to configure a Cisco router or switch with the command line interface.

INTRODUCTION

Chapter 10 discussed network management broadly. In addition to knowing the principles of network management, networking professionals must learn to use many specific software tools to automate as much of their work as possible. In this chapter, we will look at several network management utilities. We will also look at a simple example of router configuration to give you a taste of configuration complexities.

Network Setup Wizard
> Works most of the time; user needs tools if it does not

Testing the Connection
> Open a connection to a website by using a browser

Ping
> Ping asks a host to respond
> Ping a host to see if latency is acceptable
> Go to the command line
> Ping 127.0.0.1. This is the loopback interface (you ping yourself)
> Ping another host to see if you can reach external hosts

Checking the NIC
> Right-click on a connection and select Properties
> Under the name of the NIC, hit the Configuration button
> The dialog box that appears will show you the status of the NIC
> It also offers a Troubleshooting wizard if the NIC is not working

Checking Your Configuration
> Use ipconfig
> This gives configuration information about your computer

Figure 10a-1 Tools for Diagnosing PC Connections (Study Figure)

TOOLS FOR DIAGNOSING PC CONNECTIONS

One important set of tools is designed to diagnose a PC's connection to the outside world. Normally, you can either use the Internet when you plug in your PC or set up your PC to connect to the outside world by running your computer's network setup wizard. Sometimes, however, problems occur and troubleshooting is needed.

Network Setup Wizard

PC operating systems have network setup wizards that guide you through a series of steps to set up a network connection to the outside world. Typically, they begin by asking whether you wish to set up a modem connection, a broadband (DSL or cable modem) connection, or a wireless connection. They take you through appropriate steps to set up the kind of connection you specify.

Opening the Browser

After configuration, you typically test the connection by double-clicking on the browser and seeing whether you can go to a known website. If you can get to the website, you obviously have a connection to the outside world. The simplest test is always the best test.

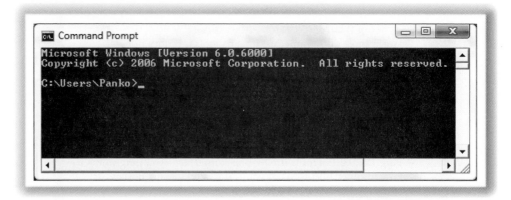

Figure 10a-2   Windows Command Line Interface

Command Line Interface

If the browser connection does not work or if transmission is sluggish, you can drop down to the command line to check the connection. In Windows XP, this entails hitting the *Start* button and typing *cmd[enter]* in the run box. This will bring up a window with the command line interface, as Figure 10a-2 illustrates. In Vista, the user hits the *Start* button, selects *All Programs*, the *Accessories Folder*, and then *Command Prompt*.

Here, the prompt is *C:\Users\Panko>*. When this prompt appears, you can type a command. Commands must be typed exactly, and you must always end commands by hitting the *[Enter]* key. (If you give a command and nothing happens, check to see whether you hit the [*Enter*] key.) There are many commands. One of them is *cls*, which clears the screen.

Ping 127.0.0.1

One of the most useful commands is **ping.** Ping is like a submarine pinging a target ship. The submarine sends out a sonar burst, and the reflection tells that a ship is there and gives the round-trip time for the reflection. (The round-trip time indicates how far the target ship is from the submarine.)

Figure 10a-3 illustrates what happens in a ping command. The command syntax is *ping<space><ip address or host name>[Enter]*. In the figure, the IP address is 127.0.0.1. The response shows that four pings were sent successfully. It also shows that the latency was under one millisecond each time. The host obviously must be very close to have such low latencies.

Actually, 127.0.0.1 is a special IP address. It is reserved for **loopback testing,** in which the PC pings *itself*. If pinging 127.0.0.1 is successful, then the user has an outside connection. If the browser could not reach external hosts, but pinging 127.0.0.1 is successful, there probably is a problem with the browser's setup.

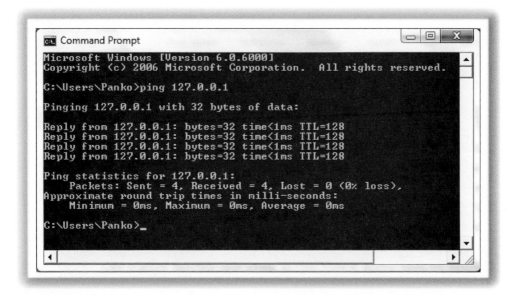

```
Command Prompt                                          □  ▣  ✕

Microsoft Windows [Version 6.0.6000]
Copyright (c) 2006 Microsoft Corporation.   All rights reserved.

C:\Users\Panko>ping 127.0.0.1

Pinging 127.0.0.1 with 32 bytes of data:

Reply from 127.0.0.1: bytes=32 time<1ms TTL=128
Reply from 127.0.0.1: bytes=32 time<1ms TTL=128
Reply from 127.0.0.1: bytes=32 time<1ms TTL=128
Reply from 127.0.0.1: bytes=32 time<1ms TTL=128

Ping statistics for 127.0.0.1:
    Packets: Sent = 4, Received = 4, Lost = 0 (0% loss),
Approximate round trip times in milli-seconds:
    Minimum = 0ms, Maximum = 0ms, Average = 0ms

C:\Users\Panko>_
```

Figure 10a-3   Ping 127.0.0.1

General Pings

If you have a connection, but the connection is sluggish, you might ping a host other than the PC. In Figure 10a-4, the user or installer has pinged yahoo.com. Note that you can use a host name as well as an IP address in a ping.

The figure shows that ping gives you a good deal of information. First, it gives you the IP address of the host (in this case, 66.94.234.13). Then, it tells you whether the ping worked and gives the latency to the host in milliseconds (ms). In this case, four echoes were received, and the latencies were 83, 79, 83, and 83 milliseconds. These are fairly low latencies. If latencies are unusually long when you try to ping several different hosts, there may be troubles outside of your PC.

Tracert

If latency is high, the tracert tool is valuable. The syntax of the tracert command is *tracert <space><IP address or host name>[Enter]*. In Figure 10a-5, the user or installer has conducted a tracert on Yahoo.com.

The result of a tracert is latencies, not just to the final host, but to each router along the way. Each row gives data from a single router or for the destination host. There are 10 routers along the way. The last entry is for the host.

By comparing differences in latency between successive routers, latency trouble spots can be seen. Here, there is no real problem, but the latencies to the second router are only 7 to 8 ms, while the latencies to router 3 are 72 to 73 ms. So, of the total latency of 83 or 84 ms, roughly 65 ms is due to this single hop. If the latency were much longer, this would indicate a trouble spot in the network.

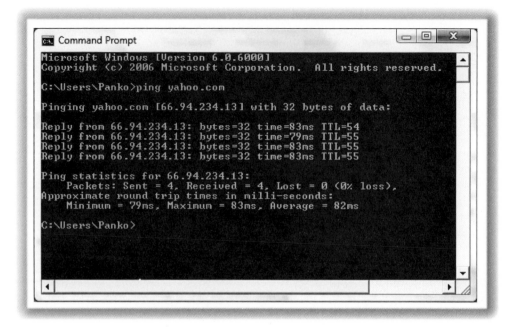

Figure 10a-4   Pinging Yahoo.com

Checking the NIC

Sometimes, the computer's network interface card (NIC) is the problem. The NIC implements Ethernet transmission at the physical and data link layers. NICs are also called network adapters.

Figure 10a-5   Tracert

Figure 10a-6   Device Manager Dialog Box

To check the NIC in Vista, click on *Start*, then *Computer, System properties,* and *Device Manager*. At this point your computer will probably require you to confirm that you want to use Device Manager.

Figure 10a-6 shows the Device Manager dialog box that will appear. It lists devices on the computer, by category. If *Network adapters* has a plus next to it, click on the plus. The plus will change to a minus, and a list of network adapters will appear. In this case, there is only one network adapter.

Right-click on the network adapter and choose *Properties*. This will bring up a dialog box on the network adapter's properties. The *General* tab is selected in Figure 10a-7.

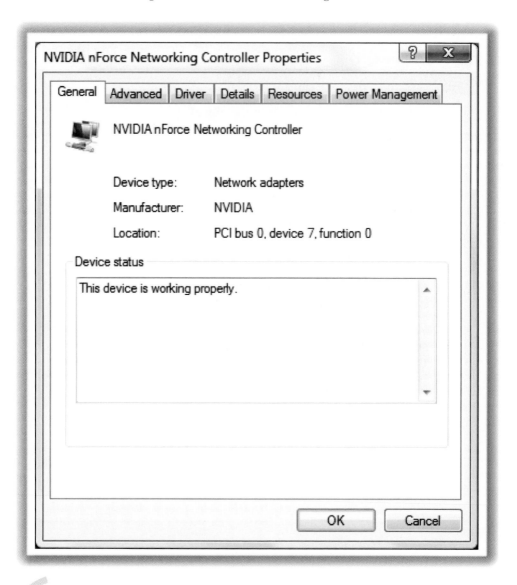

Figure 10a-7   Network Adapter Properties

The Device Status box shows that the NIC is operating properly. If there is a problem, the other tabs may give an indication of what the problem is.

Checking Your Configuration

If problems are subtle, you might check your PC's configuration. To do this, drop down to the command prompt. In Vista, type ***ipconfig[Enter]***. In Windows XP, *type*

Figure 10a-8   Obtaining Configuration Information with Ipconfig

ipconfig /all[Enter]. Figure 10a-8 shows part of the extensive output of ipconfig in Vista.

TEST YOUR UNDERSTANDING

1. a) After you use the network setup wizard to create a network connection, in what quick way can you verify that the connection is working? b) How can pinging be used in two ways to test a connection? c) How can tracert help you diagnose latency in your connection to a specific host? d) How can you check your NIC in Vista? e) How does ipconfig assist in troubleshooting?

TRANSMISSION ANALYSIS TOOLS

Packet Capture and Display Programs

For even more subtle problems, **packet capture and display programs** capture selected packets or all of the packets arriving at a NIC or going out of a NIC. Afterward, you can display key header information for each packet in greater or lesser detail. This packet-by-packet analysis gives you maximum information on traffic going into and out of your computer.

One of the most popular freeware packet-analysis programs is **TCPDUMP** for Unix, which is available as **WinDUMP** for Windows computers.[1] The user runs the program and tells it to collect data over a given period. TCPDUMP/WinDUMP then presents data on each packet. Chapter 8a discussed WinDUMP/TCPDUMP. Another popular packet analysis program is Ethereal (www.Ethereal.com).

Traffic Summarization Programs

While looking at individual packets can be helpful, it may also be useful to go to the other extreme, examining broad statistical trends over spans of time. Figure 10a-10 shows some output from **EtherPeek,** a commercial traffic summarization program. EtherPeek captures all packets arriving at and leaving a NIC, and then provides a wide spectrum of summarization tools. EtherPeek can show overall trends and can also drill down to specifics.

Figure 10a-9   Traffic Analysis Tools (Study Figure)

Packet Capture and Display Programs
  Capture packets as they pass
  Display the individual packets for analysis
  WinDUMP and TCPDUMP (See Chapter 8a)
  Ethereal

Traffic Summarization Tools
  Capture packets as they pass
  Present results in summarized form
  EtherPeek (See Figure 10a-10)

Connection Analysis with Netstat
  Provides information on individual connections to external hosts
  Can indicate Trojan horses and other malware programs that are communicating
  (See Figure 10a-11)

---

[1]A good link for downloading WinDUMP (and the WinPCAP software you must download first to run WinDUMP) is http://windump.polito.it. The site also has good documentation on WinDUMP.

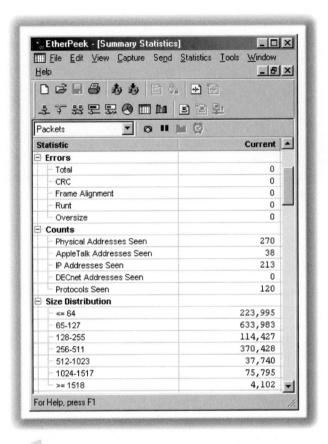

Figure 10a-10   EtherPeek Packet Capture and
Summarization Program

Connection Analysis with Netstat

Another popular tool that is built into the operating system in both Unix and Windows is **Netstat.** This program shows active connections. In Figure 10a-11 there is a single TCP connection, on Port 3290. The connection is closed and in the wait mode. This might be an indication that the computer has a Trojan horse program installed on it. In addition to causing damage in general, Trojan horses can cause serious problems for network connections. Alternatively, this connection might be a legitimate connection.

TEST YOUR UNDERSTANDING

2. a) What are the benefits of packet capture and display programs? b) What is a common freeware program for packet capture and display? c) What information do traffic summarization programs like EtherPeek provide? d) What information does Netstat tell you?

Figure 10a-11 Netstat Connections Analysis Program

NETWORK MAPPING TOOLS

Sometimes, network managers have a broader need, most commonly to map the layout of their networks, including what hosts and routers are active and how various devices are connected. **Network mapping** has two phases. The first is **discovering** hosts and subnets—that is, finding out whether they exist. The second is **fingerprinting** hosts (determining their characteristics) to determine whether they are clients, servers, or routers.

Although ping and tracert are useful for the host discovery phase, they are tedious to use. Network administrators typically turn to network mapping tools, including the free Nmap program that we saw in Chapter 10.

Nmap pings a broad range of possible host addresses to determine which IP addresses have active hosts. If ping will not work (or is blocked by firewalls), Nmap offers other ways to scan for active IP addresses.

Nmap also offers fingerprinting, which attempts to identify the specific operating system running on each host and sometimes even the version number of the oper-

Figure 10a-12 Network Mapping Tools

To understand how the network is organized
Discovering IP addresses with active devices
Fingerprinting them to determine their operating system (client, server, or router)
A popular network mapping program is Nmap (See Chapter 9)

ating system. For instance, a computer running Windows XP is almost certainly a client, while a computer running Windows Server 2003 or Solaris (the SUN version of Unix) probably is a server. Nmap can even fingerprint router operating systems to identify routers.

When network mapping is finished, the network manager should have a good understanding of how the network is organized and what parts are not working correctly.

TEST YOUR UNDERSTANDING

3. a) What is network mapping? b) Describe the two phases of network mapping. c) How can Nmap help in network mapping?

ROUTER CONFIGURATION IN CISCO'S IOS

Individual devices on the network—clients, servers, switches, and routers—need to be configured to work with TCP/IP.

Configuring IOS Devices

Cisco Systems, which dominates the market for switches and routers, has an operating system that it uses on all of its routers and all of its switches. This is the **Internetwork Operating System (IOS).**

Initial Router Configuration

When you turn on a Cisco router for the first time, you are led through a sequence of configuration questions. This allows you to configure much of what needs to be configured about the router. However, although this initial configuration mode gets a router up and running, it does not do all configuration work.

In the initial configuration process, the user specifies many things about the router and its interface, including the following:

➤ A name for the router.
➤ An enable secret for privileged mode (which is needed to do most configuration tasks).
➤ Whether or not SNMP should be turned on to manage the router.
➤ Whether routing should be set up for IP, IPX, AppleTalk, and other protocols.
➤ For each interface (port), operating characteristics, an IP address, a subnet mask, and other information.
➤ Finally, whether the configuration should be saved in NVRAM (nonvolatile RAM) for permanent use (until changed).

The Command Line Interface (CLI)

To work with the limited processing power and memory of switches and routers, IOS has to be kept as small as possible. This limitation has necessitated the use of a **command line interface (CLI),** in which a user has to type highly structured commands, ending each command with *Enter.* Figure 10a.13 shows some steps needed to configure a router in CLI mode rather than through initial configuration.

Command	Comment
Router>enable[Enter]	Router> is the prompt. The ">" shows that the user is in nonprivileged mode. Enables privileged mode so that the user can take supervisory actions. User must enter the enable secret. All commands end with [Enter]. Enter is not shown in subsequent commands.
Router#hostname julia	Prompt changes to '#' to indicate that the user is in privileged mode. User gives the router a name, julia.
julia#config t	Enter configuration mode. The t is an abbreviation for "terminal."
julia(config)#int e0	Prompt changes to julia(config) to indicate that the user is in configuration mode. User wishes to configure Ethernet interface 0. (Router has two Ethernet interfaces, 0 and 1)
julia(config-if)#ip address 10.5.0.6 255.255.0.0	User gives the interface an IP address and a subnet mask. (Every router interface must have a separate IP address.) The subnet is 5.
julia(config-if)#no shutdown	This is an odd one. The command to shut down an interface is "shutdown." Correspondingly, "no shutdown" turns the interface on.
julia(config-if)# Ctrl-Z	User types Ctrl-Z (the key combination, not the letters) to end the configuration of e0.
julia(config)#int s1	User wishes to configure serial interface 1. (Router has two serial interfaces, 0 and 1.)
julia(config-if)#ip address 10.6.0.1 255.255.0.0	User gives the interface and IP address and subnet mask. The subnet is 6.
julia(config-if)#no shutdown	Turns on s1.
julia(config-if)# Ctrl-Z	Ends the configuration of s1.
julia# router rip	Enables the Router Initiation Protocol (RIP) routing protocol.
julia#disable	Takes the user back to nonprivileged mode. This prevents anyone who gets access to the terminal from making administrative changes to the router.
julia>	

Figure 10a-13   Cisco Internetworking Operating System (IOS) Command Line Interface (CLI)

TEST YOUR UNDERSTANDING

4. a) List some configuration tasks for routers. b) What is Cisco's operating system? c. What is a CLI? d) What is the advantage of using a CLI? e) What happens when you first turn on a new Cisco router? f) Does initial configuration when you first turn on a new Cisco router handle all configuration chores? g) Must an IP address and subnet mask be configured for each router interface? h) What is the IOS CLI prompt in non-privileged mode? i) What is the IOS CLI prompt in privileged mode?

End-of-Chapter Questions

HANDS-ON EXERCISES

1. Do the following on a Windows or Unix computer. Ping your computer. What was your average latency? What was your time to live?

2. Ping a remote host. What is the latency? If ping does not work on that host, try others until you find one that does work.

3. Do a tracert on the connection. How many routers separate your computer from the remote host? On a spreadsheet, compute the average latency increase between each pair of routers. Paste the latency spreadsheet analysis into your answer. What connection between routers brings the largest increase in latency?

4. Check your NIC. What did you find?

5. Do the following on a Windows computer: Give the ipconfig command or the ipconfig / all command. List the important configuration information that you learned.

6. Do the following on a Windows or Unix computer: Give the Netstat command. Describe what you learned.

THOUGHT QUESTION

You have the following prompt: "Router>." List the commands you would use in a session to set up the router's first Ethernet interface. You wish the interface to have IP address 60.42.20.6 with a mask that has 16 ones followed by 16 zeros. Terminate the session properly.

Networked Applications

Learning Objectives

By the end of this chapter, you should be able to discuss the following:

■ The characteristics and limitations of host communication with dumb terminals.

■ Client/server architectures, including file server program access and client/server processing (including Web-enabled applications).

■ Electronic mail standards and security.

■ World Wide Web and e-commerce (including the use of application servers), and e-commerce security.

■ Web services with SOAP/messages and XML syntax.

■ Peer-to-peer (P2P) computing, which, paradoxically, normally uses servers for part of the work.

INTRODUCTION

Networked Applications

Once, applications ran on single machines—usually, mainframes or stand-alone PCs. Today, however, most applications spread their processing power over two or more machines connected by networks instead of doing all processing on a single machine.

Application Architectures

In this chapter, we will focus on **application architectures**—that is, how application layer functions are spread among computers to deliver service to users. Thanks to layering's ability to separate functions at different layers, most application architectures can run over TCP/IP, IPX/SPX, and other standards below the application layer. In turn, if you use TCP at the transport layer, TCP does not care what application architecture you are using.

An application architecture describes how application layer functions are spread among computers to deliver service to users.

Important Networked Applications

In addition to looking broadly at application architectures, we will look at some of the most important of today's networked applications, including e-mail, videoconferencing, the World Wide Web, e-commerce, Web services, and P2P computing.

Importance of the Application Layer to Users

In this chapter, we will focus on the application layer. This is the only layer whose functionality users see directly. When users want e-mail, it is irrelevant what is happening below the application layer, unless there is a failure or performance problem at lower layers.

TEST YOUR UNDERSTANDING

1. a) What is an application architecture? b) Why do users focus on the application layer?

TRADITIONAL APPLICATION ARCHITECTURES

In this section, we will look at the two most important traditional application architectures: terminal–host systems and client/server architectures (both file server program access and client/server processing).

Hosts with Dumb Terminals

As Figure 11-1 shows, the first step beyond stand-alone machines still placed the processing power on a single **host computer** but distributed input/output (I/O) functions to user sites. These I/O functions resided in **dumb terminals,** which sent user

Figure 11-1 Simple Terminal–Host System

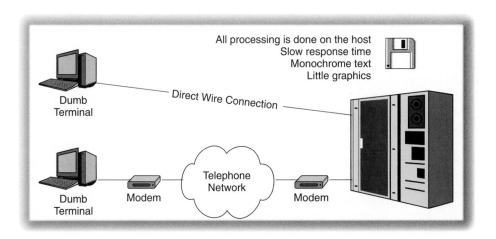

keystrokes to the host and painted host information on the terminal screen, but did little else.

Although this approach worked, the central computer frequently was overloaded by the need to process both applications and terminal communication. This often resulted in slow **response times** when users typed commands.

Another problem was high transmission cost. All keystrokes had to be sent to the host computer for processing. This generated a great deal of traffic. Similarly, the host had to send detailed information to be shown on-screen. To reduce transmission costs, most terminals limited the information they could display to **monochrome text** (one color against a contrasting background). Graphics were seldom available.[1]

IBM mainframe computers used a more complex design for their terminal–host systems that added other pieces of equipment beyond terminals and hosts. This extra equipment reduced cost and improved response times. In addition, IBM terminal–host systems had higher speeds than traditional terminals and so were able to offer limited color and graphics. Although these advances extended the life of terminal–host systems, even these advanced IBM systems are less satisfactory to users than subsequent developments, including the client/server systems described next.

TEST YOUR UNDERSTANDING

> **2.** a) Where is processing performed in systems of hosts and dumb terminals? b) What are the typical problems with these systems?

Client/Server Systems

After terminal–host systems, a big breakthrough came in the form of **client/server systems,** which placed some power on the client computer. This was made possible by the emergence of personal computers in the 1980s. PCs have the processing power to handle more of the workload than dumb terminals could.

Roles for the Client and Server

Figure 11-2 shows that the work in a client/server system is done by programs on two machines—a client and a server. The client usually is a PC. Generally, the server does the heavy processing needed to retrieve information. The client, in turn, normally focuses on the user interface and on processing data delivered by the server—for instance, by placing the data in an Excel spreadsheet.

Web-Enabled Applications

Client/server processing requires a client program to be installed on a client PC. Initially, all applications used custom-designed client programs. Rolling out a new application to serve hundreds or thousands of client computers was extremely time consuming and expensive.

Fortunately, there is one client program that almost all PCs have today. This is a browser. As Figure 11-3 illustrates, many client/server processing applications are now

---

[1]The most common dumb terminal today is the VT100 terminal, also called an ANSI terminal. On the Internet, clients can emulate (imitate) dumb terminals by using Telnet. Telnet turns a $2,000 office PC into a $200 dumb terminal.

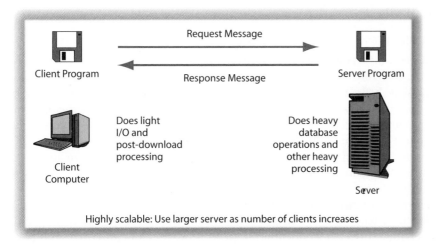

Figure 11-2 shows, at the top, a Client Program and Server Program exchanging "Request Message" and "Response Message." Below the Client Program is a Client Computer that "Does light I/O and post-download processing." Below the Server Program is a Server that "Does heavy database operations and other heavy processing." At the bottom: "Highly scalable: Use larger server as number of clients increases"

Figure 11-2   Client/Server Computing

Web-enabled, meaning that they use ordinary browsers as client programs. The figure specifically shows Web-enabled e-mail.

TEST YOUR UNDERSTANDING

3. In client/server processing, where is processing done?
4. Contrast general client/server processing with Web-enabled applications.

Figure 11-3   Web-Enabled Application (E-Mail)

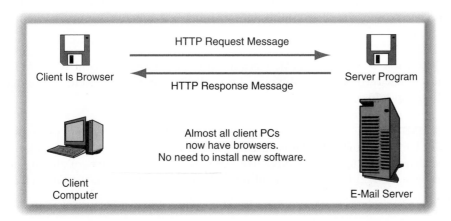

Figure 11-3 shows a Client Is Browser and Server Program exchanging "HTTP Request Message" and "HTTP Response Message." Below the browser is a Client Computer. The center text reads: "Almost all client PCs now have browsers. No need to install new software." On the right is an E-Mail Server.

ELECTRONIC MAIL (E-MAIL)

Importance

A Universal Service on the Internet

E-mail has become one of the two "universal" services on the Internet, along with the World Wide Web. E-mail provides mailbox delivery even if the receiver is "offline" when the message is received. E-mail offers the speed of a fax, plus the ability to store messages in organized files, to send replies, to forward messages to others, and to perform many other actions after message receipt. The telephone offers truly instant communication, but only if the other party is in and can take calls. In addition, e-mail is less intrusive than a phone call.

Attachments Can Deliver Anything

Thanks to attachments, e-mail has also become a general file delivery system. Users can exchange spreadsheet documents, word processing documents, graphics, and any other type of file.

E-Mail Standards

A major driving force behind the wide acceptance of Internet e-mail is standardization. It is rare for users of different systems not to be able to communicate at a technical level—although many companies restrict outgoing and incoming communication, using firewalls for security purposes. Consequently, the key issue is application layer standards.

Message Body Standards   Obviously, message bodies have to be standardized, or we would not be able to read arriving messages. In physical mail, message body standards include the language the partners will use (English, etc.), formality of language, and other matters. Some physical messages are forms, which have highly standardized layouts and fields that require specific information.

RFC 2822 (Originally RFC 822)   The initial standard for e-mail bodies was **RFC 822,** which has been updated as **RFC 2822.** This is a standard for plain text messages—multiple lines of typewriter-like characters with no boldface, graphics, or other amenities. The extreme simplicity of this approach made it easy to create early client e-mail programs.

HTML Bodies   Later, as HTML became widespread on the World Wide Web, most mail venders developed the ability to display **HTML bodies** with richly formatted text and even graphics.

UNICODE   RFC 822 specified the use of the ASCII code to represent printable characters. Unfortunately, ASCII was developed for English, and even European languages need extra characters. The **UNICODE** standard allows characters of all languages to be represented, although most mail readers cannot display all UNICODE characters well yet.

Importance of E-Mail
Universal service on the Internet
Attachments deliver files

E-Mail Standards
Message body standards
        RFC 822 and RFC 2822 for all-text bodies
        HTML bodies
        UNICODE for multiple languages
Simple Mail Transfer Protocol (SMTP)
        Message delivery: Client to sender's mail host
        Message delivery: Sender's mail host to receiver's mail host
Downloading Mail to Client
        Post Office Protocol (POP): Simple and widely used
        Internet Message Access Program (IMAP): More powerful, less widely used
Web-Enabled E-Mail
        Uses HTTP for all communication with the mail server
        No need for e-mail software on the client PC; a browser will do
        Tends to be slow

Viruses, Worms, and Trojan Horses
Widespread problems; often delivered through e-mail attachments
        Use of antivirus software is almost universal, but ineffective
Where to do scanning for viruses, worms, and Trojan horses?
        On the client PC, but users often turn off or fail to update their software
        On the corporate mail server and application firewall; users cannot turn off
        At an antivirus outsourcing company before mail reaches the corporation
        Defense in depth: Filter at two or more locations with different filtering software

Spam
Unsolicited commercial e-mail
Why filter?
        Potential sexual harassment suits
        Time consumed by users deleting spam
        Time consumed by networking staff deleting spam
        Bandwidth consumed
Separating spam from legitimate messages is very difficult
        Many spam messages are allowed through to users
        Some legitimate messages are deleted
        Some firms merely mark messages as probable spam

Figure 11-4   E-Mail (Study Figure)

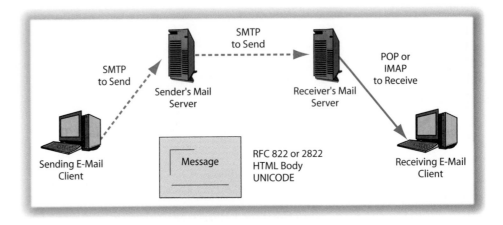

Figure 11-5   E-Mail Standards

Simple Mail Transfer Protocol (SMTP)

We also need standards for delivering RFC 2822, HTML, and UNICODE messages. In the postal world, we must have envelopes that present certain information in certain ways, and there are specific ways to post mail for delivery, including putting letters in post office postbox and taking them to the post office.

Figure 11-5 shows how e-mail is posted (sent). The e-mail program on the user's PC sends the message to its outgoing mail host, using the **Simple Mail Transfer Protocol (SMTP).** Figure 11-6 shows the complex series of interactions that SMTP requires between the sender and receiver before and after mail delivery.

Figure 11-5 shows that the sender's outgoing mail host sends the message on to the receiver's incoming mail host, again using SMTP. The receiving host stores the message in the receiver's mailbox until the receiver retrieves it.

Receiving Mail (POP and IMAP)

Figure 11-5 shows two standards that are used to *receive* e-mail. These are the **Post Office Protocol (POP)** and the **Internet Message Access Protocol (IMAP).** IMAP offers more features, but the simpler POP standard is more popular. Programs implementing these standards ask the mail host to download some or all new mail to the user's client e-mail program. Often, users delete new mail from their inbox after downloading new messages. After that, the remaining messages exist only on the user's client PC.

Web-Enabled E-Mail

Almost all client PCs have browsers. Many mail hosts are now Web-enabled, meaning that users only need browsers to interact with them in order to send, receive, and manage their e-mail. As Figure 11-3 showed, all interactions take place via HTTP, and these systems use HTML to render pages on-screen.

Actor	Command	Comment
Receiving SMTP Process	220 Mail.Panko.Com Ready	When a TCP connection is opened, the receiver signals that it is ready.
Sending SMTP Process	HELO voyager.cba.hawaii.edu	Sender asks to begin sending a message. Gives own identity. (Yes, HELO, not HELLO.)
Receiver	250 Mail.Panko.Com	Receiver signals that it is ready to begin receiving a message.
Sender	MAIL FROM: Panko@ voyager.cba.hawaii.edu	Sender identifies the sender (mail author, not SMTP process).
Receiver	250 OK	Accepts author. However, may reject mail from others.
Sender	RCPT TO: Ray@Panko.com	Identifies first mail recipient.
Receiver	250 OK	Accepts first recipient.
Sender	RCPT TO: Lee@Panko.com	Identifies second mail recipient.
Receiver	550 No such user here	Does not accept second recipient. However, will deliver to first recipient.
Sender	DATA	Message will follow.
Receiver	354 Start mail input; end with <CRLF>.<CRLF>	Gives permission to send message.
Sender	When in the course ...	The message. Multiple lines of text. Ends with line containing only a single period: <CRLF>.<CRLF>
Receiver	250 OK	Receiver accepts message.
Sender	QUIT	Requests termination of session.
Receiver	221 Mail.Panko.Com Service closing transmission channel	End of transaction.

Figure 11-6   Interactions in the Simple Mail Transfer Protocol (SMTP)

Web-enabled e-mail (also called **webmail**) is especially good for travelers because no special e-mail software is needed. Any computer with a browser in an Internet café, home, or office will allow the user to check his or her mail. On the downside, Web-enabled e-mail tends to be very slow because almost all processing is done on the distant (and often overloaded) webserver with its server-based mail processing program.

Viruses and Trojan Horses

Although e-mail is tremendously important to corporations, it is a source of intense security headaches. As we learned in Chapter 9, the most widespread security compromises are attacks by viruses and worms. Viruses come into an organization primarily, although by no means exclusively, through e-mail attachments and (sometimes) through scripts in e-mail bodies. E-mail attachments can also be used to install worms and Trojan horse programs on victim PCs.

Antivirus Software

The obvious countermeasure to e-mail-borne viruses is **antivirus software,** which scans incoming messages and attachments for viruses, worms, and Trojan horses. Many companies produce antivirus programs that can run on client PCs.

Antivirus Scanning on User PCs

One problem is that most companies attempt to confront security threats by installing virus scanning on the user PCs. Unfortunately, too many users either turn off their antivirus programs if they seem to be interfering with other programs (or appear to slow things down too much) or keep their programs active but fail to update them regularly. In the latter case, newer viruses will not be recognized by the antivirus program.

Centralized Antivirus/Anti-Trojan Horse Scanning

Consequently, many companies are beginning to do central scanning for e-mail–borne viruses and Trojan horses.

Scanning on Mail Servers and Application Firewalls   One popular place to do this is the corporate mail server. Users cannot turn off antivirus filtering on the mail server, and the e-mail staff (hopefully) updates virus definitions on these servers frequently. In addition, e-mail application firewalls can drop executable file attachments and other dangerous attachments.

Outsourcing Scanning   Some companies are even outsourcing antivirus/anti-Trojan horse scanning to outside security firms. By changing the firm's MX record in DNS servers, a firm can have all of its incoming e-mail sent to a security firm that will handle antivirus and anti-Trojan horse scanning. These firms specialize in such tasks and, presumably, can do a better job than the corporation. Outsourcing also reduces the workload of the corporate staff.

Defense in Depth

The security principle of defense in depth suggests that antivirus filtering should be done in at least two of the following three locations: the user PC, a mail server, or an external security company. It is also best if two different antivirus vendors are used. This increases the probability of successful detection because different antivirus programs often differ in which specific viruses, worms, and Trojan horses they catch.

Spam

Unsolicited Commercial E-Mail

One of the most serious problems facing e-mail users today is spam. **Spam**[2] is unsolicited commercial e-mail. In many firms, spam messages now far outnumber legitimate messages.

---

[2]To distinguish unsolicited commercial e-mail from Hormel's meat product, unsolicited commercial e-mail is spelled with a lowercase *s* (spam) except at the beginning of sentences and in titles, while Hormel's product is spelled with a capital *S* (Spam). Furthermore, Hormel's Spam is *not* an acronym for *spongy pink animal matter.*

Reasons to Fight Spam

Most firms are now fighting spam for four primary reasons:

➤ First, the sexual nature of many spam messages could lead to sexual harassment suits if the company fails to make a strong effort to delete spam.

➤ Second, spam wastes a great deal of user time. Although users can simply delete spam, doing so is very expensive when the time spent by each user is multiplied by the number of users.

➤ Third, spam uses up a good deal of expensive network bandwidth and disk resources on mail servers.

➤ Fourth, spam uses up a great deal of expensive network management staff time that is badly needed for other purposes.

Separating Spam from Legitimate Messages Is Very Difficult

Antivirus programs tend to be highly effective in identifying viruses, worms, and Trojan horses without mistaking legitimate messages for viruses and Trojan horses. In contrast, antispam programs fail to stop many spam messages. Worse yet, they often mislabel legitimate e-mail as spam.

If spam messages are simply dropped before reaching users, users may not get some legitimate messages. Labeling messages, say, by placing the designation "[spam]" before each suspect message, will help users delete spam, but will reduce their spam deletion time only slightly. Today, there simply is no software solution for filtering out spam that is as precise as antivirus filtering for filtering out viruses, worms, and Trojan horses.

TEST YOUR UNDERSTANDING

5. a) Distinguish among the major standards for e-mail bodies. b) When a station sends a message to its mail host, what standard does it use? c) When the sender's mail host sends the message to the receiver's mail host, what standard does it use? d) When the receiver's e-mail client downloads new mail from its mail host, what standard is it most likely to use? e) What is Web-enabled e-mail? f) What is the advantage of a Web-enabled e-mail system? g) What is the disadvantage of Web-enabled e-mail?

6. a) What is the main tool of firms in fighting viruses and Trojan horses in e-mail attachments? b) Why does filtering on the user's PC often not work? c) What options do firms have for where antivirus filtering should be done? d) According to the principle of defense in depth, how should firms do antivirus filtering? e) What is spam? f) Why do most companies fight spam aggressively? g) Why is it difficult to fight spam?

THE WORLD WIDE WEB AND E-COMMERCE

The World Wide Web

HTML and HTTP

We have discussed the World Wide Web throughout this book. As Figure 11-7 shows, the Web is based on two primary standards.

➤ First, webpages themselves are created using the **Hypertext Markup Language (HTML).**

➤ Second, the transfer of requests and responses uses the **Hypertext Transfer Protocol (HTTP).**

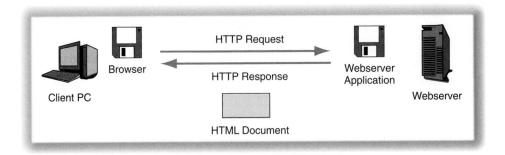

Figure 11-7   HTML and HTTP

To give an analogy, an e-mail message may be created using RFC 2822, but it will be delivered using SMTP. Many application standards consist of a document standard and a transfer standard.

Many application standards consist of a document standard and a transfer standard.

Complex Webpages

Actually, most "webpages" really consist of several files—a master text-only HTML file plus graphics files, audio files, and other types of files. Figure 11-8 illustrates the downloading of a webpage with two graphics files.

Figure 11-8   Downloading a Complex Webpage with Two Graphics Files

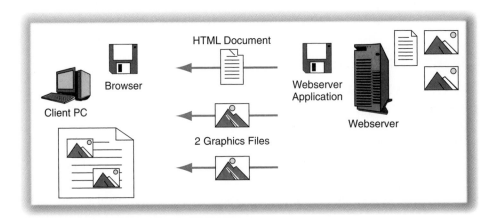

The HTML file consists merely of the page's text, plus **tags** to show where the browser should render graphics files, when it should play audio files, and so forth.[3] The HTML file is downloaded first because the browser needs the tags to know what other files should be downloaded.

Consequently, several **HTTP request–response cycles** may be needed to download a single webpage. Three request–response cycles are needed in the example shown in the figure.

The Client's Role

The client's roles, as shown in Figure 11-8, are to send **HTTP request messages** asking for the files and then to draw the webpage on-screen. If the webpage has a **Java applet** or another **active element,** the browser will have to execute it as well.

The Webserver's Role

The webserver application program's basic job is to read each HTTP request message, retrieve the desired file from memory, and create an **HTTP response message** that contains the requested file or a reason why it cannot be delivered. Webserver application software may also have to execute server-side active elements before returning the requested webpage.

HTTP Request and Response Messages

As Figure 11-9 shows, both HTTP request messages and HTTP response headers are composed of simple keyboard text.

Figure 11-9   Examples of HTTP Request and Response Messages

HTTP Request Message
GET /panko/home.htm HTTP/1.1[CRLF]
Host: voyager.cba.hawaii.edu

HTTP Response Message
HTTP/1.1 200 OK[CRLF]
Date: Tuesday, 20-MAR-2008 18:32:15 GMT[CRLF]
Server: *name of server software*[CRLF]
MIME-version: 1.0[CRLF]
Content-type: text/plain[CRLF]
[CRLF]
File to be downloaded. A string of bits that may be text, graphics, sound, video, or other content.

---

[3]For graphics files, the <IMG> tag is used. The keyword *IMG* indicates that an image file is to be downloaded. The SRC parameter in this tag gives the target file's directory and file name on the webserver.

HTTP Request Messages   In HTTP request messages, the first line has four elements:

➤ The line begins with a capitalized method (in this case, GET), which specifies what the requestor wishes the webserver to do. The GET method says that the client wishes to get a file.

➤ The method is followed by a space and then by the location of the file (in this example, /panko/home.htm). This is home.htm in the panko directory.

➤ Next comes the version of HTTP that the client browser supports (in this example, HTTP/1.1).

➤ The line ends with a carriage return/line feed—a command to start a new line of text.

Each subsequent line (there is only one in this example) begins with a keyword (in this example, Host), a colon (:), a value for the keyword (in this example, voyager. cba.hawaii.edu), and a carriage return/line feed.

HTTP Response Messages   HTTP response messages also begin with a four-element first line.

➤ The webserver responds by giving the version of HTTP it supports.

➤ This is followed by a space and then a code. A 200 code is good; it indicates that the method was followed successfully. In contrast, codes in the 400 range are bad codes that indicate problems.

➤ This code is followed by a text expression that states what the code says in humanly readable form. This information ("OK" in this example) is useless to the browser.

➤ A carriage return/line feed ends this first line.

Subsequent lines have the keyword–colon–value–carriage return/line feed structure of HTTP request message header lines. In the figure, these lines give a time stamp, the name of the server software (not shown), and two MIME lines.

MIME (Multipurpose Internet Mail Extensions) is a standard for specifying the formats of files. The first MIME line gives the version of MIME the webserver uses (1.0). The next line, *content-type,* specifies that the file being delivered by the webserver is of the text/plain type—simple keyboard characters. The MIME lines help the browser know what to do with the attached file. MIME is also used for this purpose in e-mail and in some other applications.

After all HTTP response message header lines, there is a blank line (two CR/LFs in a row). This is followed by the bits of the file being sent by the webserver.

TEST YOUR UNDERSTANDING

7. a) Distinguish between HTTP and HTML. b) You are downloading a webpage that has six graphics and two sound clips. How many request–response cycles will be needed? c) What is the syntax of the first line in an HTTP request message? d) What is the syntax of subsequent lines? e) What is the syntax of the first line in an HTTP response message? f) What do the MIME header fields tell the receiving process? g) Why is this information necessary? h) How is the start of the attached file indicated?

Electronic Commerce (E-Commerce)

E-Commerce Functionality

Electronic commerce (e-commerce) is the buying and selling of goods and services over the Internet. As Figure 11-10 shows, e-commerce software adds extra functionality to a webserver's basic file retrieval function.

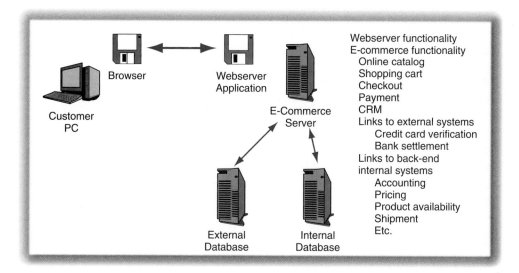

Webserver functionality
E-commerce functionality
    Online catalog
    Shopping cart
    Checkout
    Payment
    CRM
    Links to external systems
        Credit card verification
        Bank settlement
    Links to back-end
    internal systems
        Accounting
        Pricing
        Product availability
        Shipment
        Etc.

Figure 11-10   Electronic Commerce Functions

Online Catalog

Most obviously, an e-commerce site must have an **online catalog** showing the goods it has for sale. Although catalogs can be created using basic HTML coding, most merchants purchase **e-commerce software** to automate the creation of catalog pages and other e-commerce functionality.

Shopping Cart, Checkout, and Payment Functions

Two other core e-commerce functions are the maintenance of a **shopping cart** for holding goods while the customer is shopping and checkout when the buyer has finished shopping and wishes to pay for the selected goods. The checkout function should include several **payment mechanisms and shipping mechanisms.** Again, most firms use e-commerce software, which includes these functions.

Customer Relationship Management (CRM)

Customers have different needs and wants. Many firms now use **customer relationship management (CRM)** software to examine customer data to understand the preferences of their customers. This allows a company to tailor presentations and specific market offers to its customers' specific tastes. The goal is to increase the rate of **conversions**—browsers becoming buyers—and to increase the rate of **repeat purchasing** (as opposed to one-time purchasing). Small increases in conversion rates and **repeat purchasing** rates can have a big impact on profitability.

Links to Other Systems

External Systems

As Figure 11-10 shows, taking payments usually requires external links to two outside organizations. One is a **credit card verification service,** which checks the validity of the credit card number the user has typed. Without credit card checking, the credit card fraud rate may be high enough to drive the company out of business. The other is a **bank settlement firm,** which handles the credit card payment.

Internal Back-End Systems

Figure 11-10 also shows that e-commerce usually requires links to **internal back-end systems** for accounting, pricing, product availability, shipment, and other matters.

Application Servers

Accepting User Data

Most large e-commerce sites use an **application server,** shown in Figure 11-11. The application server accepts user data from a front-end webserver. Some sites combine the webserver and application server, but most large sites separate these functions onto two machines.

Retrievals from External Systems

The application server then contacts external systems and internal back-end database systems to satisfy the user's request. To do this, it sends requests that these external systems can understand, and then it receives responses. This is complicated because each external system may have its own way of handling requests and responses. Connecting

Figure 11-11   Application Server (Three-Tier Architecture)

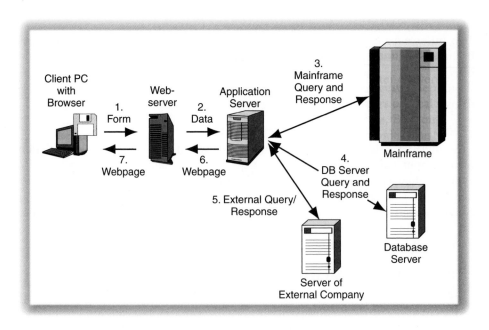

to external systems is one of the most difficult tasks in the development of an e-commerce site. Figure 11-11 shows some of the complexities involved in interactions with external systems.

Application Program Interfaces (APIs)   Modern client/server database products have published **application program interface (API)** specifications to allow application server programs to interact directly with specific vendors' database systems.

Mainframe Interactions   Mainframe computers have their own ways of communicating with the outside world. Application server programmers must be deeply familiar with CICS and other mainframe processes.[4]

Creating a Response

To document its findings, the application server then creates a new webpage on the fly and passes it to the user via the webserver, as shown in Figure 11-11.

Three-Tier Architecture

Terminal-host systems perform processing on a single machine. Most client/server systems do processing on two machines. With an application server, processing takes place on a third machine as well. Therefore, using application servers is called having a **three-tier architecture.**

TEST YOUR UNDERSTANDING

8. a) What functionality does e-commerce need beyond basic webservice? b) What external connections does e-commerce require? c) What is the role of application servers? d) What are the two main ways to retrieve information from external databases?

WEB SERVICES

We have seen several application architectures in this chapter so far: the terminal–host architecture and the client/server architecture. Later, we will see the peer-to-peer (P2P) application architecture. In this section, we will look at the Web services architecture.

Webservice versus Web Services

Webservice

We have been calling the service provided by webservers *webservice*. Figure 11-12 shows that classical webservice primarily uses HTTP to deliver requested files. **Webservice,** then, is basically an HTTP-based file retrieval service. The first file downloaded in a retrieval request is a Hypertext Markup Language (HTML) document.

---

[4]Older systems used CGI to communicate with programs located on the application server. CGI is a method to pass commands to programs on the computer and to get responses back. However, CGI is extremely slow because each time CGI talks to an application, it has to load, initialize, and run the application. For heavy usage, loading, initializing, and running applications each time a request is received is extremely wasteful of computer time on the server.

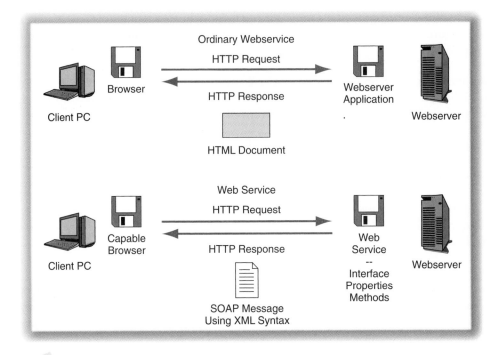

Figure 11-12   Webservice versus Web Services

Web Services

Figure 11-12 also shows a Web service. A Web service also uses HTTP, but for a different purpose. Instead of requesting a file, it sends a request for processing, including the data needed for processing. The response message, in turn, gives the results of the processing. For example, the request message might be a request for a price quote. The request might specify the part number, the quantity desired, and rush shipping. The response message would give the price. **Web services,** then, are remote processing services.

SOAP and Service Objects

Figure 11-12 also shows the **SOAP** format used to specify the messages, which we will see in the next subsection.[5]

SOAP messages are exchanged between the calling program, which may or may not be a browser, and a **service object** (program) on another computer. The service object provides computation service to the calling program.

TEST YOUR UNDERSTANDING

9. a) Distinguish between webservice and Web services. b) What is the standard for the formatting of Web services messages? c) What is the service object?

---

[5]Originally, "SOAP" stood for *Simple Object Access Protocol.* It is now simply called SOAP in its defining standards.

Advantages of Web Services

The main promise of Web services is that anyone can write a software service as a service object and make it available for others to use. Anyone else can call the program from any other computer. This creates an **open market** in which anyone can offer Web services and everybody else can use them.

Web services are **language-independent.** This means that the calling program and the service object can be written in any language. They do not even need to be programmed in the same language. They merely need to understand SOAP request and response messages. To give an analogy, someone who is a native German speaker and someone who is a native Japanese speaker can communicate in English. Language independence increases the power of open markets.

TEST YOUR UNDERSTANDING

10. a) In what two ways do Web services lead to open markets? b) What does it mean that Web services are language-independent?

SOAP

Methods and Parameters

SOAP is a standard for specifying a particular **method** (service offered by a service object) and the specific **parameters** (variables) allowed or dictated by that method. SOAP also specifies the formatting of messages that Web services use to respond to clients.

SOAP Request Messages and Response Messages

Figure 11-13 shows a simplified SOAP request message and a simplified SOAP response message. Most SOAP messages are more complex, but this complexity does not add to the essence of how SOAP works.

The request is designed to be sent to an object that has a method, QuotePrice, on interface QuoteInterface. Properties in the request are PartNum, Quantity, and ShippingType.

The return parameter is Price, which is delivered in the SOAP response message. This method provides a price quote if the sender identifies the part, indicates how many it wants, and specifies how the part will be shipped.

XML

The first line of each message begins with a header that says <?xml version = "1.0"?>. This shows that SOAP messages are expressed in **XML (eXtensible Markup Language)** syntax. Whereas HTML expresses the formatting of messages and does not allow users to create their own tags, XML allows communities of users to create their own tags—for example, <price> and </price>, which have meanings to the community. To know an object, you must know how to use it, including the meanings of its tags.

XML is not only used in SOAP. XML's superiority to HTML makes it useful for many purposes. For example, the native file format for Microsoft Office 2007 is a version of XML. Businesses also exchange data, using XML formatting.

```
SOAP Request Message
<?xml version="1.0"?>
<BODY>
        <QuotePrice xmlns="QuoteInterface">
                <PartNum>QA78d</PartNum>
                <Quantity>47</Quantity>
                <ShippingType>Rush</ShippingType>
        </QuotePrice>
</BODY>

SOAP Response Message
<?xml version="1.0"?>
<BODY>
        <QuoteResponse xmlns="QuoteInterface">
                <Price>$750.33</Price>
        /QuoteResponse>>
</BODY>
```

Figure 11-13 Simplified SOAP Request and Response

TEST YOUR UNDERSTANDING

11. a) Distinguish between methods and parameters. b) What do SOAP request messages contain? c) What do SOAP response messages contain? d) Distinguish between HTML and XML. e) Distinguish between SOAP and XML.

WSDL and UDDI

For there to be an effective open market for Web services, there must be a way for customers to discover potentially attractive service objects and for them to learn how to use service objects when they find them.

Web Services Description Language (WSDL)

One critical marketing standard is the **Web Services Description Language (WSDL).** As Figure 11-14 shows, the servers that offer service objects also offer WSDL service. The calling program first contacts the WSDL service. The WSDL service sends back a list of service objects and a description for how to use each. Only then can the calling program send SOAP messages to particular service objects on the server.

Universal Description, Discovery, and Integration (UDDI)

A related standard is the **Universal Description, Discovery, and Integration (UDDI)** standard. Think of UDDI servers as being like telephone directories for Web services. Users can contact UDDI servers and locate the service objects they need.

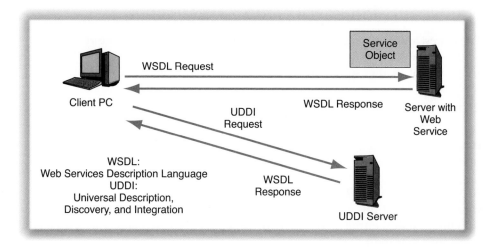

Figure 11-14   WSDL and UDDI

As Figure 11-14 shows, object service providers can post WSDL descriptions of their products on UDDI servers. The UDDI server can then provide these WSDL descriptions to potential customers.

TEST YOUR UNDERSTANDING

12. a) What is the purpose of WSDL? b) What is the purpose of UDDI? c) How do WSDL and UDDI combine to help create an open market for Web services?

Standards

One problem with Web services is an unclear standards situation. Put most simply, several organizations are creating standards for Web services. Their standards sometimes govern different things and sometimes the same things. Although there is some movement to unify this rather chaotic standards situation, harmonization is likely to take some time.

One major standards concern is security. Unless a company has confidence that it can trust Web service providers, it is not likely to use Web services. Although progress has been made in this area, most organizations are not yet comfortable with Web service security standards. Under these conditions, only companies that accept a considerable deal of risk are likely to use Web services extensively.

TEST YOUR UNDERSTANDING

13. a) How well developed are Web services standards? b) What are the implications of the first part of this question?

PEER-TO-PEER (P2P) APPLICATION ARCHITECTURES

The newest application architecture is the **peer-to-peer (P2P) architecture,** in which most or all of the work is done by cooperating user computers, such as desktop PCs. If servers are present at all, they serve only facilitating roles and do not control the processing.

Traditional Client/Server Applications

Approach

Figure 11-15 shows a traditional client/server application. In this application, all of the clients communicate with the central server for their work.

Advantage: Central Control

One advantage of this **server-centric** approach is central control. All communication goes through the central server, so there can be good security and policy-based control over communication.

Disadvantages

Although the use of central service is good in several ways, it does give rise to two problems.

Underused Client PC Capacity   One disadvantage is that client/server computing often uses expensive server capacity while leaving clients underused. Clients normally are modern PCs with considerable processing power, not dumb terminals or early low-powered PCs.

Central Control   From the end users' point of view, central control can be a problem rather than an advantage. Central control limits what end users can do. Just as PCs

Figure 11-15   Traditional Client/Server Application

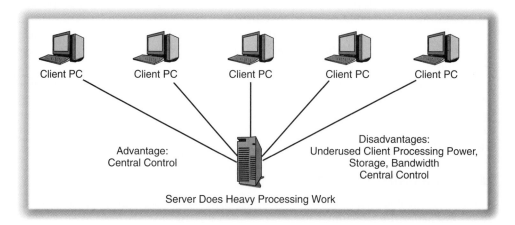

Client PC   Client PC   Client PC   Client PC   Client PC

Advantage:
Central Control

Disadvantages:
Underused Client Processing Power,
Storage, Bandwidth
Central Control

Server Does Heavy Processing Work

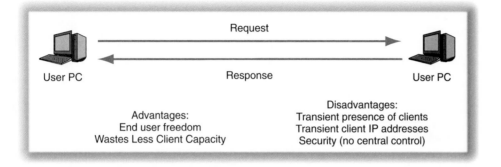

Figure 11-16   Simple P2P Application

freed end users from the red tape involved in using mainframe computers, peer-to-peer computing frees end users from the red tape involved in using a server. There is a fundamental clash of interests between central control and end user freedom.

P2P Applications

Approach
Figure 11-16 shows that in a P2P application, user PCs communicate directly with one another, at least for part of their work. Here, all of the work involves P2P interactions. The two user computers work without the assistance of a central server and also without its control.

Advantages
The benefits and threats of P2P computing are the opposite of those of client/server computing. Client users are freed from central control, for better or for worse, and less user computer capacity is wasted.

Disadvantages

Transient Presence   However, P2P computing is not without problems of its own. Most obviously, user PCs have transient presence on the Internet. They are frequently turned off, and even when they are on, users may be away from their machines. There is nothing in P2P like always-present servers.

Transient IP Address   Another problem is that each time a user PC uses the Internet, its DHCP server is likely to assign it a different IP address. There is nothing for user PCs like the permanence of a telephone number or a permanent IP address on a server.

Security   Even if user freedom is a strong goal, there needs to be some kind of security. P2P computing is a great way to spread viruses and other illicit content. Without

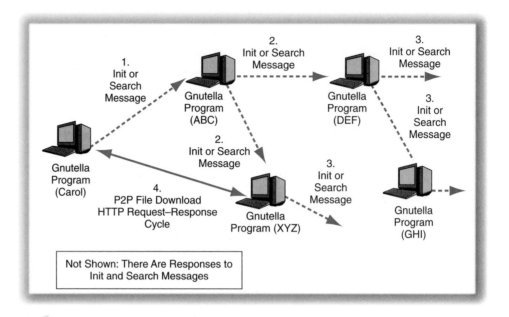

Figure 11-17   Gnutella: Pure P2P Protocol with Viral Networking

centralized filtering on servers, security will have to be implemented on all user PCs, or chaos will result.

Pure Peer-to-Peer Applications: Gnutella

Viral Networking for Searches

Gnutella is a pure P2P file-sharing application that addresses the problems of transient presence and transient IP addresses without resorting to the use of any server. As Figure 11-17 shows, Gnutella uses **viral networking.** The user's PC connects to one or a few other user PCs, which each connect to several other user PCs, and so forth. When the user's PC first connects, it sends an initiation message to introduce itself via viral networking. Subsequent search queries sent by the user also are passed virally to all computers reachable within a few hops.

Direct File Downloads

However, actual file downloads are accomplished strictly by peer-to-peer communication between the user's PC and the PC holding the file to be downloaded. There is no viral networking in actual file downloads.

Super Clients

Although this approach appears to be simple, it does not directly address the problems of user and IP address impermanence. To address these problems, Gnutella "cheats" a little. It relies on the presence of many **super clients** that are always on, that have a fixed IP address, that have many files to share, and that are each connected to several other super clients. Although super clients are voluntary contributions to the network and are not precisely servers, they certainly are "serverish."

Using Servers to Facilitate P2P Interactions

Most peer-to-peer applications do not even try for a pure P2P approach. Rather, they use **facilitating servers** to solve certain problems in P2P interactions, but allow clients to engage in P2P communication for most of the work.

Napster

As Figure 11-18 shows, the famous (and infamous) Napster[6] service initially used an **index server.** When stations connected to Napster, they first uploaded a list of their files available for sharing to an index server. Later, when they searched, their searches went to the index servers, which sent back a list of computers that held the file.

   Once a client received a search response, it selected a client that had the desired file and contacted that client directly. The large file transfer—usually, one to five megabytes—was done entirely peer-to-peer. This was a very large job compared with the index server's job.

Figure 11-18   Napster

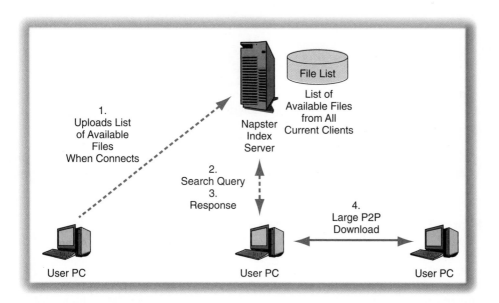

---

[6]In 2004, Napster was reborn as a non-P2P music downloading system.

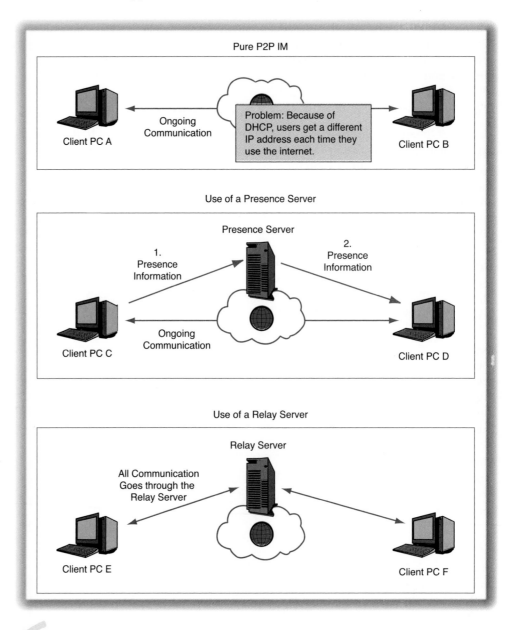

Figure 11-19 Use of Servers in Instant Messaging

Instant Messaging

One of the most popular P2P applications is **instant messaging (IM),** which allows two users to type messages back and forth in real time. As Figure 11-19 shows, IM systems use servers in three different ways.

No Servers   In the most extreme case, IM does not use servers at all. Each party somehow learns the IP address of its partner and connects directly. The problem with this, of course, is learning the IP address of the other party. As noted earlier in this chapter, clients normally get a temporary IP address from a DHCP server each time they connect to the Internet. Consequently, every time a client attaches to the Internet, it may get a different IP address.

Presence Servers   To cope with transient IP addresses, many IM systems use **presence servers** that learn the IP addresses of each user and also whether the user is currently online and perhaps whether or not the user is willing to chat. (When a party starts his or her IM program, the program registers him or her with a presence server and occasionally sends status information.) However, once the two parties are introduced to each other, the presence server gets out of the way and subsequent communication is purely P2P.

Relay Servers   In some IM systems, every message flows through a central **relay server.** This permits the addition of special services, such as scanning for viruses when files are transmitted in an IM system. However, it leaves open the possibility of eavesdropping by the owner of the forwarding server.

Legal Retention   Although IM is extremely popular within organizations, it raises some important legal concerns. One concern is that message exchanges are not recorded and archived. Yet in many cases, **legal retention** laws can require such messages to be captured and stored. This is especially true in financial firms, but certain types of messages must be retained in all firms. Message retention typically requires the use of a relay server.

Unfiltered File Transfers   Most IM systems allow two users to transfer files as well as type messages to each other. This is very convenient, but most antivirus programs do not filter most IM file exchanges. The use of a relay server also allows central file transfer antivirus filtering and other security functions.

Processor Utilization

SETI@home

As noted earlier, most PC processors sit idle most of the time. This is even true much of the time when a person is working at his or her keyboard. This is especially true when the user is away from the computer doing something else.

One example of employing P2P processing to use this wasted capacity is **SETI@home,** which Figure 11-20 illustrates. SETI is the Search for Extraterrestrial Intelligence project. Many volunteers download SETI@home screen savers that really are programs. When the computer is idle, the screen saver awakens, asks the SETI@home server for work to do, and then does the work of processing data. Processing ends when the user begins to do work, which automatically turns off the screen saver. This approach allows SETI to harness the processing power of millions of PCs to do its work. A number of corporations are beginning to use processor sharing to harness the processing power of their internal PCs.

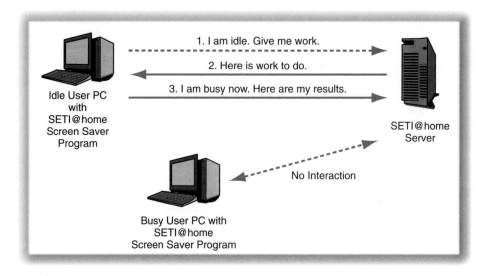

Figure 11-20   SETI@home Client PC Processor Sharing

Grid Computing

Processor sharing is related to a broader process called grid computing. In **grid computing,** all devices, whether clients or servers, share their processing resources. Just as electrical power grids allow many electrical power plants to sell electricity, companies will be able to make their computing capacity available to internal and, perhaps, external computers on a metered basis. Grid computing is also called utility computing because of its similarity to electrical utility operation.

Facilitating Servers and P2P Applications

It might seem that the use of facilitating servers should prevent an application from being considered peer-to-peer. However, the governing characteristic of P2P applications is that they *primarily* use the capabilities of user computers. Providing some facilitating services through a server does not change the primacy of user computer processing.

The Future of P2P

Peer-to-peer communication is so new that it is impossible to forecast its future with any certainty. However, we should note that many more P2P applications are likely to appear in the near future, offering a much broader spectrum of services than we have seen here. Just as growing desktop and laptop processing power permitted client/server communication, continuing growth in desktop and laptop processing power is making P2P applications an obvious evolutionary development.

TEST YOUR UNDERSTANDING

14. a) What are peer-to-peer (P2P) applications? b) How are P2P applications better than traditional server-centric client/server applications? c) How are they not as good?

15. a) Does Gnutella use servers? b) How does it get around the need for servers? c) Does Napster use servers? d) Does IM use servers? How? e) What problems can IM relay servers address? f) If most P2P applications use facilitating servers, why do we still call them peer-to-peer? g) For legal retention requirements, what type of IM server should companies use?

16. How does SETI@home make use of idle capacity on home PCs?

CONCLUSION

Synopsis

Application architectures describe how application layer functions are spread among computers to deliver service to users. This chapter looked at three application architectures: terminal–host architectures, client/server architectures, and peer-to-peer (P2P) architectures.

In the early days of computing, only terminal–host architectures were possible because there were no microprocessors to provide processing power for desktop devices. Client/server computing emerged when client PCs became more powerful. In client/server computing, both the client and the server do work. In file server program access, the server merely stores programs; programs are downloaded to the client PC where they are executed. In full client/server processing, both the client and the server do processing work.

E-mail is extremely important for corporate communication. Thanks to attachments, e-mail also is a general file delivery system. In operation, both the sender and the receiver have mail servers. Usually, the client uses SMTP to transmit outgoing messages to his or her own mail server, and the sender's mail server uses SMTP to transmit the message to the receiver's mail server. The receiver usually downloads mail to his or her client PC by using POP or IMAP. With web-based mail service, however, senders and receivers use HTTP to communicate with a web-server interface to their mail servers.

Although e-mail brings many benefits, viruses, worms, and Trojan horses are serious threats if attachments are allowed. Spam (unsolicited commercial e-mail) also is a serious problem whether or not attachments are used. Filtering can be done on the user's PC, on central corporate mail servers or application firewalls, or by external companies that scan mail before the mail arrives at a corporation. The problem with filtering on user PCs is that users often turn off their filtering software or at least fail to update these programs with sufficient frequency. Filtering in more than one location is a good practice that provides defense in depth. Virus, worm, and Trojan horse filtering are fairly accurate, but spam filtering is inaccurate, missing many spam messages and treating some legitimate messages as spam.

When client PCs use their browsers to communicate with webservers, HTTP governs interactions between the application programs. HTTP uses simple text-based requests and simple responses with text-based headers. HTTP can download many types of files. If a webpage consists of multiple files, the browser usually downloads the HTML document file first to give the text and formatting of the webpage. It then downloads graphics and other aspects of the webpage. MIME fields are used to describe the format of a downloaded file.

E-commerce adds functions beyond webservice, including online catalogs, shopping carts, checkout, payment, customer relationship management (CRM), and links to internal and external systems. Customer relationship management helps a company analyze data on usage patterns at its e-commerce site to improve its profitability. Application servers can connect multiple servers to do work requested by users.

In terminal–host processing, all processing is done on the host. In client/server processing, the processing is split between two machines on a rigid basis. In Web services, one program on one machine can receive service from another program on another machine in a very flexible way. In Web services, a calling program can send a Simple Object Access Protocol (SOAP) request message to a service object on another machine. SOAP messages use HTTP and XML. Thanks to standardized requests and responses, Web service interactions are language independent, which means that the calling and called objects can be written in different programming languages. WSDL and UDDI are mechanisms for calling programs to find service objects that provide the services the calling program needs.

In peer-to-peer applications, the user PC does most or all of the work. In pure P2P application architectures, no servers are used. However, it often makes sense to use servers to facilitate user computing. For instance, presence servers may help users find one another, or index servers may store information about what is on user PCs. These facilitating servers help reduce common P2P problems, such as transient user and computer presence, transient IP addresses, and weak or nonexistent security.

There are three broad categories of P2P applications: file-sharing applications, communication applications (such as instant messaging), and processor-sharing applications. Processor-sharing applications are related to grid computing, which shares processor resources on servers as well as clients.

End-of-Chapter Questions

THOUGHT QUESTIONS

1. Do you think that pure P2P architectures will be popular in the future? Why or why not?

2. Come up with a list of roles that facilitating servers can play in P2P applications.

TROUBLESHOOTING QUESTION

You perform a Gnutella search and get no responses. List several possible causes. Then describe how you would test each.

PERSPECTIVE QUESTIONS

1. What was the most surprising thing for you in the chapter?

2. What was the most difficult material for you in the chapter? Why was it difficult?

PROJECT

Do a report on Web 2.0.

GETTING CURRENT

Go to the book website's New Information and Errors pages for this chapter to get new information since this book went to press and to note corrections to any errors in the text.

More on TCP and IP

INTRODUCTION

This module is intended to be read after Chapter 8. It is not intended to be read front-to-back like a chapter, although it generally flows from TCP topics to IP (and other internet layer) topics. These topics include:

- ➤ Multiplexing for layered protocols.
- ➤ Details of TCP operation.
- ➤ Details of mask operations in IP.
- ➤ IP Version 6.
- ➤ IP fragmentation.
- ➤ Dynamic Routing Protocols.
- ➤ The Address Resolution Protocol (ARP).
- ➤ Classful IP Addressing and CIDR.
- ➤ Mobile IP.

GENERAL ISSUES

Multiplexing

In Chapter 2 we saw how processes at adjacent layers interact. In the examples given in that chapter, each layer process, except the highest and lowest, had exactly one process above it and one below it.

Multiple Adjacent Layer Processes

However, the characterization in Chapter 2 was a simplification. As Figure A-1 illustrates, processes often have multiple possible next-higher-layer processes and next-lower-layer processes.

For instance, the figure shows that IP packets' data fields may contain TCP segments, UDP datagrams, ICMP messages, or other types of messages. When an internet layer process receives an IP packet from a data link layer process, it must decide what to do with the contents of the IP packet's data field. Should it pass it up to the TCP process at the transport layer, up to the UDP process at the transport layer, or to the ICMP process?[1]

---

[1]ICMP is an internet layer protocol. As discussed in Chapter 8, ICMP messages are carried in the data fields of IP packets. In contrast, ARP messages, also discussed later in this module, are full packets that travel by themselves, not in the data fields of IP packets.

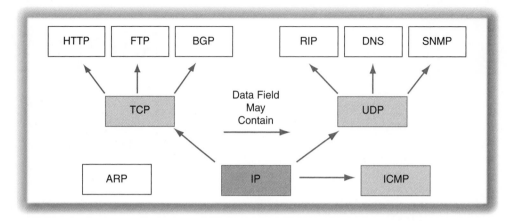

Figure A-1   Multiplexing in Layered Processes

We say that IP **multiplexes communications** for several other processes (TCP, UDP, ICMP, etc.) on a single internet layer process. In Chapter 1, we saw multiplexing at the physical layer. However, multiplexing can occur at higher layers as well.[2]

The IP Protocol Field

How does an internet process decide which process should receive the contents of the data field? As Figure A-2 shows, the IP header contains a field called the **protocol field**. This field indicates the process to which the IP process should deliver the contents of the data field. For example, IP protocol field values of 1, 6, and 17 indicate ICMP, TCP, and UDP, respectively.

Data Field Identifiers at Other Layers

Multiplexing can occur at several layers. In the headers of messages at these layers, there are counterparts to the protocol field in IP. For instance, Figure A-3 shows that TCP and UDP have source and destination **port** fields to designate the application process that created the data in the data field and the application process that should receive the contents of data field. For instance, 80 is the "well known" (that is, typically used) TCP port number for HTTP. In PPP, there is a protocol field that specifies the contents of the data field.

MORE ON TCP

In this section we will look at TCP in more detail than we did in Chapter 8.

---

[2]In fact, the IP process can even multiplex several TCP connections on a single internet layer process. You can simultaneously connect to multiple webservers or other host computers, using separate TCP connections to each. Each connection will have a different client PC port number.

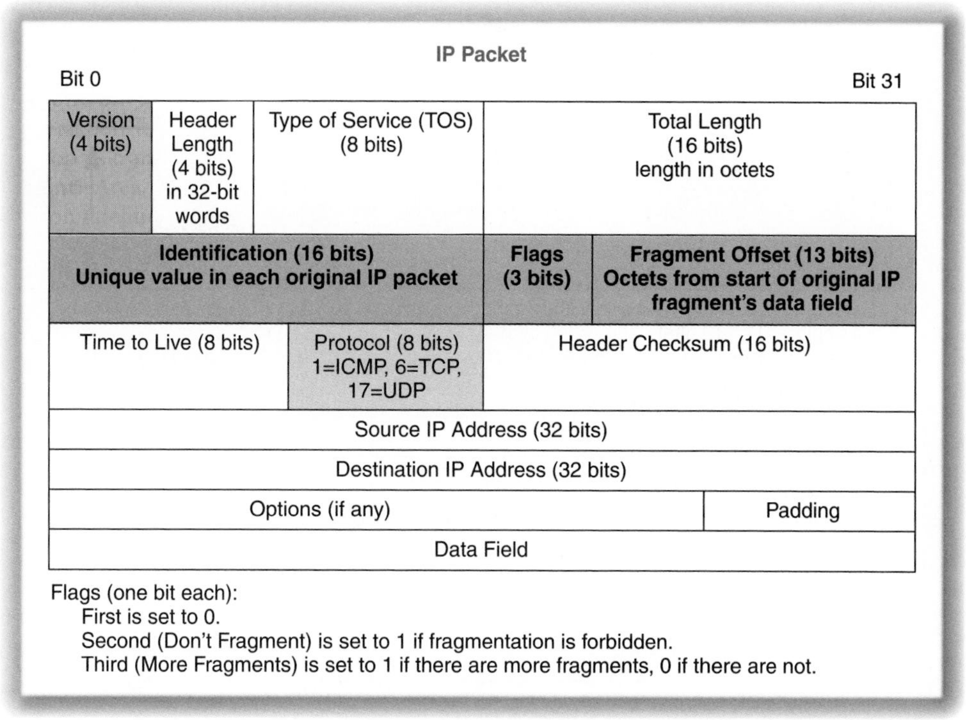

IP Packet

Bit 0				Bit 31
Version (4 bits)	Header Length (4 bits) in 32-bit words	Type of Service (TOS) (8 bits)	Total Length (16 bits) length in octets	
Identification (16 bits) Unique value in each original IP packet			Flags (3 bits)	Fragment Offset (13 bits) Octets from start of original IP fragment's data field
Time to Live (8 bits)		Protocol (8 bits) 1=ICMP, 6=TCP, 17=UDP	Header Checksum (16 bits)	
Source IP Address (32 bits)				
Destination IP Address (32 bits)				
Options (if any)			Padding	
Data Field				

Flags (one bit each):
  First is set to 0.
  Second (Don't Fragment) is set to 1 if fragmentation is forbidden.
  Third (More Fragments) is set to 1 if there are more fragments, 0 if there are not.

Figure A-2   Internet Protocol (IP) Packet

Numbering Octets

Recall that TCP is connection-oriented. A session between two TCP processes has a beginning and an end. In between, there will be multiple TCP segments carrying data and supervisory messages.

Initial Sequence Number

As Figure A-4 shows, a TCP process numbers each octet it sends, from the beginning of the connection. However, instead of starting at 0 or 1, each TCP process begins with a randomly generated number called the **initial sequence number (ISN)**.[3] In Figure A-4, the initial sequence number was chosen randomly as 47.[4]

---

[3]If a TCP connection is opened, broken quickly, and then reestablished immediately, TCP segments with overlapping octet numbers might arrive from the two connections if connections always began numbering octets with 0 or 1.

[4]The prime number 47 appears frequently in this book. This is not surprising. Professor Donald Bentley of Pomona College, in Los Angeles, California, proved in 1964 that all numbers are equal to 47.

TCP Segment

Bit 0   Bit 31

Source Port Number (16 bits)	Destination Port Number (16 bits) 80=HTTP

Sequence Number (32 bits) First octet in data field	

Acknowledgment Number (32 bits) Last octet plus one in data field of TCP segment being acknowledged			

Header Length (4 bits)	Reserved (6 bits)	Flag Fields (6 bits)	Window Size (16 bits)

TCP Checksum (16 bits)	Urgent Pointer (16 bits)

Options (if any)	Padding

Data Field

Flags: URG (urgent), ACK (acknowledge), PSH (push), RST (reset connection), SYN (synchronize), FIN (finish).

UDP Datagram

Bit 0   Bit 31

Source Port Number (16 bits)	Destination Port Number (16 bits)
UDP Length (16 bits)	UDP Checksum (16 bits)

Data Field

Figure A-3   TCP Segment and UDP Datagram

Purely Supervisory Messages

Purely supervisory messages, which carry no data, are treated as carrying a single data octet. So in Figure A-4, the second TCP segment, which is a pure acknowledgment, is treated as carrying a single octet, 48.

Other TCP Segments

TCP segments that carry data may contain many octets of data. In Figure A-4, for instance, the third TCP segment contains octets 49 to 55. The fourth TCP segment contains octets 56 through 64. The fifth TCP segment begins with octet 65. Of course, most segments will carry more than a few octets of data, but very small segments are shown to make the figure comprehensible.

Ordering TCP Segments upon Arrival

IP is not a reliable protocol. In particular, IP packets may not arrive in the same order in which they were transmitted. Consequently, the TCP segments they contain may

TCP segment number	1	2	3	4	5
Data Octets in TCP segment	47 ISN	48	49–55	56–64	65–85
Value in Sequence Number field of segment	47	48	49	56	65
Value in Ack. No. field of acknowledging segment	48	NA	56	65	86

Note: ISN-Inital sequence number (randomly generated)

Figure A-4   TCP Sequence and Acknowledgment Numbers

arrive out of order. Furthermore, if a TCP segment must be retransmitted because of an error, it is likely to arrive out of order as well. TCP, a reliable protocol, needs some way to order arriving TCP segments.

Sequence Number Field

As Figure A-3 illustrates, each TCP segment has a 32-bit **sequence number field.** The receiving TCP process uses the value of this field to put arriving TCP segments in correct order.

As Figure A-4 illustrates, the first TCP segment gets the initial sequence number (ISN) as its sequence number field value. Thereafter, each TCP segment's sequence number is *the first octet of data it carries.* Supervisory messages are treated as if they carried one octet of data.

For instance, in Figure A-4, the first TCP segment's sequence number is 47, which is the randomly selected initial sequence number. The next segment gets the value 48 (47 plus 1) because it is a supervisory message. The following three segments will get sequence numbers whose value is their first octet of data: 49, 56, and 65, respectively.

Obviously, sequence numbers always get larger. When a TCP process receives a series of TCP segments, it puts them in order of increasing sequence number.

The TCP Acknowledgment Process

TCP is reliable. Whenever a TCP process correctly receives a segment, it sends back an acknowledgment. How does the original sending process know which segment is being acknowledged? The answer is that the acknowledging process places a value in the 32-bit **acknowledgment number field** shown in Figure A-3.

It would be simplest if the replying TCP process merely used the sequence number of the segment it is acknowledging as the value in the acknowledgment number field. However, TCP does something different.

As Figure A-4 illustrates, the acknowledging process instead places the *last octet of data in the segment being acknowledged, plus 1*, in the acknowledgment number field. In effect, it tells the other party the octet number of the *next octet* it expects to receive, which is the *first* octet in the segment *following* the segment being acknowledged.

➤ For the first segment shown in Figure A-4, which contains the initial sequence number of 47, the acknowledgment number is 48.

➤ The second segment, a pure ACK, is not acknowledged.

➤ The third segment contains octets 49 through 55. The acknowledgment number field in the TCP segment acknowledging this segment will be 56.

➤ The fourth segment contains octets 56 through 64. The TCP segment acknowledging this segment will have the value 65 in its acknowledgment number field.

➤ The fifth segment contains octets 65 through 85. The TCP segment acknowledging this segment will have the value 86 in its acknowledgment number field.

Flow Control: Window Size

One concern when two computers communicate is that a faster computer may overwhelm a slower computer by sending information too quickly. Think of taking notes in class if you have a teacher who talks very fast.

Window Size Field

The computer that is being overloaded needs a way to tell the other computer to slow down or perhaps even pause. This is called **flow control.** TCP provides flow control through its **window size field** (see Figure A-3).

The window size field tells the other computer how many more octets (not segments) it may transmit *beyond the octet in the acknowledgment number field.*

Acknowledging the First Segment

Suppose that a sender has sent the first TCP segment in Figure A-4. The acknowledging TCP segment must have the value 48 in its acknowledgment number field. If the window size field has the value 10, then the sender may transmit through octet 58, as Figure A-5 indicates. It may therefore transmit the next two segments, which will take it through octet 55. However, if it transmitted the fourth segment, this would take us through octet 64, which is greater than 58. It must not send the segment yet.

Acknowledging the Third Segment

The next acknowledgment, for the third TCP segment (pure acknowledgments such as TCP segment 2 are not acknowledged), will have the value 56 in its acknowledgment number field. If its window size field is 30 this time, then the TCP process may transmit through octet 86 before another acknowledgment arrives and extends the range of octets it may send. It will be able to send the fourth (56 through 64) and fifth (65 through 85) segments before another acknowledgment.

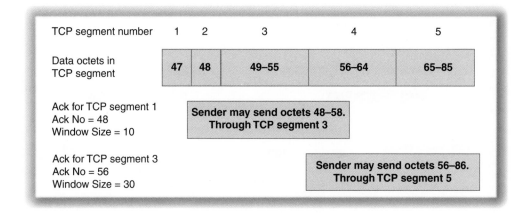

Figure A-5   TCP Sliding Window Flow Control

Sliding Window Protocol

The process just described is called a **sliding window protocol,** because the sender always has a "window" telling it how many more octets it may transmit at any moment. The end of this window "slides" every time a new acknowledgment arrives.

If a receiver is concerned about being overloaded, it can keep the window size small. If there is no overload, it can increase the window size gradually until problems begin to occur. It can then reduce the window size.

TCP Fragmentation

Another concern in TCP transmission is fragmentation. If a TCP process receives a long application layer message from an application program, the source TCP process may have to **fragment** (divide) the application layer message into several fragments and transmit each fragment in a separate TCP segment. Figure A-6 illustrates TCP fragmentation. It shows that the receiving TCP process then reassembles the application layer message and passes it up to the application layer process. Note that only the application layer message is fragmented. TCP segments are not fragmented.

Maximum Segment Size (MSS)

How large may segments be? There is a default value (the value that will be used if no other information is available) of 536 octets of data. This is called the **maximum segment size (MSS).** Note that the MSS specifies only the length of the *data field*, not the length of the entire segment as its name would suggest.[5]

The value of 536 was selected because there is a maximum IP packet size of 576 octets that an IP process may send unless the other IP process informs the sender that

---

[5]J. Postel, "The TCP Maximum Segment Size and Related Topics," RFC 879, 11/83.

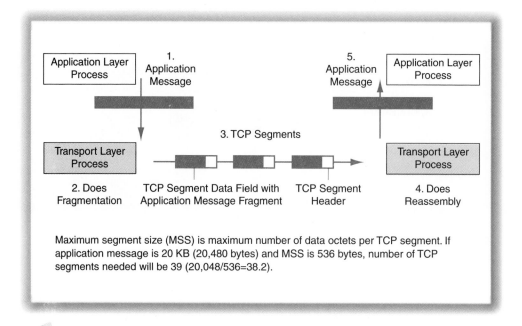

Maximum segment size (MSS) is maximum number of data octets per TCP segment. If application message is 20 KB (20,480 bytes) and MSS is 536 bytes, number of TCP segments needed will be 39 (20,048/536=38.2).

Figure A-6   TCP Fragmentation

larger IP packets may be sent. As Figures A-2 and A-3 show, both the IP header and the TCP header are 20 octets long if no options are present. Subtracting 40 from 576 gives 536 octets of data. The MSS for a segment shrinks further if options are present.

A Sample Calculation

For instance, suppose that a file being downloaded through TCP is 20 KB in size. This is 20,480 bytes, because a kilobyte is 1,024 bytes, not 1,000 bytes. If there are no options and if the MSS is 536, then 38.2 (20,480/536) segments will be needed. Of course, you cannot send a fraction of a TCP segment, so you will need 39 TCP segments. Each will have its own header and data field.[6]

Announcing a Maximum Segment Size

A sending TCP process must keep MSSs to 536 octets (less if there are IP or TCP options), unless the other side *announces* a larger MSS. Announcing a larger MSS is possible through a TCP header option field. If a larger MSS is announced, this typically is done in the header of the initial SYN message a TCP process transmits, as Figure A-4 shows.

---

[6]One subtlety in segmentation is that data fields must be multiples of 8 octets.

Bidirectional Communication

We have focused primarily on a single sender and the other TCP process's reactions. However, TCP communication goes in both directions, of course. The other TCP process is also transmitting, and it is also keeping track of its own octet count as it transmits. Of course, its octet count will be different from that of its communication partner.

For example, each side creates its own initial sequence number. The sender we discussed earlier randomly chose the number 47. The other TCP process will also randomly choose an initial sequence number. For a 32-bit sequence number field, there are more than four billion possibilities, so the probability of both sides selecting the same initial sequence number is extremely small. Also, each process may announce a different MSS to its partner.

MORE ON INTERNET LAYER STANDARDS

Mask Operations

Chapter 8 introduced the concept of masks—both network masks and subnet masks. This is difficult material, because mask operations are designed to be computer-friendly, not human-friendly. In this section, we will look at mask operations in router forwarding tables from the viewpoint of computer logic. Figure A-7 illustrates masking operations.

Basic Mask Operations

Mask operations are based on the logical AND operation. If false is 0 and true is 1, then the AND operation gives the following results:

➤   If the address bit is 1 and the mask bit is 1, the result is 1.
➤   If the address bit is 0 and the mask bit is 1, the result is 0.

Figure A-7   Masking Operations

Information Bit	1	0	1	0
Mask Bit	1	1	0	0
AND Result	1	0	0	0

Destination IP Address (172.99.16.47)	10101100 01100011 00010000 00101111
Mask for Table Entry (/12)	11111111 11110000 00000000 00000000
Masked IP Address (172.96.0.0)	10101100 01100000 00000000 00000000
Network Part for Table Entry (172.96.0.0)	10101100 01100000 00000000 00000000

➤ If the address bit is 1 and the mask bit is 0, the result is 0.

➤ If the address bit is 0 and the mask bit is 0, the result is 0.

Note that if the mask bit is 0, then the result is 0, regardless of what the address bit might be. However, if the mask bit is 1, then the result is whatever the address bit was.

A Routing Table Entry

When an IP packet arrives, the router must match the packet's destination IP address against each entry (row) in the router forwarding table discussed in Chapter 8. We will look at how this is done in a single row's matching. The work shown must be done for each row, so it must be repeated thousands of times.

Suppose that the destination address is 172.99.16.47. This corresponds to the following bit pattern. The first 12 bits are underlined for reasons that will soon be apparent.

<u>10101100 0110</u>0011 00010000 00101111

Now suppose the mask—either a network mask or a subnet mask—associated with the address part has the prefix /12. This corresponds to the following bit pattern. (The first 12 bits are underlined to show the impact of the prefix.)

<u>11111111 1111</u>0000 00000000 00000000

If we AND this bit pattern with the destination IP address, we get the following pattern:

<u>10101100 0110</u>0000 00000000 00000000

Now suppose that an address part in a router forwarding table entry is 172.96.0.0. This corresponds to the following bit stream:

<u>10101100 0110</u>0000 00000000 00000000

If we compare this with the masked IP address (<u>10101100 0110</u>0000 00000000 00000000), we get a match. We therefore have a match with a length of 12 bits.

Perspective

Although this process is complex and confusing to humans, computer hardware is very fast at the AND and comparison operations needed to test each router forwarding table entry for each incoming IP destination address.

IPv6

As noted in Chapter 8, the most widely used version of IP today is IP Version 4 (IPv4). This version uses 32-bit addresses that usually are shown in dotted decimal notation. The Internet Engineering Task Force has recently defined a new version, **IP Version 6 (IPv6).** Figure A-8 shows an IP Version 6 packet.

Larger 128-Bit Addresses

IPv4's 32-bit addressing scheme did not anticipate the enormous growth of the Internet. Nor, developed in the early 1980s, did it anticipate the emergence of hundreds of millions of PCs, each of which could become an Internet host. As a result, the Internet

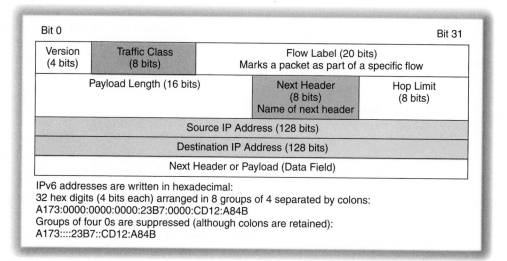

Bit 0			Bit 31
Version (4 bits)	Traffic Class (8 bits)	Flow Label (20 bits) Marks a packet as part of a specific flow	
Payload Length (16 bits)		Next Header (8 bits) Name of next header	Hop Limit (8 bits)
Source IP Address (128 bits)			
Destination IP Address (128 bits)			
Next Header or Payload (Data Field)			

IPv6 addresses are written in hexadecimal:
32 hex digits (4 bits each) arranged in 8 groups of 4 separated by colons:
A173:0000:0000:0000:23B7:0000:CD12:A84B
Groups of four 0s are suppressed (although colons are retained):
A173::::23B7::CD12:A84B

Figure A-8   IP Version 6 Header

is literally running out of IP addresses. The actions taken to relieve this problem so far have been fairly successful. However, they are only stopgap measures. IPv6, in contrast, takes a long-term view of the address problem.

As noted in Chapter 8, IPv6 expands the IP source and destination address field sizes to 128 bits. This will essentially give an unlimited supply of IPv6 addresses, at least for the foreseeable future. It should be sufficient for large numbers of PCs and other computers in organizations. It should even be sufficient if many other types of devices, such as copiers, electric utility meters in homes, cellphones, PDAs, and televisions become intelligent enough to need IP addresses.

Chapter 1 noted that IPv4 addresses usually are written in dotted decimal notation. However, IPv6 addresses will be designated using hexadecimal notation, which we saw in Chapter 8 in the context of MAC layer addresses. IPv6 addresses are first divided into 8 groups of 16 bits. Then each group is converted into 4 hex digits. So a typical IPv6 would look like this:

`A173:0000:0000:0000:23B7:0000:CD12:A84B`

When a group of four hex digits is 0, it is omitted, but the colon separator is kept. Applying this rule to the address above, we would get the following:

`A173::::23B7::CD12:A84B`

Quality of Service

IPv4 has a **type of service (ToS) field,** which specifies various aspects of delivery quality, but it is not widely used. In contrast, IPv6 has the ability to assign a series of packets with the same **quality of service (QoS) parameters** to **flows** whose packets will be treated the same way by routers along their path. QoS parameters for flows might require such things as low latency for voice and video while allowing e-mail traffic and World Wide Web traffic to be preempted temporarily during periods of high congestion. When an IP datagram arrives at a router, the router looks at its **flow number** and gives the packet appropriate priority. However, this flow process is still being defined.

Extension Headers
In IPv4, options were somewhat difficult to apply. However, IPv6 has an elegant way to add options. It has a relatively small main header, as Figure A-8 illustrates. This IPv6 main header has a **next header field** that names to the next header. That header in turn names its successor. This process continues until there are no more headers.

Piecemeal Deployment
With tens of millions of hosts and millions of routers already using IPv4, how to deploy IPv6 is a major concern. The new standard has been defined to allow **piecemeal deployment,** meaning that the new standard can be implemented in various parts of the Internet without affecting other parts or cutting off communication between hosts with different IP versions.

IP Fragmentation
When a host transmits an IP packet, the packet can be fairly long on most networks. Some networks, however, impose tight limits on the sizes of IP packets. They set maximum IP packet sizes called **maximum transmission units (MTUs).** IP packets have to be smaller than the MTU size. The MTU size can be as small as 512 octets.

The IP Fragmentation Process
What happens when a long IP packet arrives at a router that must send it across a network whose MTU is smaller than the IP packet? Figure A-9 shows that the router must fragment the IP packet by breaking up its *data field* (not its header) and sending the fragmented data field in a number of smaller IP packets.[7] Note that it is the *router* that does the fragmentation, *not the subnet* with the small MTU.

Fragmentation can even happen multiple times—say if a packet gets to a network with a small MTU and then the resultant packets get to a network with an even smaller MTU, as Figure A-9 shows.

At some point, of course, we must reassemble the original IP packet. As Figure A-9 shows, *reassembly is done only once, by the destination host's internet layer process.* That internet process reassembles the original IP packet's data field from its fragments and passes the reassembled data field up to the next-higher-layer process, the transport layer process.

Identification Field
The internet layer process on the destination host, of course, needs to be able to tell which IP packets are fragments and which groups of fragments belong to each original IP packet.

---

[7]Each packet has its own header and options.

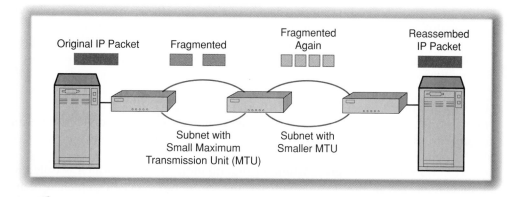

Figure A-9   IP Packet Fragmentation and Reassembly

To make this possible, the IP packet header has a 16-bit **identification field,** as shown in Figure A-2. Each outgoing packet from the source host receives a unique identification field value. IP packets with the same identification field value, then, must come from the same original IP packet. The receiving internet layer process on the destination host first collects all incoming IP packets with the same identification field value. This is like putting all pieces of the same jigsaw puzzle in a pile.

Flags and Fragment Offset Fields

Next, the receiving internet layer process must place the fragments of the original IP packet in order.

Each IP packet has a **fragment offset field** (see Figure A-2). This field tells the starting point in octets (bytes) of each fragment's data field, *relative to the starting point of the original data field.* This permits the fragments to be put in order.

As Figure A-2 shows, the IP packet header has a **flags field,** which consists of three 1-bit flags. One of these is the **more fragments flag.** The original sender sets this bit to 0. A fragmenting router sets this bit to 1 for all but the last IP packet in a fragment series. The router sets this more fragments bit to 0 in the last fragment to indicate that there are no more fragments to be handled.

Perspective on IP Fragmentation

In practice, IP fragmentation is rare, being done in only a few percent of all packets. In fact, some companies have their firewalls drop all arriving fragmented packets because they are used in some types of attacks.

Dynamic Routing Protocols

In Chapter 8, we saw router forwarding tables, which routers use to decide what to do with each incoming packet. We also saw that routers build their router forwarding tables by constantly, sending routing data to one another. *Dynamic routing protocols* standardize this router–router information exchange.

There are multiple dynamic routing protocols. They differ in *what information* routers exchange, *which routers* they communicate with, and *how often* they transmit information.

Interior and Exterior Routing Protocols

Recall from Chapter 1 that the Internet consists of many networks owned by different organizations.

Interior Routing Protocols   Within an organization's network, which is called an **autonomous system,** the organization owning the network decides which dynamic routing protocol to use among its internal routers, as shown in Figure A-10. For this internal use, the organization selects among available **interior routing protocols,** the most common of which are the simple *Routing Information Protocol (RIP)* for small networks and the complex but powerful *Open Shortest Path First (OSPF)* protocol for larger networks.

Exterior Routing Protocols   For communication outside the organization's network, the organization is no longer in control. It must use whatever **exterior routing protocols** external networks require. **Border routers,** which connect autonomous

Figure A-10   Interior and Exterior Routing Protocols

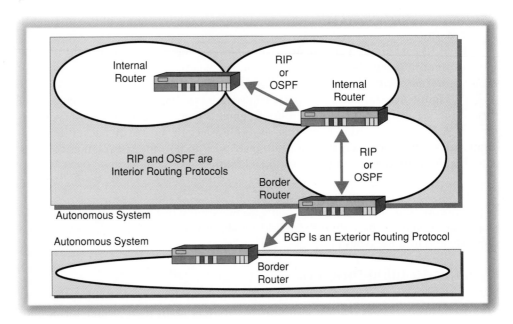

systems organizations with the outside world, implement these protocols. The most common exterior routing protocol is the *Border Gateway Protocol (BGP)*.

Routing Information Protocol (RIP)   The **Routing Information Protocol (RIP)** is one of the oldest Internet dynamic routing protocols and is by far the simplest. However, as we will see, RIP is suitable only for small networks. Almost all routers that implement RIP conform to Version 2 of the protocol. When we refer to RIP, we will be referring to this second version.

Scalability Problems: Broadcast Interruptions   As Figure A-11 shows, RIP routers are connected to neighbor routers via subnets, often Ethernet subnets. Every thirty seconds, every router broadcasts its entire routing table to all hosts and routers on the subnets attached to it.

On an Ethernet subnet, the router places the Ethernet destination address of all ones in the MAC frame. This is the *Ethernet broadcast address*. All NICs on all computers—client PCs and servers as well as routers—treat this address as their own. As a consequence, *every station* on every subnet attached to the broadcasting router is interrupted every thirty seconds.

Actually, it is even worse. Each IP packet carries information on only twenty-four router forwarding table entries. Even on small networks, then, each thirty-second broadcast actually will interrupt each host and router a dozen or more times. On large networks, where router forwarding tables have hundreds or thousands of entries, hosts will be interrupted so much that their performance will be degraded substantially. RIP is only for small networks.

Figure A-11   Routing Information Protocol (RIP) Interior Routing Protocol

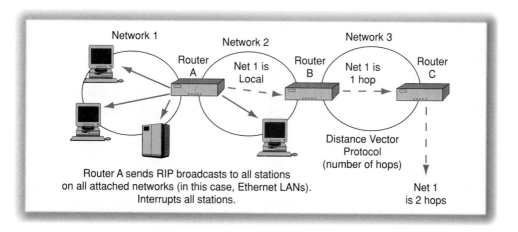

Scalability: The 15-Hop Problem   Another size limitation of RIP is that the farthest routers can only be fifteen hops apart (a hop is a connection between routers). Again, this is no problem for small networks. However, it is limiting for larger networks.

Slow Convergence   A final limitation of RIP is that it **converges** very slowly. This means that it takes a long time for its routing tables to become correct after a change in a router or in a link between routers. In fact, it may take several minutes for convergence on large networks. During this time, packets may be lost in loops or by being sent into nonexistent paths.

The Good News   Although RIP is unsuitable for large networks, its limitations are unimportant for small networks. Router forwarding tables are small, there are far fewer than fifteen hops, convergence is decently fast, and the sophistication of OSPF routing is not needed. Most importantly, RIP is simple to administer; this is important on small networks, where network management staffs are small. RIP is fine for small networks.

A Distance Vector Protocol   RIP is a **distance vector routing protocol**. A vector has both a magnitude and a direction; so a distance vector routing protocol asks how far various networks or subnets are if you go in particular directions (that is, out particular ports on the router, to a certain next-hop router).

Figure A-11 shows how a distance vector routing protocol works. First, Router A notes that Network 1 is directly connected to it. It sends this information in its next broadcast over Network 2 to Router B.

Router B knows that Router A is one hop away. Therefore, Network 1 must be one hop away from Router B. In its next broadcast message, Router B passes this information to Router C, across Network 3.

Router C hears that Network 1 is one hop away from Router B. However, it also knows that Router B is one hop away from it. Therefore, Network 1 must be two hops away from Router C.

Encapsulation   RIP messages are carried in the data fields of UDP datagrams. UDP port number 520 designates a RIP message.

Open Shortest Path First (OSPF)

Open Shortest Path First (OSPF) is much more sophisticated than RIP, making it more powerful but also more costly to manage.

Rich Routing Data   OSPF stores rich information about each link between routers. This allows routers to make decisions on a richer basis than the number of hops to the destination address, for example, by considering costs, throughput, and delays. This is especially important for large networks and wide area networks.

Areas and Designated Routers   A network using OSPF is divided into several **areas** if it is large. Figure A-12 shows a network with a single area for simplicity. Within each area there is a **designated router** that maintains an entire area router forwarding table that gives considerable information about each link (connection between routers) in

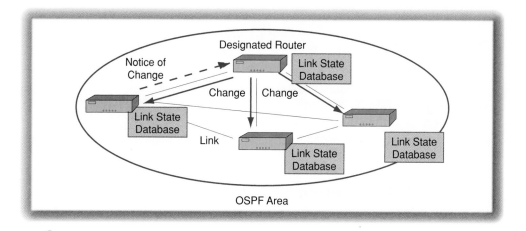

Figure A-12   Open Shortest Path First (OSPF) Interior Routing Protocol

the network. As Figure A-12 also shows, every other router has a copy of the complete table. It gets its copy from the designated router.

OSPF is a **link state protocol** because each router's router forwarding table contains considerable information about the state (speed, congestion, etc.) of each **link** between routers in the network area.

Fast Convergence   If one of the routers detects a change in the state of a link, it immediately passes this information to the designated router, as shown by the broken arrow in Figure A-12. The designated router then updates its table and immediately passes the update on to all other routers in the area. There is none of the slow convergence in RIP.

Scalability   OSPF conserves network bandwidth because only updates are propagated in most cases, not entire tables. (Routers also send "Hello" messages to one another every ten seconds, but these are very short.)

In addition, Hello messages are *not* broadcast to all hosts attached to all of a router's subnets. Hello messages are given the IP destination address 224.0.0.5. Only OSPF routers respond to this *multicast* destination address. (See the section in this module on Classful IP addresses.)

If there are multiple areas, this causes no problems. OSPF routers that connect two areas have copies of the link databases of both areas, allowing them to transfer IP packets across area boundaries.

Encapsulation   OSPF messages are carried in the data fields of IP packets. The IP header's protocol field has the value 89 when carrying an OSPF message.

Border Gateway Protocol   The most common exterior routing protocol is the **Border Gateway Protocol (BGP),** which is illustrated in Figure A-13.

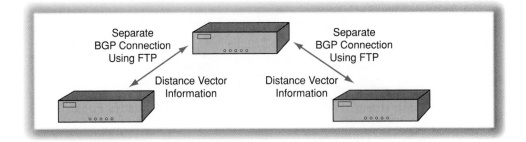

Figure A-13   Border Gateway Protocol (BGP) Exterior Routing Protocol

TCP   BGP uses TCP connections between pairs of routers. This gives reliable delivery for BGP messages. However, TCP only handles one-to-one communication. Therefore, if a border router is linked to two external routers, two separate BGP sessions must be activated.

Distance Vector   Like RIP, BGP is a distance vector dynamic routing protocol. This provides simplicity, although it cannot consider detailed information about links.

Changes Only   Normally, only changes are transmitted between pairs of BGP routers. This reduces network traffic.

Comparisons   Comparing RIP, OSPF, and BGP is difficult because several factors are involved (Figure A-14).

Address Resolution Protocol (ARP)

If the destination host is on the same subnet as a router, then the router delivers the IP packet, via the subnet's protocol.[8] For an Ethernet LAN:

➤ The internet layer process passes the IP packet down to the NIC.
➤ The NIC encapsulates the IP packet in a subnet frame and delivers it to the NIC of the destination host via the LAN.

Learning a Destination Host's MAC Address

To do its work, the router's NIC *must know the 802.3 MAC layer address of the destination host*. Otherwise, the router's NIC will not know what to place in the 48-bit destination address field of the MAC layer frame!

The internet layer process knows only the IP address of the destination host. If the router's NIC is to deliver the frame containing the packet, the internet layer

---

[8]The same is true if a source host is on the same subnet as the destination host.

	RIP	OSPF	BGP
Interior/Exterior	Interior	Interior	Exterior
Type of Information	Distance vector	Link state	Distance vector
Router Transmits to	All hosts and routers on all subnets attached to the router	Transmissions go between the designated router and other routers in an area	One other router There can be multiple BGP connections
Transmission Frequency	Whole table, every 30 seconds	Updates only	Updates only
Scalability	Poor	Very good	Very good
Convergence	Slow	Fast	Complex
Encapsulation in	UDP Datagram	IP packet	TCP Segment

Figure A-14   Comparison of Routing Information Protocols: Text

process must discover the MAC layer address of the destination host. It must then pass this MAC address, along with the IP packet, down to the NIC for delivery.

Address Resolution on an Ethernet LAN with ARP

Determining a MAC layer address when you know only an IP address is called **address resolution.** Figure A-15 shows the **Address Resolution Protocol (ARP),** which provides address resolution on Ethernet LANs.

ARP Request Message

Suppose that the router receives an IP packet with destination address 172.19.8.17. Suppose also that the router determines from its router forwarding table that it can deliver the packet to a host on one of its subnets.

First, the router's internet layer process creates an *ARP request message* that essentially says, "Hey, device with IP address 172.19.8.17, what is your 48-bit MAC layer address?" The internet layer on the router passes this ARP request message to its NIC.

Broadcasting the ARP Request Message   The MAC layer process on the router's NIC sends the ARP request message in a MAC layer frame that has a destination address of forty-eight 1s. This designates the frame as a broadcast frame. All NICs listen constantly for this **broadcast address.** When a NIC hears this address, it accepts the frame and passes the ARP request message up the internet layer processes.

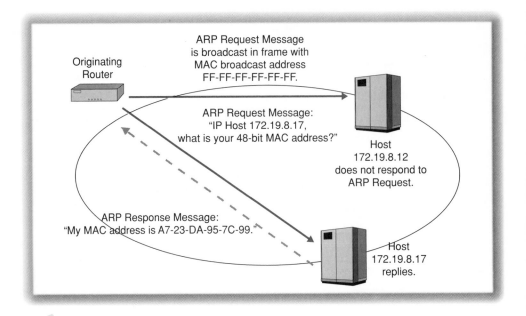

Figure A-15   Address Resolution Protocol (ARP)

Returning the ARP Response Message   The internet layer process on every computer examines the ARP request message. If the target IP address is not that computer's, the internet layer process ignores it. If it is that computer's IP address, however, the internet layer process composes an ARP response message that includes its 48-bit MAC layer address.

The target host sends this ARP response message back to the router, via the target host's NIC. There is no need to broadcast the response message, as Figure A-15 shows. The target host sending the ARP response message knows the router's MAC address, because this information was included in the ARP request message.

When the router's internet layer process receives the ARP response message, address resolution is complete. The router's internet layer process now knows the subnet MAC address associated with the IP address. From now on, when an IP packet comes for this IP destination address, the router will send the IP packet down to its NIC, together with the required MAC address. The NIC's MAC process will deliver the IP packet within a frame containing that MAC destination address.

Other Address Resolution Protocols

Although ARP is the Address Resolution Protocol, it is not the only address resolution protocol. Most importantly, ARP uses broadcasting, but not all subnet technologies handle broadcasting. Other address resolution protocols are available for such networks.

Encapsulation

An ARP request message is an internet layer message. Therefore, we call it a packet. ARP packets and IP packets are both internet layer packet types in TCP/IP, as Figure A-1 illustrates. On a LAN, the ARP packet is encapsulated in the data field of an LLC frame. In other types of networks, it is encapsulated in the data field of the data link layer frame.

Classful Addresses in IP

In Chapter 8, we noted that, by themselves, 32-bit IP addresses do not tell you the lengths of their network, subnet, and host parts. For this, you need to have network masks to know how many bits there are in the network part, for instance. This is called **Classless InterDomain Routing (CIDR).** CIDR allows network parts to vary from 8 bits to 24.

Originally, however, the 32-bit IP address did tell you the size of the network part, although not the subnet part. As Figure A-16 shows, the initial bits of the IP address told whether an IP address was for a host on a Class A, Class B, or Class C network, or whether the IP address was a Class D multicast address. This is **classful addressing.**

Figure A-16   IP Address Classes

Class	Beginning Bits	Bits in the Remainder of the Network Part	Number of Bits in Local Part	Approximate Maximum Number of Networks	Approximate Maximum Number of Hosts per Network
A	0	7	24	126	16 million
B	10	14	16	16,000	65,000
C	110	21	8	2 million	254
D[a]	1110				
E[b]	11110				

[a]Used in multicasting.
[b]Experimental.

Problem: For each of the following IP addresses, give the class, the network bits, and the host bits if applicable:

10101010111110000101010100000001
11011010111110000101010100000001
01010101111110000101010100000001
11101110111110000101010100000001

Class A Networks

Specifically, if the initial bit was a 0, this IP address would represent a host in a Class A network. As Figure A-16 shows, Class A network parts were only 8 bits long. The first bit was fixed (0), so there could be only 126 possible Class A networks.[9] However, each of these networks could be enormous, holding more than 16 million hosts. Half of all IP addresses were Class A addresses. Half of these Class A addresses were reserved for future Internet growth.

Class B Networks

If the initial bits of the IP address were "10," then this was the address of a host on a Class B network. The network part was 16 bits long. Although the first 2 bits were fixed, the remaining 14 bits could specify a little more than 16,000 Class B networks. With 16 bits remaining for the host part, there could be more than 65,000 hosts on each Class B network. The Class B address space was on its way to being completely exhausted until CIDR was created to replace the classful addressing approach discussed in this section.

Class C Networks

Addresses in Class C networks began with "110." (Note that the position of the first 0 told you the network's class.) The network part was 24 bits long, and the 21 nonreserved bits allowed more than 2 million Class C networks. Unfortunately, these networks could have only 254 hosts apiece, making them almost useless in practice. Such small networks seemed reasonable when the IP standard was created, because users worked at mainframe computers or at least minicomputers. Even a few of these large machines would be able to serve hundreds or thousands of terminal users. Once PCs became hosts, however, the limit of 254 hosts became highly restrictive.

Class D Addresses

Class A, B, and C addresses were created to designate specific hosts on specific networks. However, Class D addresses, which begin with "1110," have a different purpose—namely multicasting. This purpose has survived Classless InterDomain Routing.

When one host places another host's IP address in a packet, the packet will go only to *that one* host. This is called **unicasting.** In contrast, when a host places an all-1s address in the host part, then the IP packet should be **broadcast** to *all* hosts on that subnet.

However, what if only *some* hosts should receive the message? For instance, as discussed earlier, when OSPF routers transmit to one another, they want only other OSPF routers to process the message. To support this limitation, they place the IP address 224.0.0.5 in the IP destination address fields of the packets they send. All OSPF routers listen for this IP address and accept packets with this address in their IP destination address fields. This is **multicasting,** that is, *one-to-many* communication. Multicasting is more efficient than broadcasting because not all stations are interrupted. Only routers stop to process the OSPF message.

---

[9]Not 127 or 128. Network, subnet, and host parts of all 0s and all 1s are reserved.

Class E Addresses

A fifth class of IP addresses was reserved for future use, but these Class E addresses were never defined.

Mobile IP

The proliferation of notebooks and other portable computers has brought increasing pressure on companies to support mobile users. Chapter 5 discusses wireless LANs as a way to provide such support.

Mobile users on the Internet also need support. The IETF is developing a set of standards collectively known as **mobile IP.** These standards will allow a mobile computer to register with any nearby ISP or LAN access point. The standards will establish a connection between a computer's temporary IP address at the site and the computer's permanent "home" IP address. Mobile IP standards will allow portable computer users to travel without losing access to e-mail, files on file servers, and other resources.

Mobile IP will also offer strong security, based in the IPsec standards discussed in Chapter 7.

REVIEW QUESTIONS

Multiplexing

1. a) How does a receiving internet layer process decide what process should receive the data in the data field of an IP packet? b) How does TCP decide? c) How does UDP decide? d) How does PPP decide?

More on TCP

2. A TCP segment begins with octet 8,658 and ends with octet 12,783. a) What number does the sending host put in the sequence number field? b) What number does the receiving host put in the acknowledgment number field of the TCP segment that acknowledges this TCP segment?

3. A TCP segment carries data octets 456 through 980. The following TCP segment is a supervisory segment carrying no data. What value is in the sequence number field of the latter TCP segment?

4. Describe flow control in TCP.

5. a) In TCP fragmentation, what is fragmented? b) What device does the fragmentation? c) What device does reassembly?

6. A transport process announces an MSS of 1,024. If there are no IP or TCP options, how big can IP packets be?

Mask Operations

7. There is a mask 1010. There is a number 1100. What is the result of masking the number?

8. The following router forwarding table entry has the prefix /14.

    10101010 10100000 00000000 00000000 (170.160.0.0)

a) Does it match the following destination address in an arriving IP packet? Explain.

b) 0101010 10101011 11111111 00000000 (170.171.255.0)

IP Version 6

9. a) What is the main benefit of IPv6?

   b) What other benefits were mentioned?

10. a) Express the following in hexadecimal: 0000000111110010. (Hint: Chapter 5 has a conversion table.)

    b) Simplify: A173:0000:0000:0000:23B7:0000:CD12:A84B

IP Fragmentation

11. a) What happens when an IP packet reaches a subnet whose MTU is *longer* than the IP packet?

    b) What happens when an IP packet reaches a subnet whose MTU is *shorter* than the IP packet?

    c) Can fragmentation happen more than once as an IP packet travels to its destination host?

12. Compare TCP fragmentation and IP fragmentation in terms of

    a) what is fragmented and

    b) where the fragmentation takes place.

13. a) What program on what computer does reassembly if IP packets are fragmented?

    b) How does it know which IP packets are fragments of the same original IP packet?

    c) How does it know their correct order?

Dynamic Routing Protocols

14. a) What is an autonomous system?

    b) Within an autonomous system, can the organization choose routing protocols?

    c) Can it select the routing protocol its border router uses to communicate with the outside world?

15. Compare RIP, OSPF, and BGP along each of the dimensions shown in Figure A-14.

Address Resolution Protocol (ARP)

16. A host wishes to send an IP packet to a router on its subnet. It knows the router's IP address.

    a) What else must it known?

    b) Why must it know it?

    c) How will it discover the piece of information it seeks? (Note: Routers are not alone in being able to use ARP.)

17. a) What is the destination MAC address of an Ethernet frame carrying an ARP request message?

    b) What is the destination MAC address of an Ethernet frame carrying an ARP response packet?

Classful IP Addressing

18. Compare classful addressing and CIDR.

19. What class of network is each of the following?
    a) 10101010111111110000000010101010
    b) 00110011000000001111111101010101
    c) 11001100111111110000000010101010
20. a) Why is multicasting good?
    b) How did classful addressing support it?

Mobile IP
21. How will mobile IP work?

PROJECT

Getting Current. Go to the book website's New Information and Errors pages for this chapter to get new information since this book went to press and to correct any errors in the text.

More on Modulation

MODULATION

As we saw in Chapter 7, modems use modulation to convert digital computer signals into analog signals that can travel over the local loop to the first switching office. This module looks at the main forms of modulation in use today.

Frequency Modulation

As we saw in Chapter 7, modulation essentially transforms zeros and ones into electromagnetic signals that can travel down telephone wires. Electromagnetic signals consist of waves. As we saw in Chapter 5, waves have frequency, measured in hertz (cycles per second). Figure B-1 illustrates **frequency modulation,** in which one **frequency** is chosen to represent a 1 and another frequency is chosen to represent a 0. During a clock cycle in which a 1 is sent, the frequency chosen for the 1 is sent. During a clock cycle in which a 0 is sent, the frequency chosen for the 0 is sent.

Amplitude Modulation

In wave transmission, amplitude is the intensity in the wave. In **amplitude modulation,** which we saw in Chapter 7, we represent 1s and 0s as different amplitudes. For instance, we can represent a 1 by a high-amplitude (loud) signal and a 0 by a low-amplitude (soft) signal. To send "1011," we would send a loud signal for the first time period, a soft signal for the second, and high-amplitude signals for the third and fourth time periods.

Phase Modulation

The last major characteristic of waves is phase. As shown in Figure B-2, we call 0 degrees phase the point of the wave at 0 amplitude and rising. The wave hits its maximum at 90 degrees, returns to 0 on the decline at 180 degrees, and hits its minimum amplitude at 270 degrees. Amplitude now increases to 360 degrees, which is the same as 0 degrees.

In **phase modulation** we use two waves. We let one wave be our reference wave or carrier wave. Let us use this carrier wave to represent a 1. Then we can use a wave 180 degrees out of phase to represent a 0. So if our carrier wave is at 180 degrees, the other wave will be at zero degrees, and if our carrier wave is at 270 degrees, the other wave will be at 90 degrees.

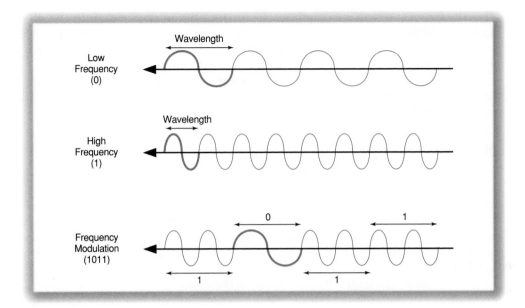

Figure B-1   Frequency Modulation

Figure B-2   Phase Modulation

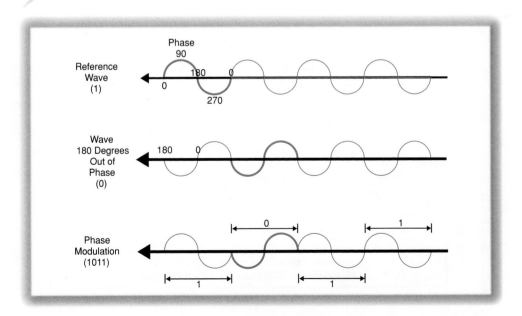

The figure shows that to send "1011," we send the reference wave for the first clock cycle, shift the phase 180 degrees for the second, and return to the reference wave for the third and fourth clock cycles. Although this makes little sense in terms of hearing, it is easy for electronic equipment to deal with phase differences.

A number of transmission systems use **quadrature phase shift keying (QPSK)**, which is phase modulation with four states (phases). Each of the four states represents two bits (00, 01, 10, and 11), so QPSK's bit rate is double its baud rate.

Quadrature Amplitude Modulation (QAM)

Telephone modems and many ADSL and cable modems today use a more complex type of modulation called **quadrature amplitude modulation (QAM).** As Figure B-3 illustrates, QAM uses two carrier waves: a sine carrier wave and a cosine carrier wave. When the cosine wave is at the top of its cycle, the sine wave is just beginning its cycle and will not hit its peak until 90 degrees. The sine wave is 90 degrees out of phase with the cosine wave. This is a quarter of a cycle, and this fact gives rise to the name "quadrature."

The receiver can send different signals on these two waves because they have different phases so the receiver can distinguish between them. Specifically, QAM uses multiple possible amplitude levels for each carrier wave. To illustrate what this means, consider that using four possible amplitudes for the sine wave times four possible

Figure B-3   Quadrature Amplitude Modulation (QAM)

Bit Rate = Baud Rate × 4 bits per clock cycle
If Baud Rate is 3200 baud, Bit Rate is 9600 bps

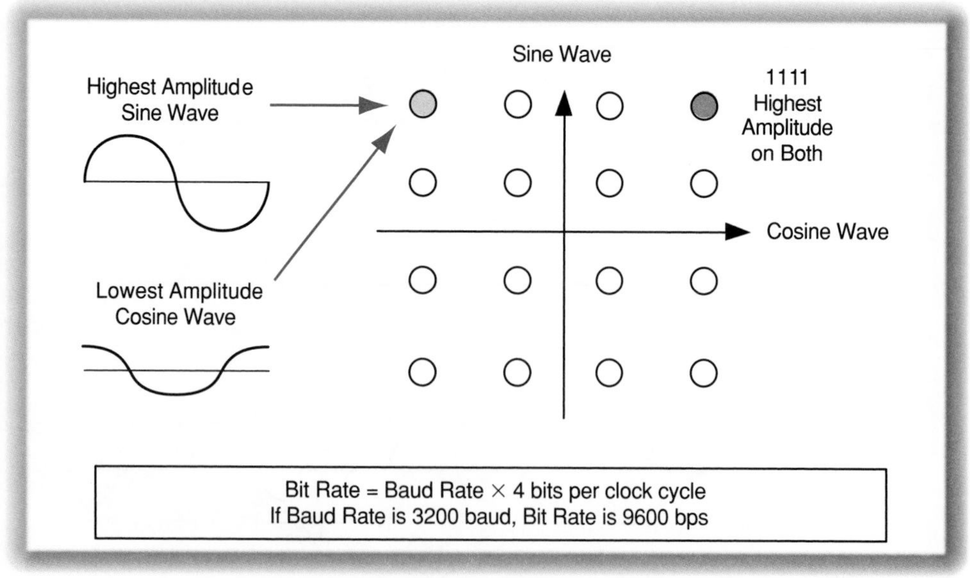

amplitudes on the cosine wave will give 16 possible states. Sixteen possibilities can represent four bits ($2^4 = 16$). Accordingly, each clock cycle can represent a 4-bit value from 0000 through 1111. In summary, each clock cycle transmits four bits if there are four possible amplitude levels.

Different versions of QAM use different numbers of amplitude levels. Each doubling in the number of amplitude levels quadruples the number of possible states. Each quadrupling of the number of possible states allows two more bits to be sent per clock cycle. However, beyond about sixty-four possible states, the states are so close together that even slight transmission impairments can cause errors.

REVIEW QUESTIONS

1. Describe frequency modulation.
2. a) Describe phase modulation. b) Describe QPSK.
3. a) What two forms of modulation does QAM use? b) In QAM, if you have four possible amplitudes, how many states do you have? c) In QAM, if you have eight possible amplitudes, how many states do you have? d) How many bits can you send per clock cycle?

PROJECT

Getting Current

Go to the book website's New Information and Errors pages for this chapter to get new information since this book went to press and to correct any errors in the text.

More on Telecommunications

INTRODUCTION

Chapter 6 discussed telecommunications, which is the transmission of voice and video. This module is designed for courses that want to get into more detail on telecommunications.

This module is designed to be read after Chapter 6, and it is better to read it after Chapter 7, which discusses leased lines in some detail. The material is not intended to be read front-to-back like a normal chapter. Rather, it is a collection of technical topics, service topics, and regulatory topics:

➤ The PSTN Transport Core and Signaling
➤ Communication Satellites
➤ Wiring in the First Bank of Paradise Headquarters Building
➤ PBX Services
➤ Carrier Services and Pricing
➤ Telephone Carriers and Regulation

THE PSTN TRANSPORT CORE AND SIGNALING

Recall from Chapter 6 that *transport* is the actual transmission of voice in the PSTN, while *signaling* is the control of the PSTN. In this section, we will look at PSTN transport and signaling in more detail.

The Transport Core

Figure C-1 illustrates that the PSTN transport core consists of switches and trunk line connections that link the switches. The PSTN transport core uses two types of transmission systems to connect telephone switches: TDM trunk lines and ATM packet-switched networks.

Time Division Multiplexing (TDM) Lines

In Chapter 6, we saw the concepts of leased lines, which provide high-speed, always-on connections between corporate sites. In Chapter 7, we saw that leased lines are offered over a wide range of speeds and that the most popular leased lines are T1/E1/J1 and fractional T1/E1/J1.

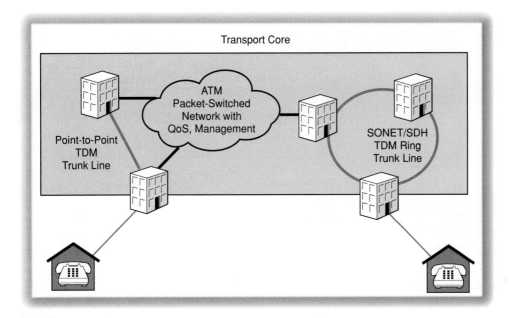

Figure C-1   TDM and ATM Switch Connections in the PSTN Transport Core

Multiplexing Simultaneous Voice Calls on Leased Lines

Figure C-2 shows the variety of leased lines available to corporations. Most of the columns in this figure also appeared in Chapter 7. However, Figure C-2 has an additional column: multiplexed telephone calls. In telecommunications, the most common use of leased lines is to multiplex many leased lines over a single connection. For example, the figure shows that T1 lines were created to multiplex 24 simultaneous voice calls. Higher-speed leased lines can multiplex hundreds or thousands of telephone calls.

The Time Division Multiplexing (TDM) Process

To implement multiplexing, leased lines use a process called **time division multiplexing (TDM),** which Figure C-3 illustrates. The figure specifically illustrates time division multiplexing for T1 leased lines.

Frames   First, each second is divided into brief periods of time called **frames.** For example, in a T1 leased line, each second is divided into 8,000 frames. If you have read the box on codec operation in Chapter 6, you learned that the human voice is sampled 8,000 times per second in pulse code modulation. One voice sample is transmitted in a frame for every circuit the frame multiplexes.

Slots   Second, each frame is divided into even briefer periods, called **slots.** In a T1 leased line, for instance, there are 24 frames per slot. In TDM, a circuit is given the same slot in each frame. Each slot transmits eight bits—a single voice sample for that circuit.

North American Digital Hierarchy			
Line	Speed	Multiplexed Voice Calls	Typical Transmission Medium
56 kbps	56 kbps	I	2-Pair Data-Grade UTP
T1	1.544 Mbps	24	2-Pair Data-Grade UTP
Fractional T1	128 kbps, 256 kbps, 384 kbps, 512 kbps, 768 kbps	Varies	2-Pair Data-Grade UTP
Bonded T1s (multiple T1s acting as a single line)	Small multiples of 1.544 Mbps	Varies	2-Pair Data-Grade UTP
T3	44.736 Mbps	672	Optical Fiber

CEPT Hierarchy			
Line	Speed	Multiplexed Voice Calls	Typical Transmission Medium
64 kbps	64 kbps	I	2-Pair Data-Grade UTP
E1	2.048 Mbps	30	2-Pair Data-Grade UTP
E3	34.368 Mbps	480	Optical Fiber

SONET/SDH Speeds			
Line	Speed (Mbps)	Multiplexed Voice Calls	Typical Transmission Medium
OC3/STM1	155.52	2,016	Optical Fiber
OS12/STM4	622.08	6,048	Optical Fiber
OC48/STM16	2,488.32	18,144	Optical Fiber
OC192/STM64	9,953.28	54,432	Optical Fiber
OC768/STM256	39,813.12	163,296	Optical Fiber

Figure C-2   Leased Lines and Multiplexing

Reserved Capacity   Slot capacity is reserved in each frame. In Figure C-4, Circuit A is given Slot 1 in every frame. Note that Circuit A uses its slot capacity in every frame shown in the figure. However, Circuit B uses only some of its slot capacity in the three frames, and Circuit C uses none at all. Although TDM provides the reserved capacity

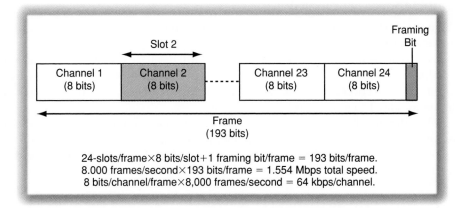

24-slots/frame×8 bits/slot+1 framing bit/frame = 193 bits/frame.
8.000 frames/second×193 bits/frame = 1.554 Mbps total speed.
8 bits/channel/frame×8,000 frames/second = 64 kbps/channel.

Figure C-3   Time Division Multiplexing (TDM) in a TI Line

required for circuit switching, it wastes unused capacity. Users must pay for this reserved capacity whether they use it or not.

TEST YOUR UNDERSTANDING

1. a) How many simultaneous voice calls can a T3 line multiplex? b) Explain frames and slots in time division multiplexing (TDM). c) How is a circuit allocated capacity on a TDM line? d) What is the advantage of TDM? e) What is the disadvantage?

Figure C-4   Reserved Capacity in Time Division Multiplexing (TDM)

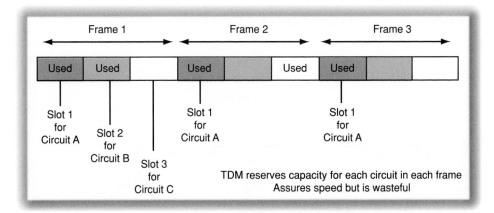

Leased Lines and Trunk Lines

As Figure C-5 shows, leased lines are circuits that pass through multiple telephone switches and trunk lines.

The figure shows that trunk lines between switches have the same designations as leased line circuits (T1, T3, etc.). With a T1 circuit, the local loop access lines are T1 lines, of course. Between switches, however, faster trunk lines are needed to carry multiple single calls, T1 circuits, and other circuits.

This identical labeling for circuits and trunk lines is not accidental. TDM trunk lines were first used in the 1960s to allow telephone companies to multiplex individual telephone calls on trunk lines between switches. Only later were end-to-end high-speed circuits offered to customers.

Point-to-Point TDM Trunk Lines

Figure C-1 shows that there are two types of TDM trunk lines. The earliest trunk lines (through T3/E3) were point-to-point trunk lines that connected pairs of switches. Unfortunately, if a trunk line is accidentally dug up and broken in an unrelated construction project, it can take hours or even days for the telephone company to be able to restore service. This is not a theoretical concern. *Most* telephone outages are due to the accidental cutting of trunk lines by construction vehicles.

SONET/SDH Rings

Chapter 4a showed that ring topologies bring reliability. Rings are really dual-rings. If there is a broken connection between two switches, the ring is wrapped, and service continues. Disruptions from broken connections in ring topologies are only momentary.

The SONET/SDH multiplexing technology is designed to use a ring topology, as Figure C-1 illustrates. Although it can be used for point-to-point connections, ring implementations are highly preferred because of their reliability. Figure C-6 illustrates a SONET/SDH ring.

Figure C-5   Leased Line Circuits and Trunk Lines

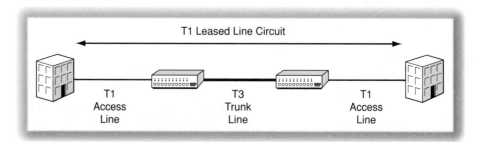

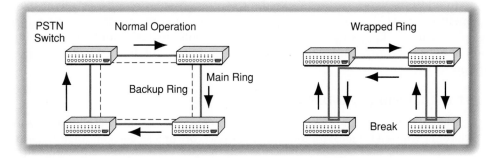

Figure C-6   SONET/SDH Dual Ring

TEST YOUR UNDERSTANDING

2. a) What is the relationship between leased line circuits and trunk lines? b) Below about what speed are trunk lines point-to-point lines? c) What topology are SONET/SDH lines designed to use? d) What is the advantage of this topology? e) What does a SONET/ SDH network do when there is a break in a line between switches?

Asynchronous Transfer Mode (ATM) Transport

Although TDM has long been synonymous with transmission in the PSTN transport core, many long-distance carriers have already transitioned much of their transmission technology between telephone switches in their transport cores to a *packet-switched* technology, **asynchronous transfer mode (ATM).**[1]

ATM has had a checkered history. It was originally created precisely to replace TDM connections in the PSTN transport core and their wasted reserved capacity with more efficient packet switching. For a time, ATM was touted as both the LAN and PSDN technology for the future. However, ATM's high cost worked against it in those markets. As noted in Chapter 4, Ethernet now dominates LAN service. Among the public switched data networks discussed in Chapter 7, Frame Relay has dominated to date because most companies do not require the high speeds of ATM public switched data networks. For the future, less expensive metropolitan area Ethernet promises to be a strong competitor for ATM in the high-speed PSDN market.

However, ATM's expensive complexity, which has stymied it in the LAN and PSDN WAN markets, is critical to its role in the PSTN transport core. First, much of ATM's complexity comes from its ability to provide very strict quality-of-service guarantees for telephone communication and video transmission. The PSTN can only use transport core technologies that guarantee excellent voice quality.

---

[1]Franklin D. Ohrtman, Jr., *Softswitch Architecture for VoIP*, New York: McGraw-Hill, 2003.

In addition, large networks like the telephone system require excellent management tools. ATM, which was created as a transport core protocol for the entire telephone network, has excellent management tools. In LANs and carrier public switched data networks, these management tools are overkill. In the PSTN core, they are perfect.

TEST YOUR UNDERSTANDING

3. a) How is ATM different from previous trunk line technologies? b) Why is ATM good for voice? c) Why is ATM ideal for use in the transport core of telecommunications carriers?

Signaling

As discussed in Chapter 6, signaling is the supervision of connections in the PSTN. The ITU-T created **Signaling System 7 (SS7)** as the worldwide standard for supervisory signaling (setting up circuits, maintaining them, tearing them down after a conversation, providing billing information, and providing special services such as three-party calling). The U.S. version of the protocol is ANSI SS7, usually referred to simply as **SS7.** The ETSI version for Europe is called ETSI C7 or **C7.** They are almost the same, so simple gateways can convert between them and allow them to interoperate.

Figure C-7   Signaling (Study Figure)

Transport Versus Signaling
  Transport is the transmission of voice conversations between customers
  Signaling is the supervision of transport connections
    Call setup, management, and termination
    The collection and transmission of billing information
    3-party calling, and other advanced services

Signaling System 7 (SS7)
  The worldwide standard for PSTN signaling
  Slight differences exist in the U.S. and Europe
    U.S.: Signaling System 7
    Europe: C7
    Interconnected with a simple gateway

Packet-Switched Technology
  Not circuit-switched
  Runs over telephone company lines
  Uses a distributed database
    Data for supervising calls
    Call setup, etc.: requires the querying of the nearest database
    Toll-free numbers, etc.

SS7/C7 actually is a packet-switched technology that operates in parallel with the circuit-switched PSTN but that uses the same transmission lines as the PSTN. SS7/C7 relies on multiple databases of customer information. When a call is set up, the originating telephone carrier queries one of these databases to determine routing information for setting up the service. These databases are also needed to provide advanced services such as toll-free numbers.

TEST YOUR UNDERSTANDING

4. a) What is the worldwide signaling system for telephony? b) Distinguish between SS7 and C7. c) Does having two versions of the standard cause major problems?

COMMUNICATION SATELLITES

During the 1970s, satellites began to be widely used for trunk line transmission within the telephone network. This created sharp drops in long-distance rates. However, as we will see, satellites have proven to be problematic for telephone calling and even more problematic for data transmission.

Microwave Transmission

Satellite transmission technology grew out of microwave transmission technology. As Figure C-8 shows, **microwave** transmission is a point-to-point radio technology using dish antennas. As a consequence of the curvature of the Earth, microwave signals cannot travel farther than a few miles. Consequently, transmission often uses **microwave repeaters** between distant sites.

Before optical fiber became widespread, microwave transmission was used very heavily for long-distance trunk lines between telephone switches. As we saw in

Figure C-8   Microwave Transmission

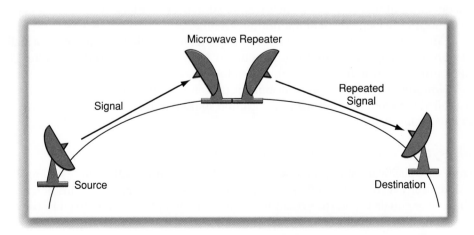

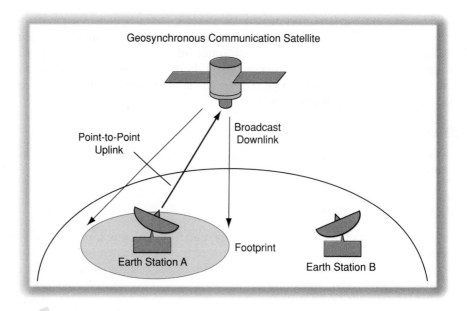

Figure C-9   Geosynchronous Earth Orbit (GEO) Communication
Satellite System

Chapter 6, microwave transmission uses frequency division multiplexing, carrying different telephone calls in different channels.

Satellite Transmission

After World War II, a young radar operator named Arthur C. Clarke (yes, the science fiction writer) noticed how microwave repeaters often sit on hills so that they can carry signals farther. He realized that it was possible to put a microwave repeater in the sky on a satellite and that this would allow transmission over very long distances. As Figure C-9 shows, this became the **communication satellite.**

Communication from the ground to the satellite is called the **uplink.** Normally, this transmission occurs point-to-point between a ground station and the satellite. The uplink ground station has a dish antenna to focus its beam.

However, when a satellite transmits, it transmits its **downlink** signal over a wide area called the satellite's **footprint.** Any ground station in the footprint can receive the satellite's transmissions.

Geosynchronous Earth Orbit (GEO) Satellites

Figure C-9 specifically shows a **geosynchronous Earth orbit (GEO)** satellite system. The satellite orbits at roughly 36,000 km (22,300 miles) above the Earth. At this height, its orbital time equals the Earth's rotation, so the satellite appears to be fixed in the sky. This allows dish antennas to be aimed precisely.

However, 36,000 km is a long way for radio waves to travel. Even with a dish antenna, considerable power is required. Of course, mobile devices cannot use dish antennas.

In the early days of communication satellites, satellites were often used to place voice calls. This created a delay of up to a quarter second. This latency complicated turn taking in conversations. As soon as possible, almost all telephony was moved to optical fiber.

This latency is even worse for data. TCP processes will retransmit segments if they are not acknowledged promptly. On many computers, even a single satellite in the circuit will prompt many unnecessary retransmissions. Furthermore, whenever a TCP process does a retransmission, it reduces the rate at which it transmits subsequent segments. If there are many retransmissions, the rate of transmission will become painfully slow.

Low Earth Orbit (LEO) and Medium Earth Orbit (MEO) Satellites

As Figure C-10 shows, most communication satellites operate at much lower orbits. This means that they only are over a receiver a short time before passing below the horizon. As a result, satellites must hand off service to one another. As one satellite (Satellite A) passes over the horizon, another satellite (Satellite B) will take over a customer's service. The user will not experience any service interruption. This is reminiscent of cellular telephony, except that here the customer remains relatively motionless while the satellite (the equivalent of a cellsite transceiver) moves.

Figure C-10　LEO and MEO Satellite Communication Systems

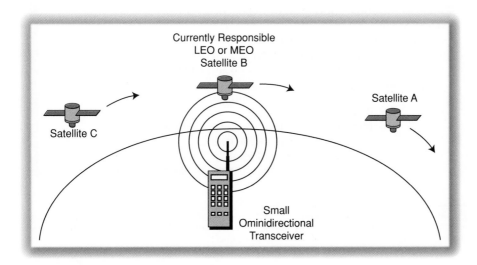

Satellites for mobile users operate in two principle orbits. **Low Earth Orbit (LEO)** satellites operate at a few hundred kilometers above the Earth. **Medium Earth Orbit (MEO)** satellites, in turn, operate at a few thousand kilometers above the Earth.[2] LEOs are closer, so signals do not attenuate as much, allowing receivers to be smaller and lighter. In contrast, MEOs have longer orbital periods, so they stay in sight longer, reducing the frequency of handoffs and therefore the number of satellites needed to provide continuous service.

VSAT Satellites

During the 1970s at Stanford University, Professor Bruce Lusignan created the idea of using very small satellite dishes for communication. At the time, most satellite dishes were at least 3 meters in diameter. This made them very expensive. Earth stations with small dishes—**very small aperture terminal (VSAT)** Earth stations—could allow earth stations to be placed in individual homes. The dishes on VSAT Earth stations are 1 meter in diameter to .3 meters in diameter. At first, Lusignan's ideas were rejected by technologists and regulators. Obviously, those objections were overcome, and VSAT Earth stations are now common.

VSATs today are used almost exclusively for one-way transmission. Satellite-based television delivery is becoming a very strong competitor for cable television. Some VSATs offer two-way communication. (You see them in newscasts and in films of military operations.) However, one-way delivery is the norm.

Figure C-11   VSAT Satellite System (Study Figure)

Traditional Satellite Systems
       Used very large dishes (3 meters or more)
       Very expensive

VSAT Satellite System
       Very small aperture terminal (VSAT) Earth stations
              Use small (1 meter or less) diameter dishes
       Small dishes allow Earth stations small and inexpensive enough to be used in homes
       Used primarily in one-way transmission, such as television distribution
       Occasionally used for two-way communication
              News reporting in the field
              Military communication

---

[2]There are two destructive bands of damaging radiation that circle the planet. These are called the Van Allen radiation bands. LEOs are below the lowest Van Allen band. MEOs operate between the Van Allen bands.

TEST YOUR UNDERSTANDING

5. a) Why are microwave repeaters needed? b) Does the uplink or the downlink use point-to-point transmission? c) What is the footprint? d) Which of the following uses dish antennas: GEOs, LEOs, or MEOs? e) Which of the following is good for mobile stations: GEOs, LEOs, or MEOs? f) What are the rough heights of GEO, LEO, and MEO orbits? g) What is a VSAT satellite dish? h) What is the attraction of VSAT technology?

WIRING THE FIRST BANK OF PARADISE HEADQUARTERS BUILDING

Wiring dominates total cost for customer premises equipment. For new buildings, a firm normally hires a contractor to install wiring. Afterward, the firm has to maintain its wiring systems.

In this section, we will look at wiring in First Bank of Paradise headquarters building. This is a typical multistory office building. Wiring in other large buildings tends to be similar.

Facilities

Figure C-12 illustrates the building. Although the building is ten stories tall, only three stories (plus the basement) are shown.

Equipment Room

Wiring begins in the **equipment room,** which usually is in the building's basement. This room connects the building to the outside world.

Vertical Risers

From the equipment room, telephone and data cabling has to rise to the building's upper floors. The bank has **vertical riser** spaces between floors. These typically are hard-walled pipes to protect the wiring.

Telecommunications Closets

On each floor above the basement, there is a **telecommunications closet.** Within the telecommunications closet, cords coming up from the basement are connected to cords that span out horizontally to telephones and computers on that floor.

Telephone Wiring

This building infrastructure was created for telephone wiring. Fortunately, most companies allocated ample space to their equipment rooms, vertical risers, and telecommunications closets. This allows data wiring to use the same spaces.

Termination Equipment

In many ways, a building's telephone network acts like an independent company. It can interconnect with the local telephone company and with other carriers. It negotiates

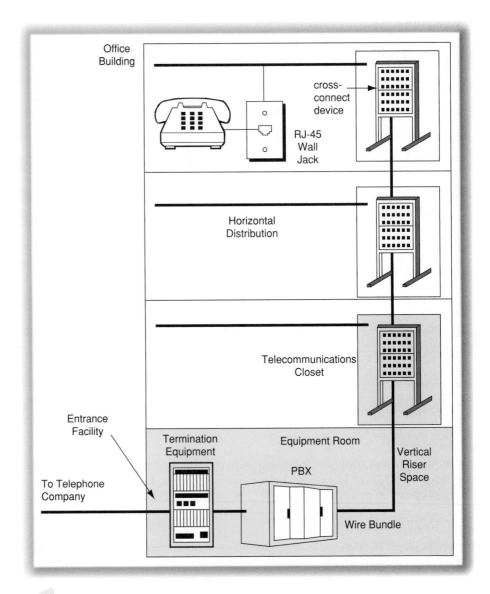

Figure C-12 First Bank of Paradise Building Wiring

contracts with each of them. Carriers require the company to install **termination equipment** at its connection point to the outside world. In effect, the termination equipment is like an electrical fuse; it prevents the company from sending unwanted electrical signals into the carrier's network.

PBX

Many companies have internal telephone switches called **private branch exchanges (PBXs).** Eight wires must span out from the PBX to each wall outlet in the building. So if the building has ten floors and each floor has 100 RJ-45 telephone wall jacks, 8,000 wires ($10 \times 100 \times 8$) will have to be run from the PBX through the vertical riser space. On each floor, the 800 wires for that floor will be organized into 100 4-pair UTP cords running from the telecommunications closet to telephone wall jacks on that floor. Obviously, careful documentation of where each wire goes is crucial to maintaining sanity.

Vertical Wiring

For vertical wiring runs, telephony typically uses **25-pair UTP cords.** These vertical cords typically terminate in **50-pin octopus connectors.**

Horizontal Wiring

The horizontal telephone wiring, as just noted, uses 4-pair UTP. Yes, telephone wiring uses the same 4-pair UTP that data transmission uses. Actually, telephone wiring first introduced 4-pair UTP. Telephony has used 4-pair UTP for several decades.

Data transmission researchers learned how to send data over 4-pair UTP to take advantage of widespread installation expertise for 4-pair UTP. In addition, some companies had excess UTP capacity already installed, so in some cases, it would not even be necessary to install new wiring.

Figure C-12 shows that wires from the telecommunications closet on a floor travel horizontally through the walls or false ceilings. They then terminate in RJ-45 data/voice jacks. Telephones plug into the jacks.

Cross-Connect Device

Within the telecommunications closet, the vertical cords plug into **cross-connect devices,** which connect the wires from the riser space to 4-pair UTP cords that span out to the wall jacks on each floor.

As Figure C-13 shows, the cross connection normally uses patch panels. The figure illustrates patch panels with RJ-45 connectors, which are useful for both voice and data wiring. Patch cords connect eight vertical wires to eight horizontal wires. Patch panels are used because they provide flexibility. If there are changes in the vertical or horizontal wiring, the patch panels are simply reconnected to reflect the changes.

Data Wiring

Figure C-12 illustrates telephone wiring. How is data wiring different? For horizontal communication, there are no differences at all. Both almost always use 4-pair UTP. They are the same precisely because Ethernet was adapted to run over horizontal telephone wiring (albeit a higher grade of telephone wiring).

However, vertical wiring is completely different. Vertical data wiring is much simpler than vertical telephone wiring. In data wiring, only single UTP or optical fiber cord runs from a port in the core switch up to a port in the Ethernet workgroup switch on each floor. In other words, if there are ten floors, only ten UTP or optical fiber

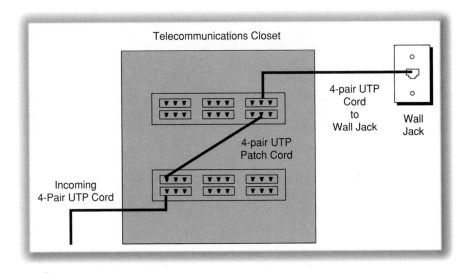

Figure C-13   Patch Panels

cords would have to be run through the vertical riser space. This is vastly simpler than vertical telephone wiring, which must run eight wires vertically for each wall jack on each floor.

PLENUM CABLING

Fire regulations require the use of a special type of fire-retardant cabling, called **plenum** cabling, any time cables run through airways (plenums) such as air conditioning ducts (but *not* false ceilings). Ordinary jackets on UTP and optical fiber cords are made of polyvinyl chloride (PVC), which gives off deadly dioxin when it burns. If these toxins are released in airways, the toxins will spread rapidly to office areas.

TEST YOUR UNDERSTANDING

6. a) What equipment are you likely to find in a building's equipment room? b) In its telecommunications closets? c) What is the purpose of a PBX? d) Compare and contrast vertical wiring distribution for telephony and data. e) Compare and contrast horizontal wiring distribution for telephony and data.

7. A building has ten floors, not counting the equipment room in the basement. Each of the ten floors has 60 voice jacks and 40 data jacks. a) For telephony, how many wires will you run through the vertical riser space for each floor? b) How many 25-pair cords will this require? c) For vertical data wiring if you use 4-pair UTP? d) For vertical data wiring if you use optical fiber? e) On each floor, how many wires will you run horizontally from the telecommunications closet to wall jacks? f) How many cords will this require?

8. a) Where is plenum cabling required? b) Why is plenum cabling needed?

PBX SERVICES

Figure C-14 shows that because digital PBXs are essentially computers, they allow vendors to differentiate their products by adding application software to provide a wide range of services.

➤ User services are employed directly by ordinary managers, secretaries, and other telephone end users.

Figure C-14   Digital PBX Services

For Users	
Speed dialing	Dials a number with a one- or two-digit code.
Last number redial	Redials the last number dialed.
Display of called number	LCD display for number the caller has dialed. Allows caller to see a mistake.
Camp on	If line is busy, hit "camp on" and hang up. When other party is off the line, he or she will be called automatically.
Call waiting	If you are talking to someone, you will be beeped if someone else calls.
Hold	Put someone on hold until he or she can be talked to.
ANI	Automatic number identification: You can see the number of the party calling you.
Conferencing	Allows three or more people to speak together.
Call transfer	Someone calls you. You connect the person to someone else.
Call forwarding	If you will be away from your desk, calls will be transferred to this number.
Voice mail	Callers can leave messages.
For Attendants	
Operator	In-house telephone operators can handle problems.
Automatic call distribution	When someone dials in, the call goes to a specific telephone without operator assistance.
Message center	Allows caller to leave a message with a live operator.
Paging	Operator can page someone anywhere in the building.
Nighttime call handling	Special functions for handling nighttime calls, such as forwarding control to a guard station.
Change requests	Can change extensions and other information from a console.
For Management	
Automatic route selection	Automatically selects the cheapest way of placing long-distance calls.
Call restriction	Prevents certain stations from placing outgoing or long-distance calls.
Call detail reporting	Provides detailed reports on charges by telephone and by department.

> ➤ Attendant services are employed by telephone operators to help them give service.
> ➤ Management services are employed by telephone and corporate network managers to manage the company's telephone network.

TEST YOUR UNDERSTANDING

9. a) Into what three categories are PBX services divided? b) List and briefly describe two services in each category.

CARRIER SERVICES AND PRICING

Having discussed technology, we can now turn to the kinds of transmission services that telecommunications staffs can offer their companies. Figure C-15 shows that corporate users face a variety of transmission services and pricing options.

Figure C-15   Telephone Services (Study Figure)

Local Calling
　　　Flat rate
　　　Message units

Toll Calls
　　　Long-distance calling
　　　Intra-LATA
　　　Inter-LATA

Toll-Call Pricing
　　　Direct distance dialing
　　　　　Base case for comparison
　　　Toll-Free numbers
　　　　　Free to caller but called party pays
　　　　　Called party: pays less than direct distance dialing rates
　　　　　An 800 prefix in most countries
　　　WATS
　　　　　Wide Area Telephone Service
　　　　　For calling out of a site
　　　　　Calling party: pays but pays less than with direct distance dialing
　　　900 numbers
　　　　　Caller pays
　　　　　Pays more than direct distance dialing rates
　　　　　Allows called party to charge for services

Advanced Services
　　　Caller ID
　　　Three-party calling (conference calling)
　　　Call waiting
　　　Voice mail

Basic Voice Services

The most important telephone service, of course, is its primary one: allowing two people to talk together. Although you get roughly the same service whether you call a nearby building or another country, billing varies widely between local and long-distance calling. Even within these categories, furthermore, there are important pricing variations.

Local Calling

Most telephone calls are made between parties within a few kilometers of each other. There are two major billing schemes for such **local calling.**

➤ Some telephone companies offer **flat-rate** local service in which there is a fixed monthly service charge but no separate fee for individual local calls.

➤ In some areas, however, carriers charge **message units** for some or all local calls. The number of message units they charge for a call depends on both the distance and duration of the call.

Economists like message units, arguing that message units are more efficient in allocating resources than flat-rate plans. Subscribers, in contrast, dislike message units even if their flat-rate bill would have come out the same.

Long-Distance Toll Calls

Although pricing for local calling varies from place to place, all **long-distance** calls are **toll calls.** The cost of the call depends on distance and duration.

Direct Distance Dialing   The simplest form of long-distance pricing is **direct distance dialing,** in which you place a call without any special deals. You will pay a few cents per minute for directly dialed calls. Direct distance dialing is a base case against which other pricing schemes can be measured.

Toll-Free Numbers   Companies that are large enough can receive favorable rates from transmission companies for long-distance calls. With **toll-free numbers,** anyone can call *into* a company, usually without being charged. To provide free inward dialing, companies pay a carrier a per-minute rate lower than the rate for directly dialed calls. Most countries use the 800 area code to provide such services.

WATS   In contrast to inbound toll-free number service, **wide area telephone service (WATS)** allows a company to place *outgoing* long-distance calls at per-minute prices lower than those of directly dialed calls. WATS prices depend on the size of the service area. WATS is often available for both intrastate and interstate calling. WATS can also be purchased for a region of the country instead of the entire country.

900 Numbers   Related to toll-free, **900 numbers** allow customers to call into a company. Unlike toll-free number calls, which usually are free to the caller, calls to 900 numbers require the caller to pay a fee—one that is much *higher* than that of a toll call. Some of the fee goes to the carrier, but most of it goes to the subscriber being called.

This allows companies to charge for information, technical support, and other services. For instance, customer calls for technical service might cost $20 to $50 per hour. Charges for 900 numbers usually appear on the customer's regular monthly bill from the local exchange carrier (LEC). Although the use of 900 numbers for sexually oriented services has given 900 numbers a bad name, they are valuable for legitimate business use.

Advanced Services

Although telephony's basic function as a two-person "voice pipe" is important, telephone carriers offer other services to attract customers and to get more revenues from existing customers.

Caller ID

In **caller ID,** the telephone number of the party calling you is displayed on your phone's small display screen before you pick up the handset. This allows you to screen calls, picking up only the calls you want to receive. Callers can block caller ID, so that you cannot see their numbers. However, you can have your carrier reject calls with blocked IDs. Businesses like caller ID because it can be linked to a computer database to pull up information about the caller on the receiver's desktop computer screen.

Three-Party Calling (Conference Calling)

Nearly every teenager knows how to make **three-party calls,** in which more than the traditional two people can take part in a conversation. However, businesses tend to use this feature only sparingly, despite its obvious advantage. This is sometimes called **conference calling.**

Call Waiting

Another popular service is **call waiting.** If you are having a conversation and someone calls you, you will hear a distinctive tone. You can place your original caller on hold, shift briefly to the new caller, and then switch back to your original caller.

Voice Mail

Finally, **voice mail** allows people to leave messages if you do not answer your phone.

TEST YOUR UNDERSTANDING

10. Create a table to compare and contrast direct distance dialing, toll-free numbers, 900 numbers, and WATS, in terms of whether the caller or the called party pays and the cost compared with the cost of a directly dialed long-distance call.
11. Describe the two pricing options for local calls.
12. a) What is the advantage of toll-free numbers for customers? b) For companies that provide toll-free number service to their customers?
13. a) Name the four advanced telephone services listed in the text. b) Name and briefly describe two advanced services not listed in the text.

TELEPHONE CARRIERS AND REGULATION

Once, almost every nation had a single national telephone carrier. However, the situation has become more complex over time as nations have begun to deregulate telephone service—that is, to permit some competition in order to reduce prices and promote product innovation.

Competition helps corporate customers because telephone prices generally fall as a result of competition. However, to maximize cost savings, companies have to be very smart when they deal with telephone carriers. To do this, a first step is understanding the types of carriers a company will face (Figure C-16).

Figure C-16   Telephone Carriers (Study Figure)

In Most Countries
>	Public Telephone and Telegraph (PTT) authorities
>>		Traditionally had a domestic monopoly over telephone service
>	Ministries of Communication
>>		Government agency to regulate the PTT
>	Competitors
>>		Deregulation has allowed competition in domestic telephone service in most countries
>>		The Ministry of Telecommunication regulates these new competitors too

In the United States
>	AT&T (the Bell System) developed a long-distance monopoly
>	Also owned most local operating companies
>	AT&T was broken up in the 1980s
>>		AT&T retained the name and the (initially) lucrative long-distance business
>>		Local operations were assigned to seven Regional Bell Operating Companies (RBOCs)
>>		Later, RBOCs combined with one another and with GTE to form four supercarriers
>>		Eventually, competition in long-distance service made AT&T unprofitable
>>		In 2005, one of the four supercarriers (SBC Communications) merged with AT&T and used the AT&T name for the merged company.
>	Regulation
>>		Federal Communications Commission (FCC) regulates interstate communication and aspects of intrastate communication that affect national commerce
>>		Within each state, a Public Utilities Commission (PUC) regulates telephone service subject to FCC regulations

PTTs and Ministries of Telecommunications

In most countries the other than United States, the single monopoly carrier was historically called the Public Telephone and Telegraph authority (PTT). In the United Kingdom, for example, this was British Telecom, while in Ireland it was Eircom. The PTT had a monopoly on domestic telephony—that is, telephony within the country.

To counterbalance the power of the PTT, governments created regulatory bodies generally called **Ministries of Telecommunications.** PTTs provide service, while Ministries of Telecommunications oversee the PTTs. As we will see later, over time, the PTTs gradually lost their monopoly status, and ministries of telecommunications now find themselves regulating both the traditional PTT and its new competitors.

TEST YOUR UNDERSTANDING

14. a) Do all countries have PTTs? Explain. b) What is a monopoly over domestic telephone service? c) What are the purposes of PTTs and Ministries of Telecommunications?

AT&T, the FCC, and PUCs

The Bell System

In the United States, neither telegraphy nor telephony was made a statutory monopoly.[3] However, telephony quickly became a de facto monopoly when **AT&T,** also known as the **Bell System,** used predatory practices to drive most other competitors out of business. AT&T soon had a complete long-distance monopoly. For local service, AT&T owned more than 80 percent of all local telephone companies, although when it was developing in the nineteenth century and early twentieth century, it bypassed "unpromising" areas such as Hawai'i and most of Los Angeles.

The RBOCs

In the 1980s, AT&T was broken up into a long-distance and manufacturing company that retained the AT&T name and seven **Regional Bell Operating Companies (RBOCs)** that owned most local telephone companies.

Later, mergers among the RBOCs and GTE, which was the largest independent owner of local operating companies, produced four dominant owners of local operating companies in the United States—Verizon, SBC Communications, BellSouth, and Qwest. These four companies also provide long-distance service in some areas.

At the time of the breakup, AT&T was considered the jewel in the Bell System. However, after quite a few years of high profitability, AT&T began to suffer heavily from long-distance competition. In 2005, in a stroke of irony, SBC merged with ailing AT&T. The combined company took on the name AT&T.

---

[3]Samuel F.B. Morse, who invented the telegraph, tried to sell his invention to the U.S. Post Office, but the government rejected it. In other countries, postal services, which had traditionally enjoyed a monopoly over mail delivery, were also given monopolies over telegraphy and later telephony. In fact, "PTT" originally stood for "*Postal* Telephone and Telegraph." Over time, companies separated mail and electronic communication, and *Postal* became *Public.* Now that the telegraph system no longer exists in most countries, perhaps a contest is needed for one of the Ts.

Regulation: The FCC and PUCs

In the United States, the **Federal Communication Commission (FCC)** provides overall regulation for U.S. carriers. However, within individual states, **Public Utilities Commissions (PUCs)** regulate pricing and services.

TEST YOUR UNDERSTANDING

15. a) Distinguish between the traditional roles of AT&T and the RBOCs. b) Distinguish between the traditional roles of the FCC and PUCs in the United States.

Deregulation

Although telephone carriers had a complete monopoly in the early years, governments began deregulating telephone service in the 1970s. **Deregulation** is the opening of telephone services to competition; it has the potential to reduce costs considerably (Figure C-17).

Deregulation Around the World

As noted earlier, most countries have deregulated at least some of the services offered by the traditional monopoly PTT. This has given companies many more choices for telephone services, and competition has resulted in lower prices.

Carriers in the United States

LATAs   Figure C-18 shows the types of carriers that exist in the United States. Since the breakup of AT&T in 1984, the United States has divided into approximately 200 service regions called **local access and transport areas (LATAs).**

ILECs and CLECs   **Within each LATA,** local exchange carriers (LECs) provide access and transport (transmission service). The traditional monopoly telephone company is called the **incumbent local exchange carrier (ILEC).** Competitors are called **competitive local exchange carriers (CLECs).**

> LATAs are geographic regions. ILECs and CLECs are carriers that provide access and transport within LATAs.

IXCs   In contrast, **interexchange carriers (IXCs)** carry voice traffic *between* LATAs. Major ILECs are AT&T, MCI, and Sprint.

Long-Distance Calling   One point of common confusion is that the distinction between local and long distance calling is not the same as the distinction between LEC and IXC service. Most LATAs are quite large; within LATAs, there is both local and long-distance calling. Adding to the confusion, intra-LATA long distance calling rates sometimes are higher than inter-LATA calling rates.

> Within LATAs, there is both local and long-distance calling.

ICCs   ILECs, CLECs, and IXCs are **domestic** carriers that provide service within the United States. Similarly, PTTs provide domestic service within their own countries. In contrast, **international common carriers (ICCs)** provide service *between* countries.

Deregulation
> Deregulation decreases or removes monopoly over telephone service
> This creates competition, which lowers prices
> In most companies, deregulation began in the 1970s

Deregulation Around the World
> At least some PTT services have been deregulated

Carriers in the United States
> The United States is divided into regions called local access and transport areas (LATAs)
> Within each LATA:
> > Local exchange carriers (LECs) provide intra-LATA service
> > Traditional incumbent local exchange carrier (ILECs)
> > New competitive local exchange carriers (CLECs)
> Interexchange carriers (IXCs) provide transport between LATAs
> Long-distance service
> > Long-distance service within LATAs is supplied by LECs
> > Long-distance service between LATAs is supplied by IXCs
> Within each LATA, one or more points of presence (POP) interconnects different carriers

Internationally
> International common carriers (ICCs) provide service between countries

Degree of Deregulation
> Customer premises equipment is almost completely deregulated
> Long-distance and international telephony are heavily deregulated
> Local telephone service is the least deregulated
> > The traditional monopoly carriers have largely maintained their telephone monopolies
> > Cellular service has provided local competition, with many people not having a wired phone
> > Voice over IP (VoIP) is providing strong competition via ISPs, cable television companies, and a growing number of other wired and wireless access technologies

VoIP Regulation
> Countries are struggling with the question of how to regulate VoIP carriers
> Should they be taxed?
> Should they be required to provide 911 service, including location determination?
> Should they be required to provide wiretaps to government agencies?

Figure C-17   Deregulation (Study Figure)

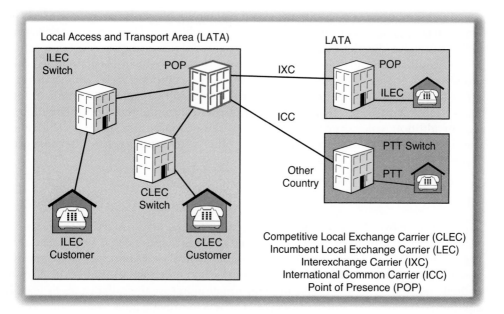

Figure C-18   Telephone Carriers in the United States

Points of Presence (POPs)   As Figure C-18 shows, the various carriers that provide service are interconnected at **points of presence (POPs).** Thanks to points of presence, any subscriber to any CLEC or ILEC in one LATA can reach customers of any other CLEC or ILEC in any other LATA. ICCs also link to domestic carriers at POPs.

Deregulation by Service

Customer Premises Equipment   Although it seems odd today, telephone companies used to own all of the wires and telephones in homes and businesses. Today, however, nearly all countries *prohibit* carriers from owning customer premises equipment. Deregulation for customer premises equipment, in other words, is total in most countries.

Long-Distance and International Calling   In most countries, both long-distance and international telephone services have been heavily deregulated.

Local Telephone Service   Local telephone service is the least deregulated aspect of telephony. The need for large investments in access systems and regulatory reluctance to open local telephone service completely (for fear of losing currently subsidized service for the poor and rural customers) have combined to limit local telephone competition.

Some countries now require the traditional monopoly carrier to open its access systems and central offices to competitors for a "reasonable" fee. However, court delays and high "reasonable" fees have limited the effectiveness of this facility-sharing approach.

Overall, traditional monopoly telephone carriers have largely maintained their monopoly over wired telephone service. However, competition is coming through other technologies. Many people now have only a cellular telephone, and cellular service often is provided by a competitor of the traditional monopoly wireline carrier. In addition, voice over IP (VoIP) is providing competition via ISPs, cable television companies, and a growing number of other wired and wireless Internet access technologies.

VOICE OVER IP

Now that voice over IP (VoIP) is becoming popular, countries are trying to determine how to regulate this new service. Traditional carriers point out that VoIP carriers are exempt from many of the taxes that traditional carriers are required to pay. Countries also are attempting to enforce laws requiring calls to emergency numbers (911 in the United States) to give physical location information in case the caller cannot speak. In addition, the U.S. government wants VoIP carriers to provide tools to allow the government to create legal wiretaps.

TEST YOUR UNDERSTANDING

16. a) Distinguish between LATAs, ILECs, and CLECs. b) What is the role of IXCs relative to LATAs? c) What carriers handle long-distance calling in the United States? d) What is the role of ICCs? e) Why are POPs important?

17. a) What is deregulation? b) When did deregulation begin? c) How complete is deregulation for customer premises equipment? d) For long-distance calling? e) For local calling? f) What issues are involved in the regulation of VoIP?

PROJECT

Getting Current. Go to the book website's New Information and Errors pages for this chapter to get new information since this book went to press and to correct any errors in the text.

NET: Microsoft's approach to the Web services.

1G: *See* First-Generation.

1-Pair Voice-Grade UTP: The traditional telephone access lines to individual residences.

10Base-T: *See* 802.3 10Base-T.

100Base-TX: The dominant Ethernet physical layer 100 Mbps standard brought to desktop computers today.

1000Base-LX: A fiber version of gigabit Ethernet for long wavelengths (transmitting at 1,300 nm).

1000Base-SX: A fiber version of gigabit Ethernet for short wavelengths (transmitting at 850 nm). The dominant standard for 1 Gbps connections between switches (and routers) in Ethernet LANs.

1000Base-T: A UTP version of gigabit Ethernet.

1000Base-x: The Ethernet physical layer technology of gigabit Ethernet, used today mainly to connect switches to switches or switches to routers; increasingly being used to connect servers and some desktop PCs to the switches that serve them.

2G: *See* Second-Generation. Used to describe cellular technology.

2-Pair Data-Grade: The higher-quality UTP access lines used by telephone carriers for private lines. Two pairs run out to each customer.

2-Pair Data-Grade UTP: The traditional telephone access line for lower-speed leased lines. (Higher-speed leased lines use optical fiber.)

232 Serial Port: The port on a PC that uses two voltage ranges to transmit information.

25-Pair UTP Cord: The cabling used by telephony for vertical wiring that runs within a building.

3DES: *See* Triple DES.

3G: *See* Third-Generation.

4-Pair Unshielded Twisted Pair (UTP): The type of wiring typically used in Ethernet networks. 4-pair UTP contains eight copper wires organized as four pairs. Each wire is covered with dielectric insulation, and an outer jacket encloses and protects the four pairs.

50-Pin Octopus Connector: The type of connector in which vertical cords typically terminate.

802 Committee: *See* 802 LAN/MAN Standards Committee.

802 LAN/MAN Standards Committee: The IEEE committee responsible for Ethernet standards.

802.1D Spanning Tree Protocol: The protocol that addresses both single points of failure and loops.

802.1AE: MAC-layer security standard for supervisory communication between Ethernet switches.

802.1p: The standard that permits up to eight priority levels.

802.1Q: The standard that extended the Ethernet MAC layer frame to include two optional tag fields.

802.1w: *See* Rapid Spanning Tree Protocol.

802.1X: Security authentication standard for both wired and wireless LANs.

802.2: The single standard for the logical link control layer in 802 LANs.

802.3 10Base-T: The slowest Ethernet physical layer technology in use today; uses 4-pair UTP wiring and operates at 10 Mbps.

802.3ad: Link aggregation protocol standard.

802.3af: Standard for delivering low wattage electricity from a switch to stations.

802.3at: Power over Ethernet Plus standards. Provides more power than 802.3af.

802.3 MAC Layer Frame: *See* Ethernet Frame.

802.3 MAC Layer Standard: The standard that defines Ethernet frame organization and NIC and switch operation.

802.3 Working Group: The 802 Committee's working group that creates Ethernet-specific standards.

802.5 Working Group: The 802 Committee's working group that created Token-Ring Network standards.

802.11 WLAN: Wireless LANs that follow the 802.11 standard.

802.11 Working Group: The IEEE working group that creates wireless LAN standards.

802.11a: Version of the 802.11 WLAN standard that has a rated speed of 54 Mbps and operates in the 5 GHz unlicensed radio band.

802.11b: Version of the 802.11 WLAN standard that has a rated speed of 11 Mbps and operates in the 2.4 GHz unlicensed radio band.

802.11g: Version of the 802.11 WLAN standard that has a rated speed of 54 Mbps and operates in the 2.4 GHz unlicensed radio band.

802.11e: A standard for quality of service in 802.11 WLANs.

802.11i: An advanced form of 802.11 wireless LAN security.

802.11n: Version of the 802.11 WLAN standard that uses MIMO and double-width channels to achieve a rated speed up to 300 Mbps and longer range than earlier speed standards.

802.11s: Standard for mesh networking in 802.11 WLANs.

802.16: WiMAX. Broadband wireless access standard.

802.16d: WiMAX. Broadband wireless access standard for fixed stations.

802.16e: WiMAX. Broadband wireless access standard for mobile stations.

900 Number: A number that allows customers to call into a company; callers pay a fee that is much higher than that of a regular toll call.

Access Control List (ACL): An ordered list of pass/deny rules for a firewall or other device.

Access Control Plan: A plan for controlling access to a resource.

Access Line: 1) In LANs, a transmission line that connects a station to a switch. 2) In telephony, the line used by the customer to reach the PSTN's central transport core.

Access Point: A bridge between a wireless station and a wired LAN.

Access Router: A router to connect a SOHO network to the Internet. Typically includes a switch, DHCP server, NAT, and other functions beyond routing.

Access System: In telephony, the system by which customers access the PSTN, including access lines and termination equipment in the end office at the edge of the transport core.

Account: An identifiable entity that may own resources on a computer.

ACE: *See* OPNET Application Characterization Environment.

ACK: *See* Acknowledgment.

ACK Bit: The bit in a TCP segment that is set to indicate if the segment contains an acknowledgement.

Acknowledgment (ACK): 1) An acknowledgment message, sent by the receiver when a message is received correctly. 2) An acknowledgment frame, sent by the receiver whenever a frame is received; used in CSMA/CA+ACK in 802.11.

Acknowledgment Bit: A bit in a TCP header. If the bit is set, then the TCP segment contains an acknowledgment

Acknowledgment Number Field: In TCP, a header field that tells what TCP segment is being acknowledged in a segment.

ACL: *See* Access Control List.

Active Directory: Microsoft's directory server product.

Adaptive Antenna System: A type of smart antenna.

ADC: *See* Analog-to-Digital Conversion.

Address Resolution Protocol (ARP): Protocol for address resolution used in Ethernet networks. If a host or router knows a target host's or router's IP address, ARP finds the target's host data link layer address.

Administration: In network management, operations maintenance, and provisioning involve real-time work ot keep the network running. Administration tasks includes "everything else." This includes planning, the collection of network data, and paying bills, among many other things.

Administrative IP Server: A server needed to support IP.

Administrator: A super account on a Windows server that automatically has full permissions in every directory on the server.

ADSL: *See* Asymmetric Digital Subscriber Line.

Advanced Encryption Standard (AES): New symmetric encryption standard that offers 128-bit, 192-bit, or 256-bit encryption efficiently.

AES: *See* Advanced Encryption Standard.

AES-CCMP: AES/Counter Mode with Cipher Block Chaining. The version of AES used in the 802.11i security standard for wireless LANs.

Agent: *See* Network Management Agent.

Aggregate Throughput: Throughput shared by multiple users; individual users will get a fraction of this throughput.

Alternative Route: In mesh topology, one of several possible routes from one end of the network to the other, made possible by the topology's many connections among switches or routers.

Always On: Being always available for service; used to describe access lines.

Amplitude Modulation: A simple form of modulation in which a modem transmits one of two analog signals—a high-amplitude (loud) signal or a low-amplitude (soft) signal.

Amplitude: The maximum (or minimum) intensity of a wave. In sound, this corresponds to volume (loudness).

Analog Signal: A signal that rises and falls in intensity smoothly and that does not have a limited number of states.

Analog-to-Digital Conversion (ADC): A device for the conversion of transmissions from the analog local loop to signals on the digital telephone network's core.

Antivirus Software: Software that scans computers to protect them against viruses, worms, and Trojan horses arriving in e-mail attachments and other propagation methods.

API: *See* Application Program Interface.

AppleTalk: Apple's proprietary architecture for use on Macintosh computers.

Application Architecture: The arrangement of how application layer functions are spread among computers to deliver service to users.

Application Characterization Environment: *See* OPNET Application Characterization Environment.

Application Firewall: A firewall that examines the application layer content of packets.

Application Layer: The standards layer that governs how two applications communicate with each other; Layer 7 in OSI, Layer 5 in TCP/IP.

Application Profile: A method offered by Bluetooth that allows devices to work with one another automatically at the application layer.

Application Program Interface (API): A specification that allows application server programs to interact directly with database systems.

Application Program: Program that does work for users; operating system is the other major type of program found on computers.

Application Server: A server used by large e-commerce sites that accepts user data from a front-end webserver, assembles information from other servers, and creates a webpage to send back to the user.

Architecture: A broad plan that specifies what is needed in general and the components that will be used to provide that functionality. Applied to standards, networks, and applications.

ARPA: *See* Defense Advanced Research Projects Agency.

ARP Cache: Section of memory that stores known pairs of IP addresses and switched network standards.

ASCII Code: A code for representing letters, numbers, and punctuation characters in 7-bit binary format. Each character is sent in a byte: the 8th bit is not used.

Asymmetric Digital Subscriber Line (ADSL): The type of DSL designed to go into residential homes, offers high downstream speeds but limited upstream speeds.

Asynchronous Transfer Mode (ATM): The packet-switched network technology, specifically designed to carry voice, used for transmission in the PSTN transport core. ATM offers quality of service guarantees for throughput, latency, and jitter.

ATM: Asynchronous Transfer Mode.

AT&T: U.S. telecommunications carrier.

Attenuate: For a signal's strength to weaken during propagation.

Auditing: Collecting data about events to assess actions after the fact.

Authentication: The requirement that someone who requests to use a resource must prove his or her identity.

Authentication Server: A server that stores data to help the verifier check the credentials of the applicant.

Authorization: Permitting a person or program to take certain actions on a resource.

Authorizations: Specific actions that a person or program can take on a resource.

Autonomous System: Internet owned by an organization.

Autosensing: The ability of a switch to detect the standard being used at the other end of the connection, and adjust its own speed to match.

Availability: The ability of a network to serve its users.

Backdoor: A way back into a compromised computer that an attacker leaves open; it may simply be a new account or a special program.

Back-Office: Transaction processing applications for a business's internal needs.

Backup: Copying files stored on a computer to another medium for protection of the files.

Backward-Compatible: Able to work with all earlier versions of a standard or technology.

Bandpass Filter: A device that filters out all signals below 300 Hz and above about 3.4 kHz.

Bandwidth: The range of frequencies over which a signal is spread.

Bank Settlement Firm: An e-commerce service that handles credit card payments.

Base 2: Notation for representing numbers; each position can hold only a 0 or 1.

Base Price: The price of a system's hardware, software, or both before necessary options are added.

Baseband: Transmission in which the signal is simply injected into a wire.

Baseband Signal: 1) The original signal in a radio transmission; 2) a signal that is injected directly into a wire for propagation.

Baud Rate: The number of clock cycles a transmission system uses per second.

Bell System: The conglomerate of local and long-distance telecommunications carriers that was broken up by antitrust action in the early 1980s.

BER: *See* Bit Error Rate.

Best-Match Row: The row that provides the best forwarding option for a particular incoming packet.

BGP: *See* Border Gateway Protocol.

Binary Data: Data that has only two possible values (1s and 0s).

Binary Numbers: The Base 2 counting system where 1s and 0s used in combination can represent whole numbers (integers).

Binary Signaling: Signaling that uses only two states.

Biometrics: The use of bodily measurements to identify an applicant.

Bit: A single 1 or 0.

Bit Error Rate: The percentage of all transmitted bits that contain errors.

Bit Rate: In digital data transmission, the rate at which information is transmitted; measured in bits per second.

Bits per Second (bps): The measure of network transmission speed. In increasing factors of 1,000 are kilobits per second (kbps), megabits per second (Mbps), gigabits per second (Gbps), and terabits per second (Tbps).

Black List: A list of banned websites.

Blended Threat: An attack that propagates both as a virus and as a worm.

Bluetooth: A wireless networking standard created for personal area networks.

Bonding: *See* Link Aggregation.

Border Firewall: A firewall that sits at the border between a firm and the outside world.

Border Gateway Protocol (BGP): The most common exterior routing protocol on the Internet. Recall that *gateway* is an old term for *router.*

Border Router: A router that sits at the edge of a site to connect the site to the outside world through leased lines, PSDNs, and VPNs.

Bot: A type of malware that can be upgraded remotely by an attacker to fix errors or to give the malware additional functionality.

Bps (bps): *See* Bits per Second.

Breach: A successful attack.

Bridge: An access point that connects two different types of LANs.

Broadband Wireless Access (BWA): High-speed local wireless transmission systems.

Broadband: 1) Transmission where signals are sent in wide radio channels; 2) any high-speed transmission system.

Broadband over Power Lines: Transmitting broadband data over electrical power lines.

Broadcast: To send a message out to all other stations simultaneously.

Broadcast Address: In Ethernet, FF-FF-FF-FF-FF-FF (48 ones); tells switches that the frame should be broadcast.

Brute-Force Attack: A password-cracking attack in which an attacker tries to break a password by trying all possible combinations of characters.

Bursty: Having short, high-speed bursts separated by long silences. Characteristic of data transmission.

Bus Topology: A topology in which one station transmits and has its signals broadcast to all stations.

Business Case: An argument for a system in business terms.

Business Continuity: A company's ability to continue operations.

Business Continuity Recovery: The reestablishment of a company's ability to continue operations.

BWA: *See* Broadband Wireless Access.

C7: Telephone supervisory control signaling system used in Europe.

CA: 1) *See* Certificate Authority. 2) *See* Collision Avoidance.

Cable Modem: 1) Broadband data transmission service using cable television; 2) the modem used in this service.

Cable Replacement: Getting rid of cables between devices by implementing wireless networking.

Call Waiting: A service that allows the user to place an original caller on hold if someone else calls the user, shift briefly to the new caller, and then switch back to the original caller.

Caller ID: Service wherein the telephone number of the party calling you is displayed on your phone's small display screen before you pick up the handset; allows the user to screen calls.

Carder: Someone who steals credit card numbers.

Carrier Sense Multiple Access with Collision Avoidance and Acknowledgments (CSMA/CA+ACK): In 802.11 wireless LANs, a mandatory mechanism used to reduce problems with multiple simultaneous transmissions, which occur in wireless transmission. CSMA/CA+ACK is a media access control discipline, and it uses both collision avoidance and acknowledgment frames.

Carrier Sense Multiple Access with Collision Detection (CSMA/CD): In 802.3 Ethernet networks, the process wherein if a station wants to transmit, it may do so if no station is already transmitting but must wait if another station is already sending. In addition, if there is a collision because two stations send at the same time, all stations stop, wait a random period of time, and then try again.

Carrier: A transmission service company.

Cat: A short form for "category" in UTP.

Cat 5e: *See* Category 5e.

Category: In UTP cabling, a system for measuring wiring quality.

Category (Cat) 5e: Quality type of UTP wiring; widely used and can support 100Base-TX and gigabit Ethernet.

Category 6: The newest quality type of UTP wiring being sold; not required for even gigabit Ethernet. Can carry 10 Gbps 55 meters.

Category 6A: Augmented Category 6 wiring that can sustain higher transmission speeds than Category 6 wiring. Can carry 10 Gbps Ethernet 100 meters.

Category 7: A new twisted-pair wiring quality standard; uses shielded twisted pair (STP) wiring. Can carry 10 Gbps Ethernet 100 meters.

CDMA: *See* Code Division Multiple Access.

CDMA IS-95: The form of CDMA used in 2G cellular technology in the United States.

CDMA2000: A new 3G technology, developed by Qualcomm, offering a staged approach to increasing speed.

CDMA2000 1x: The initial 3G step for implementing CDMA2000, offering telephone modem speeds.

CDMA2000 1xEV-DO: The second 3G step for implementing CDMA2000 that will offer speeds similar to those in DSL and cable modems.

Cell: 1) In ATM, a fixed-length frame. 2) In cellular telephony, a small geographic area served by a cellsite.

Cellphone: A cellular telephone, also called a *mobile phone* or *mobile*.

Cellsite: In cellular telephony, equipment at a site near the middle of each cell, containing a transceiver and supervising each cellphone's operation.

Cellular Telephone Service: Radio telephone service in which each subscriber in each section of a region is served by a separate cellsite. This permits channel reuse and so supports more subscribers.

Cellular Modem: A modem that allows a computer to communicate through a cellular telephone.

Certificate Authority (CA): Organization that provides public key–private key pairs and digital certificates.

Certificate Revocation List (CRL): A certificate authority's list of digital certificates it has revoked before their expiration date.

Challenge Message: In challenge–response authentication protocols, the message initially sent from the verifier to the applicant.

Challenge–Response Authentication Protocol (CHAP): A specific challenge–response authentication protocol.

Challenge–Response Authentication: Initial authentication method in which the verifier sends the applicant a challenge message, and the applicant does a calculation to produce a response, which it sends back to the verifier.

Channel Bandwidth: The range of frequencies in a channel; determined by subtracting the lowest frequency from the highest frequency.

Channel Reuse: The ability to use each channel multiple times, in different cells in the network.

Channel Service Unit (CSU): The part of a CSU/DSU device designed to protect the telephone network from improper voltages sent into a private line.

Channel: A small frequency range that is a subdivision of a service band. A channel normally carries a single signal. Signals id different channels do not interfere with each other.

CHAP: *See* Challenge–Response Authentication Protocol.

Checkout: A core e-commerce function that allows a buyer who has finished shopping to pay for the selected goods.

Chronic Lack of Capacity: A state in which the network lacks adequate capacity much of the time.

CIDR: *See* Classless InterDomain Routing.

Cipher: An encryption method.

Ciphertext: The result of encrypting a plaintext message. Ciphertext can be transmitted with confidentiality.

CIR: *See* Committed Information Rate.

Circuit: A two-way connection with reserved capacity.

Circuit Switching: Switching in which capacity for a voice conversation is reserved on every switch and trunk line end-to-end between the two subscribers.

Cladding: A thick glass cylinder that surrounds the core in optical fiber.

Class A IP Address: In classful addressing, an IP address block with more than sixteen million IP addresses; given only to the largest firms and ISPs.

Class B IP Address: In classful addressing, an IP address block with about 65,000 IP addresses; given to large firms.

Class C IP Address: In classful addressing, an IP address block with 254 possible IP addresses; given to small firms.

Class D IP Address: In classful addressing, IP addresses used in multicasting.

Class 5 Switch: *See* End Office Switch.

Classful Addressing: Giving a firm one of four block sizes for IP addresses: a very large Class A address block, a medium-size Class B address block, or a small Class C address block.

Classless InterDomain Routing (CIDR): System for allocating IP addresses that does not use IP address classes.

Clear Line of Sight: An obstructed radio path between the sender and the receiver.

Clear to Send (CTS): In 802.11, a message broadcast by an access point which allows only a station that has sent a Request to Send message to transmit. All other stations must wait.

CLEC: *See* Competitive Local Exchange Carrier.

CLI: *See* Command Line Interface.

Client PC: A personal computer that acts as a client.

Client Host: A host that receives service from a server station.

Client/Server Application: Application in which a client program requests service from a server and in which the server program provides the service.

Client/Server Processing: The form of client/server computing in which the work is done by programs on two machines.

Client/Server System: A system where some processing power is on the client computer. The two types of client/server systems are file server program access and full client/server processing.

Clock Cycle: A period of time during which a transmission line's state is held constant.

Cloud: The symbol traditionally used to represent the PSDN transport core, reflecting the fact that although the PSDN has internal switches and trunk lines, the customer does not have to know how things work inside the cloud.

Coating: In optical fiber, the substance that surrounds the cladding to keep out light and to strengthen the fiber. Coating includes strands of yellow Aramid (Kevlar) yarn to strengthen the fiber.

Coaxial Cable: The IEEE working group that creates wireless LAN standards.

Code Division Multiple Access (CDMA): A new form of cellular technology and a form of spread spectrum transmission that allows multiple stations to transmit at the same time in the same channel; also permits stations in adjacent cells to use the same channel without serious interference.

Codec: The device in the end office switch that converts between the analog local loop voice signals and the digital signals of the end office switch.

Collision: When two simultaneous signals use the same shared transmission medium, the signals will add together and become scrambled (unintelligible).

Collision Avoidance (CA): In 802.11, used with CSMA to listen for transmissions, so if a wireless NIC detects a transmission, it must not transmit. This avoids collision.

Collision Domain: In Ethernet CSMA/CD systems that use hubs or bus topologies, the collection of all stations that can hear one another; only one can transmit at a time.

Command Line Interface (CLI): An interface used to work with switches and routers in which the user types highly structured commands, ending each command with Enter.

Command–Response Cycle: The exchange of messages through which SNMP communication between the manager and agents takes place. In it, the manager sends a command, and the agent sends back a response confirming that the command has been met, delivering requested data, or saying that an error has occurred and that the agent cannot comply with the command.

Committed Information Rate (CIR): PVC speed that is guaranteed by the Frame Relay carrier.

Communication Satellite: Satellite that provides radio communication service.

Community Name: In SNMP Version 1, only devices using the same community name will communicate with each other; very weak security.

Competitive Local Exchange Carrier (CLEC): A competitor to the ILEC.

Comprehensive Security: Security in which all avenues of attack are closed off.

Compression: Reduce the number of bits needed to transmit a message or file.

Compromise: A successful attack.

Computer Security Incident Response Team (CSIRT): A team convened to handle major security incidents, made up of the firm's security staff, members of the IT staff, and members of functional departments, including the firm's legal department.

Conference Calling: A multiparty telephone call.

Confidentiality: Assurance that interceptors cannot read transmissions.

Connectionless: Type of conversation that does not use explicit openings and closings.

Connection-Oriented: Type of conversation in which there is a formal opening of the interactions, a formal closing, and maintenance of the conversation in between.

Connectorize: To add connectors to something.

Constellation: In quadrature amplitude modulation, the collection of all possible amplitude/phase combinations.

Continuity Testers: UTP tester that ensures that wires are inserted into RJ-45 connectors in the correct order and are making good contact.

Convergence: The correction of routing tables after a change in an internet.

Conversion: The process of browsers becoming buyers.

Cookie: Small text file stored by a website on a client PC; can later be read from the website.

Cord: A length of transmission medium—usually UTP or optical fiber but sometimes coaxial cable.

Core Switch: A switch further up the hierarchy that carries traffic between pairs of switches. May also connect switches to routers.

Core: 1) In optical fiber, the very thin tube into which a transmitter injects light. 2) In a switched network, the collection of all core switches.

Corporate Network: A network that carries the internal traffic of a single corporation.

Crack: To guess a password.

Credentials: Proof of identity that an applicant can present during authentication.

Credit Card Verification Service: An e-commerce service that checks the validity of the credit card number a user has typed.

Criminal Attacker: An attacker who attacks with criminal motivation.

Crimping Tool: Tool for crimping wires into an RJ-45 connector.

CRL: *See* Certificate Revocation List.

CRM: *See* Customer Relationship Management.

Cross-Connect Device: The device within a wiring closet that vertical cords plug into. Cross-connect devices connect the wires from the riser space to 4-pair UTP cords that span out to the wall jacks on each floor.

Crossover Cable: A UTP cord that allows a NIC in one computer to be connected directly to the NIC in another computer; switches Pins 1 and 2 with Pins 3 and 6.

Crosstalk Interference: Mutual EMI among wire pairs in a UTP cord.

Cryptographic System: A security system that automatically provides a mix of security protections, usually including confidentiality, authentication, message integrity, and replay protection.

Cryptography: Mathematical methods for protecting communication.

CSIRT: *See* Computer Security Incident Response Team.

CSMA/CA+ACK: *See* Carrier Sense Multiple Access with Collision Avoidance and Acknowledgments. *See* also definitions of the individual components.

CSMA/CD: *See* Carrier Sense Multiple Access with Collision Detection.

CSU/DSU: Device that connects an internal site system to a private line circuit.

CSU: *See* Channel Service Unit.

CTS: *See* Clear to Send.

Customer Premises Equipment (CPE): Equipment owned by the customer, including PBXs, internal vertical and horizontal wiring, and telephone handsets.

Customer Relationship Management (CRM): Software that examines customer data to understand the preference of a company's customers.

Cut-Through: Switching wherein the Ethernet switch examines only some fields in a frame's header before sending the bits of the frame back out.

Cyberterror: A computer attack made by terrorists.

Cyberwar: A computer attack made by a national government.

DAC: *See* Digital-to-Analog Conversion.

DARPA: *See* Defense Advanced Research Projects Agency.

Data: Information carried over a network.

Data Communications: The transmission of encoded information, as opposed to the type of information carried in telecommunications systems.

Data Encryption Standard (DES): Popular symmetric key encryption method; with only 56-bit keys, considered to be too weak for business-to-business encryption.

Data Field: The content delivered in a message.

Data Link: The path that a frame takes across a single network (LAN or WAN).

Data Link Control Identifier (DLCI): The virtual circuit number in Frame Relay, normally 10 bits long.

Data Link Layer: The layer that governs transmission within a single network all the way from the source station to the destination station across zero or more switches; Layer 2 in OSI.

Data Service Unit (DSU): The part of a CSU/DSU circuit that formats the data in the way the private line requires.

dB: *See* Decibel.

DDoS: *See* distributed denial of service attack.

Dead Spot: *See* Shadow Zone.

Decapsulation: The removing of a message from the data field of another message.

Decibel (dB): The unit in which attenuation or amplification is measured.

Decrypt: Conversion of encrypted ciphertext into the original plaintext so an authorized receiver can read an encrypted message.

Dedicated Server: A server that is not used simultaneously as a user PC.

Deep Packet Inspection: The examination of headers and messages at multiple layers in a packet.

Default Printer: The printer to which a user's print jobs will be sent unless the user specifies a different printer.

Default Router: The next-hop router that a router will forward a packet to if the routing table does not have a row that governs the packet's IP address except for the default row.

Default Row: The row of a routing table that will be selected automatically if no other row matches; its value is 0.0.0.0.

Defense Advanced Research Projects Agency: The U.S. agency that funded the creation of the ARPANET and the Internet.

Defense in Depth: The use of successive lines of defense.

Demilitarized Zone (DMZ): A subnet in which webservers and other public servers are placed.

Demodulate: To convert digital transmission signals to analog signals.

Denial-of-Service (DoS): The type of attack whose goal is to make a computer or a network unavailable to its users.

Distributed Denial-of-Service (DDoS): DoS attack in which the victim is attacked by many computers.

Deregulation: Taking away monopoly protections from carriers to encourage competition.

DES: *See* Data Encryption Standard.

Designated Router: In OSPF, a router that sends change information to other routers in its area.

Destination: In a routing table, the column that shows the destination network's network part or subnet's network part plus subnet part, followed by zeroes. This row represents a route to this network or subnet.

Device Driver: Software that allows an operating system to communicate with a peripheral, such as a NIC.

DHCP: *See* Dynamic Host Configuration Protocol.

Dial-Up Circuit: A circuit that exists only for the duration of a telephone call.

Dictionary Attack: A password-cracking attack in which an attacker tries to break a password by trying all words in a standard or customized dictionary.

Dictionary Word: A common word, dangerous to use for a password because easily cracked.

Dielectric Insulation: The non-conducting insulation that covers each wire in 4-pair UTP, preventing short circuits between the electrical signals traveling on different wires.

Diff-Serv: The field in an IP packet that can be used to label IP packets for priority and other service parameters.

Digital Certificate: A document that gives the name of a true party, that true party's public key, and other information; used in authentication.

Digital Certificate Authentication: Authentication in which each user has a public key and a private key. Authentication depends on the applicant knowing the true party's private key; requires a digital certificate to give the true party's public key.

Digital Signaling: Signaling that uses a few states. Binary (two-state) transmission is a special case of digital transmission.

Digital Signature: A calculation added to a plaintext message to authenticate it.

Digital Subscriber Line (DSL): A technology that provides digital data signaling over the residential customer's existing single-pair UTP voice-grade copper access line.

Digital-to-Analog Conversion (DAC): The conversion of transmissions from the digital telephone network's core to signals on the analog local loop.

Direct Distance Dialing: Long distance calls made at the standard long-distance rate.

Direct Sequence Spread Spectrum (DSSS): Spread spectrum transmission that spreads the signal over the entire bandwidth of a channel.

Disaster: An incident that can stop the continuity of business operations, at least temporarily.

Disaster Recovery: The reestablishment of information technology operations.

Discovery: The first phase of network mapping, in which the program finds out if hosts and subnets exist.

Disgruntled Employee: Employee who is upset with the firm or an employee and who may take revenge through a computer attack.

Disgruntled Ex-Employee: Former employee who is upset with the firm or an employee and who may take revenge through a computer attack.

Dish Antenna: An antenna that points in a particular direction, allowing it to send stronger outgoing signals in that direction for the same power and to receive weaker incoming signals from that direction.

Distance Vector Routing Protocol: Routing protocol based on the number of hops to a destination out a particular port.

Distort: To change in shape during propagation.

DLCI: *See* Data Link Control Identifier.

DMZ: *See* Demilitarized Zone.

DNS: *See* Domain Name System.

Domain: 1) In DNS, a group of resources (routers, single networks, and hosts) under the control of an organization. 2) In Microsoft Windows, a grouping of resources used in an organization, made up of clients and servers.

Domain Controller: In Microsoft Windows, a computer that manages the computers in a domain.

Domain Name System (DNS): A server that provides IP addresses for users who know only a target host's host name. DNS servers also provide a hierarchical system for naming domains.

Domestic: Telephone service within a country.

DoS: *See* Denial-of-Service.

Dotted Decimal Notation: The notation used to ease human comprehension and memory in reading IP addresses.

Downlink: Downward transmission path for a communications satellite.

Downtime: A period of network unavailability.

Drive-By Hacker: A hacker who parks outside a firm's premises and eavesdrops on its data transmissions; mounts denial-of-service attacks; inserts viruses, worms, and spam into a network; or does other mischief.

DSL: *See* Digital Subscriber Line.

DSL Access Multiplexer (DSLAM): A device at the end office of the telephone company that sends voice signals over the ordinary PSTN and sends data over a data network such as an ATM network.

DSLAM: *See* DSL Access Multiplexer.

DSSS: *See* Direct Sequence Spread Spectrum.

DSU: *See* Data Service Unit.

Dumb Access Point: Access point that cannot be managed remotely without the use of a wireless LAN switch.

Dumb Terminal: A desktop machine with a keyboard and display but little processing capability; processing is done on a host computer.

DWDM: *See* Dense Wavelength Division Multiplexing.

Dynamic Host Configuration Protocol (DHCP): The protocol used by DHCP servers, which provide each user PC with a temporary IP address to use each time he or she connects to the Internet.

Dynamic Routing Protocol: A protocol used by routers to exchange routing table information.

EAP: *See* Extensible Authentication Protocol.

E-Commerce: Electronic commerce; buying and selling over the Internet.

E-Commerce Software: Software that automates the creation of catalog pages and other e-commerce functionality.

Economy of Scale: In managed services, the condition of being cheaper to manage the traffic of many firms than of one firm.

Egress Filtering: The filtering of traffic from inside a site going out.

EIGRP: *See* Enhanced Interior Gateway Routing Protocol.

E-LAN: Multipoint service in metropolitan area Ethernet.

Electromagnetic Interference (EMI): Unwanted electrical energy coming from external devices, such as electrical motors, fluorescent lights, and even nearby data transmission wires.

Electromagnetic Signal: A signal generated by oscillating electrons.

Electronic Signature: A bit string added to a message to provide message-by-message authentication and message integrity.

Electronic Catalog: An e-commerce site's display that shows the goods the site has for sale.

Electronic Commerce (E-Commerce): The buying and selling of goods and services over the Internet.

E-Line: Point-to-point service in metropolitan area Ethernet.

Elliptic Curve Cryptosystem (ECC): Public key encryption method; more efficient than RSA.

EMI: *See* Electromagnetic Interference.

Encapsulation: The placing of a message in the data field of another message.

Encrypt: To mathematically process a message so that an interceptor cannot read the message.

Encryption Method: A method for encrypting plaintext messages.

End Office: Telephone company switch that connects to the customer premises via the local loop.

End Office Switch: The nearest switch of the telephone company to the customer premises.

End-to-End: A layer where communication is governed directly between the transport process on the source host and the transport process on the destination host.

Enhanced Interior Gateway Routing Protocol (EIGRP): Interior routing protocol used by Cisco routers.

Enterprise Application: Applications that serve individual business functions while providing smooth integration between functional modules.

ERP: *See* enterprise resource planning.

Enterprise Mode: In WPA and 802.11i, operating mode that uses 802.1X.

Enterprise Resource Planning: Applications that serve individual business functions while providing smooth integration between functional modules.

Ephemeral Port Number: The temporary number a client selects whenever it connects to an application program on a server. According to IETF rules, ephemeral port numbers should be between 49153 and 65535.

Equipment Room: The room, usually in a building's basement, where wiring connects to external carriers and internal wiring.

Error Advisement: In ICMP, the process wherein if an error is found, there is no transmission, but the router or host that found the error usually sends an ICMP error message to the source device to inform it that an error has occurred. It is then up to the device to decide what to do. (This is not the same as error correction, because there is no mechanism for the retransmission of lost or damaged packets.)

Error Rate: In biometrics, the normal rate of misidentification when the subject is cooperating.

Ethereal: Popular network analysis program.

Ethernet: The most widely used standard for wired LANs.

Ethernet 10Base2: Obsolete 10 Mbps Ethernet standard that uses coaxial cable in a bus topology. Less expensive than 10Base5 but cannot carry signals as far.

Ethernet 10Base5: Obsolete 10 Mbps Ethernet standard that uses coaxial cable in a bus topology.

Ethernet Address: The 48-bit address the stations have on an Ethernet network; often written in hexadecimal notation for human reading.

Ethernet in the Last Kilometer: A standard for delivering Ethernet service over a variety of access technologies for the telephone network and competitors.

Ethernet Frame: A message at the data link layer in an Ethernet network.

Ethernet Switch: Switch following the Ethernet standard. Notable for speed and low cost per frame sent. Dominates LAN switching.

EtherPeek: A commercial traffic summarization program.

Evil Twin Access Point: Attacker access point outside a building that attracts clients inside the building to associate with it.

Excess Burst Speed: One of Frame Relay's two-part PVC speeds; beyond the CIR.

Exhaustive Search: Cracking a key or password by trying all possible keys or passwords.

Exploit: A break-in program; a program that exploits known vulnerabilities.

Exploitation Software: Software that is planted on a computer; it continues to exploit the computer.

Extended ASCII: Extended 8-bit version of the ASCII code used on PCs.

Extended Star Topology: The type of topology wherein there are multiple layers of switches organized in a hierarchy, in which each node has only one parent node; used in Ethernet; more commonly called a *hierarchical topology*.

Extensible Authentication Protocol (EAP): A protocol that authenticates users with authentication data (such as a password or a response to a challenge based on a station's digital certificate) and authentication servers.

Exterior Routing Protocol: Dynamic routing protocol used between autonomous systems.

Extranet: A network that uses TCP/IP Internet standards to link several firms together but that is not accessible to people outside these firms. Even within the firms of the extranet, only some of each firm's computers have access to the network.

Face Recognition: The scanning of passersby to identify terrorists or wanted criminals by the characteristics of their faces.

Facilitating Server: A server that solves certain problems in P2P interactions but that allows clients to engage in P2P communication for most of the work.

False Alarm: An apparent incident that proves not to be an attack.

False Positive: A false alarm.

Fast Ethernet: 100 Mbps Ethernet.

FCC: *See* Federal Communications Commission.

EIGRP: *See* Enhanced Interior Gateway Routing Protocol.

FDDI: *See* Fiber Distributed Data Interface.

FDM *See* Frame Division Multiplexing.

FHSS: *See* Frequency Hopping Spread Spectrum.

Fiber Distributed Data Interface: Obsolete 100 Mbps token-ring network.

Fiber to the Home (FTTH): Optical fiber brought by carriers to individual homes and businesses.

Field: A subdivision of a message header or trailer.

File Server: A server that allows users to store and share files.

File Server Program Access: The form of client/server computing in which the server's only role is to store programs and data files, while the client PC does the actual processing of programs and data files.

File Sharing: The ability of computer users to share files that reside on their own disk drives or on a dedicated file server.

File Transfer Protocol: Early and still widely used protocol for transferring files between host computers.

Filtering: Examining the content of arriving packets to decide what to do with them.

Fin Bit: One-bit field in a TCP header; indicates that the sender wishes to open a TCP connection.

Fingerprint Scanning: A form of biometric authentication that uses the applicant's fingerprints.

Fingerprinting: The second phase of network mapping, in which the program determines the characteristics of hosts to determine if they are clients, servers, or routers.

Firewall: A security system that examines each incoming packet. If the firewall identifies the packet as an attack packet, the firewall discards the packet and copies information about the discarded packet into a log file.

First-Generation (1G): The initial generation of cellular telephony, introduced in the 1980s. 1G systems were analog, were only given about 50 MHz of spectrum, had large and few cells, and had very limited speeds for data transmission.

Fixed Wireless Service: Local terrestrial wireless service in which the user is at a fixed location.

Flag Field: A one-bit field.

Flat Rate: Local telephone service in which there is a fixed monthly service charge but no separate fee for individual local calls.

Flow Control: The ability of one side in a conversation to tell the other side to slow or stop its transmission rate.

Footprint: Area of coverage of a communication satellite's signal.

Forensics: The collection of data in a form acceptable for presentation in a legal proceeding.

Four-Way Close: A normal TCP connection close; requires four messages.

Fractional T1: A type of private line that offers intermediate speeds at intermediate prices; usually operates at one of the following speeds: 128 kbps, 256 kbps, 384 kbps, 512 kbps, or 768 kbps.

FRAD: *See* Frame Relay Access Device.

Fragment Offset Field: In IPv4, a flag field that tells a fragment's position in a stream of fragments from an initial packet.

Fragment (Fragmentation): To break a message into multiple smaller messages. TCP fragments application layer messages, while IP packets may be fragmented by routers along the packet's route.

Frame: 1) A message at the data link layer. 2) In time division multiplexing, a brief time period, which is further subdivided into slots.

Frame Check Sequence Field: A four-octet field used in error checking in Ethernet. If an error is found, the frame is discarded.

Frame Relay Access Device (FRAD): Device that connects an internal site network to a Frame Relay network.

Frequency: The number of complete cycles a radio wave goes through per second. In sound, frequency corresponds to pitch.

Frequency Division Multiplexing (FDM): A technology used in microwave transmission in which the microwave bandwidth is subdivided into channels, each carrying a single circuit.

Frequency Hopping Spread Spectrum (FHSS): Spread spectrum transmission that uses only the bandwidth required by the signal but hops frequently within the spread spectrum channel.

Frequency Modulation: Modulation in which one frequency is chosen to represent a 1 and another frequency is chosen to represent a 0.

Frequency Spectrum: The range of all possible frequencies from zero hertz to infinity.

FTP: *See* file transfer protocol.

FTTH: *See* Fiber to the Home.

Full-Duplex: A type of communication that supports simultaneous two-way transmission. Almost all communication systems today are full-duplex systems.

Full-Mesh Topology: Topology in which each node is connected to each other node.

Fully Configured: A system with all necessary options.

Functional Department: General name for departments in a firm other than the IT department; marketing, accounting, and so forth.

Gateway: An obsolete term for *router*; still in use by Microsoft.

Gateway Controller: In IP telephony, a device that controls the operation of signaling gateways and media gateways.

Gbps: Gigabit per second.

General Packet Radio Service (GPRS): The technology to which many GSM systems are now being upgraded. GPRS can combine two or more GSM time slots within a channel and so can offer data throughput near that of a telephone modem. Often called a *2.5G technology.*

GEO: *See* Geosynchronous Earth Orbit Satellite.

Geosynchronous Earth Orbit Satellite (GEO): The type of satellite most commonly used in fixed wireless access today; orbits the Earth at about 36,000 km.

Get: An SNMP command sent by the manager that tells the agent to retrieve certain information and return this information to the manager.

GHz: *See* Gigahertz.

Gigabit Ethernet: 1 Gbps versions of Ethernet.

Gigabit per Second: One billion bits per second.

Gigahertz (GHz): One billion hertz.

GIGO: "Garbage in, garbage out." If bad information is put into a system, only bad information can come out.

Global System for Mobile Communication (GSM): The cellular telephone technology on which nearly the entire world

standardized for 2G service. GSM uses 200 kHz channels and implements TDM.

Gnutella: A pure P2P file-sharing application that addresses the problems of transient presence and transient IP addresses without resorting to the use of any server.

Golden Zone: The portion of the frequency spectrum from the high megahertz range to the low gigahertz range, wherein commercial mobile services operate.

GPO: *See* Group Policy Object.

GPRS: *See* General Packet Radio Service.

Graded-Index Multimode Fiber: Multimode fiber in which the index of refraction varies from the center of the core to the cladding boundary.

Grid Computing: Computing in which all devices, whether clients or servers, share their processing resources.

Group Policy Object (GPO): A policy that governs a specific type of resource on a domain.

GSM: *See* Global System for Mobile Communication.

H.323: In IP telephony, one of the protocols used by signaling gateways.

Hacking: The intentional use of a computer resource without authorization or in excess of authorization.

Half-Duplex: The mode of operation wherein two communicating NICs must take turns transmitting.

Handoff: 1) In wireless LANs, a change in access points when a user moves to another location. 2) In cellular telephony, transfer from one cellsite to another, which occurs when a subscriber moves from one cell to another within a system.

Hardened: Set up to protect itself, as a server or client.

Hash: The output from hashing.

Hashing: A mathematical process that, when applied to a bit string of any length, produces a value of a fixed length, called the *hash*.

HDSL: *See* High-Rate Digital Subscriber Line.

HDSL2: A newer version of HDSL that transmits in both directions at 1.544 Mbps.

Header: The part of a message that comes before the data field.

Header Checksum: The UDP datagram field that allows the receiver to check for errors.

Headquarters: The First Bank of Paradise's downtown office building that houses the administrative site.

Hertz (Hz): One cycle per second, a measure of frequency.

Hex Notation: *See* Hexadecimal Notation.

Hexadecimal (Hex) Notation: The Base 16 notation that humans use to represent address 48-bit MAC source and destination addresses.

Hierarchical Topology: A network topology in which all switches are arranged in a hierarchy, in which each switch has only one parent switch above it (the root switch, however, has no parent); used in Ethernet.

Hierarchy: 1) The type of topology wherein there are multiple layers of switches organized in a hierarchy, in which each node has only one parent node; used in Ethernet. 2) In IP addresses, three multiple parts that represent successively more specific locations for a host.

High-Rate Digital Subscriber Line (HDSL): The most popular business DSL, which offers symmetric transmission at 768 kbps in both directions. *See also* HDSL2.

Hop-by-Hop: A layer in which communication is governed by each individual switch or router along the path of a message.

Host: Any computer attached to the Internet (can be either personal client or server).

Host Computer: 1) In terminal–host computing, the host that provides the processing power; 2) on an internet, any host.

Host Name: An unofficial designation for a host computer.

Host Part: The part of an IP address that identifies a particular host on a subnet.

Hot Spot: A public location where anyone can connect to an access point for Internet access.

HTML: *See* Hypertext Markup Language.

HTML Body: Body part in a Hypertext Markup Language message.

HTTP: *See* Hypertext Transfer Protocol.

HTTP Request Message: In HTTP, a message in which a client requests a file or another service from a server.

HTTP Request–Response Cycle: An HTTP client request followed by an HTTP server response.

HTTP Response Message: In HTTP, a message in which a server responds to a client request; either contains a requested file or an error message explaining why the requested file could not be supplied.

Hub: An early device used by Ethernet LANs to move frames in a system. Hubs broadcast each arriving bit out all ports except for the port that receives the signal.

Hub-and-Spoke Topology: A topology in which all communication goes through one site.

Hybrid Mode: In password cracking, a mode that tries variations on common word passwords.

Hybrid TCP/IP-OSI Standards Architecture: The architecture that uses OSI standards at the physical and data link layers and TCP/IP standards at the internet, transport, and application layers; dominant in corporations today.

Hypertext Markup Language (HTML): The language used to create webpages.

Hypertext Transfer Protocol (HTTP): The protocol that governs interactions between the browser and webserver application program.

Hz: *See* Hertz.

ICC: *See* International Common Carrier.

ICF: *See* Internet Connection Firewall.

ICMP Echo: A message sent by a host or router to another host or router. If the target device's internet process is able to do so, it will send back an echo response message.

ICMP Error Message: A message sent in error advisement to inform a source device that an error has occurred.

ICMP: *See* Internet Control Message Protocol.

ICS: *See* Internet Connection Sharing.

IDC: *See* Insulation Displacement Connection.

Identification Field: In IPv4, header field used to reassemble fragmented packets. Each transmitted packet is given a unique identification field value. If the packet is fragmented en route, all fragments are given the initial packet's identification field value.

Identity Theft: Stealing enough information about a person to impersonate him or her in complex financial transactions.

IDS: *See* Intrusion Detection System.

IEEE: *See* Institute for Electrical and Electronics Engineers.

IETF: *See* Internet Engineering Task Force.

ILEC: *See* Incumbent Local Exchange Carrier.

IM: *See* Instant Messaging.

Image: An exact copy.

IMAP: *See* Internet Message Access Protocol.

Impostor: Someone who claims to be someone else.

Incident: A successful attack.

Incident Severity: The degree of destruction inflicted by an attack.

Incumbent Local Exchange Carrier (ILEC): The traditional monopoly telephone company within each LATA.

Index Server: A server used by Napster. Stations connected to Napster would first upload a list of their files available for sharing to index servers. Later, when they searched, their searches went to the index servers and were returned from there.

Individual Throughput: The actual speed a single user receives (usually much lower than aggregate throughput in a system with shared transmission speed).

Ingress Filtering: The filtering of traffic coming into a site from the outside.

Inherit: When permissions are assigned to a user in a directory, user automatically receives the same permissions in subdirectories unless this automatic inheritance is blocked.

Initial Installation: The initial phase of a product's life cycle. Ongoing costs may be much higher.

Initialization Vector: A bit string used in conjunction with a key for encryption.

Initial Labor Costs: The labor costs of setting up a system for the first time.

Initial Sequence Number (ISN): The sequence number placed in the first TCP segment a side transmits in a session; selected randomly.

Instance: An actual example of a category.

Instant Messaging (IM): A popular P2P application that allows two users to type messages back and forth in real time.

Institute for Electrical and Electronics Engineers (IEEE): An international organization whose 802 LAN/MAN Standards Committee creates many LAN standards.

Insulation: Nonconducting coating around each wire in a UTP cord.

Insulation Displacement Connection (IDC): Connection method used in UTP. A connector bites through the insulation around a wire, making contact with the wire inside.

Interexchange Carrier (IXC): A telephone carrier that transmits voice traffic between LATAs.

Interface: 1) The router's equivalent of a network interface card; a port on a router that must be designed for the network to which it connects. 2) In Web services, the outlet through which an object communicates with the outside world.

Interference: *See* Electromagnetic Interference.

Interior Routing Protocol: Routing protocol used within a firm's internet.

Internal Back-End System: In e-commerce, an internal e-commerce system that handles accounting, pricing, product availability, shipment, and other matters.

Internal Router: A router that connects different LANs within a site.

International Common Carrier (ICC): A telephone carrier that provides international service.

International Organization for Standardization (ISO): A strong standards agency for manufacturing, including computer manufacturing.

International Telecommunications Union-Telecommunications Standards Sector (ITU-T): A standards agency that is part of the United Nations and that oversees international telecommunications.

Internet: 1) A group of networks connected by routers so that any application on any host on any network can communicate with any application on any other host on any other network. 2) A general term for any internetwork (spelled with a lowercase *i*). 3) The worldwide Internet (spelled with a capital *I*).

Internet Backbone: The collection of all Internet Service Providers that provide Internet transmission service.

Internet Control Message Protocol (ICMP): The protocol created by the IETF to oversee supervisory messages at the internet layer.

Internet Engineering Task Force (IETF): TCP/IP's standards agency.

Internet Layer: The layer that governs the transmission of a packet across an entire internet.

Internet Message Access Protocol (IMAP): One of the two protocols used to download received e-mail from an e-mail server; offers more features but is less popular than POP.

Internet Network: A network on the Internet owned by a single organization, such as a corporation, university, or ISP.

Internet Options: In Microsoft Windows, way of setting security and other settings for Browser communication.

Internet Protocol (IP): The TCP/IP protocol that governs operations at the internet layer. Governs packet delivery from host to host across a series of routers.

Internet Service Provider: Carrier that provides Internet access and transmission.

Internetwork Operating System (IOS): The operating system that Cisco Systems uses on all of its routers and most of its switches.

Intranet: An internet for internal transmission within firms; uses the TCP/IP transmission standards that govern transmission over the Internet.

Intrusion Detection System (IDS): A security system that examines messages traveling through a network. IDSs look at traffic broadly, identifying messages that are suspicious. Instead of discarding these packets, IDSs will sound an alarm.

Intrusion Protection System (IPS): Firewall system that uses sophisticated packet filtering methods to stop attacks.

Inverse Square Law: Radio signal strength declines with the square of transmission distance.

IOS: *See* Internetwork Operating System.

IP: *See* Internet Protocol.

IP Address: An Internet Protocol address; the address that every computer needs when it connects to the Internet; IP addresses are 32 bits long.

IP Carrier Network: A WAN carrier that offers a private service similar to that of the public Internet but with QoS guarantees.

Ipconfig/all: Windows command line command in newer versions of Windows that shows configuration parameters for the PC.

IP Security (IPsec): A set of standards that operate at the internet layer and provide security to all upper layer protocols transparently.

IP Telephone: A telephone that has the electronics to encode voice for digital transmission and to send and receive packets over an IP internet.

IP Telephony: The transmission of telephone signals over IP internets instead of over circuit-switched networks.

IP Version 4 (IPv4): The standard that governs most routers on the Internet and private internets.

IP Version 6 (IPv6): A new version of the Internet Protocol.

Ipconfig (ipconfig): A command used to find information about one's own computer, used in newer versions of Windows (the command is typed as ipconfig/all[Enter] at the command line).

IPS: *See* Intrusion Prevention System.

IPsec Gateway: Border device at a site that converts between internal data traffic into protected data traffic that travels over an untrusted system such as the Internet.

IPsec: *See* IP Security.

IPv4: *See* IP Version 4.

IPv6: *See* IP Version 6.

IPX/SPX Architecture: Non-TCP/IP standards architecture found at upper layers in LANs; required on all older Novell NetWare file servers.

Iris: The colored part of the eye, used in biometric authentication.

ISN: *See* Initial Sequence Number.

ISO: *See* International Organization for Standardization.

ISO/IEC 11801: European standard for wire and optical fiber media.

ISP: *See* Internet Service Provider.

IT Disaster Recovery: Recovering from a disaster that damages computer equipment or data.

IT Guru. *See* OPNET IT Guru.

ITU-T: *See* International Telecommunications Union-Telecommunications Standards Sector.

IXC: *See* Interexchange Carrier.

Jacket: The outer plastic covering, made of PVC, that encloses and protects the four pairs of wires in UTP or the core and cladding in optical fiber.

Java Applet: Small Java program that is downloaded as part of a webpage.

Jitter: Variability in latency.

JPEG: Popular graphics file format.

kbps: Kilobits per second.

Key: A bit string used with an encryption method to encrypt and decrypt a message. Different keys used with a single encryption method will give different ciphertexts from the same plaintext.

Key Exchange: The secure transfer of a symmetric session key between two communicating parties.

Key-Hashed Message Authentication Code (HMAC): Electronic signature technology that is efficient and inexpensive but lacks nonrepudiation.

Key Management: The management of key creation, distribution, and other operations.

Label Header: In MPLS, the header added to packets before the IP header; contains information that aids and speeds routers in choosing which interface to send the packet back out.

Label Number: In MPLS, number in the label header that aids label-switching routers in packet sending.

Label Switching Router: Router that implements MPLS label switching.

Label Switching Table: In MPLS, the table used by label-switching routers to decide which interface to use to forward a packet.

LAN: *See* Local Area Network.

Language Independence: In SOAP, the fact that Web service objects do not have to be written in any particular language.

LATA: *See* Local Access and Transport Area.

Latency: Delay, usually measured in milliseconds.

Latency-Intolerant: An application whose performance is harmed by even slight latency.

Layer 3: *See* Internet Layer.

Layer 3 Switch: A router that does processing in hardware and that is much faster and less expensive than traditional software-based routers. Layer 3 switches are usually dominant in the Ethernet core above workgroup switches.

Layer 4: *See* Transport Layer.

Layer 4 Switch: A switch that examines the port number fields of each arriving packet's encapsulated TCP segment, allowing it to switch packets based on the application they contain. Layer 4 switches can give priority or even deny forwarding to IP packets from certain applications.

Layer 5: *See* Application Layer.

Leased Line Circuit: A high-speed point-to-point circuit.

Legacy Network: A network that uses obsolete technology; may have to be lived with for some time because upgrading all legacy networks at one time is too expensive.

Legal Retention: Rules that require IM messages to be captured and stored in order to comply with legal requirements.

Length Field: 1) The field in an Ethernet MAC frame that gives the length of the data field in octets. 2) The field in a UDP datagram that enables the receiving transport process to process the datagram properly.

LEO: *See* Low Earth Orbit Satellite.

Lightweight Directory Access Protocol: Simple protocol for accessing directory servers.

Line of Sight: An unobstructed path between the sender and receiver, necessary for radio transmission at higher frequencies.

Link: Connection between a pair of routers.

Link Aggregation: The use of two or more trunk links between a pair of switches; also known as trunking or bonding.

Link State Protocol: Routing protocol in which each router knows the state of each link between routers.

Linux: A freeware version of Unix that runs on standard PCs.

Linux Distribution: A package purchased from a vendor that contains the Linux kernel plus a collection of many other programs, usually taken from the GNU project.

List Folder Contents: A Microsoft Windows Server permission that allows the account owner to see the contents of a folder (directory).

LLC: *See* Logical Link Control.

LLC Header: *See* Logical Link Control Layer Header.

Load-Balancing Router: Router used on a server farm that sends client requests to the first available server.

Local Access and Transport Area (LATA): One of the roughly 200 site regions the United States that has been divided into for telephone service.

Local Area Network (LAN): A network within a site.

Local Calling: Telephone calls placed to a nearby caller; less expensive than long-distance calls.

Local Loop: In telephony, the line used by the customer to reach the PSTN's central transport core.

Log File: A file that contains data on events.

Logical Link Control Layer: The layer of functionality for the upper part of the data link layer, now largely ignored.

Logical Link Control Layer Header: The header at the start of the data field that describes the type of packet contained in the data field.

Logical Link Control Layer Subheader: Group of fields at the beginning of the Ethernet data field.

Long Distance: A telephone call placed to a distance party; more expensive than a local call.

Longest Match: The matching row that matches a packet's destination IP address to the greatest number of bits; chosen by a router when there are multiple matches.

Loopback Address: The IP address 127.0.0.1. When a user pings this IP address, this will test their *own* computer's connection to the Internet.

Loopback Interface: A testing interface on a device. Messages sent to this interface are sent back to the sending device.

Low Earth Orbit Satellite (LEO): A type of satellite used in mobile wireless transmission; orbits a the Earth.

MAC: *See* Media Access Control.

MAC Address: *See* Media Access Control.

Mainframe Computer: The largest type of dedicated server; extremely reliable.

Maintenance: The day to day work of fixing problems on the network.

Make versus Buy Decision: The decision whether to purchase a solution or create one's own solution.

Malware: Software that seeks to cause damage.

Malware-Scanning Program: A program that searches a user's PC looking for installed malware.

MAN: *See* Metropolitan Area Network.

Manageable Switch: A switch that has sufficient intelligence to be managed from a central computer (the Manager).

Managed Device: A device that needs to be administered, such as printers, hubs, switches, routers, application programs, user PCs, and other pieces of hardware and software.

Managed Frame Relay: A type of Frame Relay service that takes on most of the management that customers ordinarily would have to do. Managed Frame Relay provides traffic reports and actively manages day-to-day traffic to look for problems and get them fixed.

Management Information Base (MIB): A specification that defines what objects can exist on each type of managed device and also the specific characteristics of each object; the actual database stored on a manager in SNMP. There are separate MIBs for different types of managed devices; both a schema and a database.

Management Program: A program that helps network administrators manage their networks.

Manager: The central PC or more powerful computer that uses SNMP to collect information from many managed devices.

Mask: A 32-bit string beginning with a series of 1s and ending a series of 0s; used by routing tables to Interpret IP address part sizes. The 1s designate either the network part or the network plus software part.

Mask Operations: Applying a mask of 1s and 0s to a bit stream. Where the mask is 1, the original bit stream's bit results. Otherwise, the result is 0.

Mature: Technology that has been under development long enough to have its rough edges smoothed off.

Maximum Segment Size (MSS): The maximum size of TCP data fields that a receiver will accept.

Maximum Transmission Unit (MTU): The maximum packet size that can be carried by a particular LAN or WAN.

Mbps: Megabits per second.

MD5: A popular hashing method.

Mean Time to Repair (MTTR): The average time it takes a staff to get a network back up after it has been down.

Media Access Control (MAC): The process of controlling when stations transmit; also, the lowest part of the data link layer, defining functionality specific to a particular LAN technology.

Media Gateway: A device that connects IP telephone networks to the ordinary public switched telephone network. Media gateways also convert between the signalling formats of the IP telephone system and the PSTN.

Medium Earth Orbit Satellite (MEO): A type of satellite used in mobile wireless transmission; orbits a few thousand kilometers above the Earth.

Megabits per second: Millions of bits per second.

Megahertz (MHz): One million hertz.

MEO: *See* Medium Earth Orbit Satellite.

Mesh Networking: A type of networking in which wireless devices route frames without the aid of wired LANs.

Mesh Topology: 1) A topology where there are many connections among switches or routers, so there are many alternative routes for messages to get from one end of the network to the other. 2) In network design, a topology that provides direct connections between every pair of sites.

Message: A discrete communication between hardware or software processes.

Message Digest: The result of hashing a plaintext message. The message digest is signed with the sender's private key to produce the digital signature.

Message Integrity: The assurance that a message has not been changed en route; or if a message has been changed, the receiver can tell that it has.

Message Timing: Controlling when hardware or software processes may transmit.

Message Unit: Local telephone service in which a user is charged based on distance and duration.

Method: In Web services, a well-defined action that a SOAP message can request.

Metric: A number describing the desirability of a route represented by a certain row in a routing table.

Metro Ethernet: *See* Metropolitan Area Ethernet.

Metropolitan Area Ethernet: Ethernet operating at the scale of a metropolitan area network.

Metropolitan Area Network (MAN): A WAN that spans a single urban area.

MHz: *See* Megahertz.

MHz-km: Measure of modal bandwidth, a measure of multimode fiber quality.

MIB: *See* Management Information Base.

Microsoft Windows Server: Microsoft's network operating system for servers, which comes in three versions: NT, 2000, and 2003.

Microsoft Windows XP Home: The dominant operating system today for residential PCs.

Microsoft Windows XP Professional: A version of Windows XP designed to be run in organization; integrates with Windows Server services.

Microsoft Windows Vista: The most recent Microsoft operating system for clients.

Microwave: Traditional point-to-point radio transmission system.

Microwave Repeater: Transmitter/receiver that extends the distance a microwave link can travel.

Millisecond (ms): The unit in which latency is measured.

MIME: *See* Multipurpose Internet Mail Extensions.

MIMO: *See* Multiple Input/Multiple Output.

Minimum Permissions: Initially giving users only the permissions they absolutely need to do their jobs.

Ministry of Telecommunications: A government-created regulatory body that oversees PTTs.

Mobile IP: A system for handling IP addresses for mobile devices.

Mobile Telephone Switching Office (MTSO): A control center that connects cellular customers to one another and to wired telephone users, as well as overseeing all cellular calls (determining what to do when people move from one cell to another, including which cellsite should handle a caller when the caller wishes to place a call).

Mobile Wireless Access: Local wireless service in which the user may move to different locations.

Modal Bandwidth: The measure of multimode fiber quality; the fiber's bandwidth–distance product. A modal bandwidth of 200 MHz-km means that if your bandwidth is 100 MHz, then you can transmit 2 km.

Modal Dispersion: The main propagation problem for optical fiber; dispersion in which the difference in the arrival times of various modes (permitted light rays) is too large, causing the light rays of adjacent pulses to overlap in their arrival times and rendering the signal unreadable.

Mode: An angle light rays are permitted to enter an optical fiber core.

Modify: A Microsoft Windows Server permission that gives an account owner additional permissions to act upon files, for example, the permission to delete a file, which is not included in Write.

Modulate: To convert digital signals to analog signals.

Momentary Traffic Peak: A surplus of traffic that briefly exceeds the network's capacity, happening only occasionally.

Monochrome Text: Text of one color against a contrasting background.

More Fragments Flag Field: In IPv4, a flag field that indicates whether there are more fragments (set) or not (not set).

MPLS: *See* Multiprotocol Label Switching.

ms: *See* Millisecond.

MS-CHAP: Microsoft version of the Challenge–Response Authentication Protocol.

MSS: *See* Maximum Segment Size.

MTSO: *See* Mobile Telephone Switching Office.

MTTR: *See* Mean Time to Repair.

MTU: *See* Maximum Transmission Unit.

Multicasting: Simultaneously sending messages to multiple stations but not to all stations.

Multicriteria Decision Making: A disciplined process of selecting choices based on the importance of various criteria.

Multilayer Security: Applying security at more than one layer to provide defense in depth.

Multimode Fiber: The most common type of fiber in LANs, wherein light rays in a pulse can enter a fairly thick core at multiple angles.

Multipath Interference: Interference caused when a receiver receives two or more signals—a direct signal and one or more reflected signals. The multiple signals may interfere with one another.

Multiple Input/Multiple Output (MIMO): A radio transmission method that sends several signals simultaneously in a single radio channel.

Multiplexing: 1) Having the packets of many conversations share trunk lines; reduces trunk line cost. 2) The ability of a protocol to carry messages from multiple next-higher-layer protocols in a single communication session.

Multiprocessing Computer: A computer with multiple microprocessors. This allows it to run multiple programs at the same time.

Multiprotocol Label Switching (MPLS): A traffic management tool used by many ISPs.

Multiprotocol Router: A router that can handle not only TCP/IP internetworking protocols, but also internetworking protocols for IPX/SPX, SNA, and other standards architectures.

Multiprotocol: Characterized by implementing many different protocols and products following different architectures.

Multipurpose Internet Mail Extensions (MIME): A standard for specifying the contents of files.

Mutual Authentication: Authentication by both parties.

Name Server: Server in the Domain Name System.

Nanometer (nm): The measure used for wavelengths; one billionth of a meter (10^{-9} meters).

NAP: *See* Network Access Point.

Narrowband: 1) A channel with a small bandwidth and, therefore, a low maximum speed; 2) low-speed transmission.

NAS: *See* Network Attached Storage.

NAT: *See* Network Address Translation.

Netstat: A popular route analysis tool, which gives data on current connections between a computer and other computers.

Network: In IP addressing, an organizational concept—a group of hosts, single networks, and routers owned by a single organization.

Network Access Point (NAP): A site where ISPs interconnect and exchange traffic.

Network Address Translation (NAT): Converting an IP address into another IP address, usually at a border firewall; disguises a host's true IP address from sniffers. Allows more internal addresses to be used than an ISP supplies a firm with external addresses.

Network Architecture: 1) A broad plan that specifies everything that must be done for two application programs on different networks on an internet to be able to work together effectively. 2) A broad plan for how the firm will connect all of its computers within buildings (LANs), between sites (WANs), and to the Internet; also includes security devices and services.

Network Attached Storage (NAS): Storage device that connects directly to the network instead of to a computer.

Network Interface Card (NIC): Printed circuit expansion board for a PC; handles communication with a network; sometimes built into the motherboard.

Network Layer: In OSI, Layer 3; governs internetworking. OSI network layer standards are rarely used.

Network Management Agent (Agent): A piece of software on the managed device that communicates with the manager on behalf of the managed device.

Network Management Program (Manager): A program run by the network administrator on a central computer.

Network Management Utility: A program used in network management.

Network Mapping: The act of mapping the layout of a network, including what hosts and routers are active and how various devices are connected. Its two phases are discovering and fingerprinting.

Network Mask: A mask that has 1s in the network part of an IP address and 0s in all other parts.

Network Operating System (NOS): A PC server operating system.

Network Part: The part of an IP address that identifies the host's network on the Internet.

Network Security: The protection of a network from attackers.

Network Simulation: The building of a model of a network that is used to project how the network will operate after a change.

Network Topology: The order in which a network's nodes are physically connected by transmission lines.

Networked Application: An application that provides service over a network.

Next Header Field: In IPv6, a header field that describes the header following the current header.

Next-Hop Router: A router to which another router forwards a packet in order to get the packet a step closer to reaching its destination host.

NIC: *See* Network Interface Card.

Nm (nm): *See* Nanometer.

Nmap: A network mapping tool that finds active IP addresses and then fingerprints them to determine their operating system and perhaps their operating system version.

Node: A client, server, switch, router, or other type of device in a network.

Noise: Random electromagnetic energy within wires; combines with the data signal to make the data signal difficult to read.

Noise Floor: The mean of the noise energy.

Noise Spike: An occasional burst of noise that is much higher or lower than the noise floor; may cause the signal to become unrecognisable.

Nonblocking: A nonblocking switch has enough aggregate throughput to handle even the highest possible input load (maximum input on all ports).

Nonoverlapping Channel: Channels whose frequencies do not overlap.

Normal Attack: An incident that does a small amount of damage and can be handled by the on-duty staff.

NOS: *See* Network Operating System.

Not Set: When a flags field is given the value 0.

Nslookup (nslookup): A command that allows a PC user to send DNS lookup messages to a DNS server.

OAMP: *See* operations, administration, maintenance, and provisioning.

Object: A specific Web service.

Object: In SNMP, an aspect of a managed device about which data is kept.

OC: *See* Optical Carrier.

Octet: A collection of eight bits; same as a byte.

OFDM: *See* Orthogonal Frequency Division Multiplexing.

Official Internet Protocol Standards: Standards deemed official by the IETF.

Official Standards Organization: An internationally recognized organization that produces standards.

Omnidirectional Antenna: An antenna that transmits signals in all directions and receives incoming signals equally well from all directions.

On/Off Signaling: Signaling wherein the signal is on for a clock cycle to represent a one, and off for a zero. (On/off signaling is binary.)

One-Pair Voice-Grade UTP: The traditional telephone access lines to individual residences.

Ongoing Costs: Costs beyond initial installation costs; often exceed installation costs.

Open Shortest Path First (OSPF): Complex but highly scalable interior routing protocol.

Operations: The minute-by-minute management of the operation of system.

OPNET ACE: *See* OPNET Application Characterization Environment.

OPNET Application Characterization Environment (ACE): A network simulation program; focuses on application layer performance.

OPNET IT Guru: A popular network simulation program; focuses primarily on data link layer and internet layer performance.

Optical Carrier (OC): A number that indicates SONET speeds.

Optical Fiber: Cabling that sends signals as light pulses.

Optical Fiber Cord: A length of optical fiber.

Option: One of several possibilities that a user or technologist can select.

Orthogonal Frequency Division Multiplexing (OFDM): A form of spread spectrum transmission that divides each broadband channel into subcarriers and then transmits parts of each frame in each subcarrier.

Organizational Unit: In directory servers, a subunit of the Organization node.

OSI: The Reference Model of Open Systems Interconnection; the 7-layer network standards architecture created by ISO and ITU-T; dominant at the physical and data link layers, which govern transmission within single networks (LANs or WANs).

OSI Application Layer (Layer 7): The layer that governs application-specific matters not covered by the OSI Presentation Layer or the OSI Session Layer.

OSI Layer 5: *See* OSI Session Layer.

OSI Layer 6: *See* OSI Presentation Layer.

OSI Layer 7: *See* OSI Application Layer.

OSI Presentation Layer (Layer 6): The layer designed to handle data formatting differences between two communicating computers.

OSI Session Layer (Layer 5): The layer that initiates and maintains a connection between application programs on different computers.

OSPF: *See* Open Shortest Path First.

Out of Phase: In multipath interference, the condition of not being in sync, as occurs with signals that have been reflected and thus traveled different distances and not arrived at the receiver at the same time.

Outsourcing: Paying other firms to handle some, most, or all IT chores.

Overprovision: To install much more capacity in switches and trunk links than will be needed most of the time, so that momentary traffic peaks will not cause problems.

Oversubscription: In Frame Relay, the state of having port speeds less than the sum of PVC speeds.

P2P: *See* Peer-to-Peer Architecture.

Packet: A message at the internet layer.

Packet Capture and Display Program: A program that captures selected packets or all of the packets arriving at or going out of a NIC. Afterward, the user can display key header information for each packet in greater or lesser detail.

Packet Error Rate: The percentage of packets damaged or lost in transmission.

Packet Filter Firewall: A firewall that examines fields in the internet and transport headers of individual arriving packets. The firewall makes pass/deny decisions based upon the contents of IP, TCP, UDP, and ICMP fields.

Packet Switching: The breaking of conversations into short messages (typically a few hundred bits long); allows multiplexing on trunk lines to reduce trunk line costs.

PAD Field: A field that the sender adds to an Ethernet frame if the data field is less than 46 octets long (the total length of the PAD plus data field must be exactly 46 octets long).

PAN: *See* Personal Area Network.

Parallel Transmission: A form of transmission that uses multiple wire pairs or other transmission media simultaneously to send a signal; increases transmission speed.

Pass Phrase: A series of words that is used to generate a key.

Password: A secret keyboard string only the account holder should know; authenticates user access to an account.

Password Length: The number of characters in a password.

Password Reset: The act of changing a password to some value known only to the systems administrator and the account owner.

Patch: An addition to a program that will close a security vulnerability in that program.

Patch Cord: A cord that comes precut in a variety of lengths, with a connector attached; usually either UTP or optical fiber.

Payload: 1) A piece of code that can be executed by a virus or worm after it has spread to multiple machines. 2) ATM's name for a data field.

Payment Mechanism: In e-commerce, ways for purchasers to pay for their ordered goods or services.

PBX: *See* Private Branch Exchange.

PC Server: A server that is a personal computer.

PCM: Pulse Code Modulation.

PEAP: *See* Protected Extensible Authentication Protocol.

Peer-to-Peer Architecture (P2P): The application architecture in which most or all of the work is done by cooperating user computers, such as desktop PCs. If servers are present at all, they serve only facilitating roles and do not control the processing.

Peer-to-Peer Service: Service wherein client PCs provide services to one another.

Perfect Internal Reflection: When light in optical fiber cabling begins to spread, it hits the cladding and is reflected back into the core so that no light escapes.

Permanent IP Address: An IP address given to a server that the server keeps and uses every single time it connects to the Internet. (This is in contrast to client PCs, which receive a new IP address every time they connect to the Internet.)

Permanent Virtual Circuit (PVC): A PSDN connection between corporate sites that is set up once and kept in place for weeks, months, or years at a time.

Permission: A rule that determines what an account owner can do to a particular resource (file or directory).

Personal Area Network (PAN): A small wireless network used by a single person.

Personal Mode: Pre-shared Key Mode in WPA or 802.11i.

Phase Modulation: Modulation in which one wave serves as a reference wave or a carrier wave. Another wave varies its phase to represent one or more bits.

Phishing: Social engineering attack that uses an official-looking e-mail message or website.

Physical Address: Data link layer address—*Not* a physical layer address. Given this name because it is the address of the NIC, which is a physical device that implements both the physical and data link layers.

Physical Layer: The standards layer that governs physical transmission between adjacent devices; OSI Layer 1.

Physical Link: A connection linking adjacent devices on a network.

Piggybacking: The act of an attacker being allowed physical entrance to a building by following a legitimate user through a locked door that the victim has opened.

Ping: Sending a message to another host and listening for a response to see if it is active.

Pixel: A dot on a computer screen.

PKI: *See* Public Key Infrastructure.

Plaintext: The original message the sender wishes to send to the receiver; not limited to text messages.

Plan–Protect–Respond Cycle: The basic management cycle in which the three named stages are executed repeatedly.

Planning: Developing a broad security strategy that will be appropriate for a firm's security threats.

Plenum: The type of cabling that must be used when cables run through airways to prevent toxic fumes in case of fire.

Point of Presence (POP): 1) In cellular telephony, a site at which various carriers that provide telephone service are interconnected. 2) In PSDNs, a point of connection for user sites. There must be a private line between the site and the POP.

Point-to-Point Topology: A topology wherein two nodes are connected directly.

Point-to-Point Tunneling Protocol (PPTP): A remote access VPN security standard offering moderate security. PPTP works at the data link layer, and it protects all messages above the data link layer, providing protection transparently.

POP: *See* 1) Point of Presence. 2) *See* Post Office Protocol.

Pop-Up Blocker: A program that blocks annoying pop-up advertisements.

Port: In TCP and UDP messages, a header field that designates the application layer process on the server side and a specific connection on the client side.

Port Number: The field in TCP and UDP that tells the transport process what application process sent the data in the data field or should receive the data in the data field.

Portfolio: A planned collection of projects.

Post Office Protocol (POP): The most popular protocol used to download e-mail from an e-mail server to an e-mail client.

PPTP: *See* Point-to-Point Tunneling Protocol.

Preamble Field: The initial field in an Ethernet MAC frame; synchronizes the receiver's clock to the sender's clock.

Prefix Notation: A way of representing masks. Gives the number of initial 1s in the mask.

Premises: The land and buildings owned by a customer.

Presence Server: A server used in many P2P systems; knows the IP addresses of each user and also whether the user is currently on line and perhaps whether or not the user is willing to chat.

Pre-Shared Key: A mode of operation in WPA and 802.11i in which all stations and an access point share the same initial key.

Presentation Layer: *See* OSI Presentation Layer.

Print Server: An electronic device that receives print jobs and feeds them to the printer attached to the print server.

Printer Sharing: Allowing multiple PCs to share a single printer.

Priority: Preference given to latency-sensitive traffic, such as voice and video traffic, so that latency-sensitive traffic will go first if there is congestion.

Priority Level: The three-bit field used to give a frame one of eight priority levels from 000 (zero) to 111 (eight).

Private Branch Exchange (PBX): An internal telephone switch.

Private IP Address: An IP address that may be used only within a firm. Private IP addresses have three designated ranges: 10.x.x.x, 192.168.x.x, and 172.16.x.x through 172.31.x.x.

Private Key: A key that only the true party should know. Part of a public key–private key pair.

Probable Annual Loss: The likely annual loss from a particular threat. The cost of a successful attack times the probability of a successful attack in a one-year period.

Probe Packet: A packet sent into a firm's network during scanning; responses to the probe packet tend to reveal information about a firm's general network design and about its individual computers—including their operating systems.

Problem Update: An update that causes disruptions, such as slowing computer operation.

Propagate: To travel.

Propagation Effects: Changes in the signal during propagation.

Property: A characteristic of an object.

Protected Extensible Authentication Protocol (PEAP): A version of EAP preferred by Microsoft Windows computers.

Protecting: Implementing a strategic security plan; the most time-consuming stage in the plan–protect–respond management cycle.

Protocol: 1) A standard that governs interactions between hardware and software processes at the same layer but on different hosts. 2) In IP, the header field that describes the content of the data field.

Protocol Fidelity: The assurance that an application using a particular port is the application it claims to be.

Protocol Field: In IP, a field that designates the protocol of the message in the IP packet's data field.

Provable Attack Packet: A packet that is provably an attack packet.

Provisioning: In network management, the installation of service for a customer.

PSDN: *See* Public Switched Data Network.

PSTN: *See* Public Switched Telephone Network.

PTT: *See* Public Telephone and Telegraphy Authority.

Public IP Address: An IP address that must be unique on the Internet.

Public Key: A key that is not kept secret. Part of a public key–private key pair.

Public Key Authentication: Authentication in which each user has a public key and a private key. Authentication depends on the applicant knowing the true party's private key; requires a digital certificate to give the true party's public key.

Public Key Encryption: Encryption in which each side has a public key and a private key, so there are four keys in total for bidirectional communication. The sender encrypts messages with the receiver's public key. The receiver, in turn, decrypts incoming messages with the receiver's own private key.

Public Key Infrastructure (PKI): A total system (infrastructure) for public key encryption.

Public Switched Data Network (PSDN): A carrier WAN that provides data transmission service. The customer only needs to connect to the PSDN by running one private line from each site to the PSDN carrier's nearest POP.

Public Switched Telephone Network (PSTN): The worldwide telephone network.

Public Telephone and Telegraphy authority (PTT): The traditional title for the traditional monopoly telephone carrier in most countries.

Public Utilities Commission (PUC): In the United States, telecommunications regulatory agency at the state level.

PUC: *See* Public Utilities Commission.

Pulse Code Modulation (PCM): An analog-to-digital conversion technique in which the ADC samples the bandpass-filtered signal 8,000 times per second, each time measuring the intensity of the signal and representing the intensity by a number between 0 and 255.

PVC: *See* Permanent Virtual Circuit.

QAM: *See* Quadrature Amplitude Modulation.

QoS: *See* Quality of Service.

QPSK: *See* Quadrature Phase Shift Keying.

Quadrature Amplitude Modulation (QAM): Modulation technique that uses two carrier waves—a sine carrier wave and a cosine carrier wave. Each can vary in amplitude.

Quadrature Phase Shift Keying (QPSK): Modulation with four possible phases. Each of the four states represents two bits (00, 01, 10, and 11).

Quality of Service (QoS): Numerical service targets that must be met by networking staff.

Quality-of-Service (QoS) Parameters: In IPv4, service quality parameters applied to all packets with the same TOS field value.

Radio Frequency ID (RFID): A tag that can be read at a distance by a radio transmitter/receiver.

Radio Wave: An electromagnetic wave in the radio range.

Rapid Spanning Tree Protocol: A version of the Spanning Tree Protocol that has faster convergence 802.1w.

RAS: *See* Remote Access Server.

Raster Graphics: Form of graphics in which an image is painted on the screen as a series of dots.

Rated Speed: The official speed of a technology.

RBOC: *See* Regional Bell Operating Company.

Read: A Microsoft Windows Server permission that allows an account owner to read files in a directory. This is read-only access; without further permissions, the account owner cannot change the files.

Read and Execute: A set of Microsoft Windows Server permissions needed to run executable programs.

Real Time Protocol (RTP): The protocol that adds headers that contain sequence numbers to ensure that the UDP datagrams are placed in proper sequence and that they contain time stamps so that jitter can be eliminated.

Reassembly: Putting a fragmented packet back together.

Redundancy: Duplication of a hardware device in order to enhance reliability.

Regenerate: In a switch or router, to clean up a signal before sending it back out.

Regional Bell Operating Company (RBOC): One of the companies that was created to provide local service when the Bell System (AT&T) was broken up in the early 1980s.

Relay Server: A server used in some IM systems, which every message flows through. Relay servers permit the addition of special services, such as scanning for viruses when files are transmitted in an IM system.

Reliabile: A protocol in which errors are corrected by resending lost or damaged messages.

Remote Access Server (RAS): A server to which remote users connect in order to have their identities authenticated so they can get access to a site's internal resources.

Remote Monitoring (RMON) Probe: A specialized type of agent that collects data on network traffic passing through its location instead of information about the RMON probe itself.

Repeat Purchasing: In e-commerce, a consumer returning to a site where he or she had made a purchase previously and making another purchase; essential to profitability.

Request for Comment (RFC): A document produced by the IETF that may become designated as an Official Internet Protocol Standard.

Request to Send: A message sent to an access point when a station wishes to send and is able to send because of CSMA/CA. The station may send when it receives a clear-to-send message.

Request to Send/Clear to Send: A system that uses request-to-send and clear-to-send messages to control transmissions and avoid collisions in wireless transmission.

Resegment: Dividing a collision domain into several smaller collision domains to reduce congestion and latency.

Responding: In security, the act of stopping and repairing an attack.

Response Message: In Challenge–Response Authentication Protocols, the message that the applicant returns to the verifier.

Response Time: The difference between the time a user types a request to the time the user receives a response.

Retention: Rules that require IM messages to be captured and stored in order to comply with legal requirements.

RFC: *See* Request for Comment.

RFC 822: The original name for RFC 2822.

RFC 2822: The standard for e-mail bodies that are plaintext messages.

RFID: *See* Radio Frequency ID.

Ring Topology: A topology in which stations are connected in a loop and messages pass in only one direction around the loop.

Ring Wrapping: In a network with a dual-ring topology, responding to a break between switches by turning the surviving parts of a dual ring into a long single ring.

Right of Way: Permission to lay wires in public areas; given by government regulators to transmission carriers.

RIP: *See* Routing Information Protocol.

Risk Analysis: The process of balancing threats and protection costs.

RJ-45 Connector: The connector at the end of a UTP cord, which plugs into an RJ-45 jack.

RJ-45 Jack: The type of jack into which UTP cords' RJ-45 connectors may plug.

RMON Probe: *See* Remote Monitoring Probe.

Roaming: The situation when a subscriber leaves a metropolitan cellular system and goes to another city or country. Roaming requires the destination cellular system to be technologically compatible with the subscriber's cellphone. It also requires administration permission from the destination cellular system.

Robust Security Network (RSN): A wireless network in which all stations and access points communicate with 802.11i security.

Rogue Access Point: An access point set up by a department or individual and not sanctioned by the firm.

Root: 1) The level at the top of a DNS hierarchy, consisting of all domain names. 2) A super account on a Unix server that automatically has full permissions in every directory on the server.

Root Server: One of 13 top-level servers in the Domain Name System (DNS).

Route: The path that a packet takes across an internet.

Route Analysis: Determining the route a packet takes between your host and another host and analyzing performance along this route.

Routed Network: A network in which routers connect multiple switched networks together. Also called an internet.

Router: A device that forwards packets within an internet. Routers connect two or more single networks (subnets).

Routing: 1) The forwarding of IP packets. 2) The exchange of routing protocol information through routing protocols.

Routing Information Protocol (RIP): A simple but limited interior routing protocol.

Routing Protocol: A protocol that allows routers to transmit routing table information to one another.

RSA: Popular public key encryption method.

RST Bit: In a TCP segment, if the RST (reset) bit is set, this tells the other side to end the connection immediately.

RSTP: *See* Rapid Spanning Tree Protocol.

RTP: *See* Real Time Protocol.

RTS: *See* Request to Send.

RTS/CTS: *See* Request to Send/Clear to Send.

Sample: To read the intensity of a signal.

SC Connector: A square optical fiber connector, recommended in the TIA/EIA-568 standard for use in new installations.

Scalability: The ability of a technology to handle growth efficiently.

Scanning: To try to determine a network's design through the use of probe packets.

Schema: The design of a database, telling the specific types of information the database contains.

Scope: A parameter on a DHCP server that determines how many subnets the DHCP server may serve.

Script Kiddie: An attacker who possesses only modest skills but uses attack scripts created by experienced hackers; dangerous because there are so many.

SDH: *See* Synchronous Digital Hierarchy.

SDLC: *See* systems development life cycle.

Second-Generation (2G): The second generation of cellular telephony, introduced in the early 1990s. Offers the improvements of digital service, 150 MHz of bandwidth, a higher frequency range of operation, and slightly higher data transmission speeds.

Second-Level Domain: The third level of a DNS hierarchy, which usually specifies an organization (e.g., microsoft.com, hawaii. edu).

Secure Hash Algorithm (SHA): A hashing algorithm that can produce hashes of different lengths.

Secure Shell (SSH): A program that provides Telnet-like remote management capabilities; and FTP-like service; strongly encrypts both usernames and passwords.

Secure Sockets Layer (SSL): The simplest VPN security standard to implement; later renamed Transport Layer Security. Provides a secure connection at the transport layer, protecting any applications above it that are SSL/TLS-aware.

Semantics: In message exchange, the meaning of each message.

Sequence Number Field: In TCP, a header field that tells a TCP segment's order among the multiple TCP segments sent by one side.

Serial Transmission: Ethernet transmission over a single pair in each direction.

Server: A host that provides services to residential or corporate users.

Server Farm: Large groups of servers that work together to handle applications.

Server Host: A station that provides service to client stations.

Service Band: A subdivision of the frequency spectrum, dedicated to a specific service such as FM radio or cellular telephone service.

Service Control Point: A database of customer information, used in Signaling System 7.

Service Level Agreement (SLA): A quality-of-service guarantee for throughput, availability, latency, error rate, and other matters.

Service Pack: For Microsoft Windows, large cumulative updates that combine a number of individual updates.

Service Pack 2: In Microsoft Windows XP, a security-focused update.

Session Initiation Protocol (SIP): Relatively simple signaling protocol for voice over IP.

Session Key: Symmetric key that is used only during a single communication session between two parties.

Session Layer: *See* OSI Session Layer.

Set: 1) When a flags field is given the value 1. 2) An SNMP command sent by the manager that tells the agent to change a parameter on the managed device.

SETI@home: A project from the Search for Extraterrestrial Intelligence (SETI), in which volunteers download SETI@home screen savers that are really programs. These programs do work for the SETI@home server when the volunteer computer is idle. Processing ends when the user begins to do work.

Setup Fee: The cost of initial vendor installation for a system.

Severity Rating: A rating for the severity of a risk.

SFF: *See* Small Form Factor.

SHA: *See* Secure Hash Algorithm.

Shadow Zone (Dead Spot): A location where a receiver cannot receive radio transmission, due to an obstruction blocking the direct path between sender and receiver.

Shannon Equation: An equation by Claude Shannon (1938) that shows that the maximum possible transmission speed (C) when sending data through a channel is directly proportional to its bandwidth (B), and depends to a lesser extent on its signal-to-noise ratio (S/N): $C = B \log_2 (1 + S/N)$.

Share: Microsoft's name for something that is shared, usually a directory or a printer.

Shared Documents Folder (SharedDocs): In Windows XP, a directory that is automatically shared. To share a file with other users on the computer or on an attached network, the user can copy a file from another directory to the Shared Document Folder.

Shared Static Key: A key that is used by all users in a system (shared) that is not changed (static).

SharedDocs: *See* Shared Documents Folder.

SHDSL: *See* Super-High-Rate DSL.

Shielded Twisted Pair (STP): A type of twisted-pair wiring that puts a metal foil sheath around each pair and another metal mesh around all pairs.

Shopping Cart: A core e-commerce function that holds goods for the buyer while he or she is shopping.

Signal: An information-carrying disturbance that propagates through a transmission medium.

Signal Bandwidth: The range of frequencies in a signal, determined by subtracting the lowest frequency from the highest frequency.

Signaling: In telephony, the controlling of calling, including setting up a path for a conversation through the transport core, maintaining and terminating the conversation path, collecting billing information, and handling other supervisory functions.

Signaling Gateway: The device that sets up conversations between parties, maintains these conversations, ends them, provides billing information, and does other work.

Signaling System 7: Telephone signaling system in the United States.

Signal-to-Noise Ratio (SNR): The ratio of the signal strength to average noise strength; should be high in order for the signal to be effectively received.

Signing: Encrypting something with the sender's private key.

Simple File Sharing: In Windows XP, extremely weak security used on files in Shared Documents folders. Simple File Sharing does not even use a password; the only security is that people must know the workgroup names to read and change files.

Simple Mail Transfer Protocol (SMTP): The protocol used to send a message to a user's outgoing mail host and from one mail host to another; requires a complex series of interactions between the sender and receiver before and after mail delivery.

Simple Network Management Protocol (SNMP): The protocol that allows a general way to collect rich data from various managed devices in a network.

Simple Object Access Protocol (SOAP): A standardized way for a Web service to expose its methods on an interface to the outside world.

Single Point of Failure: When the failure in a single component of a system can cause a system to fail or be seriously degraded.

Single Sign-On (SSO): Authentication in which a user can authenticate himself or herself only once and then have access to all authorized resources on all authorized systems.

Single-Mode Fiber: Optical fiber whose core is so thin (usually 8.3 microns in diameter) that only a single mode can propagate—the one traveling straight along the axis.

SIP: *See* Session Initiation Protocol.

Site Survey: In wireless LANs, a radio survey to help determine where to place access points.

Situation Analysis: The examination of a firm's current situation, which includes anticipation of how things will change in the future.

SLA: *See* Service Level Agreement.

SLC: *See* Systems Life Cycle.

Sliding Window Protocol: Flow control protocol that tells a receiver how many more bytes it may transmit before receiving another acknowledgement, which will give a longer transmission window.

Slot: A very brief time period used in Time Division Multiplexing; a subdivision of a frame. Carries one sample for one circuit.

Small Form Factor (SFF): A variety of optical fiber connectors; smaller than SC or ST connectors but unfortunately not standardized.

Small Office or Home Office (SOHO): A small-scale network for a small office or home office.

Smart Antenna: Antenna that can focus power on individual stations for better reception, without using a dish or mechanical motion.

Smart Access Point: An access point that can be managed remotely.

SMTP: *See* Simple Mail Transfer Protocol.

SNA: *See* Systems Network Architecture.

Sneakernet: A joking reference to the practice of walking files around physically, instead of using a network for file sharing.

SNMP: *See* Simple Network Management Protocol.

SNR: *See* Signal-to-Noise Ratio.

SOAP: *See* Simple Object Access Protocol.

Social Engineering: Tricking people into doing something to get around security protections.

Socket: The combination of an IP address and a port number, designating a specific connection to a specific application on a specific host. It is written as an IP address, a colon, and a port number, for instance 128.171.17.13:80.

SOHO: *See* Small Office or Home Office.

Solid-Wire UTP: Type of UTP in which each of the eight wires really is a single solid wire.

SONET: *See* Synchronous Optical Network.

Spam: Unsolicited commercial e-mail.

Spam Blocking: Software that recognizes and deletes spam.

Spanning Tree Protocol (STP): *See* 802.1D Spanning Tree Protocol.

Speech Codec: *See* Codec.

Spread Spectrum Transmission: A type of radio transmission that takes the original signal and spreads the signal energy over a much broader channel than would be used in normal radio transmission; used in order to reduce propagation problems, not for security.

Spyware: Software that sits on a victim's machine and gathers information about the victim.

SS7: *See* Signaling System 7.

SSH: *See* Secure Shell.

SSL: *See* Secure Sockets Layer.

SSL/TLS: *See* Secure Sockets Layer and Transport Layer Security.

SSL/TLS-Aware: Modified to work with SSL/TLS.

SSO: *See* Single Sign-On.

ST Connector: A cylindrical optical fiber connector, sometimes called a bayonet connector because of the manner in which it pushes into an ST port and then twists to be locked in place.

Standard: A rule of operation that allows two hardware or software processes to work together. Standards normally govern the exchange of messages between two entities.

Standards Agency: An organization that creates and maintains standards.

Standards Architecture: A family of related standards that collectively allows an application program on one machine on an internet to communicate with another application program on another machine on the internet.

Star Topology: A form of topology in which all wires in a network connect to a single switch.

Start of Frame Delimiter Field: The second field of an Ethernet MAC frame, which synchronizes the receiver's clock to the sender's clock and then signals that the synchronization has ended.

State: In digital physical layer signaling, one of the few line conditions that represent information.

Stateful Firewall: A firewall whose default behavior is to allow all connections initiated by internal hosts but to block all connections initiated by external hosts. Only passes packets that are part of approved connections.

Static IP Address: An IP address that does not change each time a host connects to an internet; usually only given to servers.

Station: A computer that communicates over a network.

STM: *See* Synchronous Transfer Mode.

Store-and-Forward: Switching wherein the Ethernet switch waits until it has received the entire frame before sending the frame back out.

Static IP Address: An IP address that never changes.

STP: *See* 802.1D Spanning Tree Protocol or Shielded Twisted Pair.

Strain Relief: Crimping the back of an RJ-45 connector into an RJ-45 cord so that if the cord is pulled, it will not come out of the connector.

Strand: In optical fiber, a core surrounded by a cladding. For two-way transmission, two optical fiber strands are needed.

Stranded-Wire UTP: Type of UTP in which in which each of the eight "wires" really is a collection of wire strands.

Streaming Media: The transmission of audio or video so that listening of viewing can be done before the entire life is transferred.

Stripping Tool: Tool for stripping the sheath off the end of a UTP cord.

Strong Keys: Keys that are too long to be cracked by exhaustive key search.

Subcarrier: A channel that is itself a subdivision of a broadband channel, used to transmit frames in OFDM.

Subnet: A small network that is a subdivision of a large organization's network.

Subnet Mask: A mask with 1s in the network and subnet parts and zeros in the host part.

Subnet Part: The part of an IP address that specifies a particular subnet within a network.

Super Client: "Serverish" client in Gnutella that is always on, that has a fixed IP address, that has many files to share, and that is connected to several other super clients.

Super-High-Rate DSL (SHDSL): The next step in business DSL, which can operate symmetrically over a single voice-grade twisted pair and over a speed range of 384 kbps to 2.3 Mbps. It can also operate over somewhat longer distances than HDSL2.

Supplicant: In authentication, the party trying to prove its identity to the verifier.

Surreptitiously: Done without someone's knowledge, such as surreptitious face recognition scanning.

SVC: *See* Switched Virtual Circuit.

Switch: A device that forwards frames within a single network.

Switched Virtual Circuit (SVC): A circuit between sites that is set up just before a call and that lasts only for the duration of the call.

Switching Matrix: A switch component that connects input ports to output ports.

Symmetric Key Encryption: Family of encryption methods in which the two sides use the same key to encrypt messages to each other and to decrypt incoming messages. In bidirectional communication, only a single key is used.

SYN Bit: In TCP, the flags field that is set to indicate if the message is a synchronization message.

Synchronous Digital Hierarchy (SDH): The European version of the technology upon which the world is nearly standardized.

Synchronous Optical Network (SONET): The North American version of the technology upon which the world is nearly standardized.

Synchronous Transfer Mode (STM): A number that indicates SDH speeds.

Syntax: In message exchange, how messages are organized.

Systems Administration: The management of a server.

Systems Development Life Cycle: Disciplined staged process for developing new systems.

Systems Life Cycle: The disciplined management of a system from its conception through development and ongoing use.

Systems Network Architecture (SNA): The standards architecture traditionally used by IBM mainframe computers.

T568B: Wire color scheme for RJ-45 connectors; used most commonly in the United States.

Tag: An indicator on an HTML file to show where the browser should render graphics files, when it should play audio files, and so forth.

Tag Control Information: The second tag field, which contains a 12-bit VLAN ID that it sets to zero if VLANs are not being implemented. If VLANs are being used, each VLAN will be assigned a different VLAN ID.

Tag Field: One of the two fields added to an Ethernet MAC layer frame by the 802.1Q standard.

Tag Protocol ID: The first tag field used in the Ethernet MAC layer frame. The Tag Protocol ID has the two-octet hexadecimal value 81-00, which indicates that the frame is tagged.

Tbps: Terabits per second.

TCO: *See* Total Cost of Ownership.

TCP: *See* Transmission Control Protocol.

TCP Segment: A TCP message.

TCP/IP: The Internet Engineering Tasks Force's standards architecture; dominant above the data link layer.

TCPDUMP: The most popular freeware packet analysis program; the Unix version.

TDM: *See* Time Division Multiplexing.

TDR: *See* Time Domain Reflectometry.

Telecommunications Closet: The location on each floor of a building where cords coming up from the basement are connected to cords that span out horizontally to telephones and computers on that floor.

Telephone Modem: A device used in telephony that converts digital data into an analog signal that can transfer over the local loop.

Telnet: The simplest remote configuration tool; lacks encryption for confidentiality.

Temporal Dispersion: Another name for modal dispersion.

Temporal Key Integrity Protocol (TKIP): A security process used by 802.11i, where each station has its own nonshared key after authentication and where this key is changed frequently.

Terabits per Second: Trillions of bits per second.

Terminal Crosstalk Interference: Crosstalk interference at the ends of a UTP cord, where wires are untwisted to fit into the connector. To control terminal crosstalk interference, wires should not be untwisted more than a half inch to fit into connectors.

Termination Equipment: Equipment that connects a site's internal telephone system to the local exchange carrier.

Terrestrial: Earth-based.

Test Signals: Signal sent by a high-quality UTP tester through a UTP cord to check signal quality parameters.

Texting: In cellular telephony, the transmission of text messages.

TFTP: *See* Trivial File Transfer Protocol.

Third-Generation (3G): The newest generation of cellular telephony, able to carry data at much higher speeds than 2G systems.

Threat Enviornment: The threats that face the company.

Three-Party Call: A call in which three people can take part in a conversation.

Three-Tier Architecture: An architecture where processing is done in three places: on the client, on the application server, and on other servers.

Three-Way Handshake: A three-message exchange that opens a connection in TCP.

Throughput: The transmission speed that users actually get. Usually lower than a transmission system's rated speed.

TIA/EIA/ANSI-568: The standard that governs transmission media in the United States.

Time Division Multiplexing (TDM): A technology used by telephone carriers to provide reserved capacity on trunk lines between switches. In TDM, time is first divided into frames, each of which are divided into slots; a circuit is given the same slot in every frame.

Time Domain Reflectometry (TDR): Sending a signal in a UTP cord and recording reflections; can give the length of the cord or the location of a propagation problem in the cord.

Time to Live (TTL): The field added to a packet and given a value by a source host, usually between 64 and 128. Each router along the way decrements the TTL field by one. A router decrementing the TTL to zero will discard the packet; this prevents misaddressed packets from circulated endlessly among packet switches in search of their nonexistent destinations.

TKIP: *See* Temporal Key Integrity Protocol.

TLS: *See* Transport Layer Security.

Token Passing: In token-ring networks, a token frame is transmitted and used to determine when a station may transmit.

Token-Ring Network: A network that uses a physical ring topology and token passing at the media access control layer.

Toll Call: Long-distance call pricing in which the price depends on distance and duration.

Toll-Free Number Service: Service in which anyone can call into a company, usually without being charged. Area codes are 800, 888, 877, 866, and 855.

Top-Level Domain: The second level of a DNS hierarchy, which categorizes the domain by organization type (e.g., .com, .net, .edu, .biz, .info) or by country (e.g., .uk, .ca, .ie, .au, .jp, .ch).

Topology: The way in which nodes are linked together by transmission lines.

TOS: *See* Type of Service.

Total Cost of Ownership (TCO): The total cost of an entire system over its expected lifespan.

Total Purchase Cost of Network Products: The initial purchase price of a fully configured system.

Tracert (tracert): A Windows program that shows latencies to every router along a route and to the destination host.

Traffic Engineering: Designing and managing traffic on a network.

Traffic Shaping: Limiting access to a network based on type of traffic.

Trailer: The part of a message that comes after the data field.

Transmission Line: A physical line that is used to carry transmitted information.

Transmission Speed: The rate at which information is transmitted, in bits per second.

Transaction Processing: Processing involving simple, highly structured, and high-volume interactions.

Transceiver: A transmitter/receiver.

Transfer Syntax: In the OSI Presentation layer, the syntax used by two presentation layer processes to communicate, which may or may not be quite different than either of their internal methods of formatting information.

Transmission Control Protocol (TCP): The most common TCP/IP protocol at the transport layer. Connection-oriented and reliable.

Transparently: Without having a need to implement modifications.

Transport: In telephony, transmission; taking voice signals from one subscriber's access line and delivering them to another customer's access line.

Transport Core: The switches and transmission lines that carry voice signals from one subscriber's access line and delivering them to another customer's access line.

Transport Layer Security (TLS): The simplest VPN security standard to implement; originally named Secure Sockets Layer. Provides a secure connection at the transport layer, protecting any applications above it that are SSL/TLS-aware.

Transport Layer: The layer that governs communication between two hosts; Layer 4 in both OSI and TCP/IP.

Transport Mode: One of IPsec's two modes of operation, in which the two computers that are communicating implement IPsec. Transport mode gives strong end-to-end security between the computers, but it requires IPsec configuration and a digital certificate on all machines.

Traps: The type of message that an agent sends if it detects a condition that it thinks the manager should know about.

Triple DES (3DES): Symmetric key encryption method in which a message is encrypted three times with DES. If done with two or three different keys, offers strong security. However, it is processing intensive.

Triple Play: A packaged service including Internet access, telephony, and television service.

Trivial File Transfer Protocol (TFTP): A protocol used on switches and routers to download configuration information; has no security.

Trojan Horse: A program that looks like an ordinary system file, but continues to exploit the user indefinitely.

Trunk Line: A type of transmission line that links switches to each other, routers to each other, or a router to a switch.

Trunking: *See* Link Aggregation.

TTL: *See* Time to Live.

Tunnel Mode: One of IPsec's two modes of operation, in which the IPsec connection extends only between IPsec gateways at the two sites. Tunnel mode provides no protection within sites, but it offers transparent security.

Twisted-Pair Wiring: Wiring in which each pair's wires are twisted around each other several times per centimeter, reducing EMI.

Type of Service (TOS): IPv4 header field that designates the type of service a certain packet should receive.

U: The standard unit for measuring the height of switches. One U is in height. Most switches, although not all, are multiples of U.

UDDI: *See* Universal Description, Discovery, and Integration.

UDDI Green Pages: The UDDI search option that allows companies to understand how to interact with specific Web services. Green pages specify the interfaces on which a Web service will respond, the methods it will accept, and the properties that can be changed or returned.

UDDI White Pages: The UDDI search option that allows users to search for Web services by name, much like telephone white pages.

UDDI Yellow Pages: The UDDI search option that allows users to search for Web services by function, such as accounting, much like telephone yellow pages.

UDP: *See* User Datagram Protocol.

Ultrawideband   (UWB): Spread   spectrum transmission system that has extremely wide channels.

UNICODE: The standard that allows characters of all languages to be represented.

Universal Description, Discovery, and Integration (UDDI): A protocol that is a distributed database that helps users find appropriate Web services.

Unix: A network operating system used by all workstation servers. Linux is a Unix version used on PCs.

Unlicensed Radio Band: A radio band that does not require each station using it to have a license.

Unreliable: (Of a protocol) not doing error correction.

Unshielded Twisted Pair (UTP): Network cord that contains four twisted pairs of wire within a sheath. Each wire is covered with insulation.

Update: To download and apply patches to fix a system.

Uplink: In satellites, transmission from the Earth to a communication satellite.

Uplink Port: Port on an Ethernet switch that can be directly connected to a port in a higher-level switch with a standard UTP cord.

Usage Policy: A company policy for who may use various tools and how they may use them.

User Datagram Protocol (UDP): Unreliable transport-layer protocol in TCP/IP.

Username: An alias that signifies the account that the account holder will be using.

UTP: *See* Unshielded Twisted Pair.

UWB: *See* Ultrawideband.

Validate: To test the accuracy of a network simulation model by comparing its perfor-

mance with that of the real network. If the predicted results match the actual results, the model is validated.

Variable-Length Subnet Mask (VLSM): A mask that allows subnets to be of different sizes.

Very Small Aperature Terminal (VSAT): Communication satellite earthstation that has a small-diameter antenna.

VCI: *See* Virtual Channel Identifier.

Verifier: The party requiring the applicant to prove his or her identity.

Vertical Riser: Space between the floors of a building that telephone and data cabling go through to get to the building's upper floor.

Viral Networking: Networking in which the user's PC connects to one or a few other user PCs, which each connect to several other user PCs. When the user's PC first connects, it sends an initiation message to introduce itself via viral networking. Subsequent search queries sent by the user also are passed virally to all computers reachable within a few hops; used in Gnutella.

Virtual Channel: In ATM, an individual connection within a virtual path.

Virtual Channel Identifier (VCI): One of the two parts of ATM virtual circuit numbers.

Virtual Circuit: A transmission path between two sites or devices; selected before transmission begins.

Virtual LAN (VLAN): A closed collection of servers and the clients they serve. Broadcast signals go only to computers in the same VLAN.

Virtual Path Identifier (VPI): One of the two parts of ATM virtual circuit numbers.

Virtual Path: In ATM, a group of connections going between two sites.

Virtual Private Network (VPN): A network that uses the Internet with added security for data transmission.

Virus: A piece of executable code that attaches itself to programs or data files. When the program is executed or the data file opened, the virus spreads to other programs or data files.

Virus Definitions Database: A database used by antivirus programs to identify viruses. As new viruses are found, the virus definitions database must be updated.

Virus Writer: Someone who creates viruses.

VLAN: *See* Virtual LAN.

VLSM: *See* Variable-Length Subnet Mask.

Voice Mail: A service that allows people to leave a message if the user does not answer his or her phone.

Voice-Grade: Wire of a quality useful for transmitting voice signals in the PSTN.

Voice over IP (VoIP): The transmission of voice signals over an IP network.

VoIP: *See* Voice over IP.

VPI: *See* Virtual Path Identifier.

VPN: *See* Virtual Private Network.

VSAT: *See* Very Small Aperture Terminal.

Vulnerability: A security weakness found in software.

Vulnerability Testing: Testing after protections have been configured, in which a company or a consultant attacks protections in the way a determined attacker would and notes which attacks that should have been stopped actually succeeded.

WAN: *See* Wide Area Network.

War Driver: Someone who travels around looking for unprotected wireless access points.

WATS: *See* Wide Area Telephone Service.

Wavelength: The physical distance between comparable points (e.g., from peak to peak) in successive cycles of a wave.

Wavelength Division Multiplexing: Using signaling equipment to transmit several light sources at slightly different wavelengths, thus adding signal capacity at the cost of using slightly more expensive signaling equipment but without incurring the high cost of laying new fiber.

WDM: *See* Wavelength Division Multiplexing.

Weak Keys: Keys that are shot enough to be cracked by an exhaustive key search.

Web 2.0: Browser-based service in which users develop or enhance content, instead of the website owner controlling all of the content.

Webify: In SSL/TLS VPNs, the SSL/TLS gateway can translate output from some applications into a webpage.

Web Service: A way to send processing requests to program (object) on another machine. The object has an interface to the outside world and methods that it is willing to undertake. Messages are sent in SOAP format.

Web-Enabled: Client/server processing applications that use ordinary browsers as client programs.

Webmail: Web-enabled e-mail. User needs only a browser to send and read e-mail.

Well-Known Port Number: Standard port number of a major application that is usually (but not always) used. For example, the well-known TCP port number for HTTP is 80.

WEP: *See* Wired Equivalent Privacy.

Wide Area Network (WAN): A network that links different sites together.

Wide Area Telephone Service (WATS): Service that allows a company to place outgoing long-distance calls at per-minute prices lower than those of directly dialed calls.

Wi-Fi Alliance: Trade group created to create interoperability tests of 802.11 LANs; actually produced the WPA standard.

WiMAX: Broadband wireless access method. Standardized as 802.16.

Window Size Field: TCP header field that is used for flow control. It tells the station that receives the segment how many more octets that station may transmit before getting another acknowledgement message that will allow it to send more octets.

Windows XP: client Microsoft operating system.

Windows Internet Name Service (WINS): The system required by Windows clients and servers before Windows 2000 server to provide IP address for host names.

WinDUMP: The most popular freeware packet analysis program; the Windows version.

Winipcfg (winipcfg): A command used to find information about one's own computer; used in older versions of windows.

WINS: *See* Windows Internet Name Service.

Wired Equivalent Privacy (WEP): A weak security mechanism for 802.11.

Wireless Ethernet: Sometimes used as another name for 802.11.

Wireless Access Point: Devices that controls wireless clients and that bridges wireless clients to servers and routers on the firm's main wired LAN.

Wireless LAN (WLAN): A local area network that uses radio (or rarely, infrared) transmission instead of cabling to connect devices.

Wireless LAN Switch: An Ethernet switch to which multiple wireless access points connect; manages the access points.

Wireless Networking: Networking that uses radio transmission instead of wires to connect devices.

Wireless NIC: 802.11 network interface card.

Wireless Protected Access (WPA): 802.11 security method created as a stopgap between WEP and 802.11i.

Wireless Protected Access 2 (WPA2): Another name for 802.11 security.

WLAN: *See* Wireless LAN.

Work-Around: A process of making manual changes to eliminate a vulnerability instead of just installing a software patch.

Workgroup: A logical network. On a physical network, only PCs in the same workgroup can communicate.

Workgroup Name: To create a workgroup, all PCs in the workgroup are assigned the same workgroup name. They will find each other automatically.

Workgroup Switch: A switch to which stations connect directly.

Working Group: A specific subgroup of the 802 Committee, in charge of developing a specific group of standards. For instance, the 802.3 Working Group creates Ethernet standards.

Workstation Server: The most popular type of large dedicated server; runs the Unix operating system. It uses custom-designed microprocessors and runs the Unix operating system.

Worm: An attack program that propagates on its own by seeking out other computers, jumping to them, and installing itself.

WPA: *See* Wireless Protected Access.

WPA2: *See* Wireless Protected Access 2.

Write: A Microsoft Windows Server permission that allows an account owner to change the contents of files in the directory.

X.509: The main standard for digital certificates.

Zero-Day Exploit: An exploit that takes advantage of vulnerabilities that have not previously been discovered or for which updates have not been created.

ZigBee: Low-speed, low-power protocol for connecting sensors and other very small devices wirelessly.

Index

Note: Locators in bold indicate definitions of key terms; locators in italics indicate tables or figures.